Results and Problems in Cell Differentiation

Volume 61

Series editors
Jacek Z. Kubiak, Rennes CX, France
Malgorzata Kloc, Houston, TX, USA

More information about this series at http://www.springer.com/series/400

Jean-Pierre Tassan • Jacek Z. Kubiak
Editors

Asymmetric Cell Division in Development, Differentiation and Cancer

Springer

Editors
Jean-Pierre Tassan
IGDR, UMR 6290, CNRS
Université de Rennes 1
Rennes, France

Jacek Z. Kubiak
Inst of Genetics & Dev of Rennes IG
CNRS/Unive Rennes 1 UMR 6290
Rennes Cedex, France

ISSN 0080-1844 ISSN 1861-0412 (electronic)
Results and Problems in Cell Differentiation
ISBN 978-3-319-53149-6 ISBN 978-3-319-53150-2 (eBook)
DOI 10.1007/978-3-319-53150-2

Library of Congress Control Number: 2017938011

Printed on acid-free paper

This Springer imprint is published by Springer Nature
The registered company is Springer International Publishing AG
The registered company address is: Gewerbestrasse 11, 6330 Cham, Switzerland

Contents

Chapter 1
Modeling Asymmetric Cell Division in *Caulobacter crescentus* Using a Boolean Logic Approach

Ismael Sánchez-Osorio, Carlos A. Hernández-Martínez, and Agustino Martínez-Antonio

Abstract *Caulobacter crescentus* is a model organism for the study of asymmetric division and cell type differentiation, as its cell division cycle generates a pair of daughter cells that differ from one another in their morphology and behavior. One of these cells (called stalked) develops a structure that allows it to attach to solid surfaces and is the only one capable of dividing, while the other (called swarmer) develops a flagellum that allows it to move in liquid media and divides only after differentiating into a stalked cell type. Although many genes, proteins, and other molecules involved in the asymmetric division exhibited by *C. crescentus* have been discovered and characterized for several decades, it remains as a challenging task to understand how cell properties arise from the high number of interactions between these molecular components. This chapter describes a modeling approach based on the Boolean logic framework that provides a means for the integration of knowledge and study of the emergence of asymmetric division. The text illustrates how the simulation of simple logic models gives valuable insight into the dynamic behavior of the regulatory and signaling networks driving the emergence of the phenotypes exhibited by *C. crescentus*. These models provide useful tools for the characterization and analysis of other complex biological networks.

Abbreviation

TF Transcription Factor

I. Sánchez-Osorio (✉) • C.A. Hernández-Martínez • A. Martínez-Antonio
Department of Genetic Engineering, Center for Research and Advanced Studies of the National Polytechnic Institute, Irapuato, Guanajuato CP 36821, México
e-mail: ismael.sanchez@cinvestav.mx; agustino.martinez@cinvestav.mx

J.-P. Tassan, J.Z. Kubiak (eds.), *Asymmetric Cell Division in Development, Differentiation and Cancer*, Results and Problems in Cell Differentiation 61,
DOI 10.1007/978-3-319-53150-2_1

1.1 Introduction

Caulobacter crescentus is a Gram-negative bacterium that divides asymmetrically, generating two daughter cells that develop different appendages at one of their poles at particular times during their division cycle. One newborn cell, called swarmer, develops a flagellum and a chemotactic apparatus, which enables it to move in liquid media following gradients of nutrient concentrations (Jensen 2006). The other daughter, in contrast, has a narrow extension of its cell body called stalk, which allows the cell to attach to solid surfaces. This is the only cell type capable of dividing (Wagner and Brun 2007). After acquiring enough biomass, stalked cells divide and release motile swarmer cells that will remain in that state for a certain time period (England et al. 2010). Eventually, the swarmer cells lose their flagella and acquire the stalked phenotype, generating a cyclic pattern of growth and division.

In some sense, *C. crescentus* resembles the asymmetric cell division and cell type differentiation exhibited by eukaryotes. Because of this, it is an organism that can be used as a biological model for the study of these phenomena in more complex organisms. As shown in Fig. 1.1, the cell cycle of *C. crescentus* can be

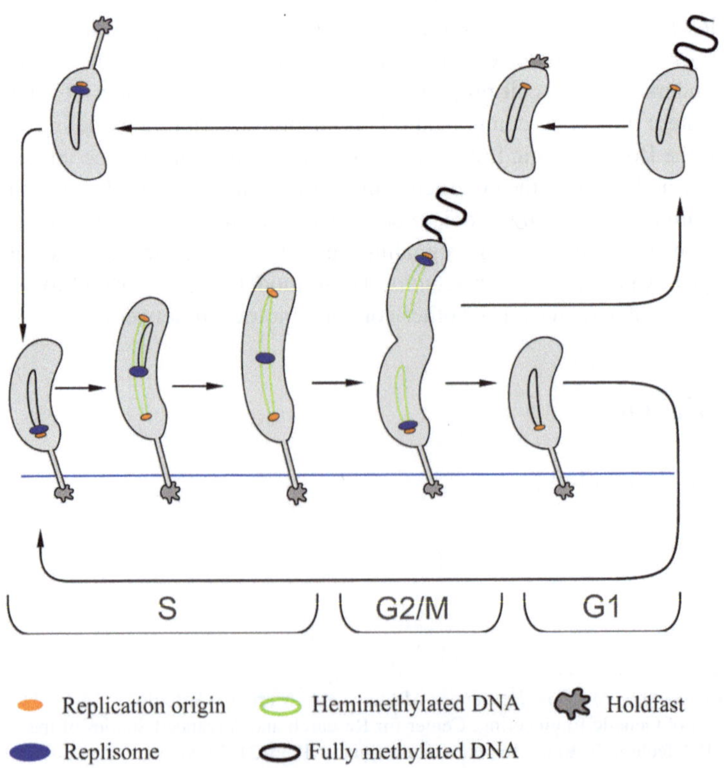

Fig. 1.1 Asymmetric division and cell cycle of *C. crescentus*. The formation of the swarmer and stalked cell types is presented in the context of cell division. The cell cycle is divided in discrete stages analogous to those of the cell division of eukaryotes

seen in terms of three stages analogous to those occurring during mitosis, namely, G1, S, and G2/M. During the G1 phase, swarmer cells eject the flagellum, synthesize the holdfast and the stalk structure, and initiate the replication of their DNA (Jensen 2006). In the S phase, when the synthesis of DNA is occurring, the newly formed DNA strands get spooled toward a pole of the cell and fold into compact chromosomal microdomains (Jensen et al. 2001). From this pre-divisional compartmentalization, by mechanisms that are still unclear, a new cell fate will emerge on each of the daughter cells (Judd et al. 2003). Finally, at the G2/M phase, *C. crescentus* divides asymmetrically into swarmer and stalked cells, and the DNA of each daughter becomes fully methylated (Reisenauer and Shapiro 2002).

From a molecular biology perspective, if we could determine the function of each of the molecules involved in the behavior displayed by *C. crescentus* during its division cycle, then we could in principle understand how this phenotype arises from the molecular constituents of the cell. However, as more genes, gene products, and other molecules were discovered, it would remain a very challenging task to extract knowledge from that information and grasp the result of the sheer number of their interactions. In consequence, it is necessary to have a way to unify that data into a coherent picture, so that this kind of biological phenomenon can be studied and understood in a more integrated manner. This integration would also require accounting for the temporal variation in the expression of all genes involved in the development of the observed phenotypes, as such dynamics may be decisive for the emergence of cell cycle properties (Ryan and Shapiro 2003). Therefore, by no longer focusing on single genes and molecules, but on global system properties arising from dynamic interactions, it may be possible to achieve a better understanding of how the asymmetric cell division of *C. crescentus* arises from its molecular cell components. This endeavor can be facilitated by the formal and precise language of mathematics.

In this chapter, we address the reconstruction, modeling, and analysis of the regulatory and signaling networks involved in the cell cycle and asymmetric division of *C. crescentus*, which we have studied in a previous work (Quiñones-Valles et al. 2014). We introduce general modeling principles that provide a foundation for the construction, simulation, and verification of simple discrete models based on the Boolean modeling framework (Thomas 1978). We argue throughout the text that, despite its limitations, this is a useful tool to explore and study the emergence of asymmetric division, as it captures some essential details of the dynamic interactions between regulatory components through the use of logical rules. These kinds of models have been successfully applied in the study of other biological phenomena, including the phenotypic transition between lysis and lysogeny in the lambda phage (Thieffry and Thomas 1995), cellular fate generation during flower development in *Arabidopsis thaliana* (Espinosa-Soto et al. 2004), and cell differentiation in lymphocytes (Mendoza 2006).

1.2 Reconstruction of the Regulatory and Signaling Network Involved in the Cell Division Cycle of *C. crescentus*

Network reconstruction consists of obtaining a comprehensive list and corresponding interconnection diagram (i.e., the topology) of all the genes, signaling molecules, and other molecular components participating in the cell cycle of *C. crescentus*. There are many methodologies to reconstruct a network (Wang and Huang 2014; Thompson et al. 2015), but all of them unavoidably require the verification of direct experimental evidence for each molecular interaction found. This is why it is advisable to retrieve data from primary literature and reconstruct a network relying as much as possible on this information. Such is the procedure we have followed earlier (Quiñones-Valles et al. 2014) and that we illustrate in Fig. 1.2.

Overall, we gathered evidence for interactions including transcriptional regulation by TFs, phosphorylation/dephosphorylation of proteins, methylation of DNA, multi-protein complex assembly, and proteolysis. From the retrieved data, it is possible to draw a graph, like the one shown in Fig. 1.3 that directly represents the interactions (edges) between genes and proteins (nodes). In the diagram, when a molecular effector promotes the transcription of a gene, it is customary to classify

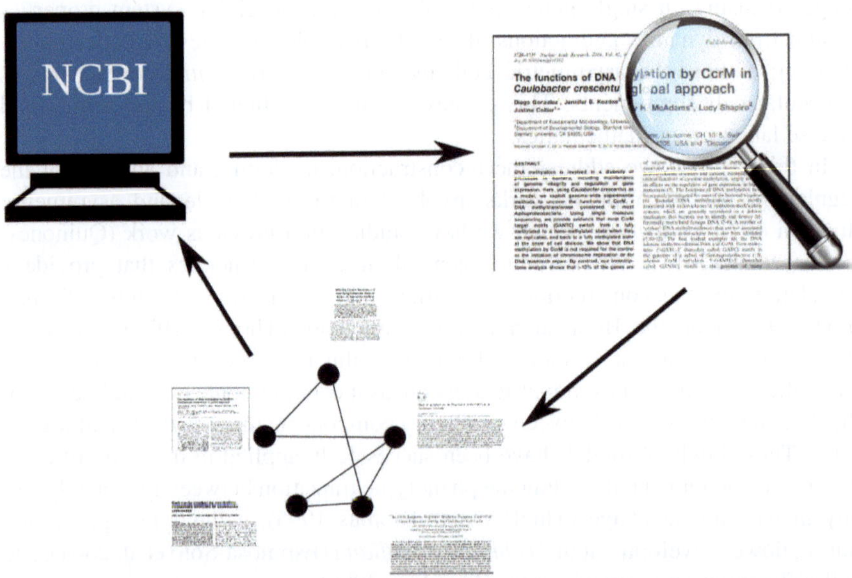

Fig. 1.2 Literature-based network reconstruction process. The iterative network reconstruction process starts from the gathering of peer-reviewed literature on the functionality of each molecular network component and their interactions. The retrieved literature is then thoroughly examined to find experimental evidence for interactions. Finally, the information is assembled into an interconnection diagram containing all components

take advantage of the available biological information to achieve further simplification. This is not strictly necessary, but it is always advisable to work with simple systems as much as possible.

In line with the above remark, the network G_2 can be divided into two simpler subsystems that differ from one another in two key aspects: first, the type of molecular interactions taking place within them, and second, the timescales on which these interactions occur. More specifically, inside the first subnetwork, named G_{2a} (Fig. 1.5a), regulatory processes mediated by transcriptional regulation and DNA methylation operate approximately on a timescale of 1–3 min (Alon 2006). In contrast, in the second subnetwork, G_{2b} (Fig. 1.5b), there is a phospho-proteolytic pathway whose reactions take place in the order of 1–100 ms. In addition to allowing for the distinction of two subnetworks within G_2, this information makes it possible to apply an approximation known as the "steady-state" assumption and omit some regulatory interactions between nodes that operate at different timescales. This strategy for system reduction is widely used in the study of biochemical reactions and enzyme kinetics (Schnell 2017).

In brief, the steady-state approximation asserts that for a dynamic system containing processes operating on well-separated timescales (e.g., enzymatic reactions, transcription, etc.), the processes with relatively fast dynamics can be assumed to have reached their equilibrium states from the perspective of the slower processes, which can be considered effectively constant for the timescales in which the faster processes operate. For the network G_2, this implies that the net effect of regulatory interactions going from the transcriptional network, G_{2a}, to the phospho-proteolytic network, G_{2b}, can be ignored for the purpose of studying the dynamics of G_{2b}. In other words, because of their relative constant behavior, regulatory interactions may be assumed to no longer influence the time variation of nodes in the phospho-proteolytic network. However, being CtrA a node shared by the two subnetworks, G_{2a} and G_{2b}, we consider it independently on each subnetwork and assign it two variable names, $CtrA_a$ for G_{2a} and $CtrA_b$ for G_{2b}.

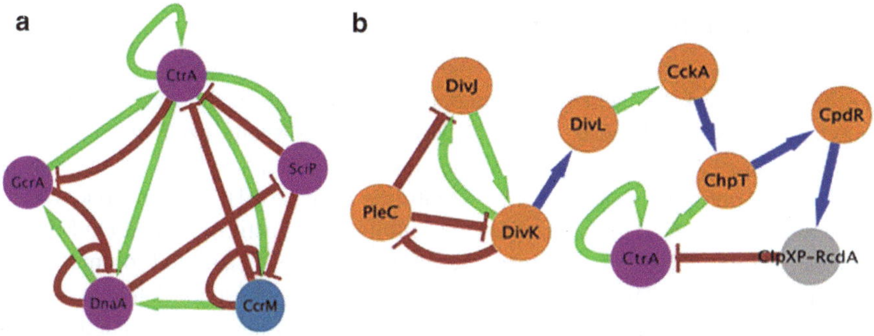

Fig. 1.5 Subnetworks G_{2a} and G_{2b}. Reduced subnetworks representing the essential elements involved in the control of cell division in *C. crescentus*. (**a**) Transcriptional subnetwork G_{2a} and (**b**) phospho-proteolytic pathway G_{2b}. The colors of *arrows* are interpreted as in Fig. 1.3. Figure modified from Quiñones-Valles et al. (2014)

Figure 1.5 shows the subnetworks G_{2a} and G_{2b} that result after applying the above simplification and approximation strategies.

1.4 Qualitative Functional Description of the Transcriptional Subnetwork G_{2a}

In the subnetwork G_{2a}, $CtrA_a$ has positive autoregulation (Fig. 1.5a), and it seems it also has a negative autoregulatory loop, but its functional operation is not currently understood (Domian et al. 1999). This TF can only be transcribed when the DNA is hemimethylated, which happens immediately after the replication process is completed. $CtrA_a$ subsequently promotes the expression of the methyl-transferase CcrM (Reisenauer and Shapiro 2002). This enzyme methylates the second strand of DNA, hence repressing the expression of $CtrA_a$ (Reisenauer et al. 1999a, b). The activity of $CtrA_a$ is regulated at a posttranslational level by the repressor SciP that sequesters the former TF, preventing it from interacting with its DNA targets (Gora et al. 2010). The transcription of SciP is promoted by $CtrA_a$ and inhibited by DnaA (Tan et al. 2010). The protein DnaA implicated in the initiation of DNA replication is subjected to several kinds of regulation; for example, its expression is positively regulated by CcrM through DNA methylation, which causes $CtrA_a$ to bind to the promoter of DnaA and initiate its transcription (Winzeler and Shapiro 1996). As the concentration of DnaA increases, it binds to its own promoter, repressing its expression via a negative feedback loop (Collier et al. 2006). Finally, the expression of the TF GcrA is repressed by $CtrA_a$ (Holtzendorff et al. 2004) and positively regulated by DnaA (Winzeler and Shapiro 1996).

1.5 Qualitative Functional Description of the Phospho-Proteolytic Signaling Subnetwork G_{2b}

In the subnetwork G_{2b}, $CtrA_b$ promotes the transcription of the sensory response regulator DivK (Laub et al. 2002). This regulator is synthesized in an inactive non-phosphorylated state that can subsequently be phosphorylated by the kinase DivJ or the phosphatase-kinase PleC (Matroule et al. 2004). The phosphorylated form of DivK can bind to the protein DivL, inhibiting its activity. When DivL is not sequestered by DivK, it binds to the phosphorylase CckA and promotes its autophosphorylation. In its phosphorylated state, CckA phosphorylates the kinase ChpT (Ryan and Shapiro 2003); however, when it is unphosphorylated, it has phosphatase activity on ChpT (Chen et al. 2009). If ChpT is phosphorylated, it activates $CtrA_b$ and CpdR by the transference of a phosphate group. Alternatively, if ChpT is unphosphorylated, it behaves as a phosphatase on CpdR (Iniesta et al. 2006). When CpdR is unphosphorylated, it facilitates the formation of the proteolytic complex ClpXP, composed of the proteases ClpX and ClpP. Together ClpXp and RcdA form a complex that degrades $CtrA_b$ (Jenal and Fuchs 1998; Chien et al. 2007; Domian et al. 1997).

1.6 General Modeling Principles

The reconstruction of the regulatory and signaling networks described above was driven by the need for integrating our current knowledge of the molecular components and the interactions involved in the asymmetric cell division of *C. crescentus*. However, a considerable difference exists between having the crude topology of such networks and the understanding of the causal and functional relationships that give rise to the two cell phenotypes exhibited by *C. crescentus*. To advance such understanding, we believe it is necessary to have appropriate conceptual frameworks that describe, in a coherent manner, the essential aspects of the multiple experimental observations on the dynamic behavior of the nodes that participate in these networks. In describing them, of course it is possible to use ordinary language, depicting their behavior in qualitative terms, as we have done in the previous section. Nevertheless, this kind of language becomes ambiguous, prone to misinterpretations, and overwhelming when describing large numbers of nodes with multiple connections. Instead, we can use the more precise language and sophisticated mode of thought of mathematical modeling, which can guide the understanding of known empirical observations and enable the exploration of diverse hypothetical conditions, leading to the generation of experimentally testable predictions.

In the context of the network we are studying, the construction of a mathematical model involves three general stages: (1) translating the topology and time variations of the nodes in the network into appropriate mathematical descriptions using assumptions and approximations based on previous knowledge; (2) studying the system in the abstract domain looking for patterns, relationships, and numerical quantities that theoretically correspond to those observed experimentally; and (3) interpreting and validating the theoretical results in terms of the biological context. To make this modeling process accessible, mathematical and models generally focus on specific aspects of the studied networks (e.g., network interconnection) while abstracting from details at other levels of resolution (e.g., TF concentrations). Figure 1.6 outlines the main components and methodological relationships involved in the model construction process.

When modeling one begins with the biological system being investigated and follows the next sequence of steps.

I→II: Description of biological components, interactions, and states in mathematical terms.

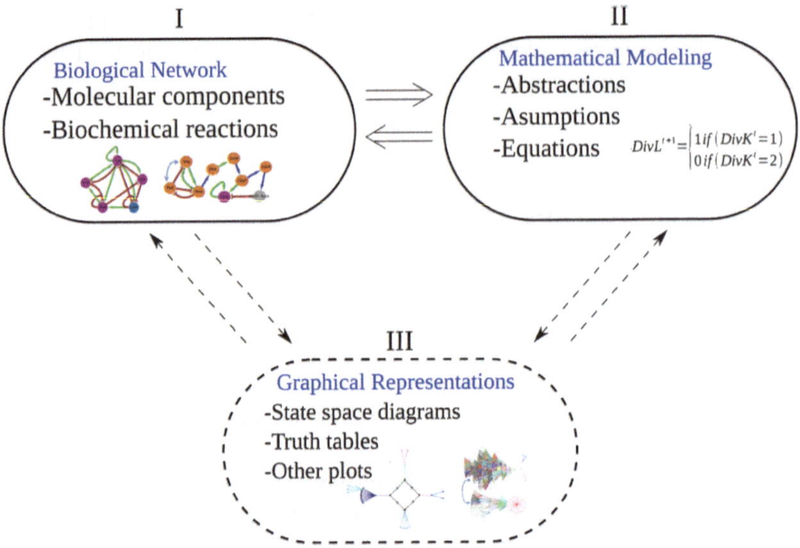

Fig. 1.6 Schematic representation showing the theoretical components of biological network modeling. In the diagram, the biological system (I) is composed of the actual asymmetric division phenomena that are affected by the regulatory and signaling networks described previously. The abstract world of the mathematical model (II) involves symbolic notational schemes that allow the time-varying interactions and dependence of the nodes in the networks to be stated with precision and economy. The auxiliary model (III) that consists of various graphical representations (schematics, diagrams, tables, plots, and other tools) helps to think about the abstract mathematical model more lucidly. *Double arrows* indicate the important transitions that are usually repeated in the construction of models

II→I: Interpretation of mathematical abstractions in terms of the biological system or phenomenon being studied. This constitutes the basic interpretation step in the modeling process.

II→III: Interpretation of mathematical notions and results in terms of visual representations (state diagrams, tables, etc.). This step constitutes a second-level interpretation in the modeling process.

III→II: Use of particular visual resources to facilitate the interpretation of the results and notions of the mathematical model that may be otherwise hard to grasp.

I↔III: Analysis of the biological system by means of graphical representations. The correlation of empirical observations with concrete visualizations helps in the translation of biological phenomena into particular scenarios in the model and facilitates the interpretation of abstract results.

In the next sections, the general modeling principles that have been introduced here will be further illustrated continuing with our study of the reconstructed network of *C. crescentus*. In particular, we will address the translation of the dynamics of the network into an appropriate mathematical description, the theoretical study of the system via simulation, and the validation of theoretical results in terms of the biological evidence.

1.7 Model Construction Using a Discrete Logical Approach

One possible way to model a biological system would be to incorporate all available biochemical details about its molecular interactions. The resulting quantitative model may reveal phenotypes that reflect in detail the dynamics of the regulatory and signaling networks of *C. crescentus* at a physicochemical level. This is a legitimate approach that has led to useful insights about the mechanisms producing the phenotypes involved in the control of the cell division cycle (Subramanian et al. 2013); unfortunately, when trying to extend this level of description to consider a larger number of genes and gene products in a network, solving the resulting model via analytical methods becomes infeasible and a numerical approach turns out computationally very demanding. From a theoretical perspective, a recurrent problem is that the number of parameters and equations grows exponentially with the number of state variables, making the analysis of the model less intuitive and its solutions hardly intelligible. From an experimental perspective, it is challenging to obtain data of molecular interactions and kinetic parameters with the spatiotemporal resolution required to build and validate these kinds of models. In consequence, this level of description becomes inaccessible for large-scale networks. This is why simpler, yet expressive, modeling approaches are required to study the behavior of complex biological systems, even at the cost of losing numerical resolution.

Perhaps the simplest way of describing the expression of a gene is to view it as a variable, $\sigma_i(t)$, with only two possible states, 0 and 1, which can be made to correspond to "low expression or inactive" and "high expression or active," respectively. These kinds of binary variables are named Boolean, after George Boole, who developed the mathematics to work with these kinds of quantities (Boole 1848). Using this conceptual framework, the dynamics of gene expression can be represented by discrete transitions between logic states throughout time, as depicted in Fig. 1.7a, b. Although at first sight these models may seem very restrictive, in the end models are only a representation of a limited range of observed facts, which fit the observed behavior of biological systems to a certain degree of approximation. From this viewpoint, it is possible to preserve a model for its usefulness, even if it never reflects exactly the observed data.

For a whole regulatory network, a handy representation for encoding the states of all its N genes, at a particular time t, is the vector notation $\boldsymbol{\sigma}(t) = [\sigma_1(t), \sigma_2(t),\ldots, \sigma_N(t)]$, in which the expression state of a gene i is denoted by $\sigma_i(t)$ (see Fig. 1.7c). Each gene i is affected by the expression of the k genes that directly regulate it. Hence, its dynamics can be specified as a function of the states of these genes, namely $\sigma_{i1}(t), \ldots, \sigma_{ik}(t)$. More precisely, the time evolution of the value σ_i can be written as $\sigma_i(t+1) = f_i(\sigma_{i1}(t), \sigma_{i2}(t), \ldots, \sigma_{ik}(t))$, where f_i is a Boolean function that

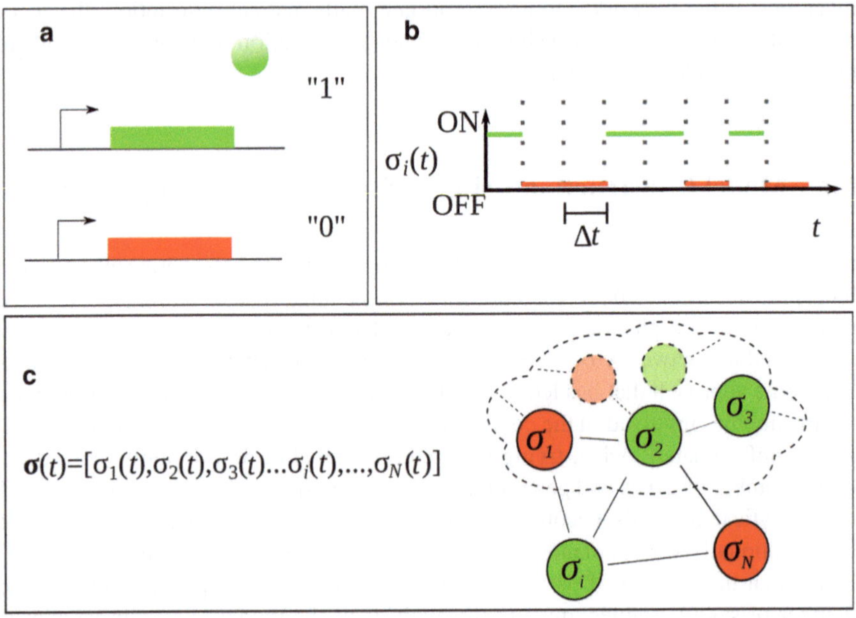

Fig. 1.7 Gene expression states modeled as logic values. (**a**) The state of a single gene is represented as "1" or "0." The active state of a gene is labeled with *green* and its repressed state is labeled with *red*. (**b**) The dynamics of gene expression is depicted throughout time. (**c**) Network state is represented as a vector

represents a regulation rule for gene *i*, as shown in Fig. 1.8. This rule originates from previous biological knowledge and determines the state of *i* at the next time step from the current states of the other genes. When considering the totality of gene states and their corresponding Boolean functions in a network, it is possible to compute global network states at uniformly spaced time steps. This synchronous updating scheme to simulate the dynamics of network states is the most common in practice, and it is the one used in this work.

The logical rule for a gene can be formulated using logic functions, such as AND, OR, and NOT. Consider, for example, the gene *gcrA* that belongs to the network G_{2a}. This gene expresses at time $t+1$ (i.e., $GcrA^{t+1}$ equals 1) when DnaA is present and $CtraA_a$ is absent, at time *t*. In contrast, it is repressed (i.e., $GcrA^{t+1}$ equals 0) if DnaA is absent or $CtrA_a$ is present. This relationship can be encoded using the AND operator, as stated in Eq. 1.1.

$$GcrA^{t+1} = \begin{cases} 1 \text{ if } \left(DnaA^t = 1\right) \text{ AND } \left(CtrA_a^t = 0\right) \\ 0 \text{ if } \left(DnaA^t = 0\right) \text{ OR } \left(CtrA_a^t = 1\right) \end{cases} \quad (1.1)$$

We show in Table 1.1 the set of logic equations for the rest of the genes in G_{2a}.

When formulating the equations for the nodes in the phospho-proteolytic network G_{2b}, it should be noticed that, besides being expressed, some signaling

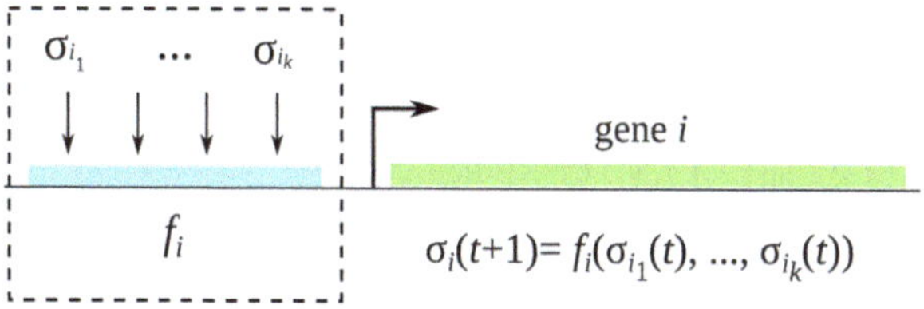

Fig. 1.8 Dynamics of gene *i* as a function of the state of its regulating nodes. In this figure, f_i represents the regulatory logic function that maps the expression state of the *k* regulating genes at time *t* to the expression of *i* at time *t* + 1

Table 1.1 Logic functions for the genes in G_{2a}

Node value at *t* + 1	Logic function
$CtrA_a^{t+1}$	1 if (($CtrA_a^t$=1) OR ($GcrA^t$=1)) AND ($CcrM^t$=0) AND ($SciP^t$=0) 0 if ($CtrA_a^t$=0 AND $GcrA^t$=0) OR $CcrM^t$=1 OR $SciP^t$=1
$GcrA^{t+1}$	1 if $DnaA^t$=1 AND $CtrA_a^t$=0 0 if $DnaA^t$=0 OR $CtrA_a^t$=1
$DnaA^{t+1}$	1 if $CtrA_a^t$=1 AND $CcrM^t$=1 AND $GcrA^t$=0 AND $DnaA^t$=0 0 if $CtrA_a^t$=0 OR $CcrM^t$=0 OR $GcrA^t$=1 OR $DnaA^t$=1
$CcrM^{t+1}$	1 if $CtrA_a^t$=1 AND $CcrM^t$=0 AND $SciP^t$=0 0 if $CtrA_a^t$=0 OR $CcrM^t$=1 OR $SciP^t$=1
$SciP^{t+1}$	1 if $CtrA_a^t$=1 AND $DnaA^t$=0 0 if $CtrA_a^t$=1 OR $DnaA^t$=1

molecules must be phosphorylated to be functional (e.g., PleC and DivK). As a consequence, the use of binary variables would fall short when representing the states of most of the nodes in G_{2b}. To overcome this limitation, it is possible to use the values 0, 1, and 2 to denote the state of regulators as "absent," "present," and "phosphorylated," respectively. However, as the modeling process favors simplicity, it is better to reduce the set of states to "functional" or "nonfunctional or absent" and accordingly use 1 or 0. This approach can be extended to represent functional states of signaling molecules that assume mutually exclusive roles, depending on whether they are phosphorylated or not. For example, we can assign the value 1 to ChpT when it acts as a phosphorylase and 0 when it behaves as a phosphatase (see Table 1.2). Given that ChpT is constitutively expressed (Chen et al. 2009), this is a valid method for model simplification. Based on the above arguments, Table 1.2 displays the resulting equations for each signaling molecule.

An alternative way of representing the logic function for a node is the use of a truth table. Such table contains the possible output states, given all the combinations of inputs coming from other regulatory nodes, as illustrated in Table 1.3 for GcrA.

Table 1.2 Logic functions for G_{2b}

Node value at $t+1$	Logic function
$CtrA_b^{t+1}$	1 if $ChpT^t$=1 AND ClpXP-$RcdA^t$=0 0 if ($ChpT^t$=1 AND ClpXP-$RcdA^t$=0) OR ($ChpT^t$=1 OR $ChpT^t$=0) AND ClpXP-$RcdA^t$=1
$DivK^{t+1}$	1 if $DivJ^t$=1 AND $PleC^t$=0 0 if $DivJ^t$=0 OR $PleC^t$=1
$PleC^{t+1}$	1 if $DivK^t$=0 0 if $DivK^t$=1
$DivJ^{t+1}$	1 if $DivK^t$=1 AND $PleC^t$=0 0 if $DivK^t$=0 OR $PleC^t$=1
$DivL^{t+1}$	1 if $DivK^t$=0 0 if $DivK^t$=1
$CckA^{t+1}$	1 if $DivL^t$=1 0 if $DivL^t$=0
$ChpT^{t+1}$	1 if $CckA^t$=1 0 if $CckA^t$=0
$CpdR^{t+1}$	1 if $ChpT^t$=1 0 if $ChpT^t$=0
ClpXP-$RcdA^{t+1}$	1 if $CpdR^t$=0 0 if $CpdR^t$=1

Table 1.3 Truth table for the expression of GcrA at time t +; 1 as a function for $CtrA_a$ and DnaA at time t.

$CtrA_a^t$	$DnaA^t$	GcrA $^{t+1}$	Biological description
0	0	0	GcrA is absent when DnaA is absent, because DnaA promotes the transcription of *gcrA*
0	1	1	GcrA is expressed when DnaA is active, because DnaA promotes the transcription of *gcrA*
1	0	0	GcrA is inhibited when CtrA is active, because CtrA inhibits the transcription of *gcrA*
1	1	0	GcrA is inhibited when CtrA is active, because CtrA inhibits the transcription of *gcrA*

1.8 Model Simulation

Having the topology of a subnetwork and the logical rules corresponding to each node, global state transitions can be computed starting at a particular initial condition. For instance, assume that at an initial time t_0 the states of the nodes CtrA, GcrA, DnaA, CcrM, and SciP in G_{2a} are given by the vector $\boldsymbol{\sigma}(t_0) = [0, 0, 1, 0, 1]$. Each position in the vector corresponds to the order in which the nodes were enumerated. Figure 1.9a shows how each node in the subnetwork changes at

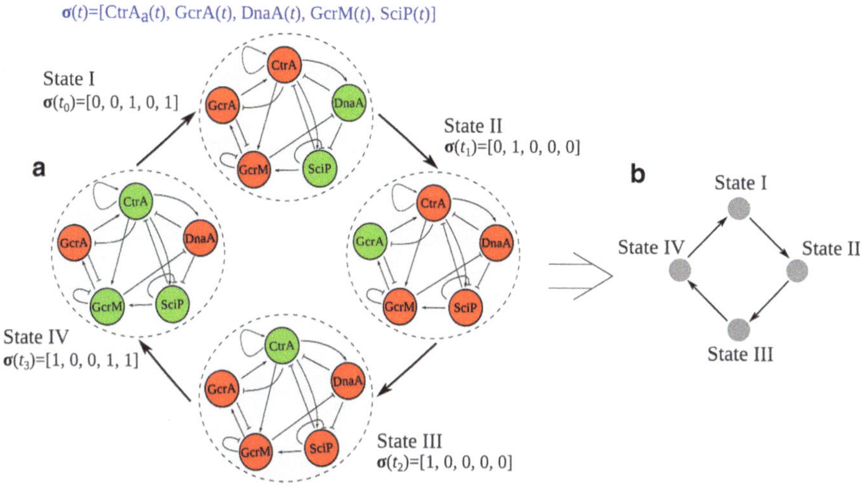

Fig. 1.9 Translation of network state transitions to an abstract state transition diagram. (**a**) The figure shows the dynamic changes in the states of the nodes corresponding to subnetwork G_{2a}. (**b**) Simplification of the dynamic state transitions displayed in (**a**)

discrete time steps according to rules about their current states and the states of their regulating nodes (see Table 1.1).

For illustrative purposes, in Fig. 1.9a the dynamics of the subnetwork G_{2a} is portrayed using state-to-state transitions. This graphical representation, however visual, gets cumbersome for large networks. Hence, to simplify the notation it is convenient to introduce the notion of a state space network (also called state transition diagram), which contains all possible combinations of network states and the possible transitions between them. The time evolution of the subnetwork can thus be viewed as trajectories in the state space network. Using this notion, a transition from state I to II in Fig. 1.9a means that the state of the biological network goes from $\boldsymbol{\sigma}(t) = [0, 0, 1, 0, 0]$ to $\boldsymbol{\sigma}(t+1) = [0, 1, 0, 0, 0]$, and this corresponds to an edge in the state transition diagram in Fig. 1.9b.

As might be expected, when the number of variables increases the manual calculation of trajectories becomes an impractical task. In consequence, software tools become imperative. In our case, we used GINsim (Chaouiya et al. 2012) to calculate the whole state space trajectories for the subnetworks G_{2a} and G_{2b}. This is a software tool explicitly designed for the automated modeling and simulation of gene regulatory networks. With this software and an appropriate visualizer, such as Cytoscape (Shannon et al. 2003), a user can compute and plot state space networks like those shown in Fig. 1.10, corresponding to the dynamics of the signaling network G_{2b} associated with the control of asymmetric cell division in *C. crescentus*. The logical models for G_{2a} and G_{2b} are available in the GINsim repository (http://ginsim.org/models_repository), under the names *g2a.zginml* and *g2b.zginml*.

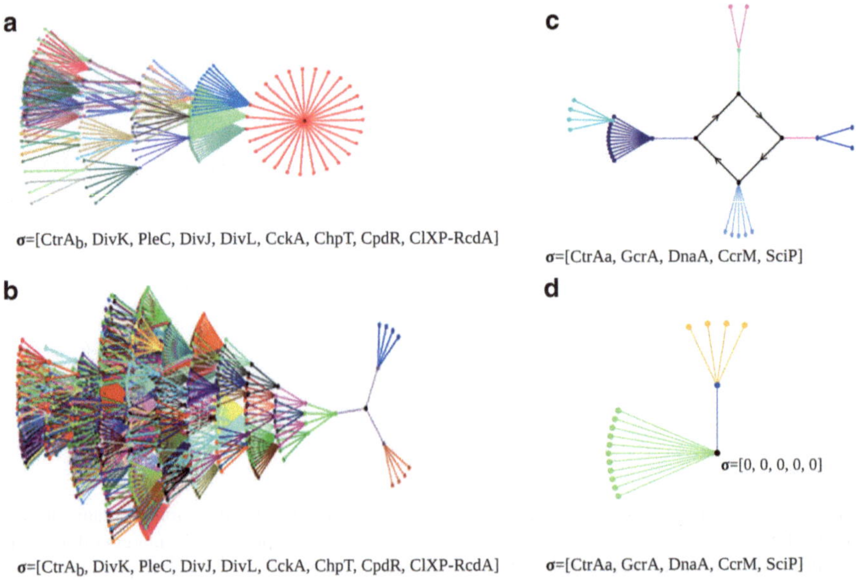

Fig. 1.10 State space network corresponding to the G_{2a} and G_{2b}. (**a**) Fraction of the state space of G_{2b} containing the point attractor corresponding to the stalked cell type. (**b**) Part of the state space of G_{2b} having the point attractor corresponding to the swarmer cell type. Both a and b form the state space of G_{2b}. (**c**) Portion of the state space of the network G_{2a} containing the cyclic attractor associated to the cell cycle. (**d**) Region of the state space of G_{2a} that converges to a point attractor whose defining values, $\boldsymbol{\sigma} = [0, 0, 0, 0, 0]$, are not attained by the real biological system. This attractor will be ignored in the interpretation of results (see Sect. 1.9). Together, (**c**) and (**d**) represent the state space of the network G_{2a}. Figure adapted from Quiñones-Valles et al. (2014)

As the reader can figure out using the logical models for G_{2a} and G_{2b}, a series of repeating network states appears after a certain simulation time. The sets of states, to which the system eventually evolves, are known as attractors. The number of states that an attractor contains determines its length (also called its period). An attractor with length greater than 1 is called a cyclic (or periodic) attractor, such as the one displayed in Fig. 1.10c. In contrast, an attractor with only one element (i.e., cycle of length one) is called a steady-state attractor or a fixed point, like the three attractors in Fig. 1.10. The simulation of the model presented in this work results in a cyclic attractor of length four and three steady-state attractors.

1.9 Interpretation and Validation of Results

The last stage of the modeling process consists of validating the model by contrasting the results obtained from the dynamic simulations of the subnetworks with experimental evidence. We mentioned in the previous section that the simulation of G_{2b} produced two fixed point attractors (see Fig. 1.10a and b). From the first point attractor shown in Fig. 1.10a, $\boldsymbol{\sigma} = [1, 0, 1, 0, 1, 1, 1, 1, 0]$, it can be inferred that the phosphorylation cascade activates $CtrA_b$ by the action of the phosphorylated form of ChpT (ChpT=1). In contrast, the state of the network in the second point attractor displayed in Fig. 1.10b, $\boldsymbol{\sigma} = [0, 1, 0, 1, 0, 0, 0, 0, 1]$, suggests that the same phosphorylation pathway inhibits the activation of $CtrA_b$, through the complementary phosphatase activity of ChpT (ChpT=0). Additionally, this form of ChpT promotes the formation of the proteolytic complex that degrades $CtrA_b$. Since the phosphorylated form of CtrA is present in the microdomain belonging to the swarmer cell type and absent in the microdomain corresponding to the stalked one, the first point attractor (Fig. 1.10a) can be made to correspond to the swarmer cell type and the second to the stalked cell type.

The simulations also show that the phosphorylation states of PleC and DivJ are mutually exclusive (i.e., either PleC=1, DivJ=0 or PleC=0, DivJ=1) and that the presence of one condition or the other determines which of the two point attractors is reached by G_{2b}. This result supports the current notion that the swarmer and stalked phenotypes arise from the spatially separated microdomains formed during the cell division of *C. crescentus*. For instance, they are consistent with the findings of Subramanian et al. (2013), whose model based on differential equations predicts a bistable behavior of PleC, acting as a phosphorylase at one pole of the cell and as a phosphatase at the other. Our results are also in agreement with the experimental observations of (Jacobs et al. 2001) who found evidence for the existence of a spatial separation of PleC and DivJ, which causes mutually exclusive conditions on their phosphorylation states at each cell pole. Furthermore, the results are coherent with the observation of spatial heterogeneity in the location of the kinases and phosphatases that mediate the asymmetric division of *C. crescentus*, more specifically, with the anchoring of PleC to the pole that gives origin to the swarmer cell, and the attachment of DivJ to the pole that corresponds to the stalked cell (Wheeler and Shapiro 1999; Viollier et al. 2002).

Complementary information on the cell cycle for each cell type can be obtained from the dynamic simulation of G_{2a}. This simulation produces one cyclic attractor of length four (see Fig. 1.11b) that resembles the gene expression pattern associated with the cell cycle of the swarmer phenotype and, in agreement with experimental evidence (Reisenauer et al. 1999a), exhibits the oscillatory behavior of the regulators DnaA, CtrA, and GcrA. Because the presented model does not reproduce the division cycle corresponding to the stalked cell type, our results account only partially for the observed phenotypes in *C. crescentus*. Despite this, the model is still useful to integrate known experimental observations within a theoretical framework, helping to think of the system in an intuitive manner.

In line with the above remark, the descriptive language provided by the Boolean formalism can be reasonably used to enumerate the empirically observed sequence of expression states for the division cycle of the stalked cell type, even when not predicted by the model. According to empirical evidence (Tan et al. 2010; Holtzendorff et al. 2004), such sequence happens to be very similar to the expression pattern associated with the swarmer cell type, differing only in the expression

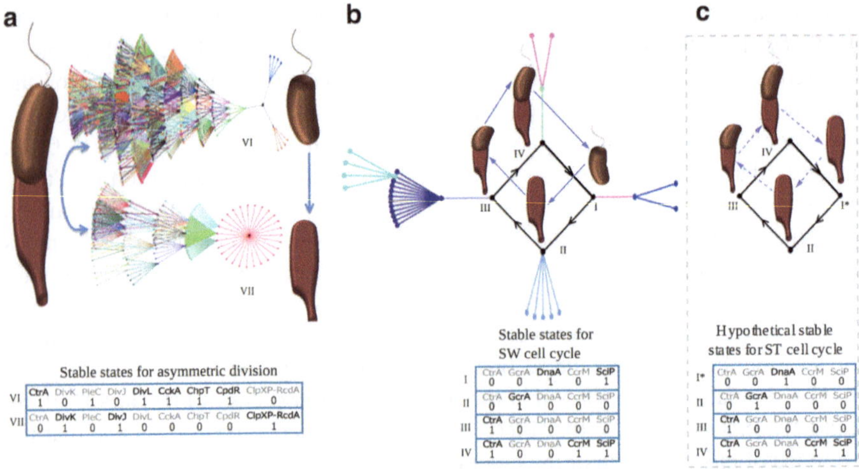

Fig. 1.11 Interpretation of network attractors in terms of cell phenotypes. (**a**) Point attractors corresponding to the states of the phosphorylation cascade of CtrA at each daughter cell, during asymmetric division, (**b**) cyclic attractor for the swarmer cell, (**c**) proposed, hypothetical cyclic attractor for the stalked cell. Adapted from Quiñones-Valles et al. (2014)

of regulators SciP and DnaA at state I (see Fig. 1.11b and c). This biological feature of the cell cycle may underlie the inability of the Boolean model presented here to predict the second attractor, which we have proposed as a hypothetical construct (Fig. 1.11c). More specifically, the deterministic and discrete nature of the model does not allow for two attractors to be different in only one state.

The aforementioned situation points, first, to the need of more expressive theoretical frameworks that address the limitations of the presented model, for example, through the incorporation of stochasticity or spatial resolution in the model (Shmulevich et al. 2002; Sanchez-Osorio et al. 2014), and second, to the iterative character of model refinement in which a simple model is usually used as an exploratory tool to gain some insight into the particular features of the studied system, and once the limitations of the current model are reached, successive rounds of model reformulation and experimental testing are applied to further improve the understanding of the phenotype of interest.

Finally, the point attractor $\boldsymbol{\sigma} = [0, 0, 0, 0, 0]$ for the network G_{2a} (see Fig. 1.10d) represent a combination of logic values that are not consistent with any observed expression pattern of the regulators being analyzed. In other words, the appearance of this attractor does not make biological sense because a situation in which all of the involved regulators (i.e., CtrA, GcrA, DnaA, CcrM, and SciP) are not expressed would correspond to a nonfunctional state of the biological system. Having no associated phenotype, such attractor has been omitted in Fig. 1.11. In the same vein, we should emphasize that models explore the whole space of states that are attainable in the abstract domain, regardless its biological feasibility. Thus, it is important to verify the validity of each of the results produced by a model and discard some of them when they are inconsistent with observed phenomena.

1.10 Conclusion

Logical modeling is a simple, yet powerful, framework to study the dynamics of regulatory networks. The emergence of complex phenotypes such as asymmetric cell division is the outcome of many interactions between genes, gene products, and small molecules that mutually affect each other or interconvert through physico-chemical reactions. While acknowledging the enormous complexity of such systems, we illustrated in this chapter that even very simple models and their simulations can provide useful insight into the mechanisms that determine the functions of cells and ultimately the emergence of biological phenotypes. In the same vein, this chapter highlighted a systematic method for model development and analysis that combined knowledge from different disciplines. We showed that well-established tools of the logic modeling approach could reduce the complexity of model development, give valuable and rapid insight into the dynamic behavior of regulatory and signaling networks, and provide practical tools for the characterization of other complex biological networks. Furthermore, they represent the simplest level of abstraction that can be used to capture gene network dynamics and explain experimental observations in a highly intuitive way. Logical modeling may thus serve as a first approximation toward the understanding of more complex mechanisms driving cell differentiation in eukaryotes.

Acknowledgements We thank the anonymous reviewer whose suggestions helped improve and clarify this manuscript and Claudine Chaouiya for critically reading the manuscript and helping with the submission of the model to the GINsim repository. We also thank Laura Carolina Diaz de León and Jaime Obeb Cervantes Pérez for their assistance with the edition of figures. IS-O and CAH-M thank the National Council of Science and Technology (CONACYT México) for postdoctoral and master fellowships, respectively.

References

Alon U (2006) An introduction to systems biology: design principles of biological circuits. CRC press, Boca Raton

Boole G (1848) The calculus of logic. Camb Dublin Math J 3(1848):183–198

Chaouiya C, Naldi A, Thieffry D (2012) Logical modelling of gene regulatory networks with GINsim. In: van Helden J, Toussaint A, Thieffry D (eds) Bacterial molecular networks: methods and protocols. Humana Press, New York, pp 463–479

Chen YE, Tsokos CG, Biondi EG, Perchuk BS, Laub MT (2009) Dynamics of two Phosphorelays controlling cell cycle progression in *Caulobacter crescentus*. J Bacteriol 191(24):7417–7429

Chien P, Perchuk BS, Laub MT, Sauer RT, Baker TA (2007) Direct and adaptor-mediated substrate recognition by an essential AAA+ protease. Proc Natl Acad Sci 104(16):6590–6595

Collier J, Murray SR, Shapiro L (2006) DnaA couples DNA replication and the expression of two cell cycle master regulators. EMBO J 25(2):346–356

Domian IJ, Quon KC, Shapiro L (1997) Cell type-specific phosphorylation and proteolysis of a transcriptional regulator controls the G1-to-S transition in a bacterial cell cycle. Cell 90 (3):415–424

Domian IJ, Reisenauer A, Shapiro L (1999) Feedback control of a master bacterial cell-cycle regulator. Proc Natl Acad Sci 96(12):6648–6653

England JC, Perchuk BS, Laub MT, Gober JW (2010) Global regulation of gene expression and cell differentiation in Caulobacter crescentus in response to nutrient availability. J Bacteriol 192(3):819–833

Espinosa-Soto C, Padilla-Longoria P, Alvarez-Buylla ER (2004) A gene regulatory network model for cell-fate determination during Arabidopsis thaliana flower development that is robust and recovers experimental gene expression profiles. Plant Cell 16(11):2923–2939

Gora KG, Tsokos CG, Chen YE, Srinivasan BS, Perchuk BS, Laub MT (2010) A cell-type-specific protein-protein interaction modulates transcriptional activity of a master regulator in Caulobacter crescentus. Mol cell 39(3):455–467

Holtzendorff J, Hung D, Brende P, Reisenauer A, Viollier PH, McAdams HH, Shapiro L (2004) Oscillating global regulators control the genetic circuit driving a bacterial cell cycle. Science 304(5673):983–987

Iniesta AA, McGrath PT, Reisenauer A, McAdams HH, Shapiro L (2006) A phospho-signaling pathway controls the localization and activity of a protease complex critical for bacterial cell cycle progression. Proc Natl Acad Sci 103(29):10935–10940

Jacobs C, Hung D, Shapiro L (2001) Dynamic localization of a cytoplasmic signal transduction response regulator controls morphogenesis during the Caulobacter cell cycle. Proc Natl Acad Sci 98(7):4095–4100

Jenal U, Fuchs T (1998) An essential protease involved in bacterial cell cycle control. EMBO J 17 (19):5658–5669

Jensen RB (2006) Coordination between chromosome replication, segregation, and cell division in Caulobacter crescentus. J Bacteriol 188(6):2244–2253

Jensen RB, Wang SC, Shapiro L (2001) A moving DNA replication factory in Caulobacter crescentus. EMBO J 20(17):4952–4963

Judd EM, Ryan KR, Moerner WE, Shapiro L, McAdams HH (2003) Fluorescence bleaching reveals asymmetric compartment formation prior to cell division in Caulobacter. Proc Natl Acad Sci 100(14):8235–8240

Laub MT, Chen SL, Shapiro L, McAdams HH (2002) Genes directly controlled by CtrA, a master regulator of the Caulobacter cell cycle. Proc Natl Acad Sci 99(7):4632–4637

Matroule JY, Lam H, Burnette DT, Jacobs-Wagner C (2004) Cytokinesis monitoring during development: rapid pole-to-pole shuttling of a signaling protein by localized kinase and phosphatase in Caulobacter. Cell 118(5):579–590

McAdams HH, Shapiro L (2009) System-level design of bacterial cell cycle control Harley. FEBS Lett 583:3984–3991

Mendoza L (2006) A network model for the control of the differentiation process in Th cells. Biosystems 84(2):101–114

Naldi A, Remy E, Thieffry D, Chaouiya C (2009) A reduction of logical regulatory graphs preserving essential dynamical properties. In: Degano P, Gorrieri R (eds) Computational methods in systems biology. Springer, Berlin, pp 24–280

Quiñones-Valles C, Sánchez-Osorio I, Martínez-Antonio A (2014) Dynamical Modeling of the Cell Cycle and Cell Fate Emergence in *Caulobacter crescentus*. PloS One 9(11):e111116

Quon KC, Yang B, Domian IJ, Shapiro L, Marczynski GT (1998) Negative control of bacterial DNA replication by a cell cycle regulatory protein that binds at the chromosome origin. Proc Natl Acad Sci 95(1):120–125

Reisenauer A, Shapiro L (2002) DNA methylation affects the cell cycle transcription of the CtrA global regulator in Caulobacter. EMBO J 21(18):4969–4977

Reisenauer A, Kahng LS, McCollum S, Shapiro L (1999a) Bacterial DNA methylation: a cell cycle regulator? J Bacteriol 181(17):5135–5139

Reisenauer A, Quon K, Shapiro L (1999b) The CtrA response regulator mediates temporal control of gene expression during the Caulobacter cell cycle. J Bacteriol 181(8):2430–2439

Ryan KR, Shapiro L (2003) Temporal and spatial regulation in prokaryotic cell cycle progression and development. Annu Rev Biochem 72(1):367–394

Sanchez-Osorio I, Ramos F, Mayorga P, Dantan E (2014) Foundations for modeling the dynamics of gene regulatory networks: a multilevel-perspective review. J Bioinform Comput Biol 12 (1):1330003

Schnell S (ed) (2017) Modeling biochemical reactions: steady-state approximation. Wiley, New York

Shannon P, Markiel A, Ozier O, Baliga NS, Wang JT, Ramage D, Amin N, Schwikowski B, Ideker T (2003) Cytoscape: a software environment for integrated models of biomolecular interaction networks. Genome Res 13(11):2498–2504

Shmulevich I, Dougherty ER, Kim S, Zhang W (2002) Probabilistic Boolean networks: a rule-based uncertainty model for gene regulatory networks. Bioinformatics 18(2):261–274

Subramanian K, Paul MR, Tyson JJ (2013) Potential role of a bistable histidine kinase switch in the asymmetric division cycle of Caulobacter crescentus. PLoS Comput Biol 9(9):e1003221

Tan MH, Kozdon JB, Shen X, Shapiro L, McAdams HH (2010) An essential transcription factor, SciP, enhances robustness of Caulobacter cell cycle regulation. Proc Natl Acad Sci 107 (44):18985–18990

Thieffry D, Thomas R (1995) Dynamical behaviour of biological regulatory networks—II. Immunity control in bacteriophage lambda. Bull Math Biol 57(2):277–297

Thomas R (1978) Logical analysis of systems comprising feedback loops. J Theor Biol 73 (4):631–656

Thomas R, D'Ari R (eds) (1990) Biological feedback. CRC-Press, Boca Raton

Thompson D, Regev A, Roy S (2015) Comparative analysis of gene regulatory networks: from network reconstruction to evolution. Annu Rev Cell Dev Biol 31:399–428

Viollier PH, Sternheim N, Shapiro L (2002) A dynamically localized histidine kinase controls the asymmetric distribution of polar pili proteins. EMBO J 21(17):4420–4428

Wagner JK, Brun YV (2007) Out on a limb: how the Caulobacter stalk can boost the study of bacterial cell shape. Mol Microbiol 64(1):28–33

Wang YR, Huang H (2014) Review on statistical methods for gene network reconstruction using expression data. J Theor Biol 362:53–61

Wheeler RT, Shapiro L (1999) Differential localization of two histidine kinases controlling bacterial cell differentiation. Mol Cell 4(5):683–694

Winzeler E, Shapiro L (1996) A novel promoter motif for Caulobacter cell cycle-controlled DNA replication genes. J Mol Biol 264(3):412–425

Chapter 2
Spatiotemporal Models of the Asymmetric Division Cycle of *Caulobacter crescentus*

Kartik Subramanian and John J. Tyson

Abstract The spatial localization of proteins within the cytoplasm of bacteria is an underappreciated but critical aspect of cell cycle regulation for many prokaryotes. In *Caulobacter crescentus*—a model organism for the study of asymmetric cell reproduction in prokaryotes—heterogeneous localization of proteins has been identified as the underlying cause of asymmetry in cell morphology, DNA replication, and cell division. However, significant questions remain. Firstly, the mechanisms by which proteins localize in the organelle-free prokaryotic cytoplasm remain obscure. Furthermore, how variations in the spatial and temporal dynamics of cell fate determinants regulate signaling pathways and orchestrate the complex programs of asymmetric cell division and differentiation are subjects of ongoing research. In this chapter, we review current efforts in investigating these two questions. We describe how mathematical models of spatiotemporal protein dynamics are being used to generate and test competing hypotheses and provide complementary insight about the control mechanisms that regulate asymmetry in protein localization and cell division.

2.1 Asymmetry in Protein Localization and Cell Division in Prokaryotes

Unlike its eukaryotic counterpart, the bacterial cytoplasm has traditionally been regarded as "well mixed," with limited need for organizing cellular components into spatial niches. However, this view of a homogeneously distributed bacterial cytoplasm has always been suspect, given the commonly observed asymmetric localization of the flagellum and virulence factors in chemotactic and pathogenic bacteria, respectively (Goldberg et al. 1993; Steinhauer et al. 1999). Moreover, the

K. Subramanian (✉)
Laboratory of Systems Pharmacology, Department of Systems Biology, Harvard Medical School, Boston, MA 02115, USA
e-mail: skartik@vt.edu

J.J. Tyson
Department of Biological Sciences, Virginia Tech, Blacksburg, VA 24061, USA

J.-P. Tassan, J.Z. Kubiak (eds.), *Asymmetric Cell Division in Development, Differentiation and Cancer*, Results and Problems in Cell Differentiation 61,
DOI 10.1007/978-3-319-53150-2_2

prevalence of asymmetric division and differentiation in bacteria—of which *Bacillus subtilis* (Wu and Errington 2003; dos Santos et al. 2012) is a well-known example—has always alluded to the presence of regulatory mechanisms that rely on spatial cues to orchestrate the cell cycle. Advances in microscopy over the past two decades have provided an unprecedented view of a highly organized landscape of proteins (Shapiro et al. 2009; Goley et al. 2007), RNA (Nevo-Dinur et al. 2012), and chromosomes (Thanbichler and Shapiro 2006a) within the cytoplasm of bacteria. Furthermore, analogues of all three major cytoskeletal elements—actin filaments, microtubules, and intermediate filaments—have all been observed to occupy spatial niches in prokaryotes (Charbon et al. 2009). These observations have spurred research into the mechanisms of cell division in prokaryotes that exhibit some form of asymmetry, such as *E. coli* (Winkler et al. 2010), *B. subtilis*, *V. cholerae* (Yamaichi et al. 2012), and *S. flexneri* (Goldberg et al. 1993; Steinhauer et al. 1999). However, it is in the aquatic, nonpathogenic bacterium *Caulobacter crescentus* that asymmetric protein localization and the resultant asymmetric division cycle are most pronounced. In addition to division (Goley et al. 2011), protein localization in *Caulobacter* also affects a range of physiological processes such as growth (Aaron et al. 2007; Kühn et al. 2010), cell shape (Charbon et al. 2009; Jin and Sun 2009), differentiation (Lam et al. 2003), motility (Briegel et al. 2008), and stringent response (Boutte et al. 2012; Henry and Crosson 2013). These readily observable physiological traits in an experimentally tractable organism have established *Caulobacter* as the model organism for the study of asymmetric division in prokaryotes.

Here, we will highlight some basic principles that have emerged from studies of asymmetric cell division in *Caulobacter*. Firstly, we discuss the concept of self-organization of cell fate determinants in generating asymmetry in the spatial distribution of cell components. Next, we discuss how self-organization—which is otherwise a spontaneous process dependent on cell length—is brought under cell cycle control and coupled with differentiation events. Finally, we will discuss how asymmetric localization of protein kinases, and the subsequent establishment of phosphorylation gradients, is leveraged by the cell to drive morphogenesis, as well as asymmetry in DNA replication, in cell division, and in the distribution of cell appendages. In each case, we will describe dynamical mathematical models that we have developed (Subramanian et al. 2013, 2014, 2015) to investigate hypotheses about the molecular mechanisms driving asymmetric cell cycle events.

There is minimal overlap among molecular players that regulate cell fate asymmetry in prokaryotes and eukaryotes; homologs of the Par family are a rare example of proteins common to both domains. However, the basic principles of the underlying molecular mechanisms of cell cycle asymmetry share many similarities. For instance, Par proteins are required for asymmetric protein distributions in developmental systems (Knoblich 2014) and in *Caulobacter* (Shebelut et al. 2010); reaction-diffusion systems for self-organization are thought to be responsible for localization patterns in *C. elegans* (Daniels et al. 2009, 2010) and in *E. coli* (Hale et al. 2001; Howard and Kruse 2005); and spatial regulation of kinase activity has been central to understanding asymmetric cell division and development in

D. melanogaster and *C. elegans* (Knoblich 2014) and in *Caulobacter* (Chen et al. 2011). The tubulin homolog FtsZ dynamically organizes into a ringlike structure close to mid-cell to initiate the process of cytokinesis (Aaron et al. 2007; Meinhardt and de Boer 2001; Thanbichler and Shapiro 2006b). MreB, which is homologous to actin filaments, regulates the process of cell elongation (Gitai et al. 2005; Takacs et al. 2010), while the intermediate filament Crescentin that gives *Caulobacter* its distinctive crescent shape shares many properties with eukaryotic intermediate filament. These include features such as spontaneous self-assembly, lateral interactions with its filamentous form, and interaction with actin filaments (Charbon et al. 2009). Given the potential similarities in mechanisms, and the experimental tractability of prokaryotes, it is conceivable that fundamental research in *Caulobacter* may reveal molecular mechanisms that drive asymmetric division and differentiation in stem cells and other developmental programs in eukaryotes.

2.2 *Caulobacter crescentus* Is a Model Organism for the Study of Asymmetry

The natural niche of *Caulobacter crescentus* is an oligotrophic, aquatic environment. As a means to restrict competition for limited resources, the bacterium has evolved an asymmetric cell division cycle. The sessile parent cell, which is attached to substratum in the environment by a stalk and holdfast, divides asymmetrically to give birth to a sessile daughter cell, as well as a motile daughter cell that is equipped with a flagellum and pili for swimming away from its place of birth (Poindexter 1981). The sessile "stalked" cell is replication competent, whereas the motile "swarmer" cell is reproductively quiescent. The holdfast, found at the tip of the stalked cell, is a polysaccharide gel that enables the stalked *Calulobacter* cell to adhere itself to almost all biotic and abiotic surfaces, such as Teflon, plastic, as well as other bacterial cells. These properties result in the stalked cell being fixed irreversibly to substratum and absorbing nutrients, despite the shear stress of flowing currents encountered in its aquatic environment (Mitchell and Smit 1990; Ausmees and Jacobs-Wagner 2003; Curtis and Brun 2010). The irreversible attachment of stalked cells may explain the requirement for a dimorphic life cycle where an obligate swarmer cell that is morphologically and functionally different is necessary. When a swarmer cell finds itself in some other favorable location, it can differentiate into a stalked cell and initiate its own rounds of cell division. Asymmetric division of stalked cells and differentiation of swarmer cells into stalked cells are the fundamental events of the life cycle of *Caulobacter* that enable this organism to survive in nutrient-limiting conditions (Curtis and Brun 2010) (Fig. 2.1).

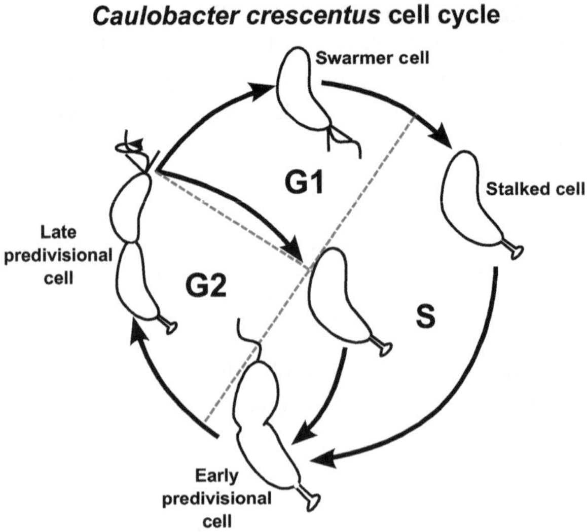

Fig. 2.1 Life cycle of *Caulobacter crescentus*. The motile swarmer cell contains a flagellum and pili, and it cannot undergo DNA replication (G1 phase). The stalked cell remains attached to substratum by its stalk and undergoes active DNA replication (S phase). The predivisional cell is equipped with a stalk at one end and a flagellum/pili at the other. It is preparing for cell division (G2 phase)

2.3 Role of Computational Models in Investigating the Cell Cycle Regulatory Network

Technological advances in microscopy have provided unprecedented insight into the localization of proteins in the bacterial cytoplasm (Sliusarenko et al. 2011). Using a combination of imaging and genetic manipulations, cell biologists are now able to infer causal links between protein localization and cell behavior. However, the mechanistic underpinnings remain elusive. How do proteins recognize cell poles and consistently choose one pole over the other? What are the mechanisms that govern temporally varying localizations of proteins? How is information about protein locations integrated into signaling and cell fate determination? Based on observations made on wild-type and mutant cells, molecular biologists have put forward hypotheses of the underlying molecular mechanisms with the aid of probable network models or schematics of the signaling system. Reasoning over network models without the aid of computational tools can be quite insightful as long as the number of distinct molecular players is limited, and the combinatorial complexity of the control mechanisms is not too great. However, we are now aware of dozens of genes and proteins taking part in cell cycle control of *Caulobacter* (Werner et al. 2009), and the complexity of the signaling network is daunting. It is improbable that one can predict—in the absence of computational tools—how mechanisms of such complexity will respond under a variety of natural and artificial conditions, in wild-type cells and mutant strains. An alternative approach is to leverage the benefits of executable dynamical computational models that can be rigorously tested in a transparent and reliable manner as compared to manual reasoning with large static schematic networks. Competing hypotheses, suggested by experimentalists and theoreticians, can be tested in an unbiased fashion.

Furthermore, computational models can be decomposed into modules that can be tested independently, to identify components critical to a certain physiological outcome. Computational models that successfully reproduce many known cellular behaviors can be used to predict outcomes under novel conditions. These predictions can, in turn, be tested experimentally. The iterative process of modeling and experimentation is of significant utility in the investigation of molecular mechanisms.

2.4 Dynamic Self-organization of Proteins During the *Caulobacter* Cell Cycle

The establishment of cell polarity is a critical step in the developmental programs of many organisms. The underlying cause of polarity is the movement of proteins to specific subcellular locations. The number of proteins that are known to localize within cells is a steadily growing list (Shapiro et al. 2009; Goley et al. 2007; Werner et al. 2009; Rudner and Losick 2010; Stekhoven et al. 2014). In most cases, the explanation for a protein localizing at a specific site is because it binds to an "upstream" protein that was previously localized at that site. An important question therefore is how the protein at the top of the hierarchy—referred to henceforth as a landmark protein—localizes in the absence of further-upstream factors. In *Caulobacter*, TipN and PodJ are two examples of proteins that are required for localization of other proteins (Lam et al. 2006; Huitema et al. 2006; Hinz et al. 2003; Curtis et al. 2012), yet the mechanism of their own localization remains unclear. In the subsequent section, we focus on the landmark protein called PopZ that is of significant interest not only because it shows dynamic self-assembly but also because it shows a range of functions that are important for the asymmetric division cycle (Ebersbach et al. 2008; Bowman et al. 2008, 2010). PopZ is a scaffolding protein not only for other proteins but also for anchoring chromosomes via interaction with the ParA–ParB replication origin complex. The polymeric protein PopZ localizes at the old pole of swarmer cells and later assumes bipolar localization shortly after the cell loses its flagellum and transitions to the stalked phenotype. Subsequent to cell division, each daughter cell inherits one focus of PopZ at their respective old poles (Fig. 2.2). PopZ polymers form scaffolds that serve as a hub to bind proteins. In addition, PopZ also anchors the chromosome (Ebersbach et al. 2008; Bowman et al. 2008). What are the factors responsible for restricting PopZ only to the poles, and what drives a second focus of PopZ to emerge at the new pole of stalked cells? There are competing but incomplete answers to these questions, none of which provides a completely satisfactory explanation for PopZ localization (Ebersbach et al. 2008; Bowman et al. 2010; Laloux and Jacobs-Wagner 2013; Ptacin et al. 2014). Suggestions include (a) localization of PopZ at the poles as a result of directed transport on MreB filaments (Bowman et al. 2008); (b) a DNA occlusion mechanism, in which PopZ

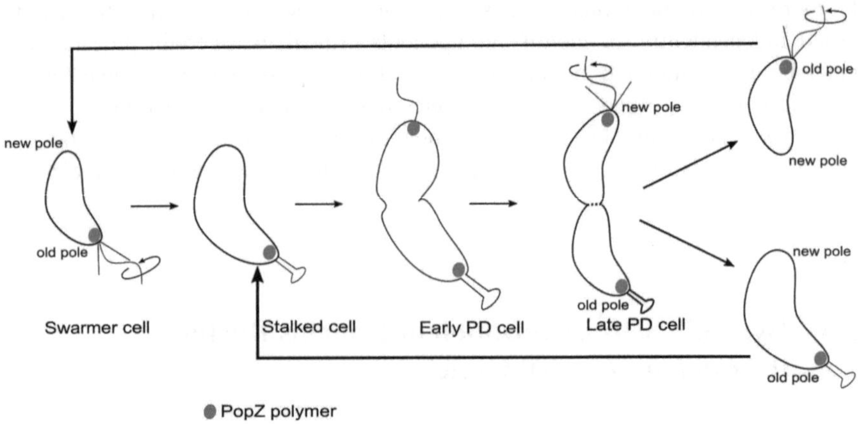

Fig. 2.2 Schematic representation of the dynamic localization of PopZ polymer during the *Caulobacter crescentus* cell cycle. PopZ polymers are present only at the "old" pole of swarmer cells. After the swarmer-to-stalked transition, PopZ polymers appear at the "new" pole, resulting in a bipolar distribution. When the cell divides, the site of cytokinesis is now designated as the "new" pole for each daughter cell. Both swarmer and stalked daughter cells inherit the PopZ polymers present at the old pole of the cell cytoplasm

polymers aggregate in DNA-free regions (Ebersbach et al. 2008; Saberi and Emberly 2010); and (c) a ParA-dependent mechanism, where ParA is suggested to nucleate the formation of PopZ polymers (Laloux and Jacobs-Wagner 2013). Based on the observed patterns of PopZ distribution during the course of the cell cycle, we speculated that a Turing-type mechanism of protein self-assembly might provide a coherent explanation for PopZ localization. In the next section, we review the basic concept of Turing-type pattern formation and the adaptations to the classical Turing mechanism that are required to explain cell cycle-dependent localization of PopZ polymers in *Caulobacter*.

2.5 Concept of Turing Mechanism

Spatial pattern formation is evident at all levels of biological organization, from single cells to whole organisms to population of interacting organisms. To achieve such complex organization at the cellular level, macromolecules have to be distributed in a highly organized, dynamic, and reproducible fashion. A fundamental question, therefore, is how molecules are able to "sense" their location and get distributed in an ordered manner. This problem of spatial self-organization in living systems was first approached by Alan Turing in his classical 1952 paper "On the chemical basis of morphogenesis" (Turing 1952). Turing, with some clarifying help from subsequent mathematical biologists (Segel and Jackson 1972), showed that a system of reacting and diffusing chemicals could generate spontaneous, stable patterns if they satisfied the following conditions: one component must be self-activating (autocatalytic) and slowly diffusing, and another rapidly diffusing component must inhibit the self-activating process. Turing proposed that such reaction-diffusion systems could account for the self-organizing patterns of morphogen gradients observed in developing embryos.

By a reaction-diffusion system, we mean a system of coupled partial differential equations such as (Eq. 2.1):

$$\frac{\partial C_1}{\partial t} = R_1(C_1, C_2) + D_1 \frac{\partial^2 C_1}{\partial x^2} \tag{2.1a}$$

$$\frac{\partial C_2}{\partial t} = R_2(C_1, C_2) + D_2 \frac{\partial^2 C_2}{\partial x^2} \tag{2.1b}$$

These equations describe the dynamics of two species, C_1 and C_2. For any species C_i, the rate of change of its concentration $C_i(x, t)$ is informed by the reactions it participates in (R_i) and its diffusion constant D_i. The reaction term is an abstraction for multiple types of molecular interaction such as synthesis, degradation, posttranslational modifications, and multimerization. Two basic reaction-diffusion schemes generate Turing patterns (Gierer and Meinhardt 1972; Meinhardt 1982): the Activator-Inhibitor Production scheme (A-IP), in which the slow-diffusing autocatalytic species (A) activates the rapidly diffusing species (I), which in turn inhibits the production of the activator; and the Activator-Substrate Depletion (A-SD) scheme, in which the slow-diffusing autocatalytic species (A) accumulates by consuming the fast-diffusing species (S), which may hence be regarded as a substrate for the activator (Fig. 2.3a). In either scenario, the activator has to diffuse significantly more slowly than the inhibitor/substrate ($D_\mathrm{i} \gg D_\mathrm{a}$, or $D_\mathrm{s} \gg D_\mathrm{a}$). When the appropriate conditions are met, the Turing instability will result in a pattern where the activator concentration is high within a spatial niche, surrounded by regions of lower activator concentration, because the surrounding regions have either higher inhibitor concentration or lower substrate concentration that cannot sustain growth of the activator. An important distinction between the classical A-IP and A-SD schemes is that the resultant patterns are phase shifted. In the case of A-IP, the activator peak is at the boundaries of the spatial domain, while in the A-SD mechanism, the activator tends to occupy the center of the spatial domain (Fig. 2.3b). In either case, spatial patterns develop within confined domains, and the distance between any two peaks or troughs of the distribution is a characteristic wavelength (λ_o) that is a function of the reaction rate constants and the diffusion constants (Segel and Jackson 1972).

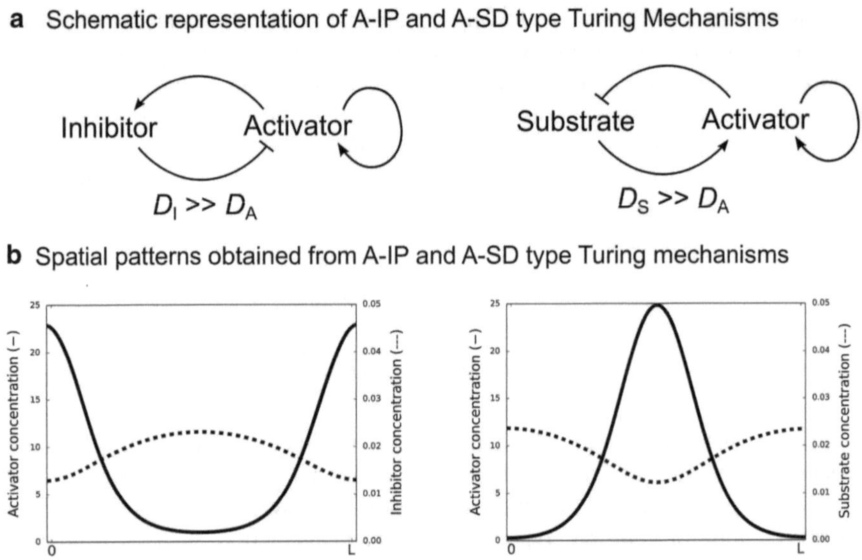

Fig. 2.3 Turing mechanisms can be broadly classified as Activator-Inhibitor Production (A-IP) or Activator-Substrate Depletion (A-SD). (**a**) In the two-chemical systems of both the AIP and A-SD type Turing mechanisms, the activator species is the one which has a slower diffusion rate, and whose level increases in an autocatalytic manner. In the A-SD mechanism, the activator level increases by the consumption of the substrate. Hence, as compared to the A-IP mechanism, the activator is thought of as inhibiting the substrate, or the substrate facilitates the production of the activator. (**b**) Spatial patterns obtained from Turing mechanisms have a characteristic wavelength λ_o. For a domain of size $L = \lambda_o$, A-IP mechanisms give rise to a pattern where two half-peaks of activator are obtained at the boundaries of the domain. In comparison, classical A-SD mechanisms give rise to patterns where the peak of activator level is at the *center*, and the troughs are at the boundaries

2.6 A Modified A-SD Can Generate Patterns with an Activator Peak at the Center of the Cell

The A-IP model results in the generation of bipolar patterns, similar to those observed in the case of PopZ. However, the underlying reaction mechanisms and diffusion timescales that form the basis of the A-IP model differ from the inherent biochemical mechanism of PopZ polymer formation. For instance, the activator in the A-IP model is a slow-moving species that produces the rapidly diffusing inhibitor. In the case of PopZ, the polymer is the slow-moving form while the monomer is the faster diffusing species. The PopZ polymer does not “activate” the monomer, neither does the monomer “inhibit” the growth of the polymer. Hence, the A-IP model is not a valid model abstraction of the PopZ reaction-diffusion system. Instead, we find that the biochemical properties of PopZ polymerization are congruent with an A-SD type Turing pattern. PopZ monomers are rapidly diffusing substrates, and PopZ polymers are the slow-moving activators that grow by consuming PopZ monomers in an autocatalytic process. However, the peaks of a typical A-SD Turing mechanism are at the center of the domain, whereas PopZ shows polar localization in *Caulobacter*. Hence, the classical A-SD equations, as developed by Gierer and Meinhardt (Gierer and Meinhardt 1972; Meinhardt and Gierer 2000), cannot directly explain the observed localization pattern of PopZ at the poles of the cell. We were motivated to modify the original A-SD mechanism in order to account for peaks at the poles of the cell. The Gierer and Meinhardt equations (Eq. 2.2) have only a single term for the generation of activators, namely through the consumption of substrates in an autocatalytic fashion. Hence, the activator peak is constantly moving toward the regions of higher substrate concentration. When the activator peak reaches the center of the cell, the concentration of substrate is equal on either side, resulting in the activator forming a stable focus there.

$$\frac{\partial a}{\partial t} = k_{aut} \cdot a^2 \cdot s - k_{\text{deg}} \cdot a + D_a \cdot \frac{\partial^2 a}{\partial x^2} \tag{2.2a}$$

$$\frac{\partial s}{\partial t} = k_{syn} - k_{\text{deg}} \cdot s - k_{aut} \cdot a^2 \cdot s + D_s \cdot \frac{\partial^2 s}{\partial x^2} \tag{2.2b}$$

We proposed that an additional source of de novo (non-autocatalytic) activator production from substrate would allow for the appearance of activator at new sites (Subramanian et al. 2014). The de novo term thus describes the production of activator from an initial pool of substrates, while the autocatalytic term describes the extension of activator by the addition of new monomeric subunits (Eq. 2.3):

$$\frac{\partial a}{\partial t} = k_{aut} \cdot a^2 \cdot s - k_{\text{deg}} \cdot a + k_{dnv} \cdot s - k_{as} \cdot a + D_a \cdot \frac{\partial^2 a}{\partial x^2} \tag{2.3a}$$

$$\frac{\partial s}{\partial t} = k_{\text{syn}} - k_{\text{deg}} \cdot s - k_{aut} \cdot a^2 \cdot s - k_{dnv} \cdot s + k_{as} \cdot a + D_s \cdot \frac{\partial^2 s}{\partial x^2} \tag{2.3b}$$

To investigate the role of the de novo term, we first simulated a cell of length $L \cong \lambda_{\text{o}}$, i.e., a length that can accommodate either a central peak or two half-peaks at the poles. First, we simulated the scenario of the classical A-SD equations. (To do this, we set the rate constants for de novo production k_{dnv} and for depolymerization k_{as} to zero.) As expected, stable peaks were obtained at the center of the spatial domain. Next, we studied the modified model with increasing values for de novo production. For small values of the de novo rate constant k_{dnv}, we found that central peaks were obtained (Fig. 2.4; left panel). However, beyond a critical value of k_{dnv}, we found peaks that were either localized at the poles or the center, depending on the initial location of the activator. If the activator was abundant at the center to begin with, it stayed at the center. However, if the activator was most abundant at one of the poles at beginning of the simulation, a bipolar distribution of peaks was obtained (Subramanian et al. 2014). In the case of *Caulobacter* PopZ polymers, an initial polar localization is inherited from the parental cell, thus excluding the

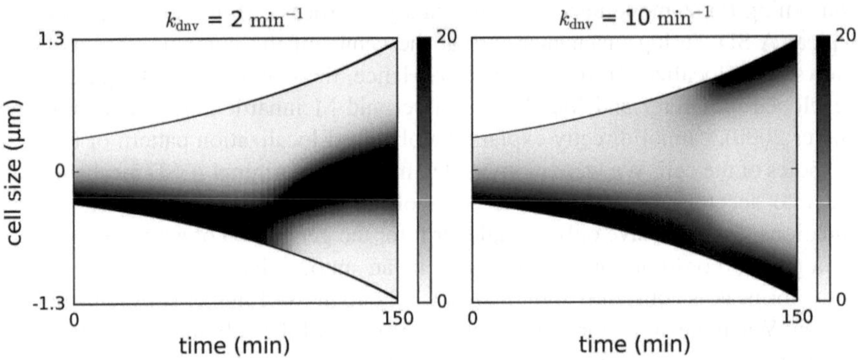

Fig. 2.4 Space–time plot of activator production in a growing cell. At $L(t = 0) = 1.3$ μm, the cell in the simulation was initialized with a bias of activator at one pole. The *color bars* indicate the concentration gradient of the activator (in dimensionless units) from low (*white*) to high (*black*). In the *left panel*, the de novo synthesis rate (k_{dnv}) was set to 2 min^{-1}, while in the *right panel*, $k_{dnv} = 10$ min^{-1}

possibility of mid-cell localization. Therefore, our modified A-SD model predicts that PopZ will consistently show a bipolar pattern of localization.

2.7 A Modified A-SD Turing Mechanism Is Sufficient to Explain Dynamic PopZ Localization

This pattern of activator localization is reminiscent of the spatiotemporal localization of PopZ. In smaller swarmer cells, PopZ is present at one of the poles of the cell; however, as the cell grows larger into a stalked cell, two foci are found at each pole. Moreover, the biochemistry of PopZ polymer growth is consistent with an A-SD-type Turing mechanism, where PopZ monomers are the substrate, and the polymers are the activator species. PopZ forms branched polymers, which have the potential to grow in an autocatalytic manner. It is reasonable to assume that the polymers diffuse much slower than the monomers. Therefore, we applied the modified A-SD-type reaction-diffusion equations to the scenario of growth of PopZ polymers. We tuned the parameters of the model such that the critical wavelength is approximately the size of the *Caulobacter* cell when bipolar patterns arise, i.e., $\lambda_o \cong 2$ μm. Our model simulation faithfully reproduces the temporal dynamics of polymer localization in a *Caulobacter* cell that doubles in size roughly every 2 h. In swarmer cells of length 1.3 μm, only a single PopZ focus is found at the old pole. As the in silico cell grows past the characteristic wavelength λ_o, a second focus spontaneously appears at the opposite pole (Fig. 2.4; right panel). Furthermore, if we allow the cell to grow well past the size at which wild-type cells normally divide, additional foci appear, each separated by a distance equal to the characteristic wavelength. This simulation result is consistent with observations in filamentous *Caulobacter* cells where many PopZ foci appear at evenly distributed intervals (Ebersbach et al. 2008). Further, the bipolar PopZ pattern with an optimal characteristic wavelength is consistently observed for a wide range of biochemical rate constants.

2.8 Coupling Bipolar Localization with Cell Cycle Events

As observed in Fig. 2.4, spatial patterns that arise from Turing mechanisms are dependent on cell length. Indeed, the hypothesis of a cell length-dependent pattern of self-assembly is consistent with observed localization of PopZ not only in normal and filamentous *Caulobacter* cells but also in smaller *E. coli* cells, where only a single focus is observed when PopZ is expressed exogenously (Ebersbach et al. 2008). However, experiments have shown that cell length is not the only determining factor for establishing bipolar PopZ localization. When DNA replication is blocked, PopZ fails to appear at the new pole even if the cell continues to grow normally (Bowman et al. 2010), suggesting that the G1-to-S transition serves as a checkpoint for new-pole PopZ localization. One study suggests a role for ParA as a nucleating factor for PopZ localization. Localization of ParA at the new pole requires the translocation of the newly replicated chromosome. The hypothesis that emerged from these observations is that DNA replication and translocation are required to localize ParA at the new pole, where it nucleates the aggregation of a new focus of PopZ (Laloux and Jacobs-Wagner 2013). In contrast, a different study shows that PopZ mutants that are unable to bind ParA retain their ability to form bipolar foci, suggesting ParA is not essential for bipolar localization (Ptacin et al. 2014). Since DNA replication seems important for bipolar localization, and since mRNA in *Caulobacter* seems to move slower than previously suggested (Montero Llopis et al. 2010), we suggested that limited dispersal of mRNA from the location of the *popZ* gene might restrict the spatial localization of PopZ polymers to sites close to the gene itself. In light of these varying opinions, it seems fair to say that the mechanism bringing PopZ accumulation under cell cycle control remains ambiguous.

To investigate the relevance of ParA as a nucleating factor for PopZ localization, we extended our equations to include a ParA-dependent polymerization term. In addition, we also modeled replication and translocation of *popZ* genes and dispersal of mRNA (Fig. 2.5). We used the extended Turing model of PopZ localization to simulate alternate scenarios. Thus far, our simulation results suggest that although not critical, the localization of ParA and that of the *popZ* gene in proximity to the new pole play important roles in enhancing the robustness of PopZ bipolarity and bringing it under cell cycle control. In the case that the *popZ* gene is normally replicated and translocated in the absence of ParA, bipolar localization is retained but delayed. Bipolarity is altogether lost if translocation of the *PopZ* gene is also inhibited.

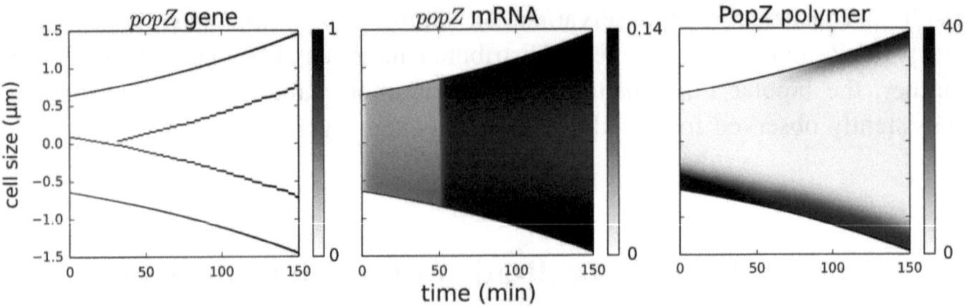

Fig. 2.5 Space–time plot of the spatiotemporal distribution of the popZ gene, mRNA, and PopZ polymer. $L(t = 0) = 1.3$ μm, only a single *popZ* gene is present at mid-cell. The initial bias in the PopZ polymer is at one of the poles. At $t = 40$ min, a second *popZ* gene is introduced to represent DNA replication. The positions of the genes are updated during the simulation. For the *popZ* gene, colors indicate the presence (*black*) or absence (*white*) of the gene. For the *popZ* mRNA and PopZ polymer, colors represent concentration gradient (in dimensionless units) from low (*white*) to high (*black*)

Given the conflicting conclusion on the role of additional factors in PopZ localization, additional experimental and modeling studies would be required to obtain a complete understanding of how the otherwise spontaneous unipolar-to-bipolar transitions in PopZ localization is brought under cell cycle control. It is likely that control of the Turing pattern by cell cycle events is not reliant on just a single factor, but a combination of cell length, ParA-dependent nucleation, and the number and positions of *popZ* genes.

2.9 The Turing Mechanism of Protein Localization Is Not Broadly Applicable

The Turing mechanism provides a theoretical framework to explain how periodic patterns develop in space. Turing patterns have been shown to reproduce observed spatial patterns across many scales of biology: the spatial distribution of Min proteins in *E. coli*, the pattern of pigmentation in skin, and the population dynamics of animals (Kondo and Miura 2010). However, experimental validation of Turing patterns is challenging and still under active consideration. More recently, it has been shown that the Turing mechanisms may not be responsible for patterns observed in stem cell proliferation and morphogenesis (Kunche et al. 2016). Its relevance in other systems remains debatable. In short, Turing mechanisms are not a universal explanation for autonomous pattern formation, and further investigation, using a combination of experiments and simulation, is required to understand the relevance of Turing-type mechanisms in self-organization of proteins in pro- and eukaryotes alike.

2.10 Asymmetry in the Localization of Bifunctional Protein Kinases Drive Differentiation and Asymmetric Cell Division

A characteristic feature of many bacteria are two-component signaling systems, comprised of histidine kinase and their response regulators (Stock et al. 2000). A two-component system is an elegant stimulus–response mechanism, in which the histidine kinase perceives signals from the extracellular environment and undergoes activation by autophosphorylation. The activated kinase in turn phosphorylates and activates a downstream response regulator that effects an appropriate physiological response (Stewart 2010). Bacterial two-component systems have been studied extensively to understand how cells adapt to their environment. The range of physiological responses includes differentiation, growth arrest, and chemotaxis (Stewart 2010; Wylie et al. 1988; Lin et al. 2010). A less understood aspect, however, is how bacteria encode and integrate spatial information in order to regulate the activity of signaling pathways that govern cell fate.

In the preceding section, we described how landmark proteins, like PopZ polymers, assemble at the cell pole in a semiautonomous, cell cycle-regulated manner. PopZ is a scaffolding protein that serves as an anchor for many proteins, including two-component histidine kinases such as DivJ and CckA (Ebersbach et al. 2008). In the following sections, we will describe how differential localization of the kinases and their binding partners affects kinase activities and, as a consequence, the state of each signaling pathway.

2.11 The Swarmer-to-Stalked and G1-to-S Transitions Are Coupled by Cross Talk Between Two-Signal Transduction Modules

An important event in the cell cycle of *Caulobacter* is the transition of the motile swarmer cell to a stalked cell. This transition is characterized by two changes; a morphological change, where the flagellum and pili are replaced by a stalk, and a biochemical change, where an impediment to DNA replication is lifted, allowing the cell to transition from G1 to S phase. The DivJ-PleC-DivK-PleD signal transduction module controls the morphological transition (Paul et al. 2008; Wheeler and Shapiro 1999), while the DivL-CckA-ChpT-CtrA signal transduction module controls the G1-to-S transition (Angelastro et al. 2010; Chen et al. 2009). The two modules are coupled by phosphorylated DivK binding and inhibiting DivL (Tsokos et al. 2011) (Fig. 2.6a). DivJ, PleC, and CckA are histidine kinases, while DivK and CtrA are response regulators. DivL is an exception in that it is a pseudo-histidine kinase (Childers et al. 2014), where the histidine residue is replaced by a tyrosine. It is therefore referred to as a tyrosine kinase (Wu et al. 1999; Sciochetti et al. 2005;

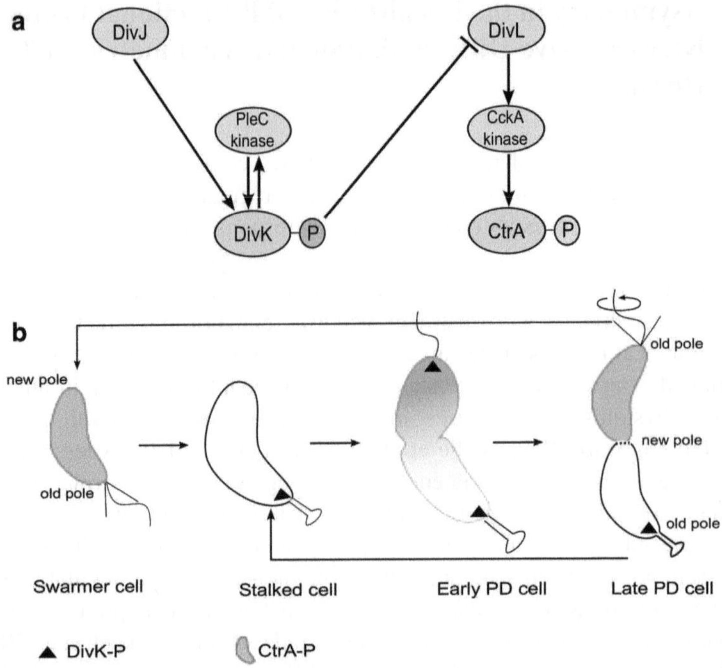

Fig. 2.6 Conundrum of simultaneous phosphorylation states of CtrA and DivK in predivisional cells. (**a**) A schematic of the coupled two-component signal transduction systems. *Lines with arrows* indicate activation, while lines with *blunt ends* represent inhibition. (**b**) Spatiotemporal distribution of DivK~P and CtrA~P during the *Caulobacter* cell cycle

Pierce et al. 2006). Although its kinase function is nonessential, DivL plays an important role in maintaining the activity of CckA as a kinase (Reisinger et al. 2007). At the molecular level, swarmer and stalked cells can be distinguished by which response regulator—DivK or CtrA—is phosphorylated. In the swarmer cell, DivK and PleD are unphosphorylated, while CtrA is phosphorylated. PleD is a diguanylate cyclase that signals the formation of the stalk via cyclic-di-GMP (Aldridge et al. 2003). However, it has to be phosphorylated in order to be active (Paul et al. 2007); hence, the swarmer cell retains its flagellum and pili instead of a stalk. CtrA is a master regulator, which, upon phosphorylation, binds the origin of replication on DNA and inhibits replication (Wortinger et al. 2000). Thus, swarmer cells—abundant in CtrA~P—are blocked in G1 phase. For the transition into a stalked cell to occur, PleD has to be phosphorylated so that it can signal formation of the stalk, while CtrA has to be dephosphorylated so that the cell can initiate DNA replication.

PleC is thought to function primarily as a phosphatase acting on DivK~P (Tsokos et al. 2011; Matroule et al. 2004), whereas CckA is thought to be primarily a kinase acting on CtrA. However, both PleC and CckA are bifunctional histidine kinases that can switch between phosphatase and kinase forms (Paul et al. 2008;

Chen et al. 2009). The relevance of the dual functions of these enzymes in cell cycle regulation is now coming to the fore. For instance, in predivisional cells, new-pole CckA is a kinase, while old-pole CckA is a phosphatase. The opposing functions at the two poles create a gradient of phosphorylated CtrA across the cell (Chen et al. 2011). As a result of this gradient in the predivisional cell, the swarmer daughter cell inherits CtrA~P and enters G1 phase, while the stalked daughter cell inherits inactive CtrA and continues DNA replication. PleC's bifunctional nature has been observed only in vitro, and its relevance to cell cycle regulation is debated. The reason for this lack of clarity is that it is not currently possible to distinguish between the functional states of histidine kinases in vivo, although efforts to develop methods are under way. Typically, change in function of bifunctional kinases is brought about by the binding of extracellular ligands. In the case of PleC, however, its own response regulator DivK brings about the change from phosphatase to kinase (Paul et al. 2008). Such interactions, where a response regulator binds to and changes the function of the histidine kinase, have, to our knowledge, been observed only in the case of the *Caulobacter* PleC-DivK two-component system. Furthermore, in vitro experiments show that DivK, irrespective of its phosphorylation status, can upregulate the kinase activity of PleC. Given that the total level of DivK remains constant throughout the cell cycle, it is interesting that PleC is believed to become a kinase only in the stalked cell.

2.12 The PleC-DivJ-DivK Module Confers Bistability to the Differentiation Program in *Caulobacter*

We formulated a mathematical model (Subramanian et al. 2013) to gain insight into the potential mechanism by which a response regulator can upregulate the activity of its histidine kinase and to understand the factors responsible for restricting the DivK-dependent upregulation of PleC kinase to the stalked stage. The model is composed of elementary chemical reactions that describe the binding of response regulator DivK (unphosphorylated) and DivK~P to the histidine kinase PleC in its phosphatase or kinase forms. As it must be, the model is consistent with the principles of allostery, thermodynamics, and detailed balance. Since DivK is a product with respect to the phosphatase form of PleC, it has lower binding affinity than DivK~P does to the phosphatase form. In contrast, DivK~P is a substrate for PleC phosphatase and therefore has a higher relative binding affinity. The relative differences in binding affinities might prevent unphosphorylated DivK from upregulating PleC kinase in the swarmer cell, unless DivK is overexpressed. If the cellular level of total DivK is such that only phosphorylated DivK can upregulate PleC kinase, then a positive feedback loop will prevail, whereby DivK~P upregulates PleC kinase, and PleC kinase in turn phosphorylates more DivK. Positive feedback loops are a common occurrence in eukaryotes and

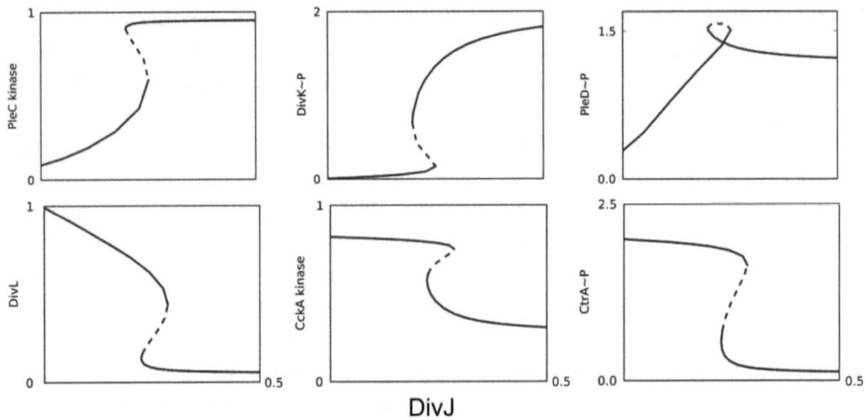

Fig. 2.7 DivJ initiates the PleC phosphatase-to-kinase transition. The plots show signal–response curves (one-parameter bifurcation diagrams) for the steady-state levels of PleC kinase, DivK~P, PleD~P, DivL, CckA kinase, and CtrA~P. The *solid lines* show stable steady states, while the *dashed lines* represent unstable steady states

prokaryotes alike as a mechanism to regulate critical transitions in a robust and switch-like manner. In the context of *Caulobacter*, the differentiation from a motile but quiescent swarmer cell to an immotile replicative stalked cell is an outcome of the cross talk between two-component signal transduction pathways, localization dynamics, and the proteolytic machinery. Our simulations show how DivJ can act as a cell fate determinant that causes a surge in PleC kinase activity, provided DivJ concentration rises above a critical threshold (a "bifurcation point"). DivK~P and PleD~P levels also increase, while CckA kinase and CtrA~P levels drop (Fig. 2.7). The positive feedback loop can generate a bistable response to DivJ activity, suggesting robust transitions that are insensitive to fluctuations in active DivJ.

2.13 Conundrum of Phosphorylation States of DivK and CtrA in the Predivisional Stage

The schematic in Fig. 2.6a, showing that DivK~P inhibits DivL, suggests that DivK and CtrA cannot be phosphorylated at the same time. Indeed, this inhibitory mechanism is critical to the swarmer-to-stalked differentiation that takes place as a result of changes in the phosphorylation of DivK and CtrA. Yet, in the late predivisional stage, both CtrA and DivK are phosphorylated (Fig. 2.6b). How can CtrA be phosphorylated in the predivisional stage when the DivK-PleC-DivK pathway is still active? Recent experimental observations have shown that DivL, which is inactivated during the swarmer-to-stalked transition, is reactivated in the predivisional stage (Chen et al. 2011). It is yet unclear how DivK~P-dependent inactivation of DivL is overcome only in the predivisional stage of the cell cycle, although some hypotheses have been put forward (Chen et al. 2011).

2.14 Contrasting Models of PleC Function

One hypothesis suggests that PleC, co-localized with DivL at the new pole of predivisional cells, is a phosphatase that continuously dephosphorylates DivK~P, thus protecting DivL by "inhibitor dephosphorylation" (Chen et al. 2011; Tsokos et al. 2011). However, the functional status of PleC is in itself unclear, with some studies alluding that it is a kinase (Paul et al. 2008; Thanbichler 2009; Jenal and Galperin 2009), while others considering it to be a phosphatase (Matroule et al. 2004). The notion that PleC is a kinase comes from in vitro studies that show DivK~P as an allosteric modulator of PleC kinase activity (Paul et al. 2008). Under this scenario, DivK~P can upregulate kinase activity of PleC, suggesting that PleC in predivisional stage is a kinase. Indeed, our own modeling studies, described in the previous section, demonstrated that PleC could be a kinase in predivisional cells.

The location of PleC changes from the old pole in the swarmer cell to the new pole in the stalked and predivisional cells. This event may be of significance to the question of PleC activity. The two alternate possibilities are that PleC reverts back to a phosphatase in the predivisional cell or retains its kinase activity (Fig. 2.8). Although one can now monitor localization of proteins, resolving the functional status of a protein kinase or phosphatase is not yet possible experimentally. Therefore, a mathematical model that incorporates spatial information may be useful to resolve this conundrum by testing each hypothesis across a range of contexts (for example, observed mutant states) and determining which hypothesis, if any, is most consistent with the observed facts.

2.15 Spatiotemporal Model of the Two-Component Network for Testing and Discriminating Alternate Hypotheses

In Fig. 2.7, we demonstrated how PleC shifts from a phosphatase to a kinase as a function of the cell fate determinant DivJ. The model only looked at the temporal aspects of the two-component systems during the swarmer and stalked stages of the cell cycle. To address the important question of reactivation of DivL in the predivisional cell, we extended our temporal model to include spatial localization as well (Subramanian et al. 2015). Our first goal was to investigate how the change in location of PleC from old to new pole affects its activity. If it were to revert back to a phosphatase, then the "inhibitor dephosphorylation" hypothesis is plausible and

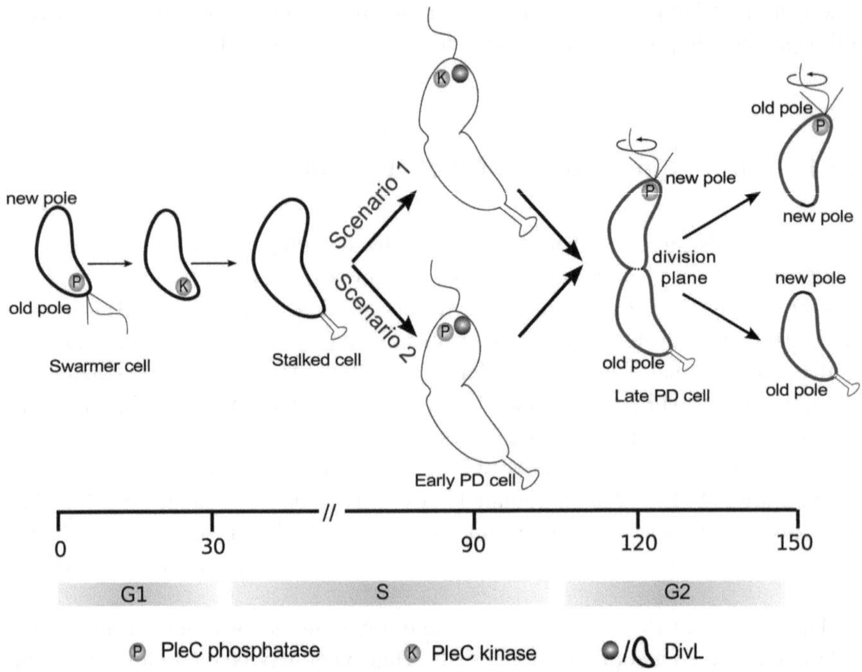

Fig. 2.8 Two scenarios for the function of PleC (kinase or phosphatase) in the early predivisional (PD) cell. Spatiotemporal dynamics of PleC [*light circles* marked as K (kinase) or P (phosphatase)] and DivL (*dark circles* for polar localization or *dark boundaries* for membrane localization) during the cell cycle under two scenarios. In scenario 1, PleC is a kinase in early PD cells. In scenario 2, PleC is a phosphatase in early PD cells

could be further tested. Alternatively, if PleC remained a kinase, then we would use the model to test alternate hypotheses of how DivL is reactivated.

A protein called PodJ is required for the correct localization of PleC and DivL (Curtis et al. 2012; Chen et al. 2006). The spatiotemporal dynamics of PodJ localization is akin to PleC, i.e., it localizes at the old (flagellar) pole in swarmer cells, and shifts to the new pole in stalked and predivisonal cells. In the absence of PodJ, PleC is delocalized. In contrast, DivL is found to localize preferentially at the opposing stalked pole (Curtis et al. 2012), indicating the presence of additional mechanisms for DivL localization. The reason for preferential localization of PodJ to the new pole in stalked cells is unclear, although TipN may be a potential landmark protein that determines its localization (Lawler and Brun 2007). CckA, another protein of significant interest, is thought to localize at the poles by binding to PopZ (Ebersbach et al. 2008). However, the localization profiles of PopZ and CckA do not always overlap during the cell cycle, suggesting the role of yet unidentified proteins required for recruiting CckA to the poles (Angelastro et al. 2010). Since the detailed mechanisms underlying the spatial and temporal distribution of PleC, DivL, and CckA are not completely understood, localization of these proteins in our model was enforced, based on experimentally observed patterns (Lam et al. 2003; Wheeler and Shapiro 1999; Angelastro et al. 2010) rather than predicted by the model itself. In stalked cells, PleC and DivK briefly co-localize, and DivK~P produced by DivJ upregulates the kinase form of PleC (Paul et al. 2008). Consistent with our temporal model, the extended spatiotemporal model showed that PleC shifts from a phosphatase to a kinase in stalked cells. However, PleC and DivJ are at opposite poles in the predivisional cells on account of PleC relocating to the new pole. Our simulations (Subramanian et al. 2015) showed that the functional state of PleC in predivisional cells was now sensitive to the diffusion rate of DivK. For slow-diffusing DivK (~0.1 $\mu m^2\ min^{-1}$), PleC reverted back to a phosphatase. For higher rates of diffusion ($>1\ \mu m^2\ min^{-1}$), PleC retained kinase activity on account of allosteric upregulation by DivK~P (Fig. 2.9a). Fluorescence-loss-in-photobleaching (FLIP) experiments show that DivK shuttles from pole to pole in about 5 s, indicating a diffusion rate in the range 20–100 $\mu m^2\ min^{-1}$ (Matroule et al. 2004). Therefore, within physiological estimates of diffusion rates, the spatiotemporal model supports the hypothesis that PleC is a kinase in predivisional cells. Only after cytokinesis separates DivJ and PleC into separate compartments does PleC revert back to a phosphatase.

If PleC in the predivisional cell is not a phosphatase that dephosphorylates the inhibitor DivK~P, then how is DivL maintained in the active state? Simulations from both the temporal-only and the spatiotemporal models showed that DivK~P, due to its role of an allosteric activator, remains bound to the kinase form of PleC (Fig. 2.9b) (Subramanian et al. 2013, 2015). These results are consistent with FRET microscopy experiments that show DivK and PleC being part of a complex in the new pole. Based on these observations, we proposed that an additional role of PleC kinase in the predivisional cell might be to bind and sequester DivK~P. By varying the localization pattern of PleC kinase in our model, we came to the conclusion that sequestration rather than dephosphorylation of the inhibitor may result in the activation of DivL and the establishment of the CtrA~P gradient that is essential for asymmetric cell division (Fig. 2.9c). Our spatiotemporal model simulations are qualitatively consistent with the observed phenotypes of *pleC* depletion mutants and *divK* overexpressing strains (Subramanian et al. 2013, 2015). If either PleC kinase or DivL were mislocalized, our simulations show that DivK~P would not be effectively sequestered away from DivL. In these circumstances, DivK~P binds and co-localizes with DivL, preventing its activation. The "inhibitor sequestration" hypothesis (Subramanian et al. 2015), which awaits experimental verification, is an example of a model-driven explanation of plausible molecular mechanisms that underlie cell fate asymmetry.

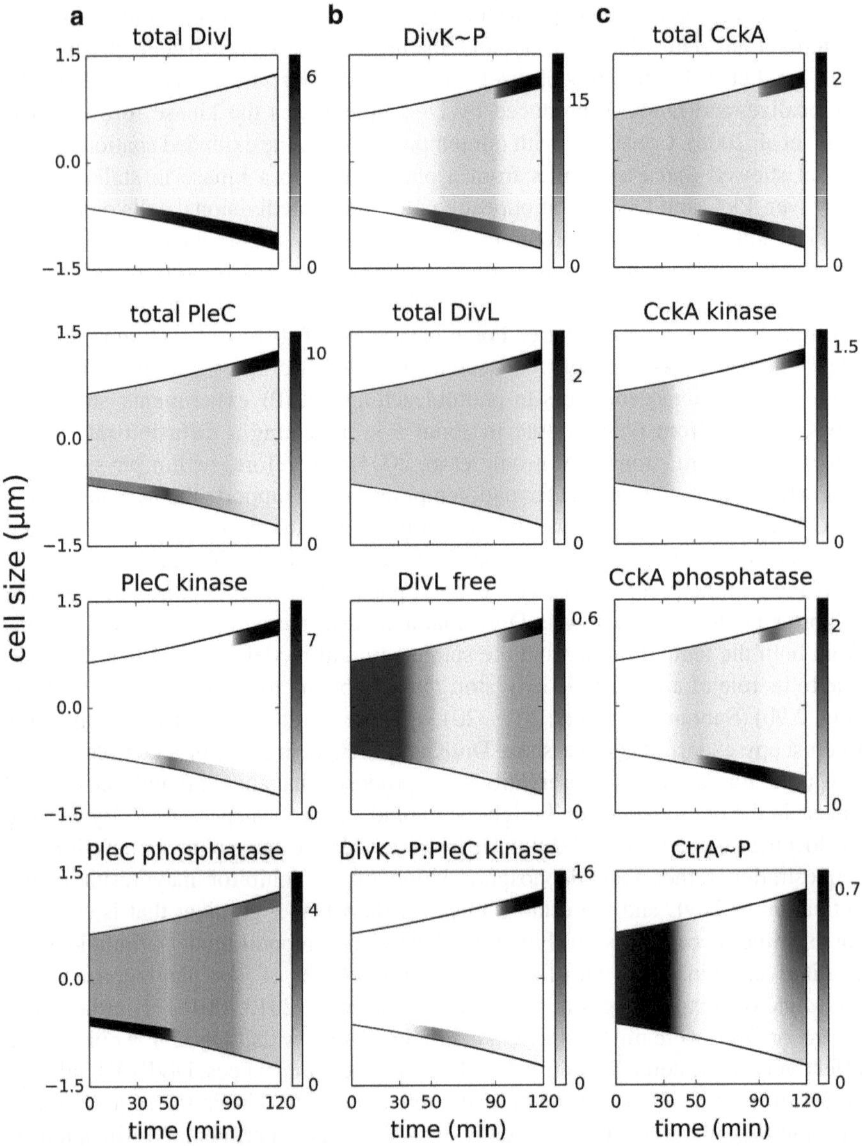

Fig. 2.9 Co-localization of PleC kinase and DivL in the predivisional cell is required for the reactivation of DivL. The plots show the change in activity and localization of the kinases and their substrates during the cell cycle (prior to cytokinesis). Colors indicate concentration gradients from minimum (*white*) to maximum (*black*). Mechanisms of DivJ, PleC, and CckA localization are not well understood, so the localization of these proteins is externally enforced in the model. At $t = 30$ min, DivJ is localized to the old pole. PleC is first localized at the old pole ($t = 0$ to $t = 50$ min) and then shifted to the new pole for the remainder of the simulation. CckA and DivL are uniformly distributed in swarmer cells. In the predivisional cell, CckA is localized at both poles, while DivL is localized only at the new pole

2.16 Discussion

Caulobacter crescentus continues to serve as an ideal model organism to study and understand asymmetric protein localization, cell division, and differentiation. Many cell cycle genes are shared between *Caulobacter* and other alpha-proteobacteria (Hallez et al. 2004; Brilli et al. 2010). These include the nitrogen-fixing bacterium *Sinorhizobium meliloti*, the plant pathogen *Agrobacterium tumefaciens*, and mammalian pathogens *Rickettsia rickettsii* and *Brucella abortus*. The basic principles that have emerged from studies in *Caulobacter* have been valuable in understanding the cell cycle of these related species that are of importance to medicine and agriculture. Turing mechanisms for self-organization of landmark proteins, the asymmetric localization of bifunctional histidine kinases, allosteric regulation of enzymes by their substrates, bistable morphogenesis, and the importance of spatial dynamics in regulating signal transduction are some examples of the elegant mechanisms in place for executing asymmetric cell division processes. The striking parallels between prokaryotes and eukaryotes with respect to proposed molecular mechanisms of asymmetric cell division are exciting, and they demonstrate the relevance of investigating bacterial systems to gain insight into the inherently more complex, regulatory mechanisms of eukaryotes.

Despite the advances in our understanding, several key questions and challenges remain. The mechanism of how PopZ localization is brought under cell cycle control needs further investigation. Although our extended Turing model offers a satisfactory and cohesive explanation, key assumptions and conclusions will have to be experimentally validated. PopZ plays a role in localizing DivJ and CckA, but the spatiotemporal profiles of the three proteins are dissimilar, indicating that additional mechanisms regulate their localization. Even less is known about how PleC shifts from the old to new pole (Viollier et al. 2002). *Caulobacter* also possesses a complex proteolytic machinery that complements phosphorylation dynamics in regulating cell differentiation and division. ClpXP is the primary effector of the proteolytic complex that degrades key cell cycle regulator such as CtrA, CpdR, KidO, and SspB. Although significant strides have been made in understanding how substrate specificity of ClpXP is modulated (Abel et al. 2011; Iniesta et al. 2006; Gora et al. 2013), little is known about the factors responsible for the dynamic localization of ClpXP during the cell cycle. The role of environmental signals in regulating the asymmetric division cycle and differentiation program is an active area of research. Recently, many of the proteins responsible for coupling metabolic control and stringent response with the cell cycle machinery are being identified and characterized (Boutte et al. 2012; Beaufay et al. 2016). Understanding the mechanisms by which these proteins transmit environmental information to cell fate pathways is of significant interest.

The technological innovations that have enabled the investigation of protein localization in prokaryotes continue to advance at a rapid pace and are becoming available to the larger experimental community. Quantitative live cell imaging at high throughput allows monitoring proteins for hundreds of cell generations

(Sliusarenko et al. 2011; Iyer-Biswas et al. 2014). High-resolution imaging studies have revealed that over 300 proteins in the *Caulobacter* proteome are capable of localization (Werner et al. 2009; Christen et al. 2010). These technologies, in combination with conventional approaches of flow cytometry, western blots, and population growth studies, are now providing large amounts of high-quality, quantitative data that will enable the comprehensive mapping of the intricate network of molecular interactions responsible for asymmetry. Molecular, single-cell, and population level data can be integrated by computational biologists to build multi-scale, highly constrained, predictive models of the *Caulobacter* cell cycle. Importantly, the wealth of data makes computational models both necessary and significantly more reliable for data analysis, hypotheses discrimination, providing explanations of molecular–level mechanisms, and making predictions. The inherent stochasticity in molecular level dynamics that translate to variable cell division times (Lin et al. 2010; Iyer-Biswas et al. 2014) can also be modeled with state-of-the-art algorithms for spatial stochastic simulations. An integrated approach of combining the best of experimental and computational tools will be required to shed light on the outstanding questions about regulatory mechanisms not only in *Caulobacter crescentus* but also for other organisms exhibiting asymmetric division cycles.

Acknowledgements The work on mathematical models presented in this chapter (Subramanian et al. 2013, 2014, 2015) was funded by the National Science Foundation (Division of Mathematical Sciences-1225160). Ongoing investigation of the Caulobacter crescentus cell cycle is currently being funded by the National Science Foundation grant (MCB-1613741). Subramanian is currently a postdoctoral fellow in the Sorger Lab at Harvard Medical School (Funding no: GM107618).

References

Aaron M, Charbon G, Lam H, Schwarz H, Vollmer W, Jacobs-Wagner C (2007) The tubulin homologue FtsZ contributes to cell elongation by guiding cell wall precursor synthesis in Caulobacter crescentus. Mol Microbiol 64:938–952

Abel S, Chien P, Wassmann P, Schirmer T, Kaever V, Laub MT, Baker TA, Jenal U (2011) Regulatory cohesion of cell cycle and cell differentiation through interlinked phosphorylation and second messenger networks. Mol Cell 43:550–560

Aldridge P, Paul R, Goymer P, Rainey P, Jenal U (2003) Role of the GGDEF regulator PleD in polar development of Caulobacter crescentus. Mol Microbiol 47:1695–1708

Angelastro PS, Sliusarenko O, Jacobs-Wagner C (2010) Polar localization of the CckA histidine kinase and cell cycle periodicity of the essential master regulator CtrA in Caulobacter crescentus. J Bacteriol 192:539–552

Ausmees N, Jacobs-Wagner C (2003) Spatial and temporal control of differentiation and cell cycle progression in Caulobacter crescentus. Annu Rev Microbiol 57:225–247

Beaufay F, De Bolle X, Hallez R (2016) Metabolic control of cell division in α-proteobacteria by a NAD-dependent glutamate dehydrogenase. Commun Integr Biol 9:e1125052

Boutte CC, Henry JT, Crosson S (2012) ppGpp and polyphosphate modulate cell cycle progression in Caulobacter crescentus. J Bacteriol 194:28–35

Bowman GR, Comolli LR, Zhu J, Eckart M, Koenig M, Downing KH, Moerner WE, Earnest T, Shapiro L (2008) A polymeric protein anchors the chromosomal origin/ParB complex at a bacterial cell pole. Cell 134:945–955

Bowman GR, Comolli LR, Gaietta GM, Fero M, Hong S-H, Jones Y, Lee JH, Downing KH, Ellisman MH, McAdams HH, Shapiro L (2010) Caulobacter PopZ forms a polar subdomain dictating sequential changes in pole composition and function. Mol Microbiol 76:173–189

Briegel A, Ding HJ, Li Z, Werner J, Gitai Z, Dias DP, Jensen RB, Jensen GJ (2008) Location and architecture of the Caulobacter crescentus chemoreceptor array. Mol Microbiol 69:30–41

Brilli M, Fondi M, Fani R, Mengoni A, Ferri L, Bazzicalupo M, Biondi EG (2010) The diversity and evolution of cell cycle regulation in alpha-proteobacteria: a comparative genomic analysis *BMC*. Syst Biol 4:52

Charbon G, Cabeen MT, Jacobs-Wagner C (2009) Bacterial intermediate filaments: in vivo assembly, organization, and dynamics of crescentin. Genes Dev 23:1131–1144

Chen JC, Hottes AK, McAdams HH, McGrath PT, Viollier PH, Shapiro L (2006) Cytokinesis signals truncation of the PodJ polarity factor by a cell cycle-regulated protease. Eur Mol Biol Organ J 25:377–386

Chen YE, Tsokos CG, Biondi EG, Perchuk BS, Laub MT (2009) Dynamics of two Phosphorelays controlling cell cycle progression in Caulobacter crescentus. J Bacteriol 191:7417–7429

Chen YE, Tropini C, Jonas K, Tsokos CG, Huang KC, Laub MT (2011) Spatial gradient of protein phosphorylation underlies replicative asymmetry in a bacterium. Proc Natl Acad Sci USA 108:1052–1057

Childers WS, Xu Q, Mann TH, Mathews II, Blair JA, Deacon AM, Shapiro L (2014) Cell fate regulation governed by a repurposed bacterial histidine kinase. PLoS Biol 12:e1001979

Christen B, Fero MJ, Hillson NJ, Bowman G, Hong S-H, Shapiro L, McAdams HH (2010) High-throughput identification of protein localization dependency networks. Proc Natl Acad Sci USA 107:4681–4686

Curtis PD, Brun YV (2010) Getting in the loop: regulation of development in Caulobacter crescentus. Microbiol Mol Biol Rev 74:13–41

Curtis PD, Quardokus EM, Lawler ML, Guo X, Klein D, Chen JC, Arnold RJ, Brun YV (2012) The scaffolding and signalling functions of a localization factor impact polar development. Mol Microbiol 84:1–24

Daniels BR, Perkins EM, Dobrowsky TM, Sun SX, Wirtz D (2009) Asymmetric enrichment of PIE-1 in the *Caenorhabditis elegans* zygote mediated by binary counterdiffusion. J Cell Biol 184:473–479

Daniels BR, Dobrowsky TM, Perkins EM, Sun SX, Wirtz D (2010) MEX-5 enrichment in the *C. elegans* early embryo mediated by differential diffusion. Development 137:2579–2585

dos Santos VT, Bisson-Filho AW, Gueiros-Filho FJ (2012) DivIVA-mediated polar localization of ComN, a posttranscriptional regulator of bacillus subtilis. J Bacteriol 194:3661–3669

Ebersbach G, Briegel A, Jensen GJ, Jacobs-Wagner C (2008) A self-associating protein critical for chromosome attachment, division, and polar organization in caulobacter. Cell 134:956–968

Gierer A, Meinhardt H (1972) A theory of biological pattern formation. Kybernetik 12:30–39

Gitai Z, Dye NA, Reisenauer A, Wachi M, Shapiro L (2005) MreB actin-mediated segregation of a specific region of a bacterial chromosome. Cell 120:329–341

Goldberg MB, Bârzu O, Parsot C, Sansonetti PJ (1993) Unipolar localization and ATPase activity of IcsA, a Shigella flexneri protein involved in intracellular movement. J Bacteriol 175:2189–2196

Goley ED, Iniesta AA, Shapiro L (2007) Cell cycle regulation in Caulobacter: location, location, location. J Cell Sci 120:3501–3507

Goley ED, Yeh YC, Hong SH, Fero MJ, Abeliuk E, Mcadams HH, Shapiro L (2011) Assembly of the Caulobacter cell division machine. Mol Microbiol 80:1680–1698

Gora KG, Cantin A, Wohlever M, Joshi KK, Perchuk BS, Chien P, Laub MT (2013) Regulated proteolysis of a transcription factor complex is critical to cell cycle progression in Caulobacter crescentus. Mol Microbiol 87:1277–1289

Hale CA, Meinhardt H, de Boer PA (2001) Dynamic localization cycle of the cell division regulator MinE in *Escherichia coli*. EMBO J 20:1563–1572
Hallez R, Bellefontaine A-F, Letesson J-J, De Bolle X (2004) Morphological and functional asymmetry in alpha-proteobacteria. Trends Microbiol 12:361–365
Henry JT, Crosson S (2013) Chromosome replication and segregation govern the biogenesis and inheritance of inorganic polyphosphate granules. Mol Biol Cell 24:3177–3186
Hinz AJ, Larson DE, Smith CS, Brun YV (2003) The Caulobacter crescentus polar organelle development protein PodJ is differentially localized and is required for polar targeting of the PleC development regulator. Mol Microbiol 47:929–941
Howard M, Kruse K (2005) Cellular organization by self-organization: mechanisms and models for Min protein dynamics. J Cell Biol 168:533–536
Huitema E, Pritchard S, Matteson D, Radhakrishnan SK, Viollier PH (2006) Bacterial birth scar proteins mark future flagellum assembly site. Cell 124:1025–1037
Iniesta AA, McGrath PT, Reisenauer A, McAdams HH, Shapiro L (2006) A phospho-signaling pathway controls the localization and activity of a protease complex critical for bacterial cell cycle progression. Proc Natl Acad Sci USA 103:10935–10940
Iyer-Biswas S, Wright CS, Henry JT, Lo K, Burov S, Lin Y, Crooks GE, Crosson S, Dinner AR, Scherer NF (2014) Scaling laws governing stochastic growth and division of single bacterial cells. Proc Natl Acad Sci USA 111:15912–15917
Jenal U, Galperin MY (2009) Single domain response regulators: molecular switches with emerging roles in cell organization and dynamics. Curr Opin Microbiol 12:152–160
Jin SK, Sun SX (2009) Morphology of Caulobacter crescentus and the mechanical role of crescentin. Biophys J 96:L47–L49
Knoblich JA (2014) Asymmetric cell division: recent developments and their implications for tumour biology. Nat Rev Mol Cell Biol 11:849–860. Europe PMC Funders Group
Kondo S, Miura T (2010) Reaction-diffusion model as a framework for understanding biological pattern formation. Science 329:1616–1620
Kühn J, Briegel A, Mörschel E, Kahnt J, Leser K, Wick S, Jensen GJ, Thanbichler M (2010) Bactofilins, a ubiquitous class of cytoskeletal proteins mediating polar localization of a cell wall synthase in Caulobacter crescentus. EMBO J 29:327–339
Kunche S, Yan H, Calof AL, Lowengrub JS, Lander AD (2016) Feedback, lineages and self-organizing morphogenesis. PLOS Comput Biol 12:e1004814
Laloux G, Jacobs-Wagner C (2013) Spatiotemporal control of PopZ localization through cell cycle-coupled multimerization. J Cell Biol 201:827–841
Lam H, Matroule J-Y, Jacobs-Wagner C (2003) The asymmetric spatial distribution of bacterial signal transduction proteins coordinates cell cycle events. Dev Cell 5:149–159
Lam H, Schofield WB, Jacobs-Wagner C (2006) A landmark protein essential for establishing and perpetuating the polarity of a bacterial cell. Cell 124:1011–1023
Lawler ML, Brun YV (2007) Advantages and mechanisms of polarity and cell shape determination in Caulobacter crescentus. Curr Opin Microbiol 10:630–637
Lin Y, Crosson S, Scherer NF (2010) Single-gene tuning of Caulobacter cell cycle period and noise, swarming motility, and surface adhesion. Mol Syst Biol 6:445
Matroule J-Y, Lam H, Burnette DT, Jacobs-Wagner C (2004) Cytokinesis monitoring during development; rapid pole-to-pole shuttling of a signaling protein by localized kinase and phosphatase in Caulobacter. Cell 118:579–590
Meinhardt H (1982) Models of biological pattern formation. Research Gate, pp 1–10
Meinhardt H, de Boer PA (2001) Pattern formation in *Escherichia coli*: a model for the pole-to-pole oscillations of Min proteins and the localization of the division site. Proc Natl Acad Sci USA 98:14202–14207
Meinhardt H, Gierer A (2000) Pattern formation by local self-activation and lateral inhibition. Bioessays 22:753–760
Mitchell D, Smit J (1990) Identification of genes affecting production of the adhesion organelle of Caulobacter crescentus CB2. J Bacteriol 172:5425–5431

Montero Llopis P, Jackson AF, Sliusarenko O, Surovtsev I, Heinritz J, Emonet T, Jacobs-Wagner C (2010) Spatial organization of the flow of genetic information in bacteria. Nature 466:77–81

Nevo-Dinur K, Govindarajan S, Amster-Choder O (2012) Subcellular localization of RNA and proteins in prokaryotes. Trends Genet 28:314–322

Paul R, Abel S, Wassmann P, Beck A, Heerklotz H, Jenal U (2007) Activation of the diguanylate cyclase PleD by phosphorylation-mediated dimerization. J Biol Chem 282:29170–29177

Paul R, Jaeger T, Abel S, Wiederkehr I, Folcher M, Biondi EG, Laub MT, Jenal U (2008) Allosteric regulation of histidine kinases by their cognate response regulator determines cell fate. Cell 133:452–461

Pierce DL, O'Donnol DS, Allen RC, Javens JW, Quardokus EM, Brun YV (2006) Mutations in DivL and CckA rescue a divJ null mutant of Caulobacter crescentus by reducing the activity of CtrA. J Bacteriol 188:2473–2482

Poindexter JS (1981) The caulobacters: ubiquitous unusual bacteria. Microbiol Rev 45:123–179

Ptacin JL, Gahlmann A, Bowman GR, Perez AM, von Diezmann ARS, Eckart MR, Moerner WE, Shapiro L (2014) Bacterial scaffold directs pole-specific centromere segregation. Proc Natl Acad Sci USA 111:E2046–E2055

Reisinger SJ, Huntwork S, Viollier PH, Ryan KR (2007) DivL performs critical cell cycle functions in caulobacter crescentus independent of kinase activity. J Bacteriol 189:8308–8320

Rudner DZ, Losick R (2010) Protein subcellular localization in bacteria. Cold Spring Harb Perspect Biol 2:a000307

Saberi S, Emberly E (2010) Chromosome driven spatial patterning of proteins in bacteria. Briggs JM (ed). PLoS Comput Biol 6:e1000986

Sciochetti SA, Ohta N, Newton A (2005) The role of polar localization in the function of an essential Caulobacter crescentus tyrosine kinase. Mol Microbiol 56:1467–1480

Segel LA, Jackson JL (1972) Dissipative structure: an explanation and an ecological example. J Theor Biol 37:545–559

Shapiro L, McAdams HH, Losick R (2009) Why and how bacteria localize proteins. Science 326:1225–1228

Shebelut CW, Guberman JM, Van Teeffelen S, Yakhnina AA, Gitai Z (2010) Caulobacter chromosome segregation is an ordered multistep process. Proc Natl Acad Sci USA 107:14194–14198

Sliusarenko O, Heinritz J, Emonet T, Jacobs-Wagner C (2011) High-throughput, subpixel precision analysis of bacterial morphogenesis and intracellular spatio-temporal dynamics. Mol Microbiol 80:612–627

Steinhauer J, Agha R, Pham T, Varga AW, Goldberg MB (1999) The unipolar Shigella surface protein IcsA is targeted directly to the bacterial old pole: IcsP cleavage of IcsA occurs over the entire bacterial surface. Mol Microbiol 32:367–377

Stekhoven DJ, Omasits U, Quebatte M, Dehio C, Ahrens CH (2014) Proteome-wide identification of predominant subcellular protein localizations in a bacterial model organism. J Proteomics 99:123–137

Stewart RC (2010) Protein histidine kinases: assembly of active sites and their regulation in signaling pathways. Curr Opin Microbiol 13:133–141

Stock AM, Robinson VL, Goudreau PN (2000) Two-component signal transduction. Annu Rev Biochem 69:183–215

Subramanian K, Paul MR, Tyson JJ (2013) Potential role of a bistable histidine kinase switch in the asymmetric division cycle of Caulobacter crescentus. PLoS Comput Biol 9:e1003221

Subramanian K, Paul MR, Tyson JJ (2014) De novo production of Turing activator generates polarity in pattrn formation. In: Proceedings of the Evry Sping School on Modeling Complex Biological systems in the context of Genomics, Evry, France, pp 131–142

Subramanian K, Paul MR, Tyson JJ (2015) Dynamical localization of DivL and PleC in the asymmetric division cycle of Caulobacter crescentus: a theoretical investigation of alternative models. PLoS Comput Biol 11:e1004348

Takacs CN, Poggio S, Charbon G, Pucheault M, Vollmer W, Jacobs-Wagner C (2010) MreB drives de novo rod morphogenesis in Caulobacter crescentus via remodeling of the cell wall. J Bacteriol 192:1671–1684

Thanbichler M (2009) Spatial regulation in Caulobacter crescentus. Curr Opin Microbiol 12:715–721

Thanbichler M, Shapiro L (2006a) Chromosome organization and segregation in bacteria. J Struct Biol 156:292–303

Thanbichler M, Shapiro L (2006b) MipZ, a spatial regulator coordinating chromosome segregation with cell division in Caulobacter. Cell 126:147–162

Tsokos CG, Perchuk BS, Laub MT (2011) A dynamic complex of signaling proteins uses polar localization to regulate cell fate asymmetry in Caulobacter crescentus. Dev Cell 20:329–341

Turing AM (1952) The chemical basis of morphogenesis. Philos Trans R Soc B Biol Sci 237:37–72

Viollier PH, Sternheim N, Shapiro L (2002) Identification of a localization factor for the polar positioning of bacterial structural and regulatory proteins. Proc Natl Acad Sci USA 99:13831–13836

Werner JN, Chen EY, Guberman JM, Zippilli AR, Irgon JJ, Gitai Z (2009) Quantitative genome-scale analysis of protein localization in an asymmetric bacterium. Proc Natl Acad Sci USA 106:7858–7863

Wheeler RT, Shapiro L (1999) Differential localization of two histidine kinases controlling bacterial cell differentiation. Mol Cell 4:683–694

Winkler J, Seybert A, König L, Pruggnaller S, Haselmann U, Sourjik V, Weiss M, Frangakis AS, Mogk A, Bukau B (2010) Quantitative and spatio-temporal features of protein aggregation in *Escherichia coli* and consequences on protein quality control and cellular ageing. EMBO J 29:910–923

Wortinger M, Sackett MJ, Brun YV (2000) CtrA mediates a DNA replication checkpoint that prevents cell division in Caulobacter crescentus. Eur Mol Biol Organ J 19:4503–4512

Wu LJ, Errington J (2003) RacA and the Soj-Spo0J system combine to effect polar chromosome segregation in sporulating Bacillus subtilis. Mol Microbiol 49:1463–1475

Wu J, Ohta N, Zhao J-L, Newton A (1999) A novel bacterial tyrosine kinase essential for cell division and differentiation. Proc Natl Acad Sci USA 96:13068–13073

Wylie D, Stock A, Wong CY, Stock J (1988) Sensory transduction in bacterial chemotaxis involves phosphotransfer between Che proteins. Biochem Biophys Res Commun 151:891–896

Yamaichi Y, Bruckner R, Ringgaard S, Moll A, Cameron DE, Briegel A, Jensen GJ, Davis BM, Waldor MK (2012) A multidomain hub anchors the chromosome segregation and chemotactic machinery to the bacterial pole. Genes Dev 26:2348–2360

Chapter 3
Intrinsic and Extrinsic Determinants Linking Spindle Pole Fate, Spindle Polarity, and Asymmetric Cell Division in the Budding Yeast *S. cerevisiae*

Marco Geymonat and Marisa Segal

Abstract The budding yeast *S. cerevisiae* is a powerful model to understand the multiple layers of control driving an asymmetric cell division. In budding yeast, asymmetric targeting of the spindle poles to the mother and bud cell compartments respectively orients the mitotic spindle along the mother–bud axis. This program exploits an intrinsic functional asymmetry arising from the age distinction between the spindle poles—one inherited from the preceding division and the other newly assembled. Extrinsic mechanisms convert this age distinction into differential fate. Execution of this program couples spindle orientation with the segregation of the older spindle pole to the bud. Remarkably, similar stereotyped patterns of inheritance occur in self-renewing stem cell divisions underscoring the general importance of studying spindle polarity and differential fate in yeast. Here, we review the mechanisms accounting for this pivotal interplay between intrinsic and extrinsic asymmetries that translate spindle pole age into differential fate.

Abbreviations

+TIP	microtubule plus-end tracking protein
aMT	astral microtubule
APC	adenomatous polyposis coli
CDK	cyclin-dependent kinase
FEAR	Fourteen Early Anaphase Release
GAP	GTPase-Activating Protein
GDI	Guanyl Dissociation Inhibitor
GEF	Guanyl Nucleotide Exchange Factor
MEN	Mitotic Exit Network
PAK	p21-activated kinase

M. Geymonat • M. Segal (✉)
Department of Genetics, University of Cambridge, Downing Street, Cambridge CB2 3EH, UK
e-mail: ms433@cam.ac.uk

J.-P. Tassan, J.Z. Kubiak (eds.), *Asymmetric Cell Division in Development, Differentiation and Cancer*, Results and Problems in Cell Differentiation 61,
DOI 10.1007/978-3-319-53150-2_3

SPB	spindle pole body
SPB_{bud}	SPB destined to the daughter cell
SPOC	Spindle Position Checkpoint
γTC	gamma-tubulin complex

3.1 Introduction

Asymmetric cell divisions orchestrate chromosomal segregation along an axis of cell polarity in order to impart distinctive cell fates to the resulting daughter cells (Li 2013). This calls for an elaborate interplay between temporal and spatial asymmetries that determine the targeting of the poles of the mitotic spindle to opposite cell ends along this axis. Intracellular or extracellular landmarks may impose further constraints to the specification of the cell polarity axis, thus promoting cell division patterning. For example, internal or external cues may pattern asymmetric stem cell divisions that balance self-renewing and differentiating progenies either by autonomous polarization in order to partition intracellular fate factors asymmetrically (Chia et al. 2008; Januschke et al. 2013) or by changing the identity of the daughter cell exiting the stem cell microenvironment (Wang et al. 2009; Yamashita et al. 2007). It is therefore of great interest to achieve an integral view of the mechanisms underpinning such critical decision-making events.

Model systems offer unique insight into this fundamental problem. The budding yeast *Saccharomyces cerevisiae* is one of the best unicellular paradigms to explore the integral program producing an asymmetric cell division. In yeast, mother cells divide in a highly polarized manner to give rise to smaller daughter (bud) cells. Furthermore, cells designate the site for a new bud in reference to the previous division site based on distinctive molecular landmarks that pattern cell divisions characteristically in haploid versus diploid cells. In turn, those landmarks instruct the cell polarizing machinery to establish a singular axis of cell polarity—the mother–bud axis—through a set of highly conserved molecular effectors that organize a polarized cytoskeleton. Yeast cells thus determine the position of the future division site, the bud neck, coincident with the establishment of an axis of cell polarity (Bi and Park 2012; Juanes and Piatti 2016). Finally, polarized cytoskeletal organization sets the stage for differential gene expression programs in mother and daughter cells, a classic paradigm for exploring determination of cell lineages (Haber 2012).

Yeast cells adhere to the fundamental premise that spindle orientation must be coupled to the axis of cell polarity in a cell dividing asymmetrically (Segal and Bloom 2001). Shortly after assembly, the mitotic spindle becomes aligned along the mother–bud axis and later elongates across the bud neck during anaphase. Moreover, a surveillance mechanism known as the Spindle Position Checkpoint (SPOC) delays mitotic exit until one pole of the spindle has correctly translocated into the bud, effectively enforcing temporal and spatial coordination between chromosomal segregation and cytokinesis (Caydasi and Pereira 2012). Throughout, spindle morphogenesis and position are controlled from the poles by the spindle pole bodies

(SPBs), equivalent to the animal centrosome (Winey and Bloom 2012; Conduit et al. 2015). Budding yeast cells undergo closed mitosis with SPBs inserted in the persisting nuclear envelope. The nuclear face of the SPB organizes the spindle microtubules. In addition, astral microtubules (aMTs) derived from the SPB cytoplasmic face position the spindle pole through dynamic interactions with cortical factors (Huisman and Segal 2005).

Yeast spindle orientation is linked to a stereotyped pattern of spindle pole inheritance (Pereira et al. 2001). Conservative SPB duplication gives rise to a new SPB next to the old SPB from the preceding division (Winey and Bloom 2012). Yeast cells exploit this intrinsic age distinction to target the old SPB to the bud through the interplay between age-dependent built-in asymmetric function and extrinsic spatial cues. Importantly, the yeast system has effectively predicted similar asymmetries built into the centrosome cycle of higher eukaryotes (Yamashita et al. 2007; Wang et al. 2009; Rebollo et al. 2007).

Here, we will explore the integration of multiple layers of control that drive asymmetric cell division in yeast and converge to enforce the position of the mitotic spindle along the mother–bud axis. Several excellent reviews focus in depth on each of these cellular processes (Howell and Lew 2012; Moseley and Goode 2006; Winey and Bloom 2012; Bi and Park 2012; Weiss 2012). We will therefore attempt here an overview of the mechanisms for the establishment of cell polarity and landmark events of the yeast spindle pathway and proceed to elaborate on inherent spindle pole asymmetries and the instructive roles played by cell polarity in translating spindle pole age into differential fate.

3.2 Establishment of a Cell Polarity Axis and Bud Singularity

S. cerevisiae cells orient the new bud with respect to the previous division site in a stereotypical and distinctive manner in haploid and diploid cells (Fig. 3.1). Haploid cells select a new budding site adjacent to the previous (axial pattern) while diploid cells alternate the use of distal (opposite to the recent division site) and proximal ends of the cell (bipolar pattern). Instructive landmarks deposited in previous cell cycles determine these choices (Pringle et al. 1995; Bi and Park 2012). Importantly, genetic analysis demonstrates that these programs may be abrogated without impeding the formation of the bud. Instead, cells may still exploit stochastic fluctuations to undergo symmetry breaking leading to the random orientation of the new bud (Howell and Lew 2012). Feedback mechanisms play central roles in amplifying and suppressing these differences, so that despite spontaneous polarization, singularity is achieved and only one bud emerges (Howell et al. 2009, 2012; Wu et al. 2015).

Axial or bipolar landmarks instruct the local activation of the GTPase Rsr1, through the antagonistic roles of the Guanyl nucleotide Exchange Factor (GEF)

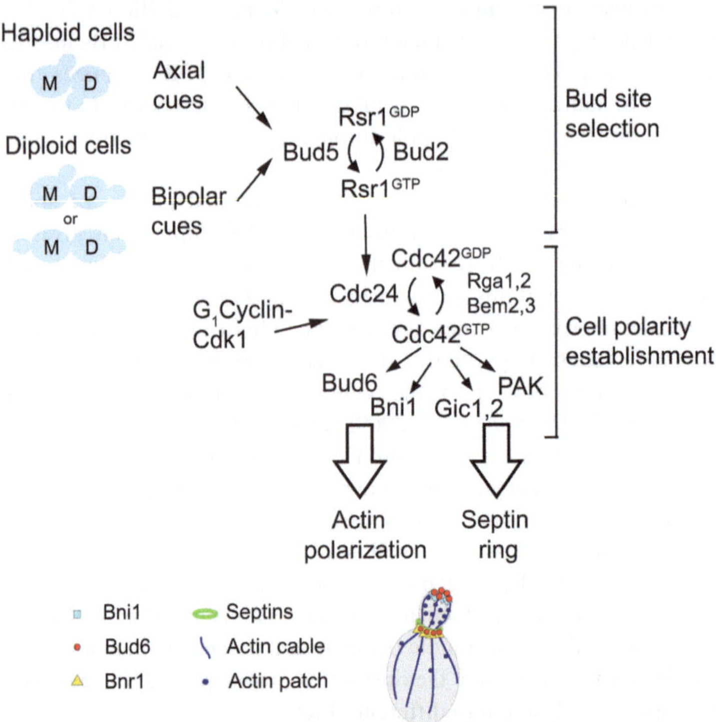

Fig. 3.1 Cell division patterning and the establishment of the mother–bud axis in *S. cerevisiae*. A simplified overview of the key players linking the program of bud site selection with the establishment of cell polarity (reviewed in Bi and Park 2012). Landmarks deposited in a previous cell cycle instruct the local burst of Cdc42 activity. Positional and cell cycle controls converge on Cdc24, the GEF for Cdc42. Downregulation is controlled by several GAPs. Active Cdc42^{GTP} recruits a number of effectors including the components of the "polarisome"—Bni1, Spa2, Pea2, and Bud6 that organize polarized actin cables from the incipient bud tip, evidencing an axis of cell polarity. Cdc42^{GTP} also promotes the assembly of the septin ring, which constitutes a scaffold supporting the recruitment of further polarity determinants and is essential to set up cytokinesis. After bud emergence, a second formin, Bnr1, nucleates polarized actin cables from the bud neck. In addition, the two formins dictate the cortical partitioning of Bud6 between the bud tip and bud neck

Bud5 and the GTPase-Activating Protein (GAP) Bud2. In turn, Rsr1^{GTP} recruits Cdc24, a GEF that promotes the local activation of the master regulator Cdc42 (Pringle et al. 1995), a critical requirement for the establishment of cell polarity (Bi and Park 2012; Woods et al. 2015; Howell and Lew 2012). Moreover, the activity of Rga1, a GAP for Cdc42, establishes an exclusion zone that prevents a new budding site from overlapping with the previous site (Tong et al. 2007). In addition, a separate mechanism centered on Gps1 (GTPase-mediated polarity switch 1) coordinates Rho1- and Cdc42-dependent signaling during cytokinesis to prevent the emergence of the new bud from clashing with ongoing events at the cell division site (Meitinger et al. 2013). Gps1 remains at the bud neck during cytokinesis for continued Rho1-dependent polarization during septation while preventing Cdc42 activation via Nba1, a protein directly interacting with Gps1 (Meitinger et al. 2014). However, Gps1 does not shift to the newly selected budding site so that it no longer inhibits bud emergence promoted by locally active $Cdc42^{GTP}$. Importantly, Nba1 is part of a protein complex remaining subsequently at cytokinesis remnants to preclude repeated use of former budding sites. Interestingly, inactivation of this mechanism has been linked to impaired nuclear segregation and shorter yeast cell life span (Meitinger et al. 2014).

Bud site selection landmarks are insufficiently focused at the cortex to restrict polarization to one point. Indeed, as is the case in cells in which the bud site selection system has been abrogated, feedback mechanisms centered on the recruitment of an activating module containing the scaffold protein Bem1, Cdc24, and an effector p21-activated kinase (PAK) may amplify GDP to GTP exchange on nearby Cdc42 molecules (Kozubowski et al. 2008; Howell and Lew 2012; Howell et al. 2012). At the same time, after initial bias is set by local activation of Rsr1, feedback helps resolve incipient polarity clusters into a single site by competition for those and other polarity factors (Wu et al. 2013, 2015). Crucially, rewiring cells to deliver Bem1 through the secretory system (staging actin-dependent feedback) allows polarization but undermines competition and resolution of polarity clusters, resulting in cells breaching singularity and producing two buds (Howell et al. 2009). The generation of an axis of cell polarity is linked to cell cycle progression as it requires activation of G_1 cyclin-dependent kinase (CDK) upon execution of START—the point for commitment into a new round of mitotic division (Hartwell 1974). G_1-CDK is required for polarizing Cdc24, Cdc42, and effector proteins, but the mechanistic details are unclear (reviewed in Howell et al. 2012).

Localized activation of Cdc42 triggers the recruitment of multiple interconnected effectors leading to actin-polarized organization and the assembly of a scaffold at the bud neck—the septin ring (Bridges and Gladfelter 2015) by separate pathways. These two systems contribute to the proper shaping of the bud and demark and confine mother and daughter cortical domains with distinctive identities. These will play key extrinsic roles in asymmetric marking leading to SPB differential fate (see below). The characteristic shape of yeast cells is promoted by the polarization of the actin cytoskeleton, while microtubules do not contribute to cell shape (Palmer et al. 1992; Huffaker et al. 1988). Actin filaments are organized into polarized actin patches that control endocytosis and linear actin cables that deliver exocytic vesicles and other components to sites of growth (Adams and Pringle 1984; Goode et al. 2015; Moseley and Goode 2006). Mother cells experience limited change in size, with the polarity machinery directing cell growth to the bud. At late stages in the cell cycle, polarity is redirected to the mother–bud neck to promote cytokinesis by the assembly of a contractile actomyosin ring and deposition of a septum (Howell and Lew 2012; Bi and Park 2012; Weiss 2012).

Actin cables are essential for mitotic spindle orientation within a temporal window ending shortly before anaphase (Theesfeld et al. 1999). They are nucleated from the bud tip by the formin Bni1 (Pruyne et al. 2002; Evangelista et al. 2002; Sagot et al. 2002; Imamura et al. 1997), a member of the Diaphanous family

(Chesarone et al. 2010). A second Diaphanous-related formin, Bnr1, organizes actin cables from the bud neck that extend into the mother cell after bud emergence (Pruyne et al. 2004; Buttery et al. 2007; Juanes and Piatti 2016). At least one of the yeast formins is essential for viability. A second cell polarity determinant recruited to the incipient bud is Bud6, a bifunctional protein acting at the interface between the actin and microtubule cytoskeletal systems (Delgehyr et al. 2008; Amberg et al. 1997; Segal et al. 2000a). Bud6 stimulates actin-cable assembly by binding formins through a C-terminal domain (Graziano et al. 2013; Delgehyr et al. 2008). Interaction with formins also governs the temporal partition of Bud6 between the bud tip and the bud neck following bud emergence (Delgehyr et al. 2008), a process additionally controlled by the axial determinant Bud3 that accumulates at the bud neck in late S phase (Segal et al. 2000a; Lord et al. 2000). Localization of Bnr1, Bud3, and Bud6 at the bud neck requires an intact septin ring (Huisman et al. 2004; Pruyne et al. 2004; Chant et al. 1995). As described below, a separate N-terminal region of Bud6 is directly involved in a program for aMT capture that hinges on Bud6 temporal partition to promote spindle polarity.

3.3 Intrinsic and Extrinsic Asymmetries Coupling the Spindle Pathway with Cell Polarity

Yeast cells single one SPB to establish contacts with the bud while confining the second SPB to the mother cell. This critical hallmark of the spindle pathway, the establishment of spindle polarity, directs the orientation of the mitotic spindle along the mother–bud axis. As summarized in Fig. 3.2 and described below, spindle polarity results from intricate cross talk between intrinsic asymmetries arising from the SPB duplication cycle and extrinsic mechanisms relaying instructions from the cell cortex to designate the pole that will enter the bud.

3.3.1 Built-In Asymmetry in the SPB Duplication Cycle: Old Versus New SPB

Unbudded yeast cells in G_1 contain a single SPB with its adjoining half-bridge inherited from the preceding division (Fig. 3.3). The SPB is a three-layered structure consisting of an inner plaque facing the nucleus, a central plaque rooted in the nuclear envelope throughout the cell cycle, and an outer plaque facing the cytoplasm. The SPB organizes separately nuclear and aMTs from the inner and outer plaques, respectively (Byers 1981; Byers and Goetsch 1975). SPB duplication is primed by the formation of a satellite at the distal end of the asymmetrically extended half-bridge in late G_1. With the execution of START and G_1-CDK activation, the satellite expands and a "new" SPB assembles side by side to the

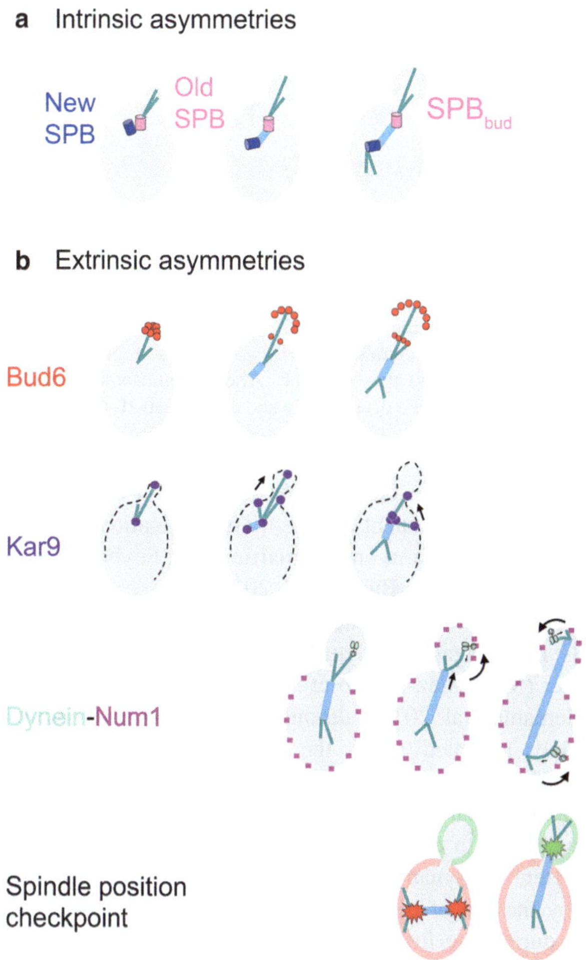

Fig. 3.2 Interplay between intrinsic and extrinsic mechanisms determines asymmetric spindle pole fate. (**a**) Intrinsic asymmetry built into the SPB duplication cycle. The old SPB can contact the bud by its existing aMTs while the new pole delays aMT organization until a spindle has formed. This asymmetry primes the inheritance of the old SPB by the bud. (**b**) Extrinsic asymmetries instructing SPB fate. Asymmetric marking of the bud cortex by Bud6 promotes aMT-mediated capture of the old SPB. Bud6 later accumulates at the bud neck confining the new SPB to the mother cell. Kar9 guides aMT plus ends toward the bud along actin cables. Kar9 is progressively polarized to deliver a single SPB to the bud. Force generation by the minus end-directed motor dynein is temporally asymmetric although not directly involved in assigning SPB fate. Dynein is first recruited to aMTs from the SPB_{bud}, while its cortical anchor Num1 initially decorates the mother cell cortex. Once Num1 gains access to the bud, dynein may be off-loaded to generate pulling force from the bud. Later, force generation redistributes to both poles (reviewed in Moore et al. 2009b). Antagonistic SPOC determinants are compartmentalized in the mother and bud cells and modulate key targets at spindle poles that will mark the SPB that enters the bud once it escapes negative regulation within the mother cell. This system operates irrespective of intrinsic SPB history, as it is always the pole ultimately translocating into the bud that becomes labeled by these components (Pereira et al. 2001)

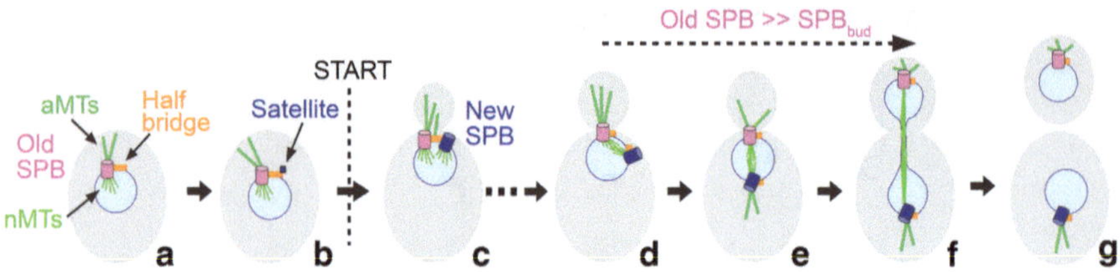

Fig. 3.3 Functional asymmetry links SPB age and fate. Cells in G_1 have a single SPB inherited from the preceding cell cycle (**a**). At the distal end of the adjoining bridge, satellite formation primes the new SPB (**b**). After START, the new SPB assembles (**c**) but lacks aMTs at the onset of spindle assembly (**d**) and stays in the mother cell. Upon spindle alignment, the old SPB is set to enter the daughter cell, i.e., it becomes the SPB_{bud} (**e**). Spindle elongation is coupled to the translocation of the SPB_{bud} across the bud neck. After chromosomal segregation and nuclear division, the cell proceeds to spindle disassembly and cytokinesis (**f–g**)

"old" SPB with the bridge in between. Additionally, a cycle of phosphorylation centered on the bridge component Sfi1 restricts duplication to once per cell cycle (Winey and Bloom 2012; Adams and Kilmartin 1999; Seybold et al. 2015; Elserafy et al. 2014; Avena et al. 2014; Burns et al. 2015).

The idea of conservative SPB duplication has been challenged by the observation that SPBs may engage in dynamic exchange and symmetric addition of components later (Menendez-Benito et al. 2013; Winey and Bloom 2012; Yoder et al. 2003; Erlemann et al. 2012), although the conservative nature of duplication per se is widely accepted as is the presence of a stereotyped pattern of age-dependent inheritance despite this dynamic backdrop (Winey and Bloom 2012; Adams and Kilmartin 1999; Pereira et al. 2001). For these reasons, here "old" and "new" refer to chronological history in the SPB duplication cycle. In addition, the extent of dynamic exchange might have been overstated in some studies either by the use of overexpression (Yoder et al. 2003) or from tracking protein history during the first cell cycle after return to growth from stationary phase (Menendez-Benito et al. 2013), when bulk remodeling of the SPB structure may occur (Byers and Goetsch 1975).

Conservative duplication paves the way for SPB age-dependent functional asymmetry. The old SPB is already competent to initiate contacts with the incipient bud by its existing aMTs, thus priming spindle polarity. By contrast, the new SPB only acquires aMTs later (Shaw et al. 1997; Segal et al. 2000b; Juanes et al. 2013). At the onset of spindle assembly, the bridge is severed and SPBs separate, each retaining a half-bridge. While the old SPB stays dynamically targeted to the bud, the new SPB, devoid of aMTs, moves away and remains confined in the mother cell. Asymmetric contacts, underscoring spindle polarity, direct the alignment of the spindle along the mother–bud axis prior to chromosomal segregation across the bud neck (Fig. 3.3). Polarity thus configured commits the old SPB to become the SPB_{bud}, i.e., the SPB destined to the daughter cell (Huisman and Segal 2005; Pereira et al. 2001). S-phase Clb5-CDK enforces the temporal asymmetry in aMT organization linked to SPB age through unknown targets (Huisman et al. 2007;

Segal et al. 2000b). Upon loss of Clb5-CDK, SPB asymmetry is abolished—both SPBs contain aMTs at the onset of spindle assembly and develop contacts with the bud cell cortex that drag the entire mitotic spindle across the bud neck, a lethal event highly penetrant in diploid cells (Segal et al. 1998).

A body of work has led to important advances regarding the spatial and temporal control of microtubule nucleation along the yeast cell cycle. Moreover, they convey crucial structural insight into aMT asymmetry and the basis for the inheritance of the old SPB by the bud. Microtubule nucleation in yeast requires the association of the conserved γ-tubulin complex (γTC) to the SPB (Lin et al. 2015). The yeast γTC consisting of Tub4 (γ-tubulin), Spc98, and Spc97 is recruited to docking sites at the nuclear and cytoplasmic faces of the SPB through separate receptors—Spc110 and Spc72, respectively (Geissler et al. 1996; Knop et al. 1997; Knop and Schiebel 1997, 1998; Pereira et al. 1998; Schiebel 2000; Marschall et al. 1996; Lin et al. 2011; Sobel and Snyder 1995). Cell cycle control of aMT organization in relation to spindle assembly may involve phosphorylation of multiple targets at the SPBs, but the molecular details are incompletely understood (Keck et al. 2011; Rock et al. 2013; Lin et al. 2011, 2014; Huisman et al. 2007).

In agreement with live imaging studies (Shaw et al. 1997; Huisman et al. 2007; Segal et al. 2000b; Ten Hoopen et al. 2012), three-dimensional ultrastructural analyses point to asymmetric aMT organization early in the spindle pathway. aMT nucleation sites occur preferentially at the SPB outer plaque throughout the cell cycle. However, aMTs emerge from both the outer plaque and the (half) bridge of the old SPB during G_1 (Byers and Goetsch 1975). Nucleation from these two sites depends on Spc72 binding to Nud1 or Kar1, respectively (Pereira et al. 1999; Gruneberg et al. 2000). During a G_1 arrest imposed by mating pheromone, Spc72 binds exclusively to Kar1 at the half-bridge to support karyogamy (nuclear fusion) and diploid formation. However, cells proceeding unperturbed through G_1 maintain Spc72 localization at the outer plaque (Pereira et al. 1999; Juanes et al. 2013), demonstrating the inherent ability of the old SPB to retain Spc72 and nucleation competence in cycling cells from both locations. After SPB duplication (Fig. 3.3), nuclear microtubules emanate from each of the side-by-side SPBs while aMTs continue to emerge from the bridge and the old SPB outer plaque (Byers and Goetsch 1975; McIntosh and O'Toole 1999; O'Toole et al. 1999). Indeed, tomography model reconstructions point to aMTs originating from one SPB outer plaque only and the bridge (McIntosh and O'Toole 1999), supporting temporal asymmetry in aMT organization.

Structural and functional asymmetry is intrinsically linked to the SPB cycle. Quantitative live imaging analysis has showed that γTCs at the SPB outer plaque favor the old SPB, a bias resulting from the acquisition of the γTC receptor Spc72, by the new SPB outer plaque after spindle assembly, while other core components of the inner, central, and outer plaques are already present at similar levels after separation of the new and old SPB (Huisman et al. 2007; Juanes et al. 2013; Yoder et al. 2003; Erlemann et al. 2012). Forcing the symmetric recruitment of Spc72 by fusion to a core component of the SPB outer plaque drives aMT organization at both SPBs from the outset and randomizes SPB identity, demonstrating that temporally asymmetric aMT nucleation directed by Spc72 links SPB history and fate (Juanes et al. 2013). Spc72 asymmetric recruitment and the pattern of SPB inheritance are retained in *kar1Δ15* cells (Juanes et al. 2013) in which Spc72 localization to the bridge is abrogated (Pereira et al. 1999) proving that asymmetric function is inherent to the SPB outer plaque.

Bulk γTC bias to the old SPB is also apparent at early stages of spindle assembly (Juanes et al. 2013). It is estimated that haploid cells contain ~22 nuclear microtubules and 3–5 aMTs per SPB. Yet, the precise timing for acquisition and saturation of nucleation capacity as well as the actual partition of γTC at the two faces of the SPB remains unknown (Erlemann et al. 2012). Crucially, three-dimensional ultrastructure analysis of early stages in spindle assembly also points to an asymmetric distribution of nuclear microtubules from the respective poles (O'Toole et al. 1997, 1999; Winey et al. 1995). In a remarkable parallel, centrosome age also governs the asymmetric behavior of spindle microtubules in human cells (Gasic et al. 2015), in addition to previously noted centrosome asymmetries in self-renewing asymmetric cell divisions (Pereira and Yamashita 2011).

In conclusion, maturation of the new SPB inner and the outer plaques proceeds sequentially. As a result, spindle assembly precedes the acquisition of nucleation competence by the new SPB outer plaque and therefore only the old SPB can connect with the bud through existing aMTs during spindle assembly. This stages the establishment of spindle polarity and the inheritance of the old SPB by the bud.

3.3.2 Extrinsic Asymmetries: Instructive Roles of Cortical Determinants in aMT Capture

Extrinsic controls for SPB fate involve cortical components operating at the interface between microtubule and actin cytoskeletal systems (Fig. 3.2) in association with a variety of motor proteins and microtubule plus-end tracking proteins or +TIPs (Akhmanova and Steinmetz 2015; Moore and Cooper 2010). These molecular partnerships generate further layers of control that govern distinctive modes of aMT–cortex interaction (Fig. 3.4) positioning the spindle poles (Yeh et al. 2000; Adames and Cooper 2000; Ten Hoopen et al. 2012; Carminati and Stearns 1997; Segal et al. 2002), although the basis for their stage-specific prevalence along the cell cycle is poorly understood.

Two polarity determinants, Bud6 and Kar9, act through separate mechanisms (Fig. 3.2) to establish SPB asymmetric fate (Huisman et al. 2004). Bud6 plays a dual role through distinct molecular partners in actin organization and aMT capture at the cell cortex to prime spindle polarity by engaging the aMTs of the old SPB with the bud (Huisman et al. 2004; Delgehyr et al. 2008; Huisman and Segal 2005; Segal et al. 2002; Ten Hoopen et al. 2012). Kar9 guides spindle orientation by linking aMTs to myosin-based transport along actin cables (Pearson and Bloom 2004). Kar9 progressive polarization consolidates the identity of the bud-ward SPB

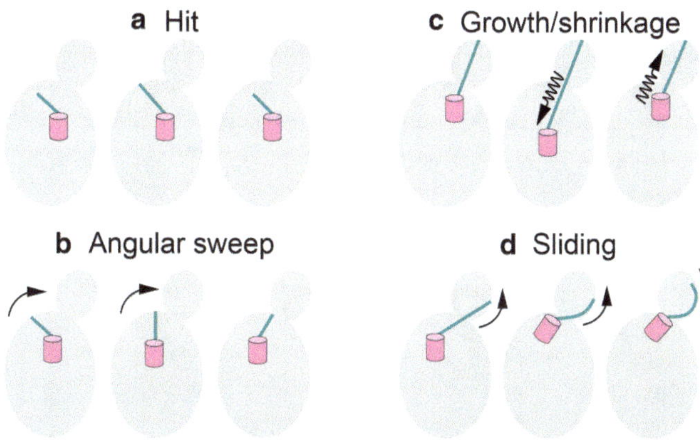

Fig. 3.4 Distinct modes of aMT plus end–cortex interaction involved in positioning the spindle poles in yeast. (**a**) Hits represent transient contacts with the cell cortex during cycles of aMT growth and shortening; (**b**) angular sweeping reflects aMT plus-end guidance toward the bud; (**c**) growth and shrinkage at the cell cortex, with aMT plus end stationary at the site of contact, cause the associated movement of the spindle pole away or toward the point of contact, respectively; (**d**) aMT lateral sliding creates pulling force for insertion of the spindle pole into the bud neck

(Cepeda-Garcia et al. 2010; Liakopoulos et al. 2003; Maekawa and Schiebel 2004; Huisman et al. 2004; Segal and Bloom 2001; Hotz et al. 2012a). How Kar9 polarized marking and SPB inheritance are unified (i.e., Kar9 bias linked to SPB age) remains an outstanding question. Finally, insertion of the spindle across the bud neck is powered by off-loading of cytoplasmic dynein onto its cortical anchor Num1, also subject to asymmetric distribution although not directly contributing to designation of the SPB_{bud}. Instead, spatial confinement of the motor and anchor secures initial pulling force into the bud during spindle elongation. Force generation later redistributes to bring each spindle pole close to the respective ends of the cell (reviewed in Moore et al. 2009b). While dynein inactivation compromises spindle position, it does not prevent the asymmetric partition of aMT–cortex contacts or the polarized marking of the SPB_{bud} by Kar9 (Cepeda-Garcia et al. 2010).

3.3.2.1 Mechanism for aMT Capture Centered on Bud6

Bud6 exerts an instructive role by sequentially directing aMTs to the bud tip and the bud neck (Fig. 3.5). As part of the polarisome, Bud6 marks the incipient bud once a new axis of cell polarity is specified (Bi and Park 2012); it spreads on the growing bud cell cortex and additionally accumulates at the bud neck during spindle assembly. At mitotic exit, Bud6 redistributes to form a double ring at the division site. Mother and daughter cells retain one of these rings that progressively disappear as Bud6 becomes recruited to the next budding site (Amberg et al. 1997; Segal et al. 2000a). Dynamic aMT contacts with Bud6 foci occur with remarkable precision throughout the cell cycle (Segal et al. 2002). Upon exit from mitosis, existing aMTs

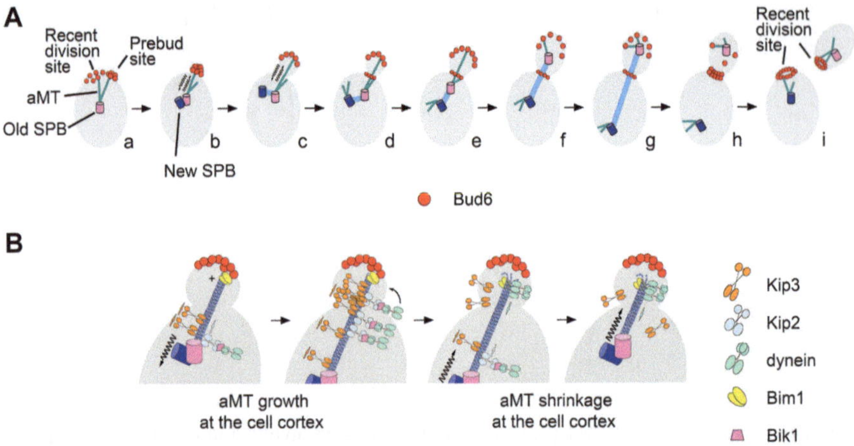

Fig. 3.5 Program of aMT–cortex interactions instructed by Bud6 that primes the commitment of the SPB_{bud}. (**A**) Landmark events in the spindle pathway in relation to Bud6-mediated aMT capture. (**a**) aMT shrinkage during contacts with the former division site marked by Bud6 repositions the SPB. Capture then shifts to the incipient bud. (**b**) Cycles of growth and shrinkage link the old SPB with the bud, maintaining the duplicated SPBs facing the bud neck. (**c**) At the onset of SPB separation and spindle assembly, the new SPB lacks aMTs and cannot establish interactions with the bud. (**d**) Then, Bud6 accumulates at the bud neck creating a new area of cortical capture that confines the new SPB to the mother cell. Bud6 also plays an instructive role in the progressive polarization of Kar9 to the old SPB, which continues to guide aMTs from this pole to the bud along actin cables (not depicted; see below). (**B**) Proposed mechanism for aMT growth and shrinkage at Bud6 cortical foci. Bud6 at the bud cortex captures aMTs via the plus-end binding protein Bim1. Cortical capture centered on Bud6 requires, in addition, the concerted action of two microtubule motor proteins—the kinesin-8 Kip3 and the minus end-directed motor dynein. During growth at the cortex upon Bud6 interaction with Bim1, Kip3 accumulates at the aMT plus end. Kip3 exhibits plus end-directed motor activity, and enrichment at plus ends induces microtubule length-dependent depolymerization. Dynein is delivered to aMT plus ends or is recruited by Bik1 (the yeast CLIP170), a cargo of the kinesin-7 Kip2, alongside other factors (reviewed in Moore et al. 2009b). Dynein tethers the receding aMT plus end to the cell cortex thus coupling aMT depolymerization with SPB movement toward the site of contact. aMTs encountering instead the cell cortex away from Bud6 foci would not be sufficiently stabilized and would undergo catastrophe

drive the old SPB toward the recent division site. Contacts then shift toward the emerging bud. Balanced cycles of aMT growth and shrinkage at the bud cell cortex, cued by Bud6, position the duplicated SPBs facing the bud neck in preparation for spindle assembly (Segal et al. 2002, 2000b; Adames and Cooper 2000). After SPB separation, Bud6 appearance at the bud neck imposes a barrier confining the new SPB to the mother cell. Spindle alignment along the mother–bud axis follows, with the old SPB becoming the SPB_{bud} (Fig. 3.5). Bud6 continues to attract aMT contacts throughout anaphase and mitotic exit, leading to the characteristic repositioning of the SPBs near the division site after cytokinesis (Huisman et al. 2004; Segal et al. 2000a, 2002; Ten Hoopen et al. 2012).

Bud6 functions in actin organization and cortical aMT capture through separate domains (Delgehyr et al. 2008). As described in Sect. 1, a C-terminal domain activates actin cable organization through interaction with formins (Graziano et al. 2013). In addition, formins determine the temporality of Bud6 partition between the bud tip and bud neck instructing spindle polarity (Delgehyr et al. 2008; Segal et al. 2000a). By contrast, a Bud6 N-terminal region captures aMTs by binding the C-terminal cargo domain of the +TIP Bim1 (Delgehyr et al. 2008; Huisman et al. 2004; Segal et al. 2002; Ten Hoopen et al. 2012), the yeast homologue of human EB1(Akhmanova and Steinmetz 2015). Astral MTs contacting Bud6 sites prolong their interaction with the cell cortex and typically couple MT growth and shrinkage with SPB movement (Segal et al. 2002; Ten Hoopen et al. 2012). Indeed, the precision of aMT interactions with Bud6 foci and the ensuing shrinkage and growth at the cell cortex are abolished by a *bim1Δ* mutation. Conversely, a Bud6 mutant unable to bind Bim1 supports correct actin organization but cannot promote aMT capture (Segal et al. 2002; Ten Hoopen et al. 2012).

Bim1 at growing aMT plus ends stabilizes growth at the cell cortex upon encountering Bud6 cortical foci (Fig. 3.5). This is antagonized by the kinesin-8 Kip3, a length-dependent microtubule depolymerase (Gupta et al. 2006; Su et al. 2011; Ten Hoopen et al. 2012). Depolymerization may terminate the cortical contact. Yet, a depolymerizing aMT may also couple SPB movement toward the site of contact when the receding plus end is tethered to the cell cortex. This requires minus end-directed dynein persisting at the aMT plus end during contact (Ten Hoopen et al. 2012). The involvement of dynein in Bud6-dependent aMT shrinkage, as revealed by live imaging studies, is surprising given the accepted view that dynein promotes aMT sliding to power translocation of one spindle pole across the bud neck in anaphase, with Num1 as its cortical anchor (Moore et al. 2009b). Furthermore, Num1 is dispensable for aMT shrinkage at the cell cortex (Ten Hoopen et al. 2012) while dynein is essential (Carminati and Stearns 1997). Similarly, "barrier-attached" dynein exerts pulling force by interacting with shrinking microtubules in vitro (Laan et al. 2012). In conclusion, Bud6 captures aMTs via Bim1, while the concerted action of Kip3 and dynein promotes aMT shrinkage driving the SPB closer to the point of contact.

Astral MT capture is redirected to mirror Bud6 abnormal distribution in formin mutants, emphasizing Bud6's instructive role in orienting aMT attachments. While *bni1Δ* mutant cells continue to tether the duplicated SPBs to the incipient bud (Segal et al. 2000a), they exhibit a tendency for symmetric interactions from both SPBs with the bud neck due to excessive and premature accumulation of Bud6 there, which perturbs SPB identity (Delgehyr et al. 2008; Segal et al. 2000a; Yeh et al. 2000). Conversely, a *bnr1Δ* mutation selectively decreases Bud6 buildup at the bud neck and associated aMT attachments. In turn, defective preanaphase spindle retention leads to wide spindle oscillations along the mother–bud axis (Delgehyr et al. 2008). Thus, formins are not directly involved in aMT capture beyond their contribution to Bud6 cortical partition, with Bud6 linking aMT plus ends to the cortex via Bim1.

3.3.2.2 Kar9-Mediated Guidance of aMTs to the Bud

A second mechanism of extrinsic control enforcing SPB asymmetric fate involves Kar9, a protein that guides aMTs toward the bud along polarized actin tracks generated from the bud tip and bud neck by Bni1 and Bnr1, respectively, in partnership with Bud6 (Graziano et al. 2013; Delgehyr et al. 2008; Pruyne et al. 2004).

Kar9 translocates from the SPB to aMT plus ends by binding Bim1 (Fig. 3.6). At the same time, Kar9 interacts with the cargo domain of the type V myosin Myo2, thus acting as a bridge to deliver aMT plus ends to the bud through myosin-based transport (Hwang et al. 2003; Yin et al. 2000; reviewed in Pearson and Bloom 2004). Kar9-based guidance induces characteristic angular movement of aMTs of constant length (Fig. 3.4). These movements are coupled to rotation that brings the spindle in closer alignment with the mother–bud axis (Liakopoulos et al. 2003; Huisman et al. 2004). As the bud grows, transport sustains targeting of aMTs through the bud neck to be engaged with shrinkage at the bud cell cortex via Bud6 (Huisman et al. 2004; Ten Hoopen et al. 2012). Upon completion of a delivery cycle and disengagement from Myo2, Kar9 returns to the SPB on the receding aMT (Liakopoulos et al. 2003; Huisman et al. 2004; Cuschieri et al. 2006; Cepeda-Garcia et al. 2010). Close to anaphase onset, spindle orientation becomes insensitive to actin perturbation (Theesfeld et al. 1999) as dynein-dependent force takes over and promotes the insertion of the spindle into the bud neck (Moore et al. 2009b).

Kar9 is present at both poles at the onset of SPB separation, but is progressively polarized marking the SPB_{bud} after spindle assembly. As a result, Kar9 selectively guides aMT plus ends to the bud from one pole, enforcing spindle polarity (Cepeda-Garcia et al. 2010; Huisman et al. 2004; Liakopoulos et al. 2003; Maekawa and Schiebel 2004). How Kar9 polarization is achieved is a mystery, although it is generally accepted that Kar9 localization requires Bim1 and intact aMTs (Pearson and Bloom 2004; Maekawa and Schiebel 2004; Cepeda-Garcia et al. 2010). Multiple processes have been evoked, but their effect on Kar9 polarity

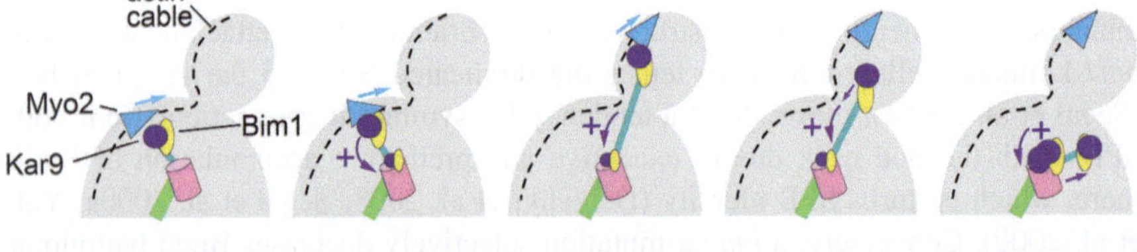

Fig. 3.6 Kar9 circuit during aMT delivery. Kar9 accumulates at aMT plus ends bound to the +TIP Bim1. While bound to plus ends, Kar9 may act as cargo of the type V myosin Myo2. Transport efficiently guides aMT plus ends toward the bud. Following a cycle of aMT plus-end delivery, Kar9 returns to the SPB on a receding aMT. The establishment of a feedback loop reinforcing Kar9 recruitment to the pole engaged by iterative cycles of transport may provide the basis for linking initial age-dependent aMT asymmetry and Kar9 bias to consolidate bud-ward fate of the old SPB

has been the topic of contrasting reports (Table 3.1). Importantly, their respective impact on Kar9–Bim1 complex formation, its dynamics, and any relationship to Myo2 transport and Kar9 bias linked to fate remain obscure. Further links have been proposed between Kar9 retention by the old SPB and the Mitotic Exit Network (MEN; see below) (Hotz et al. 2012a; Scarfone et al. 2015), although

Table 3.1 Summary of studies exploring the basis for Kar9-polarized localization

Proposed contribution	Reference
Phosphorylation by cyclin-dependent kinase	
Critical for asymmetric recruitment. Unphosphorylatable Kar9 localizes symmetrically. Cyclin mutants exhibit excess symmetry (Note: assayed in the presence of microtubule poisons)	Liakopoulos et al. (2003)
Unphosphorylatable Kar9 or cyclin mutants only show modest effect on symmetry	Maekawa and Schiebel (2004); Moore and Miller (2007); Cepeda-Garcia et al. (2010)
Kar9 localizes CDK complexes asymmetrically at aMT plus ends	Maekawa and Schiebel (2004)
CDK controls Kar9 partitioning between SPB and aMT plus end with concomitant impact on aMT dynamics, not symmetry	Maekawa and Schiebel (2004); Moore and Miller (2007)
Effect of cytoskeletal poisons	
Dismissed	Liakopoulos et al. (2003)
Actin poisons induce absolute symmetry independent of the morphogenesis checkpoint	Cepeda-Garcia et al. (2010)
Microtubule poisons induce Kar9 symmetry independent of the spindle assembly checkpoint (SAC)	Cepeda-Garcia et al. (2010); Maekawa and Schiebel (2004)
Link between Clb5-CDK, the SAC, and the MEN	Hotz et al. (2012b)
Effect of Myo2 inactivation	
No effect	Liakopoulos et al. (2003)
myo2ts inactivation induces absolute symmetry. Myo2 mutants retaining Kar9 binding support asymmetric loading. Thus, Kar9 polarization is linked to its ability to act as cargo in Myo2-based transport	Cepeda-Garcia et al. (2010)
Other	
SUMO-targeted ubiquitylation. Modulation of Kar9 turnover may impact asymmetric loading	Schweiggert et al. (2016)
The MEN kinase Dbf2 promotes Kar9 bias to the old SPB at metaphase. SPOC mutants retain Kar9 polarity	Hotz et al. (2012a)
aMT asymmetric organization determines Kar9 bias to the old SPB at the onset of spindle assembly	Juanes et al. (2013)
Mutations increasing Tem1 symmetry impair Kar9 polarity	Scarfone et al. (2015)

such mechanisms may not explain the apparent bias of Kar9 for the old pole already at the onset of SPB separation (Juanes et al. 2013).

3.3.2.3 Instructive Role of Cell Polarity in Kar9 Polarization and the Pattern of SPB Inheritance

As an alternative view, a cytoskeletal-centric proposal may provide a basis for unifying Kar9 asymmetric marking and the choice of the old SPB as SPB_{bud}. At the heart of this proposal is Kar9's inherent ability to build polarity through positive feedback (Fig. 3.6) set by Kar9 recycling along aMTs to the pole engaged in Myo2-based transport (Cepeda-Garcia et al. 2010; Cuschieri et al. 2006). A number of observations support this proposal:

- Kar9 polarity depends on cell polarity determinants, intact actin cables, and microtubules. Moreover, polarity is selectively disabled by *myo2* alleles preventing Kar9 binding to the Myo2 cargo domain, showing that engagement in Myo2-dependent delivery of aMT plus ends instructs and maintains Kar9-polarized localization to the SPB_{bud} (Cepeda-Garcia et al. 2010). Recycling based on Kar9 return on a depolymerizing aMT after a transport event increases localization at the SPB engaged in delivery (Cepeda-Garcia et al. 2010; Huisman et al. 2004; Liakopoulos et al. 2003; Cuschieri et al. 2006).
- Kar9 favors the old SPB at the onset of spindle assembly. Yet, forced aMT symmetry through constitutive tethering of Spc72 abolishes Kar9 bias to the old SPB. In that scenario, Kar9 still breaks symmetry driving spindle alignment, although cells no longer link that choice to the old SPB (Juanes et al. 2013).
- Actin depolymerization causes Kar9 symmetry. Actin poison washout allows Kar9 to stochastically rebuild polarity, with the identity of the SPB_{bud} randomized (Cepeda-Garcia et al. 2010). Similarly, cells may regain Kar9 polarity after transient microtubule depolymerization. Recovery is delayed in cell polarity mutants, reflecting the role of cell polarity in promoting symmetry breaking via actin-based transport (Cepeda-Garcia et al. 2010). Microtubule poisons also randomize SPB identity (Pereira et al. 2001; Cepeda-Garcia et al. 2010).

Taken together, asymmetric SPB function in wild-type yeast cells primes Kar9 bias to the old SPB through its existing aMTs, an event superimposed to the temporal program for partitioning aMT–cortex interactions cued by Bud6-dependent capture. Moreover, Bud6 stimulates formin-dependent organization of actin cables generating the tracks for Kar9-dependent delivery of aMTs. Kar9 continued engagement in cycles of aMT delivery to the bud reinforces asymmetric marking throughout the spindle pathway, thus consolidating spindle polarity. In this way, built-in asymmetric aMT organization linked to Spc72 acquisition by the new SPB represents the most upstream structural asymmetry coupling SPB age, Kar9 marking, and fate (Juanes et al. 2013). The importance of Bud6 in setting in motion the extrinsic program is underscored by the behavior of Bud6$\Delta^{229-549}$, a separation-of-function mutant unable to bind Bim1 but supporting correct actin organization.

This mutant preserves Kar9-dependent delivery along actin cables but no longer supports growth and shrinkage at the bud cell cortex. Thus, Bud6 binding to Bim1 is critical for aMT dynamics associated with Bud6 foci without relation to aMT guidance by Kar9. Furthermore, this mutant decreases the fidelity of inheritance of the old SPB by the bud (Ten Hoopen et al. 2012). Taken together, Bud6 and Kar9 cooperate by distinct mechanisms to set aside the old SPB for bud-ward fate upon establishment of spindle polarity.

3.4 Coupling Successful Chromosomal Segregation with Cell Division

The Spindle Position Checkpoint (SPOC) is a surveillance mechanism by which *S. cerevisiae* monitors that chromosomal segregation between the mother and daughter cells has been accomplished. The SPOC delays mitotic progression until the elongated spindle intersects the mother–bud neck with one SPB translocated into the bud. This delay is imposed by inhibition of the Mitotic Exit Network (MEN), a signal transduction cascade that otherwise targets the phosphatase Cdc14 for release from the nucleolus to cause reversal of phosphorylation by CDK. Cdc14 undergoes release at two points in the cell cycle by the action of two interconnected pathways. First, transient release occurs following anaphase onset triggered by the Fourteen Early Anaphase Release (FEAR) pathway. Later, activation of the MEN causes persistent release of Cdc14 that promotes two sets of events converging on CDK inactivation (for a comprehensive review, see Weiss 2012), a prerequisite for mitotic exit and cytokinesis (Fig. 3.7).

Localization of MEN components to the SPB is essential for its function (Valerio-Santiago and Monje-Casas 2011). Moreover, SPOC and MEN components exhibit an asymmetric bias toward the SPB_{bud} (Pereira et al. 2002; Yoshida et al. 2002; Visintin and Amon 2001). Importantly, this bias may be uncoupled from SPB identity by age. Indeed, transient treatment with microtubule poisons erases SPB history without preventing SPOC and MEN components from associating asymmetrically with the SPB that ultimately enters the bud after the spindle reforms (Pereira et al. 2001). The significance of this localization bias for activation of the MEN remains unclear.

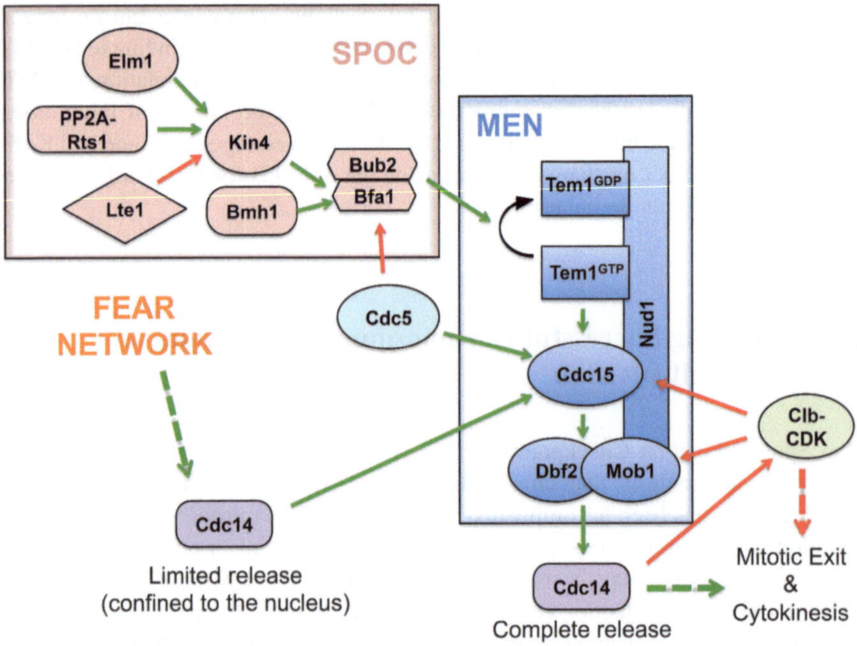

Fig. 3.7 Key relationships between the FEAR network, the MEN, and the SPOC. Two separate pathways control Cdc14 activation. Anaphase onset triggers the Cdc14 Early Anaphase Release network (FEAR network, components not depicted) followed by the essential Mitotic Exit Network (MEN) that promotes progression into G_1. The Spindle position checkpoint (SPOC) translates positional information via the $GAP^{Bfa1\text{-}Bub2}$ to control the most upstream element of the MEN, the GTPase Tem1. Phosphorylation of Bfa1 by the protein kinase Kin4 protects $GAP^{Bfa1\text{-}Bub2}$ from inhibition by the polo-like kinase Cdc5. In turn, Lte1 inhibits Kin4 to allow downregulation of the $GAP^{Bfa1\text{-}Bub2}$ when the SPB_{bud} enters the daughter cell. Thus, upon correct translocation of a spindle pole across the bud neck, $Tem1^{GTP}$ triggers activation of the MEN kinase cascade consisting of Cdc15 and Mob1-Dbf2, leading to full release of Cdc14. Cross talk between pathways based on antagonism between Cdc14 and Clb-CDK (mitotic CDK) is shown. Cdc5 plays multiple roles in controlling the MEN: it inhibits $GAP^{Bfa1\text{-}Bub2}$ and activates Cdc15. Cdc5 has also been placed in the FEAR pathway (not depicted). Additionally, another component of the FEAR network, the phosphatase Cdc55 (PP2A), may further modulate Bfa1 and Mob1 (Baro et al. 2013). *Green* and *red solid arrows* indicate direct positive or negative role, respectively

3.4.1 The SPOC: Antagonistic Players Set a Mother–Daughter Boundary for Spatial Control of Tem1 Activity

3.4.1.1 Modulators of the Tem1 GTPase Cycle

At the molecular level, the SPOC regulates the MEN by inhibiting its most upstream element, the GTPase Tem1, via an unconventional two-component GAP composed of regulatory and catalytic subunits—Bfa1 and Bub2, respectively

(Alexandru et al. 1999; Fraschini et al. 1999; Li 1999; Hu et al. 2001; Geymonat et al. 2002). Since no corresponding GEF has been identified (Yoshida et al. 2003; Geymonat et al. 2009), it is believed that Tem1 activation relies mainly on downregulation of the GAP and spontaneous GDP–GTP exchange. Localization of Bfa1 and Bub2 to the SPB is interdependent and essential for $GAP^{Bfa1\text{-}Bub2}$ function. Thus, either a *bfa1Δ* or a *bub2Δ* mutation is sufficient to abrogate the SPOC (Pereira et al. 2000). Yet, both proteins play mechanistically distinct roles. Indeed, in vitro studies revealed that Bfa1 binds directly to Tem1 to exert Bub2-independent nucleotide-stabilizing activity, making Bfa1 more like a Guanyl Dissociation Inhibitor (GDI), while Bub2 is responsible for GTPase stimulation (Geymonat et al. 2002; Fraschini et al. 2006). At the same time, Bfa1 is also important for Tem1 localization (Pereira et al. 2000). In either *bfa1Δ* or *bub2Δ* mutant cells, Tem1 concentration at the SPBs is drastically reduced, demonstrating that Tem1 has two modes of SPB binding, one of which depends upon Bfa1 (Rock and Amon 2011; Caydasi et al. 2012).

Bfa1 is a cell cycle-dependent phosphoprotein. The yeast polo-like kinase Cdc5 phosphorylates Bfa1 and inhibits its GAP activity (Hu et al. 2001; Geymonat et al. 2003). Accordingly, downregulation of GAP activity by Cdc5 may increase $Tem1^{GTP}$ on the SPB_{bud}, to allow exit from mitosis (Hu et al. 2001; Rock and Amon 2011; Valerio-Santiago and Monje-Casas 2011).

The protein kinase Kin4 is the central player in the SPOC linking GAP activity to the position of the spindle poles. Kin4 is localized to the mother cell cortex, the mother-directed SPB, and the bud neck (D'Aquino et al. 2005; Pereira and Schiebel 2005). At the SPB, Kin4 phosphorylates Bfa1 impeding inhibitory phosphorylation by Cdc5; hence, active GAP holds the MEN inactive (Maekawa et al. 2007). Phosphorylation at the kinase activation loop by the bud neck-associated protein kinase Elm1 promotes Kin4 activity (Caydasi et al. 2010; Moore et al. 2010). At the same time, the phosphatase PP2A-Rts1 maintains Kin4 in a hypo-phosphorylated form essential for its localization at the mother cortex and SPB (Chan and Amon 2009; Caydasi et al. 2010). Both *elm1Δ* and *rts1Δ* mutant strains are SPOC deficient demonstrating the impact of Kin4 activity and localization control on checkpoint proficiency.

As stated above, a GEF for Tem1 has not been identified. Early studies implicated Lte1, a protein containing a GEF-like C-terminal domain, as a MEN regulator. Yet, the molecular nature of its role has remained elusive for a long time. Lte1 is localized exclusively to the bud cell cortex (Bardin et al. 2000; Hofken and Schiebel 2002; Jensen et al. 2002), and its overexpression or leakage into mother cells in mutant backgrounds impairs SPOC-dependent arrest (Bardin et al. 2000; Pereira et al. 2000; Castillon et al. 2003). Accordingly, early models postulated Lte1 as the GEF for Tem1. In this view, translocation of the SPB_{bud} into the daughter compartment would allow Tem1 to encounter its GEF, thus triggering MEN activation (Bardin et al. 2000; Pereira et al. 2000). However, GEF activity has never been demonstrated in vitro, and instead, it has been shown that the putative GEF domain binds Ras, an interaction responsible for the cortical localization of Lte1 (Yoshida et al. 2003; Seshan and Amon 2005; Geymonat et al. 2009).

Interestingly, despite sequence homology to GEFs, Lte1 appears to act as an effector for Ras since it binds preferentially Ras^{GTP}, in a phospho-dependent manner (Yoshida et al. 2003; Seshan and Amon 2005; Geymonat et al. 2010). Finally, recent studies have showed that Lte1 functions in the SPOC by antagonizing Kin4 as described below.

3.4.1.2 SPOC Dynamics: Polarization of the GAP-Tem1 Module and GAP Downregulation

In cells with correctly aligned preanaphase spindles, $GAP^{Bfa1\text{-}Bub2}$ and Tem1 are the first elements to display a strong bias for the SPB_{bud} (Caydasi and Pereira 2009; Monje-Casas and Amon 2009). In cells with misaligned spindles, $GAP^{Bfa1\text{-}Bub2}$ localizes at intermediate levels but symmetrically (Fig. 3.8). According to photobleaching experiments, symmetric distribution at the poles is linked to a more dynamic association between Bfa1 and SPBs. By contrast, unperturbed cells with correctly aligned spindles show relatively stable association between Bfa1 and the SPB, irrespective of cell cycle stage or whether the SPB_{bud} is still in the mother cell or across the bud neck, correlating with Bfa1 asymmetry (Caydasi and Pereira 2009).

Dynamic association between $GAP^{Bfa1\text{-}Bub2}$ and the SPB is modulated by Kin4-dependent phosphorylation of Bfa1. Either *KIN4* deletion or cancelling Kin4 phosphorylation sites in Bfa1 increases Bfa1 stable association with SPBs even in misaligned spindles, while overexpression of Kin4, but not $Kin4^{kd}$ (kinase-dead), enhances dynamics (Caydasi and Pereira 2009). Preventing Kin4-dependent phosphorylation and the ensuing increase in Bfa1 dynamics at the SPB impairs checkpoint performance (D'Aquino et al. 2005; Maekawa et al. 2007; Pereira and Schiebel 2005). Interestingly, tethering Bfa1 or Tem1 to a core component of the outer plaque of the SPB also disrupts the SPOC without affecting the MEN per se (Caydasi and Pereira 2009; Scarfone et al. 2015). These findings have been validated by quantitative analysis and mathematical modeling pointing to the importance of $GAP^{Bfa1\text{-}Bub2}$ dynamic shuttling to the cytoplasm for complete inhibition of Tem1 and checkpoint proficiency (Caydasi et al. 2012). Another study shows that phosphorylated Bfa1 binds Bmh1, a 14–3-3 yeast homologue, which enhances Bfa1 dynamics onto the SPB. Indeed, *bmh1Δ* cells retain asymmetric marking by Bfa1 even in the presence of mispositioned spindles and are therefore checkpoint defective (Caydasi et al. 2014). Taken together, Kin4 phosphorylation of Bfa1 plays a dual role in the SPOC to accomplish inhibition of Tem1: it protects Bfa1 from the inhibitory phosphorylation by Cdc5 and increases dynamic shuttling of $GAP^{Bfa1\text{-}Bub2}$ between SPB and cytoplasm by creating a docking site for Bmh1.

In turn, antagonistic interplay between Kin4 and Lte1 generates boundaries between mother and daughter compartments that offer a spatial context for surveillance of the position of the spindle poles. Indeed, Lte1 influences the asymmetric distribution of $GAP^{Bfa1\text{-}Bub2}$ (Geymonat et al. 2009) by binding directly to Kin4

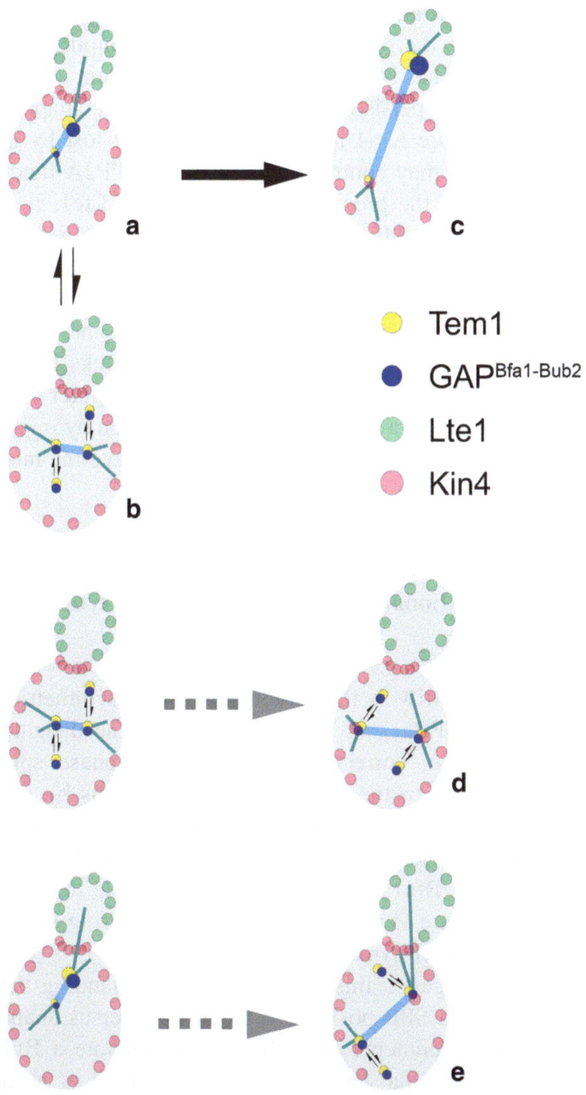

Fig. 3.8 Relative impact of asymmetric aMT contacts and spindle alignment on Tem1-GAP$^{Bfa1\text{-}Bub2}$ module dynamics and localization. Asymmetric aMT contacts underscoring spindle polarity in preanaphase cells correlate with polarized localization of the Tem1-GAP$^{Bfa1\text{-}Bub2}$ module (**a**). Before establishment of spindle polarity, cells exhibit Tem1-GAP$^{Bfa1\text{-}Bub2}$ symmetry (**b**)—typically observed in a small fraction of wild-type cells at this stage. As cells proceed to elongate the spindle across the bud neck during anaphase, localization of the Tem1-GAP$^{Bfa1\text{-}Bub2}$ module to the SPB_{bud} is further enhanced (**c**) after transit from the Kin4- into the Lte1-containing compartment. Mutant cells that fail to polarize the spindle prior to anaphase, e.g., *kar9Δ*, will proceed to elongate the spindle off axis within the mother cell in the absence of asymmetric contacts (**d**). By contrast, inactivation of dynein abrogates pulling force to translocate the SPB_{bud} across the bud neck while preserving spindle polarity. This results in elongated spindles in the mother cell that retain the asymmetric attachments (**e**). In both cases, aMTs may grow to reach the bud neck as well. Regardless of the partition of aMT contacts, in the presence of elongated spindles confined to the mother cell, the SPOC is active: Kin4 is recruited to

(Bertazzi et al. 2011; Falk et al. 2011). Lte1 binds and clearly inhibits Kin4 kinase activity in vitro. However, the situation must be more complicated in vivo as no difference in Kin4 activity has been observed in *lte1Δ* mutant cells. Instead, Kin4 localization was affected by persisting on the SPB_{bud} during anaphase. Moreover, deletion of *LTE1* can restore Kin4 localization otherwise lost by inactivation of Rts1, suggesting that Lte1 may contribute to counteract PP2A function (Bertazzi et al. 2011). Regardless of the precise molecular role of Lte1, it is clear that it opposes Kin4 and allows the MEN to be activated once the SPB_{bud} escapes the mother compartment. In this "zone model" (Chan and Amon 2010), the mother compartment represents a MEN-inhibitory environment where Kin4 is fully active and gives a "wait" signal to mitotic progression while the bud compartment represents a MEN-activating environment where Lte1 blocks Kin4 allowing the cell to exit mitosis after the SPB_{bud} gains access to the daughter cell. Overall, the SPOC enforces tight control of a critical cell cycle transition by imposing a double requirement for activation of Tem1—the SPB_{bud} must leave the mother compartment *and* enter the bud to allow mitotic exit.

3.4.1.3 Translating Positional Information into SPB Asymmetric Marking

A hallmark of spindle polarity is the establishment of asymmetric aMT attachments that set aside the SPB_{bud}, i.e., the pole to enter the daughter cell during anaphase. As stated above, correctly polarized preanaphase spindles already exhibit a $GAP^{Bfa1\text{-}Bub2}$ bias to the SPB_{bud} (Fig. 3.8a versus b), despite the fact that both SPBs coexist in the mother compartment and are therefore potentially under the control of Kin4. Spindle elongation across the bud neck enhances this asymmetry by reinforcing Bfa1 at the SPB_{bud} once outside the mother compartment (Fig. 3.8c). Yet, failure to translocate the SPB_{bud} during spindle elongation triggers Kin4-dependent Bfa1 symmetry (Fig. 3.8d–e).

These observations suggest a missing link for understanding the relationship between Bfa1 partition and SPB position in metaphase versus anaphase. In other words, polarized aMT attachments are sufficient to generate a Bfa1 bias to the SPB_{bud} at metaphase (Fig. 3.8a versus b), while the effect of asymmetric aMT attachments may be superseded by the two SPBs remaining in the same compartment during anaphase (Fig. 3.8d–e). How are these contrasting behaviors explained on the basis of the central role of Kin4? What is the basis for discriminating the SPB_{bud} while confined in the mother compartment at metaphase and why this

Fig. 3.8 (continued) both SPBs and elicits increased dynamics and symmetric association of the Tem1-$GAP^{Bfa1\text{-}Bub2}$ module. Under these conditions, aMT contacts might play a role in sustaining SPOC-dependent arrest

distinction does not persist in anaphase? Is the cell sensing asymmetric aMT attachments or absolute SPB position? How is this information transduced to Kin4?

Kin4 binds to Spc72 at the SPB outer plaque (Maekawa et al. 2007) and may sense information emanating directly from aMTs. Accordingly, preanaphase spindle rotation causing exchange of SPB identity redirected Bfa1 bias to the SPB contacting the bud cortex. Thus, aMT–cortex interactions (rather than SPB age) may be important to determine $GAP^{Bfa1\text{-}Bub2}$ asymmetry (Adames et al. 2001; Monje-Casas and Amon 2009; Pereira et al. 2001). Furthermore, in anaphase cells with correctly aligned spindles, the severing of the aMTs linking the SPB_{bud} to the bud cortex induces a temporal loss of Bfa1 localization, while no change is observed if the spindle midzone or the aMTs in the mother compartment were severed (Monje-Casas and Amon 2009). Conversely, it has been proposed that cells with misaligned spindles may sustain SPOC activation through aMTs penetrating the bud neck (Adames et al. 2001; Castillon et al. 2003) as their severing by laser ablation induced premature mitotic exit (Moore et al. 2009a). Yet, a more recent study reported that inappropriate mitotic exit in cells with misoriented spindles did not correlate with the presence of aMTs reaching into the bud (Falk et al. 2016). Moreover, laser ablation experiments performed in this study indicated that aMT severing to abrogate contacts with the bud is insufficient to provoke mitotic exit in the presence of mispositioned spindles.

In conclusion, aMTs promote spindle orientation and may also relay signals to sense the position of the spindle poles (presumably by contacting cortical components marking the bud neck, the mother or the daughter cell) and/or promote polarization of SPOC components. However, the significance of these aMT-mediated polarity for SPOC proficient arrest remains unclear. Nevertheless, several cortical proteins including Elm1 (mentioned above), Bud6 (Nelson and Cooper 2007; Huisman et al. 2004), and the septin ring component Cdc10 (Castillon et al. 2003) are required for SPOC proficiency.

Crucially, the penetration of the SPB into the Lte1-controlled bud environment signals an irreversible green light for mitotic exit and cytokinesis. Moreover, the study of cells engineered to contain two spindles—one misaligned and one correctly elongated across the bud neck—revealed that entry of one SPB into the bud was sufficient to trigger a dominant activating signal resulting in mitotic exit and the disassembly of both spindles (Falk et al. 2016; Gryaznova et al. 2016).

3.4.2 *Is Polarization to the SPB_{bud} of Kar9 and Bfa1 Mechanistically Linked?*

Actin and septin ring integrity is required to maintain cell morphogenesis (Bi and Park 2012) and to delimit two compartments with distinctive cortical identities—the mother cell and the bud. In turn, asymmetric attachments to the respective compartments established by the mother or bud-ward pole may determine polarized loading of checkpoint regulators such as Bfa1 (Monje-Casas and Amon 2009). Is this scenario applicable to Kar9? The contrasting behavior of Kar9 and Bfa1 in response to polarity disruption or spindle alignment argues against this possibility. By extension, it points to independent pathways polarizing SPOC/MEN components and Kar9.

Perturbation of actin organization by drugs or mutations disrupting cell polarity abolishes Kar9 asymmetry by a mechanism that can be separated from the effects of actin perturbation on spindle orientation per se. Indeed, preanaphase cells retain spindle orientation upon treatment with actin poisons despite actin cable loss (Theesfeld et al. 1999) and Kar9 symmetry. By contrast, Bfa1 polarization to the SPB_{bud}, *conditioned to spindle orientation*, is retained under the same conditions (Cepeda-Garcia et al. 2010). In turn, rescue of spindle orientation by dynein-driven sliding permits polarization to the SPB_{bud} of Bfa1 (Caydasi and Pereira 2009; Monje-Casas and Amon 2009) but not Kar9 (Cepeda-Garcia et al. 2010). Thus, extrinsic control of Bfa1 polarity still persists despite polarity loss eliciting Kar9 symmetry. It follows that Kar9 symmetry is not increased in response to misalignment per se. Moreover, Kar9 continues to polarize upon allele-specific inactivation of *MYO2*—only alleles that disengage Kar9 from the Myo2 cargo domain induce loss of asymmetry, demonstrating that the effect of actin perturbation relates directly to the ability of Kar9 to act as Myo2 cargo.

The distinctive behavior of Kar9 and Bfa1 is also evident in dynein mutant cells, which lack a key force generator to drag a pole of the spindle across the bud neck, resulting in cells exhibiting elongated spindles confined to the mother cell. These mispositioned spindles retain both Kar9-polarized localization to the SPB_{bud} and asymmetric aMT attachments. Yet, they are symmetrically marked by Bfa1 under the same conditions, thus imposing SPOC-dependent arrest as described above (Cepeda-Garcia et al. 2010).

Finally, Bfa1 polarizes proficiently in *kar9Δ* mutants and correctly senses orientation and asymmetric attachments. Thus, Bfa1 polarization cannot arise in direct response to delivery of aMT plus ends powered by Myo2-dependent transport (absent without Kar9) but instead responds to defined signals derived from asymmetric aMT contacts (mother vs. bud cell) as proposed by Monje-Casas and Amon (2009) and by spindle pole translocation during anaphase. Conversely, Kar9 localization and spindle polarity are unaffected in a *bfa1Δ* mutant (Cepeda-Garcia et al. 2010; Juanes et al. 2013). In conclusion, unconnected pathways determine Kar9 and Bfa1 polarization independent from each other.

Despite these precedents, a recent study has reported that perturbation of polarization of the Tem1-$GAP^{Bfa1\text{-}Bub2}$ module to the SPB_{bud} impaired Kar9 polarity without affecting the MEN (Scarfone et al. 2015). This might suggest a mechanistic link between the Tem1 GTPase cycle and Kar9 dynamics. Yet, it remains unclear whether this relationship is relevant to Kar9 control under unperturbed conditions. Finally, another study has pointed to links between the MEN and Kar9 (Hotz et al. 2012a). That study posits that the MEN kinase Dbf2 might phosphorylate Kar9 at metaphase (when the MEN is otherwise inactive) to enforce Kar9 marking of the old SBP, thus linking SPB age and fate. Crucially, Kar9 polarity was unaffected in SPOC mutants.

3.5 Asymmetric Cell Division—*Lessons from Yeast*

The orientation of the mitotic spindle in self-renewing stem cell divisions shares prominent features of the asymmetric cell division in yeast. In particular, cells exploit structural asymmetries built into the centrosome cycle to produce temporal patterns of microtubule organization underlying differential centrosome fate. For example, *Drosophila* male germline stem cells orient the spindle perpendicular to the periphery of a hub of somatic cells and retain the "old" centrosome at the junction between the stem cell and the hub (Yamashita et al. 2007). By contrast, in *Drosophila* neuroblasts the "new" centrosome remains targeted to the apical cortex (Januschke et al. 2011; Conduit and Raff 2010). The significance of this opposite trends remains a mystery. Male germline stem cells orient the spindle by anchoring the old centrosome to a preestablished cortical landmark. Perturbation of centrosome function disrupts differential cell fate (Yamashita et al. 2007). Yet, autonomous asymmetric cell fate in neuroblasts persists even when centrosome identity is randomized, as is the case in yeast (Pereira et al. 2001; Januschke et al. 2013). In neuroblasts, a specialized centrosome cycle is in effect based on early separation of single centrioles and the incorporation of Centrobin by the new centriole (Januschke et al. 2013). This structural distinction governs microtubule inherent asymmetry and centrosome fate. However, cross talk between the centrosome and the apical cortex effectively couples spindle orientation with the configuration of polarized cortical crescents. The instructive role of the cell cortex in symmetry breaking is also apparent in yeast (Cepeda-Garcia et al. 2010; Januschke et al. 2013).

Along the same lines, asymmetric cell divisions of neural progenitors in the mouse neocortex are coupled to the inheritance of the "old" centrosome by the progenitor cells (Wang et al. 2009). Centrosome inheritance may be randomized by depletion of ninein (a mother centriole component of subdistal appendages) with concomitant disruption of cell division patterning and loss of progenitor cells (Wang et al. 2009). Thus, the idea that a pattern of inheritance is intrinsically linked to the ability of a centrosome to retain microtubule organization may be a general principle in self-renewing asymmetric stem cell divisions and beyond. Supporting this view, Cenexin, also a component of the mother centriole subdistal appendages, is required for polarized spindle orientation supporting lumen morphogenesis (Hung et al. 2016). A future challenge is to integrate cytoskeletal systems with cell cycle control to achieve a comprehensive view of mechanisms that translate asymmetric controls into differential fate.

In that regard, yeast provides valuable insight into the potential for integration through extrinsic determinants operating at the interface between actin and microtubule cytoskeletal systems, which may transcend mechanism for spindle orientation. In yeast, the best-characterized molecular bridge between the two systems is Kar9, the accepted functional counterpart of human adenomatous polyposis coli protein, APC (Akhmanova and Steinmetz 2015). Yet, while acting as cargo for Myo2, Kar9 plays no direct roles in actin organization. By contrast, APC exhibits competing interactions with actin filaments and microtubules, pointing to potential mechanisms for modulating APC distribution in vivo. Moreover, APC stimulates actin assembly synergizing with the formin mDia1 (Okada et al. 2010). In addition, formins have been implicated as MT-stabilizing factors in a range of processes. Yet, comparable relationships have not been demonstrated in the context of spindle orientation in budding yeast. Instead, it is Bud6 that captures aMTs at the cell cortex via Bim1. At the same time, a separate domain promotes actin cable formation by formins, in a remarkable parallel to APC (Graziano et al. 2013; Delgehyr et al. 2008), a role that also ensures temporal control of Bud6 partition defining sites for cortical capture. Thus, bifunctional proteins like Bud6 or APC may exploit their dual roles to fine-tune temporal and spatially restricted function.

Acknowledgements We thank members of the Segal, Geymonat, Draviam, Lindon, and Glover laboratories for advice and fruitful discussions. Work in our laboratory has been partly supported by the Wellcome Trust and the Biotechnology and Biological Sciences Research Council.

References

Adames NR, Cooper JA (2000) Microtubule interactions with the cell cortex causing nuclear movements in *Saccharomyces cerevisiae*. J Cell Biol 149(4):863–874

Adames NR, Oberle JR, Cooper JA (2001) The surveillance mechanism of the spindle position checkpoint in yeast. J Cell Biol 153(1):159–168

Adams IR, Kilmartin JV (1999) Localization of core spindle pole body (SPB) components during SPB duplication in *Saccharomyces cerevisiae*. J Cell Biol 145(4):809–823

Adams AE, Pringle JR (1984) Relationship of actin and tubulin distribution to bud growth in wild-type and morphogenetic-mutant *Saccharomyces cerevisiae*. J Cell Biol 98(3):934–945

Akhmanova A, Steinmetz MO (2015) Control of microtubule organization and dynamics: two ends in the limelight. Nat Rev Mol Cell Biol 16(12):711–726. doi:10.1038/nrm4084

Alexandru G, Zachariae W, Schleiffer A, Nasmyth K (1999) Sister chromatid separation and chromosome re-duplication are regulated by different mechanisms in response to spindle damage. EMBO J 18(10):2707–2721. doi:10.1093/emboj/18.10.2707

Amberg DC, Zahner JE, Mulholland JW, Pringle JR, Botstein D (1997) Aip3p/Bud6p, a yeast actin-interacting protein that is involved in morphogenesis and the selection of bipolar budding sites. Mol Biol Cell 8(4):729–753

Avena JS, Burns S, Yu Z, Ebmeier CC, Old WM, Jaspersen SL, Winey M (2014) Licensing of yeast centrosome duplication requires phosphoregulation of sfi1. Plos Genet 10(10):e1004666. doi:10.1371/journal.pgen.1004666

Bardin AJ, Visintin R, Amon A (2000) A mechanism for coupling exit from mitosis to partitioning of the nucleus. Cell 102(1):21–31

Baro B, Rodriguez-Rodriguez JA, Calabria I, Hernaez ML, Gil C, Queralt E (2013) Dual regulation of the mitotic exit network (MEN) by PP2A-Cdc55 phosphatase. Plos Genet 9 (12):e1003966. doi:10.1371/journal.pgen.1003966

Bertazzi DT, Kurtulmus B, Pereira G (2011) The cortical protein Lte1 promotes mitotic exit by inhibiting the spindle position checkpoint kinase Kin4. J Cell Biol 193(6):1033–1048. doi:10.1083/jcb.201101056

Bi E, Park HO (2012) Cell polarization and cytokinesis in budding yeast. Genetics 191 (2):347–387. doi:10.1534/genetics.111.132886

Bridges AA, Gladfelter AS (2015) Septin form and function at the cell cortex. J Biol Chem 290 (28):17173–17180. doi:10.1074/jbc.R114.634444

Burns S, Avena JS, Unruh JR, Yu Z, Smith SE, Slaughter BD, Winey M, Jaspersen SL (2015) Structured illumination with particle averaging reveals novel roles for yeast centrosome components during duplication. eLife:4. doi:10.7554/eLife.08586

Buttery SM, Yoshida S, Pellman D (2007) Yeast formins Bni1 and Bnr1 utilize different modes of cortical interaction during the assembly of actin cables. Mol Biol Cell 18(5):1826–1838

Byers B (1981) Cytology of the yeast life cycle. In: Strathern JN, Jones EW, Broach JR (eds) The molecular biology of the yeast Saccharomyces: life cycle and inheritance. Cold Spring Harbor Laboratory, Cold Spring Harbor, NY, pp 59–96

Byers B, Goetsch L (1975) Behavior of spindles and spindle plaques in the cell cycle and conjugation of *Saccharomyces cerevisiae*. J Bacteriol 124(1):511–523

Carminati JL, Stearns T (1997) Microtubules orient the mitotic spindle in yeast through dynein-dependent interactions with the cell cortex. J Cell Biol 138(3):629–641

Castillon GA, Adames NR, Rosello CH, Seidel HS, Longtine MS, Cooper JA, Heil-Chapdelaine RA (2003) Septins have a dual role in controlling mitotic exit in budding yeast. Curr Biol 13 (8):654–658

Caydasi AK, Pereira G (2009) Spindle alignment regulates the dynamic association of checkpoint proteins with yeast spindle pole bodies. Dev Cell 16(1):146–156. doi:10.1016/j.devcel.2008.10.013

Caydasi AK, Pereira G (2012) SPOC alert—when chromosomes get the wrong direction. Exp Cell Res 318(12):1421–1427. doi:S0014-4827(12)00167-X [pii] 10.1016/j.yexcr.2012.03.031

Caydasi AK, Kurtulmus B, Orrico MI, Hofmann A, Ibrahim B, Pereira G (2010) Elm1 kinase activates the spindle position checkpoint kinase Kin4. J Cell Biol 190(6):975–989. doi:10.1083/jcb.201006151

Caydasi AK, Lohel M, Grunert G, Dittrich P, Pereira G, Ibrahim B (2012) A dynamical model of the spindle position checkpoint. Mol Syst Biol 8:582. doi:10.1038/msb.2012.15

Caydasi AK, Micoogullari Y, Kurtulmus B, Palani S, Pereira G (2014) The 14-3-3 protein Bmh1 functions in the spindle position checkpoint by breaking Bfa1 asymmetry at yeast centrosomes. Mol Biol Cell 25(14):2143–2151. doi:10.1091/mbc.E14-04-0890

Cepeda-Garcia C, Delgehyr N, Ortiz MA, ten Hoopen R, Zhiteneva A, Segal M (2010) Actin-mediated delivery of astral microtubules instructs Kar9p asymmetric loading to the bud-ward spindle pole. Mol Biol Cell 21(15):2685–2695

Chan LY, Amon A (2009) The protein phosphatase 2A functions in the spindle position checkpoint by regulating the checkpoint kinase Kin4. Genes Dev 23(14):1639–1649. doi:10.1101/gad.1804609

Chan LY, Amon A (2010) Spindle position is coordinated with cell-cycle progression through establishment of mitotic exit-activating and -inhibitory zones. Mol Cell 39(3):444–454. doi:10.1016/j.molcel.2010.07.032

Chant J, Mischke M, Mitchell E, Herskowitz I, Pringle JR (1995) Role of Bud3p in producing the axial budding pattern of yeast. J Cell Biol 129(3):767–778

Chesarone MA, DuPage AG, Goode BL (2010) Unleashing formins to remodel the actin and microtubule cytoskeletons. Nat Rev Mol Cell Biol 11(1):62–74

Chia W, Somers WG, Wang HY (2008) Drosophila neuroblast asymmetric divisions: cell cycle regulators, asymmetric protein localization, and tumorigenesis. J Cell Biol 180(2):267–272. doi:10.1083/jcb.200708159

Conduit PT, Raff JW (2010) Cnn dynamics drive centrosome size asymmetry to ensure daughter centriole retention in Drosophila neuroblasts. Curr Biol 20(24):2187–2192. doi:S0960-9822 (10)01521-6 [pii] 10.1016/j.cub.2010.11.055

Conduit PT, Wainman A, Raff JW (2015) Centrosome function and assembly in animal cells. Nat Rev Mol Cell Biol 16(10):611–624. doi:10.1038/nrm4062

Cuschieri L, Miller R, Vogel J (2006) Gamma-tubulin is required for proper recruitment and assembly of Kar9-Bim1 complexes in budding yeast. Mol Biol Cell 17(10):4420–4434

D'Aquino KE, Monje-Casas F, Paulson J, Reiser V, Charles GM, Lai L, Shokat KM, Amon A (2005) The protein kinase Kin4 inhibits exit from mitosis in response to spindle position defects. Mol Cell 19(2):223–234. doi:10.1016/j.molcel.2005.06.005

Delgehyr N, Lopes CS, Moir CA, Huisman SM, Segal M (2008) Dissecting the involvement of formins in Bud6p-mediated cortical capture of microtubules in *S. cerevisiae*. J Cell Sci 121 (Pt 22):3803–3814

Elserafy M, Saric M, Neuner A, Lin TC, Zhang WL, Seybold C, Sivashanmugam L, Schiebel E (2014) Molecular mechanisms that restrict yeast centrosome duplication to one event per cell cycle. Curr Biol 24(13):1456–1466. doi:10.1016/j.cub.2014.05.032

Erlemann S, Neuner A, Gombos L, Gibeaux R, Antony C, Schiebel E (2012) An extended gamma-tubulin ring functions as a stable platform in microtubule nucleation. J Cell Biol 197(1):59–74. doi:10.1083/jcb.201111123

Evangelista M, Pruyne D, Amberg DC, Boone C, Bretscher A (2002) Formins direct Arp2/3-independent actin filament assembly to polarize cell growth in yeast. Nat Cell Biol 4(3):260–269

Falk JE, Chan LY, Amon A (2011) Lte1 promotes mitotic exit by controlling the localization of the spindle position checkpoint kinase Kin4. Proc Natl Acad Sci U S A 108(31):12584–12590. doi:10.1073/pnas.1107784108

Falk JE, Tsuchiya D, Verdaasdonk J, Lacefield S, Bloom K, Amon A (2016) Spatial signals link exit from mitosis to spindle position. eLife 5. doi:10.7554/eLife.14036

Fraschini R, Formenti E, Lucchini G, Piatti S (1999) Budding yeast Bub2 is localized at spindle pole bodies and activates the mitotic checkpoint via a different pathway from Mad2. J Cell Biol 145(5):979–991

Fraschini R, D'Ambrosio C, Venturetti M, Lucchini G, Piatti S (2006) Disappearance of the budding yeast Bub2-Bfa1 complex from the mother-bound spindle pole contributes to mitotic exit. J Cell Biol 172(3):335–346. doi:10.1083/jcb.200507162

Gasic I, Nerurkar P, Meraldi P (2015) Centrosome age regulates kinetochore-microtubule stability and biases chromosome mis-segregation. eLife 4:e07909. doi:10.7554/eLife.07909

Geissler S, Pereira G, Spang A, Knop M, Soues S, Kilmartin J, Schiebel E (1996) The spindle pole body component Spc98p interacts with the gamma-tubulin-like Tub4p of *Saccharomyces cerevisiae* at the sites of microtubule attachment. Embo J 15(15):3899–3911

Geymonat M, Spanos A, Smith SJ, Wheatley E, Rittinger K, Johnston LH, Sedgwick SG (2002) Control of mitotic exit in budding yeast. In vitro regulation of Tem1 GTPase by Bub2 and Bfa1. J Biol Chem 277(32):28439–28445. doi:10.1074/jbc.M202540200

Geymonat M, Spanos A, Walker PA, Johnston LH, Sedgwick SG (2003) In vitro regulation of budding yeast Bfa1/Bub2 GAP activity by Cdc5. J Biol Chem 278(17):14591–14594. doi:10.1074/jbc.C300059200

Geymonat M, Spanos A, de Bettignies G, Sedgwick SG (2009) Lte1 contributes to Bfa1 localization rather than stimulating nucleotide exchange by Tem1. J Cell Biol 187(4):497–511. doi:10.1083/jcb.200905114

Geymonat M, Spanos A, Jensen S, Sedgwick SG (2010) Phosphorylation of Lte1 by Cdk prevents polarized growth during mitotic arrest in *S. cerevisiae*. J Cell Biol 191(6):1097–1112. doi:10.1083/jcb.201005070

Goode BL, Eskin JA, Wendland B (2015) Actin and endocytosis in budding yeast. Genetics 199 (2):315–358. doi:10.1534/genetics.112.145540

Graziano BR, Jonasson EM, Pullen JG, Gould CJ, Goode BL (2013) Ligand-induced activation of a formin-NPF pair leads to collaborative actin nucleation. J Cell Biol 201(4):595–611. doi:10.1083/jcb.201212059

Gruneberg U, Campbell K, Simpson C, Grindlay J, Schiebel E (2000) Nud1p links astral microtubule organization and the control of exit from mitosis. EMBO J 19(23):6475–6488. doi:10.1093/emboj/19.23.6475

Gryaznova Y, Koca Caydasi A, Malengo G, Sourjik V, Pereira G (2016) A FRET-based study reveals site-specific regulation of spindle position checkpoint proteins at yeast centrosomes. eLife:5. doi:10.7554/eLife.14029

Gupta ML Jr, Carvalho P, Roof DM, Pellman D (2006) Plus end-specific depolymerase activity of Kip3, a kinesin-8 protein, explains its role in positioning the yeast mitotic spindle. Nat Cell Biol 8(9):913–923

Haber JE (2012) Mating-type genes and MAT switching in *Saccharomyces cerevisiae*. Genetics 191(1):33–64. doi:10.1534/genetics.111.134577

Hartwell LH (1974) *Saccharomyces cerevisiae* cell cycle. Bacteriol Rev 38(2):164–198

Hofken T, Schiebel E (2002) A role for cell polarity proteins in mitotic exit. EMBO J 21 (18):4851–4862

Hotz M, Leisner C, Chen D, Manatschal C, Wegleiter T, Ouellet J, Lindstrom D, Gottschling DE, Vogel J, Barral Y (2012a) Spindle pole bodies exploit the mitotic exit network in metaphase to drive their age-dependent segregation. Cell 148(5):958–972. doi:S0092-8674(12)00151-1[pii] 10.1016/j.cell.2012.01.041

Hotz M, Lengefeld J, Barral Y (2012b) The MEN mediates the effects of the spindle assembly checkpoint on Kar9-dependent spindle pole body inheritance in budding yeast. Cell Cycle 11 (16):3109–3116. doi:10.4161/cc.21504

Howell AS, Lew DJ (2012) Morphogenesis and the cell cycle. Genetics 190(1):51–77. doi:10.1534/genetics.111.128314

Howell AS, Savage NS, Johnson SA, Bose I, Wagner AW, Zyla TR, Nijhout HF, Reed MC, Goryachev AB, Lew DJ (2009) Singularity in polarization: rewiring yeast cells to make two buds. Cell 139(4):731–743. doi:10.1016/j.cell.2009.10.024

Howell AS, Jin M, Wu CF, Zyla TR, Elston TC, Lew DJ (2012) Negative feedback enhances robustness in the yeast polarity establishment circuit. Cell 149(2):322–333. doi:10.1016/j.cell.2012.03.012

Hu F, Wang Y, Liu D, Li Y, Qin J, Elledge SJ (2001) Regulation of the Bub2/Bfa1 GAP complex by Cdc5 and cell cycle checkpoints. Cell 107(5):655–665

Huffaker TC, Thomas JH, Botstein D (1988) Diverse effects of beta-tubulin mutations on microtubule formation and function. J Cell Biol 106(6):1997–2010

Huisman SM, Segal M (2005) Cortical capture of microtubules and spindle polarity in budding yeast—where's the catch? J Cell Sci 118(Pt 3):463–471

Huisman SM, Bales OA, Bertrand M, Smeets MF, Reed SI, Segal M (2004) Differential contribution of Bud6p and Kar9p to microtubule capture and spindle orientation in *S. cerevisiae*. J Cell Biol 167(2):231–244

Huisman SM, Smeets MF, Segal M (2007) Phosphorylation of Spc110p by Cdc28p-Clb5p kinase contributes to correct spindle morphogenesis in *S. cerevisiae*. J Cell Sci 120(Pt 3):435–446

Hung HF, Hehnly H, Doxsey S (2016) The mother centriole appendage protein cenexin modulates lumen formation through spindle orientation. Curr Biol 26(6):793–801. doi:10.1016/j.cub.2016.01.025

Hwang E, Kusch J, Barral Y, Huffaker TC (2003) Spindle orientation in *Saccharomyces cerevisiae* depends on the transport of microtubule ends along polarized actin cables. J Cell Biol 161 (3):483–488

Imamura H, Tanaka K, Hihara T, Umikawa M, Kamei T, Takahashi K, Sasaki T, Takai Y (1997) Bni1p and Bnr1p: downstream targets of the Rho family small G-proteins which interact with profilin and regulate actin cytoskeleton in *Saccharomyces cerevisiae*. EMBO J 16(10):2745–2755

Januschke J, Llamazares S, Reina J, Gonzalez C (2011) Drosophila neuroblasts retain the daughter centrosome. Nat Commun 2:243. doi:ncomms1245 [pii] 10.1038/ncomms1245

Januschke J, Reina J, Llamazares S, Bertran T, Rossi F, Roig J, Gonzalez C (2013) Centrobin controls mother-daughter centriole asymmetry in Drosophila neuroblasts. Nat Cell Biol 15 (3):241–248. doi:ncb2671 [pii] 10.1038/ncb2671

Jensen S, Geymonat M, Johnson AL, Segal M, Johnston LH (2002) Spatial regulation of the guanine nucleotide exchange factor Lte1 in *Saccharomyces cerevisiae*. J Cell Sci 115 (Pt 24):4977–4991

Juanes MA, Piatti S (2016) The final cut: cell polarity meets cytokinesis at the bud neck in *S. cerevisiae*. Cell Mol Life Sci. doi:10.1007/s00018-016-2220-3

Juanes MA, Twyman H, Tunnacliffe E, Guo Z, ten Hoopen R, Segal M (2013) Spindle pole body history intrinsically links pole identity with asymmetric fate in budding yeast. Curr Biol 23 (14):1310–1319. doi:10.1016/j.cub.2013.05.057

Keck JM, Jones MH, Wong CCL, Binkley J, Chen DC, Jaspersen SL, Holinger EP, Xu T, Niepel M, Rout MP, Vogel J, Sidow A, Yates JR, Winey M (2011) A cell cycle phosphoproteome of the yeast centrosome. Science 332(6037):1557–1561. doi:10.1126/science.1205193

Knop M, Schiebel E (1997) Spc98p and Spc97p of the yeast gamma-tubulin complex mediate binding to the spindle pole body via their interaction with Spc110p. EMBO J 16 (23):6985–6995. doi:10.1093/emboj/16.23.6985

Knop M, Schiebel E (1998) Receptors determine the cellular localization of a gamma-tubulin complex and thereby the site of microtubule formation. EMBO J 17(14):3952–3967. doi:10.1093/emboj/17.14.3952

Knop M, Pereira G, Geissler S, Grein K, Schiebel E (1997) The spindle pole body component Spc97p interacts with the gamma-tubulin of *Saccharomyces cerevisiae* and functions in microtubule organization and spindle pole body duplication. EMBO J 16(7):1550–1564. doi:10.1093/emboj/16.7.1550

Kozubowski L, Saito K, Johnson JM, Howell AS, Zyla TR, Lew DJ (2008) Symmetry-breaking polarization driven by a Cdc42p GEF-PAK complex. Curr Biol 18(22):1719–1726. doi:10.1016/j.cub.2008.09.060

Laan L, Pavin N, Husson J, Romet-Lemonne G, van Duijn M, López MP, Vale RD, Jülicher F, Reck-Peterson SL, Dogterom M (2012) Cortical dynein controls microtubule dynamics to generate pulling forces that position microtubule asters. Cell 148(3):502–514

Li R (1999) Bifurcation of the mitotic checkpoint pathway in budding yeast. Proc Natl Acad Sci U S A 96(9):4989–4994

Li R (2013) The art of choreographing asymmetric cell division. Dev Cell 25(5):439–450. doi:10.1016/j.devcel.2013.05.003

Liakopoulos D, Kusch J, Grava S, Vogel J, Barral Y (2003) Asymmetric loading of Kar9 onto spindle poles and microtubules ensures proper spindle alignment. Cell 112(4):561–574

Lin TC, Gombos L, Neuner A, Sebastian D, Olsen JV, Hrle A, Benda C, Schiebel E (2011) Phosphorylation of the yeast gamma-tubulin Tub4 regulates microtubule function. Plos One 6 (5). doi:ARTN e19700 10.1371/journal.pone.0019700

Lin TC, Neuner A, Schlosser YT, Scharf AN, Weber L, Schiebel E (2014) Cell-cycle dependent phosphorylation of yeast pericentrin regulates gamma-TuSC-mediated microtubule nucleation. eLife 3:e02208. doi:10.7554/eLife.02208

Lin TC, Neuner A, Schiebel E (2015) Targeting of gamma-tubulin complexes to microtubule organizing centers: conservation and divergence. Trends Cell Biol 25(5):296–307. doi:10.1016/j.tcb.2014.12.002

Lord M, Yang MC, Mischke M, Chant J (2000) Cell cycle programs of gene expression control morphogenetic protein localization. J Cell Biol 151(7):1501–1512

Maekawa H, Schiebel E (2004) Cdk1-Clb4 controls the interaction of astral microtubule plus ends with subdomains of the daughter cell cortex. Genes Dev 18(14):1709–1724

Maekawa H, Priest C, Lechner J, Pereira G, Schiebel E (2007) The yeast centrosome translates the positional information of the anaphase spindle into a cell cycle signal. J Cell Biol 179 (3):423–436. doi:10.1083/jcb.200705197

Marschall LG, Jeng RL, Mulholland J, Stearns T (1996) Analysis of Tub4p, a yeast gamma-tubulin-like protein: implications for microtubule-organizing center function. J Cell Biol 134 (2):443–454. doi:10.1083/jcb.134.2.443

McIntosh JR, O'Toole ET (1999) Life cycles of yeast spindle pole bodies: getting microtubules into a closed nucleus. Biol Cell 91(4–5):305–312. doi:S0248-4900(99)80091-4 [pii]

Meitinger F, Richter H, Heisel S, Hub B, Seufert W, Pereira G (2013) A safeguard mechanism regulates Rho GTPases to coordinate cytokinesis with the establishment of cell polarity. PLoS Biol 11(2):e1001495. doi:10.1371/journal.pbio.1001495

Meitinger F, Khmelinskii A, Morlot S, Kurtulmus B, Palani S, Andres-Pons A, Hub B, Knop M, Charvin G, Pereira G (2014) A memory system of negative polarity cues prevents replicative aging. Cell 159(5):1056–1069. doi:10.1016/j.cell.2014.10.014

Menendez-Benito V, van Deventer SJ, Jimenez-Garcia V, Roy-Luzarraga M, van Leeuwen F, Neefjes J (2013) Spatiotemporal analysis of organelle and macromolecular complex inheritance. Proc Natl Acad Sci U S A 110(1):175–180. doi:1207424110 [pii] 10.1073/pnas.1207424110

Monje-Casas F, Amon A (2009) Cell polarity determinants establish asymmetry in MEN signaling. Dev Cell 16(1):132–145. doi:10.1016/j.devcel.2008.11.002

Moore JK, Cooper JA (2010) Coordinating mitosis with cell polarity: molecular motors at the cell cortex. Semin Cell Dev Biol 21(3):283–289. doi:S1084-9521(10)00021-2 [pii] 10.1016/j.semcdb.2010.01.020

Moore JK, Miller RK (2007) The CDK, Cdc28p, regulates multiple aspects of Kar9p function in yeast. Mol Biol Cell 18(4):1187–1202. doi:10.1091/mbc.E06-04-0360

Moore JK, Magidson V, Khodjakov A, Cooper JA (2009a) The spindle position checkpoint requires positional feedback from cytoplasmic microtubules. Curr Biol 19(23):2026–2030. doi:10.1016/j.cub.2009.10.020

Moore JK, Stuchell-Brereton MD, Cooper JA (2009b) Function of dynein in budding yeast: mitotic spindle positioning in a polarized cell. Cell Motil Cytoskeleton 66(8):546–555. doi:10.1002/cm.20364

Moore JK, Chudalayandi P, Heil-Chapdelaine RA, Cooper JA (2010) The spindle position checkpoint is coordinated by the Elm1 kinase. J Cell Biol 191(3):493–503. doi:10.1083/jcb.201006092

Moseley JB, Goode BL (2006) The yeast actin cytoskeleton: from cellular function to biochemical mechanism. Microbiol Mol Biol Rev 70(3):605–645

Nelson SA, Cooper JA (2007) A novel pathway that coordinates mitotic exit with spindle position. Mol Biol Cell 18(9):3440–3450. doi:10.1091/mbc.E07-03-0242

O'Toole ET, Mastronarde DN, Giddings TH Jr, Winey M, Burke DJ, McIntosh JR (1997) Three-dimensional analysis and ultrastructural design of mitotic spindles from the cdc20 mutant of *Saccharomyces cerevisiae*. Mol Biol Cell 8(1):1–11

O'Toole ET, Winey M, McIntosh JR (1999) High-voltage electron tomography of spindle pole bodies and early mitotic spindles in the yeast *Saccharomyces cerevisiae*. Mol Biol Cell 10 (6):2017–2031

Okada K, Bartolini F, Deaconescu AM, Moseley JB, Dogic Z, Grigorieff N, Gundersen GG, Goode BL (2010) Adenomatous polyposis coli protein nucleates actin assembly and synergizes with the formin mDia1. J Cell Biol 189(7):1087–1096

Palmer RE, Sullivan DS, Huffaker T, Koshland D (1992) Role of astral microtubules and actin in spindle orientation and migration in the budding yeast, *Saccharomyces cerevisiae*. J Cell Biol 119(3):583–593

Pearson CG, Bloom K (2004) Dynamic microtubules lead the way for spindle positioning. Nat Rev Mol Cell Biol 5(6):481–492

Pereira G, Schiebel E (2005) Kin4 kinase delays mitotic exit in response to spindle alignment defects. Mol Cell 19(2):209–221. doi:10.1016/j.molcel.2005.05.030

Pereira G, Yamashita YM (2011) Fly meets yeast: checking the correct orientation of cell division. Trends Cell Biol 21(9):526–533

Pereira G, Knop M, Schiebel E (1998) Spc98p directs the yeast gamma-tubulin complex into the nucleus and is subject to cell cycle-dependent phosphorylation on the nuclear side of the spindle pole body. Mol Biol Cell 9(4):775–793

Pereira G, Grueneberg U, Knop M, Schiebel E (1999) Interaction of the yeast gamma-tubulin complex-binding protein Spc72p with Kar1p is essential for microtubule function during karyogamy. EMBO J 18(15):4180–4195. doi:10.1093/emboj/18.15.4180

Pereira G, Hofken T, Grindlay J, Manson C, Schiebel E (2000) The Bub2p spindle checkpoint links nuclear migration with mitotic exit. Mol Cell 6(1):1–10

Pereira G, Tanaka TU, Nasmyth K, Schiebel E (2001) Modes of spindle pole body inheritance and segregation of the Bfa1p-Bub2p checkpoint protein complex. EMBO J 20(22):6359–6370

Pereira G, Manson C, Grindlay J, Schiebel E (2002) Regulation of the Bfa1p-Bub2p complex at spindle pole bodies by the cell cycle phosphatase Cdc14p. J Cell Biol 157(3):367–379. doi:10.1083/jcb.200112085

Pringle JR, Bi E, Harkins HA, Zahner JE, De Virgilio C, Chant J, Corrado K, Fares H (1995) Establishment of cell polarity in yeast. Cold Spring Harb Symp Quant Biol 60:729–744

Pruyne D, Evangelista M, Yang C, Bi E, Zigmond S, Bretscher A, Boone C (2002) Role of formins in actin assembly: nucleation and barbed-end association. Science 297(5581):612–615

Pruyne D, Gao L, Bi E, Bretscher A (2004) Stable and dynamic axes of polarity use distinct formin isoforms in budding yeast. Mol Biol Cell 15(11):4971–4989

Rebollo E, Sampaio P, Januschke J, Llamazares S, Varmark H, Gonzalez C (2007) Functionally unequal centrosomes drive spindle orientation in asymmetrically dividing Drosophila neural stem cells. Dev Cell 12(3):467–474

Rock JM, Amon A (2011) Cdc15 integrates Tem1 GTPase-mediated spatial signals with Polo kinase-mediated temporal cues to activate mitotic exit. Genes Dev 25(18):1943–1954. doi:10.1101/gad.17257711

Rock JM, Lim D, Stach L, Ogrodowicz RW, Keck JM, Jones MH, Wong CCL, Yates JR, Winey M, Smerdon SJ, Yaffe MB, Amon A (2013) Activation of the yeast hippo pathway by phosphorylation-dependent assembly of signaling complexes. Science 340(6134):871–875. doi:10.1126/science.1235822

Sagot I, Klee SK, Pellman D (2002) Yeast formins regulate cell polarity by controlling the assembly of actin cables. Nat Cell Biol 4(1):42–50

Scarfone I, Venturetti M, Hotz M, Lengefeld J, Barral Y, Piatti S (2015) Asymmetry of the budding yeast Tem1 GTPase at spindle poles is required for spindle positioning but not for mitotic exit. Plos Genet 11(2):e1004938. doi:10.1371/journal.pgen.1004938

Schiebel E (2000) Gamma-tubulin complexes: binding to the centrosome, regulation and microtubule nucleation. Curr Opin Cell Biol 12(1):113–118. doi:S0955-0674(99)00064-2 [pii]

Schweiggert J, Stevermann L, Panigada D, Kammerer D, Liakopoulos D (2016) Regulation of a spindle positioning factor at kinetochores by SUMO-targeted ubiquitin ligases. Dev Cell 36 (4):415–427. doi:10.1016/j.devcel.2016.01.011

Segal M, Bloom K (2001) Control of spindle polarity and orientation in *Saccharomyces cerevisiae*. Trends Cell Biol 11(4):160–166

Segal M, Clarke DJ, Reed SI (1998) Clb5-associated kinase activity is required early in the spindle pathway for correct preanaphase nuclear positioning in *Saccharomyces cerevisiae*. J Cell Biol 143(1):135–145

Segal M, Bloom K, Reed SI (2000a) Bud6 directs sequential microtubule interactions with the bud tip and bud neck during spindle morphogenesis in *Saccharomyces cerevisiae*. Mol Biol Cell 11 (11):3689–3702

Segal M, Clarke DJ, Maddox P, Salmon ED, Bloom K, Reed SI (2000b) Coordinated spindle assembly and orientation requires Clb5p-dependent kinase in budding yeast. J Cell Biol 148 (3):441–452

Segal M, Bloom K, Reed SI (2002) Kar9p-independent microtubule capture at Bud6p cortical sites primes spindle polarity before bud emergence in *Saccharomyces cerevisiae*. Mol Biol Cell 13 (12):4141–4155

Seshan A, Amon A (2005) Ras and the Rho effector Cla4 collaborate to target and anchor Lte1 at the bud cortex. Cell Cycle 4(7):940–946

Seybold C, Elserafy M, Ruthnick D, Ozboyaci M, Neuner A, Flottmann B, Heilemann M, Wade RC, Schiebel E (2015) Kar1 binding to Sfi1 C-terminal regions anchors the SPB bridge to the nuclear envelope. J Cell Biol 209(6):843–861. doi:10.1083/jcb.201412050

Shaw SL, Yeh E, Maddox P, Salmon ED, Bloom K (1997) Astral microtubule dynamics in yeast: a microtubule-based searching mechanism for spindle orientation and nuclear migration into the bud. J Cell Biol 139(4):985–994

Sobel SG, Snyder M (1995) Highly divergent gamma-tubulin gene is essential for cell growth and proper microtubule organization in *Saccharomyces cerevisiae*. J Cell Biol 131(6):1775–1788. doi:10.1083/jcb.131.6.1775

Su X, Qiu W, Gupta ML Jr, Pereira-Leal JB, Reck-Peterson SL, Pellman D (2011) Mechanisms underlying the dual-mode regulation of microtubule dynamics by Kip3/kinesin-8. Mol Cell 43 (5):751–763

Ten Hoopen R, Cepeda-Garcia C, Fernandez-Arruti R, Juanes MA, Delgehyr N, Segal M (2012) Mechanism for astral microtubule capture by cortical Bud6p priming spindle polarity in *S. cerevisiae*. Curr Biol 22(12):1075–1083. doi:10.1016/j.cub.2012.04.059

Theesfeld CL, Irazoqui JE, Bloom K, Lew DJ (1999) The role of actin in spindle orientation changes during the *Saccharomyces cerevisiae* cell cycle. J Cell Biol 146(5):1019–1032

Tong Z, Gao XD, Howell AS, Bose I, Lew DJ, Bi E (2007) Adjacent positioning of cellular structures enabled by a Cdc42 GTPase-activating protein-mediated zone of inhibition. J Cell Biol 179(7):1375–1384. doi:10.1083/jcb.200705160

Valerio-Santiago M, Monje-Casas F (2011) Tem1 localization to the spindle pole bodies is essential for mitotic exit and impairs spindle checkpoint function. J Cell Biol 192 (4):599–614. doi:10.1083/jcb.201007044

Visintin R, Amon A (2001) Regulation of the mitotic exit protein kinases Cdc15 and Dbf2. Mol Biol Cell 12(10):2961–2974

Wang X, Tsai JW, Imai JH, Lian WN, Vallee RB, Shi SH (2009) Asymmetric centrosome inheritance maintains neural progenitors in the neocortex. Nature 461(7266):947–955

Weiss EL (2012) Mitotic exit and separation of mother and daughter cells. Genetics 192 (4):1165–1202. doi:10.1534/genetics.112.145516

Winey M, Bloom K (2012) Mitotic spindle form and function. Genetics 190(4):1197–1224. doi:190/4/1197 [pii] 10.1534/genetics.111.128710

Winey M, Mamay CL, O'Toole ET, Mastronarde DN, Giddings TH Jr, McDonald KL, McIntosh JR (1995) Three-dimensional ultrastructural analysis of the *Saccharomyces cerevisiae* mitotic spindle. J Cell Biol 129(6):1601–1615

Woods B, Kuo CC, Wu CF, Zyla TR, Lew DJ (2015) Polarity establishment requires localized activation of Cdc42. J Cell Biol 211(1):19–26. doi:10.1083/jcb.201506108

Wu CF, Savage NS, Lew DJ (2013) Interaction between bud-site selection and polarity-establishment machineries in budding yeast. Philos Trans R Soc London B Biol Sci 368 (1629):20130006. doi:10.1098/rstb.2013.0006

Wu CF, Chiou JG, Minakova M, Woods B, Tsygankov D, Zyla TR, Savage NS, Elston TC, Lew DJ (2015) Role of competition between polarity sites in establishing a unique front. eLife 4. doi:10.7554/eLife.11611

Yamashita YM, Mahowald AP, Perlin JR, Fuller MT (2007) Asymmetric inheritance of mother versus daughter centrosome in stem cell division. Science 315(5811):518–521

Yeh E, Yang C, Chin E, Maddox P, Salmon ED, Lew DJ, Bloom K (2000) Dynamic positioning of mitotic spindles in yeast: role of microtubule motors and cortical determinants. Mol Biol Cell 11(11):3949–3961

Yin H, Pruyne D, Huffaker TC, Bretscher A (2000) Myosin V orientates the mitotic spindle in yeast. Nature 406(6799):1013–1015

Yoder TJ, Pearson CG, Bloom K, Davis TN (2003) The *Saccharomyces cerevisiae* spindle pole body is a dynamic structure. Mol Biol Cell 14(8):3494–3505. doi:10.1091/mbc.E02-10-0655 E02-10-0655 [pii]

Yoshida S, Asakawa K, Toh-e A (2002) Mitotic exit network controls the localization of Cdc14 to the spindle pole body in *Saccharomyces cerevisiae*. Curr Biol 12(11):944–950

Yoshida S, Ichihashi R, Toh-e A (2003) Ras recruits mitotic exit regulator Lte1 to the bud cortex in budding yeast. J Cell Biol 161(5):889–897. doi:10.1083/jcb.200301128

Chapter 4
Wnt Signaling Polarizes *C. elegans* Asymmetric Cell Divisions During Development

Arielle Koonyee Lam and Bryan T. Phillips

Abstract Asymmetric cell division is a common mode of cell differentiation during the invariant lineage of the nematode, *C. elegans*. Beginning at the four-cell stage, and continuing throughout embryogenesis and larval development, mother cells are polarized by Wnt ligands, causing an asymmetric inheritance of key members of a Wnt/β-catenin signal transduction pathway termed the Wnt/β-catenin asymmetry pathway. The resulting daughter cells are distinct at birth with one daughter cell activating Wnt target gene expression via β-catenin activation of TCF, while the other daughter displays transcriptional repression of these target genes. Here, we seek to review the body of evidence underlying a unified model for Wnt-driven asymmetric cell division in *C. elegans*, identify global themes that occur during asymmetric cell division, as well as highlight tissue-specific variations. We also discuss outstanding questions that remain unanswered regarding this intriguing mode of asymmetric cell division.

4.1 Introduction

Asymmetric cell division (ACD) that results in two daughter cells with different developmental fate potentials is required for proper fate specification and diversification of tissues during development and also regulates self-renewal of these tissues during adulthood. These decisions are coordinated across space and time by cell signaling pathways, such as the well-conserved Wnt signaling pathways. Wnt signal transduction pathways can differ considerably; however, they typically share requirements of Wnt ligand-binding receptors and coreceptors, such as Frizzled and Lrp6, to activate the phosphoprotein Dishevelled. Downstream of

A.K. Lam
Interdisciplinary Graduate Program in Molecular and Cellular Biology, University of Iowa, Iowa City, IA, USA

B.T. Phillips (✉)
Department of Biology, University of Iowa, Iowa City, IA, USA
e-mail: bryan-phillips@uiowa.edu

J.-P. Tassan, J.Z. Kubiak (eds.), *Asymmetric Cell Division in Development, Differentiation and Cancer*, Results and Problems in Cell Differentiation 61,
DOI 10.1007/978-3-319-53150-2_4

Dishevelled the pathways diverge to control several distinct intracellular pathways depending on the protein expression profiles of the recipient cells. These include the β-catenin-dependent or "canonical" Wnt signaling pathway, the Wnt planar cell polarity (PCP) pathway, and the Ca^{2+}/Calmodulin pathway (Sokol 2015; Gao and Chen 2010). Though Wnt/PCP is often associated with polarizing cells in various contexts including vertebrate convergent extension and the fly eye and wing bristle (Axelrod et al. 1996; Wehrli and Tomlinson 1998; Klein and Mlodzik 2005; Gao 2012) and the Wnt/Ca^{2+} pathway also polarizes cell movements during gastrulation (Lin et al. 2010), the Wnt/β-catenin canonical pathway regulates mother cell polarity at the time of division and subsequent asymmetry of newly formed daughter cells in addition to its well-established role in transcriptional activation, cell proliferation, differentiation, and stem cell maintenance (Habib et al. 2013; Sawa and Korswagen 2013; Murgan and Bertrand 2015; Munro and Bowerman 2009; Hardin and King 2008).

4.1.1 Canonical Wnt Signaling Stabilizes β-Catenin to Activate TCF Target Genes

The canonical Wnt pathway affects changes in cell behavior by altering gene expression through regulation of a transcriptional coactivator, β-catenin. In populations of cells that lack Wnt signal transduction, a "destruction complex" forms, consisting of two kinases, Glycogen Synthase Kinase 3β (GSK3β) and Casein Kinase Iα (CKIα), and two scaffold proteins, Axin and Adenomatous Polyposis Coli (APC), which binds and phosphorylates cytoplasmic β-catenin, triggering β-catenin ubiquitination and proteasomal degradation (Clevers and Nusse 2012; Logan and Nusse 2004; MacDonald and He 2012). In the absence of β-catenin, the DNA binding protein TCF transcriptionally represses target gene expression, but in the presence of Wnt signal transduction, TCF transcriptional repression is converted to target gene activation. This is accomplished by Wnt activation of Frizzled receptors and LRP5/6 coreceptors, which promotes the activation of Dishevelled (Dvl) and inactivates the destruction complex by physically interacting with its components, thereby inhibiting β-catenin degradation (Clevers and Nusse 2012; MacDonald and He 2012). As a result, β-catenin within the cytoplasm is stabilized and translocated into the nucleus, where it binds TCF. This binding converts TCF into a transcriptional activator and promotes the expression of TCF target genes due to β-catenin's ability to recruit transcriptional activators such as Mediator and chromatin modifiers such as CBP, Brg-1, Bcl-9, and Pygopus (Fig. 4.1a) (Hecht et al. 2000; Takemaru and Moon 2000; Barker et al. 2001; Kramps et al. 2002; Parker et al. 2002; Thompson et al. 2002; Roose and Clevers 1999; Yoda et al. 2005).

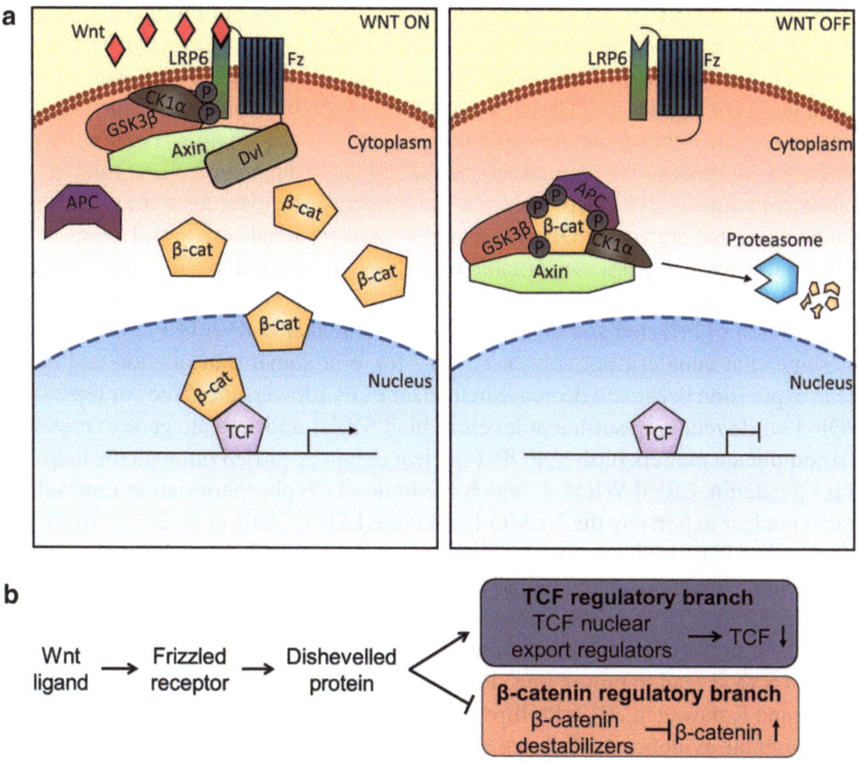

Fig. 4.1 The Wnt/β-catenin signing pathway. (**a**) Canonical Wnt signaling in mammalian systems activates TCF target gene expression by stabilizing β-catenin. (**b**) Simplified representation of the branched WβA pathway in *C. elegans* with two outputs: elevated β-catenin levels and decreased nuclear TCF

4.1.2 A Branched Wnt Pathway Controls Asymmetric Divisions in C. elegans

While a similar Wnt/β-catenin signaling pathway is found in nematodes that stabilizes the β-catenin homolog, BAR-1, to generate cell diversity in tissues including the Q neuroblast and vulva precursor cell lineages (Eisenmann et al. 1998; Gleason et al. 2006; Jiang and Sternberg 1999; Schmid and Hajnal 2015), a more broadly used Wnt/β-catenin pathway polarizes mother cells and drives serial ACD throughout development, globally generating tissue diversity. In this pathway, termed the Wnt/β-catenin Asymmetry (WβA) pathway, two additional β-catenins, SYS-1 and WRM-1, regulate the transcriptional activity and nuclear levels to the sole *C. elegans* TCF transcription factor, POP-1. After ACD, the daughter cell whose fate does not depend on Wnt signaling, the "unsignaled daughter," contains low SYS-1/β-catenin due to destruction complex activity, and, similar to the canonical Wnt pathway, the lack of nuclear β-catenin causes POP-1 to repress the transcription of Wnt target genes (Sawa and Korswagen 2013; Baldwin and Phillips 2014; Jiang and Sternberg 1999). The mother cell asymmetrically localizes the destruction complex members (e.g., APC and Axin) such that the unsignaled daughter inherits these negative regulators (Mizumoto and Sawa 2007b). Conversely, the daughter cell whose fate depends on Wnt signaling (the "signaled daughter") exhibits lower levels of these negative regulators. In the signaled daughter cell, similar to the canonical pathway, SYS-1/β-catenin accumulates in the cytoplasm, translocates to the nucleus, and converts POP-1/TCF into a transcriptional activator. However, the WβA pathway also possesses notable differences compared to the canonical pathway. In addition to the stabilization of SYS-1/β-catenin, a second mechanism downstream of Frizzled and Dvl exports excess nuclear POP-1/TCF. POP-1 export is, somewhat counterintuitively, necessary for Wnt signal transduction and target gene expression because a decrease in nuclear POP-1 lowers the "free" or repressive POP-1 while retaining sufficient levels to bind SYS-1 and activate gene expression. To complicate matters further, POP-1 nuclear export is carried out with the help of a third β-catenin, called WRM-1, which facilitates TCF phosphorylation (and subsequent nuclear export) by the NEMO-like kinase LIT-1 (Yang et al. 2011). In all, the low level of POP-1/TCF in the signaled cell increases the likelihood that most of the POP-1 in the nucleus will be bound by SYS-1/β-catenin, which is increasing in this cell, therefore activating the transcription of genes in the signaled daughter. Conversely, in the Wnt inactive ("unsignaled") cell, a high level of POP-1 and a low level of SYS-1 lead to more "free" POP-1 and transcriptional repression (Fig. 4.1b) (Sawa and Korswagen 2013; Phillips and Kimble 2009). A detailed discussion of the experimental evidence underlying this model and considerations of future challenges are presented below.

4.2 Wnt Polarizes the Endo-mesoderm Lineage

The WβA pathway appears to regulate the many ACDs of *C. elegans* embryogenesis. Embryonic blastomeres undergo multiple asymmetric divisions, each cell producing daughter cells with different developmental fates. ACDs affect differential cell fate specification as early as the first cell division, where the site of sperm entry determines the first cleavage plane and breaks symmetry by producing a larger somatic cell (AB) and a smaller germ cell (P1). However, the first incidence of Wnt-driven ACD occurs at the four-cell stage, where the posterior daughter of P1, called P2, polarizes its neighbor cell, called EMS, which divides asymmetrically to give rise to the endoderm and mesoderm lineages (Fig. 4.2a) (Munro and Bowerman 2009; Goldstein and Hird 1996). EMS polarization by P2-derived Wnt ligand has been well studied and gives excellent insight into the mechanisms of the WβA pathway.

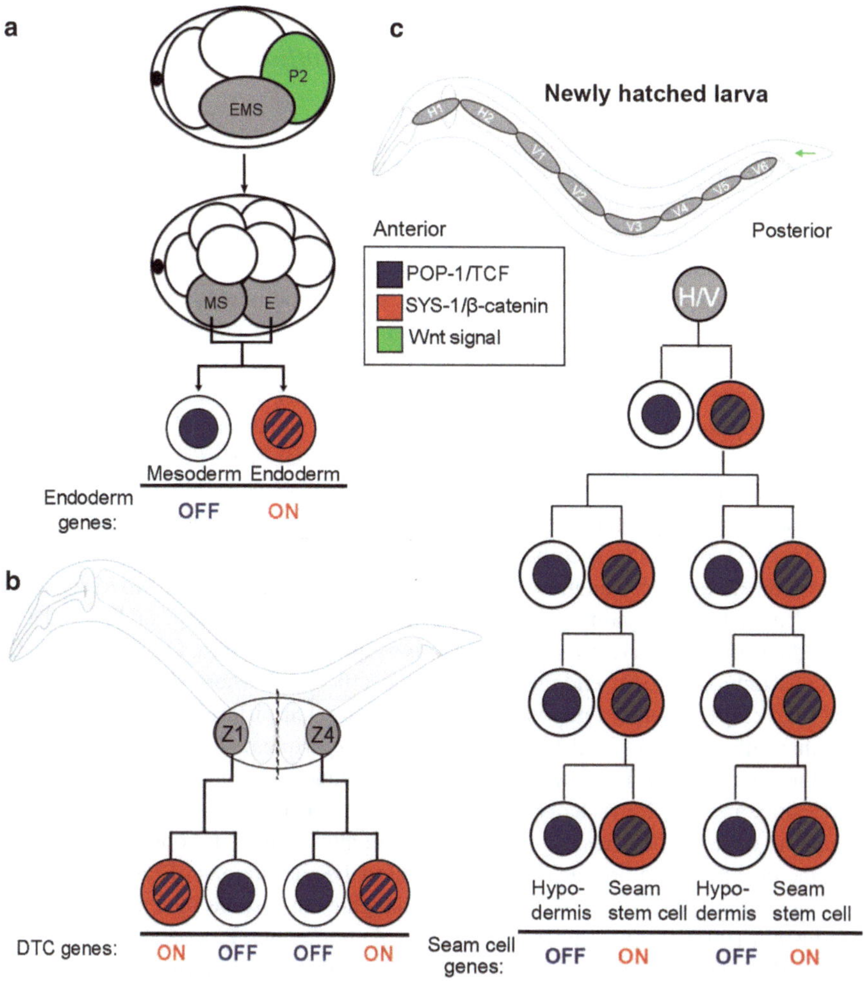

Fig. 4.2 Asymmetric cell division in *C. elegans*. (**a**) A Wnt signal from P2 polarizes EMS, which divides into endoderm (E) and mesoderm (MS) daughter cells. E displays lowered nuclear POP-1 (*blue cross-hatching*) and elevated SYS-1 compared to MS. (**b**) Z1/Z4 somatic gonadal precursor cells divide along the proximal–distal axis (*dashed line*), each asymmetrically generating a distal tip cell (DTC) lineage daughter and a proximal anchor cell potential lineage. (**c**) A newly hatched worm contains several seam cells on both sides of the worm. Most seam cells (H1–H2, V1–V6) divide in a stem cell-like manner giving rise to a seam cell daughter and a terminally differentiated hypodermal daughter at each larval molt. In each case (**a–c**), WβA signaling activates a transcriptional profile specific for the signaled daughter that is repressed in the unsignaled daughter. A symmetric cell division at L2 increases overall seam cell number to 16. Modified from Phillips and Kimble (2009)

4.2.1 Embryonic Blastomere Manipulations Demonstrate an Instructive Signal from the Germline Polarized EMS

In normal embryo development, polarized EMS divides asymmetrically to produce a posterior E cell and an anterior MS cell, where the descedants of the E cell form endoderm while the descedants of MS cell form mesoderm (Fig. 4.2a) (Rose and Gonczy 2014; Sulston and Horvitz 1977). In a technically challenging and revealing series of experiments, Goldstein (1992) studied the interaction between each embryonic cell and its closest neighbor by removing the embryo's eggshell and physically isolating the blastomeres from the 4-cell stage embryo. These studies found that the descedants of the isolated EMS mother cell do not result in two separate populations of endoderm and mesoderm populations of cells, but rather EMS divides symmetrically into duplicated mesodermal lineages, evident from the absence of gut. Since the descedants of isolated EMS cells fail to differentiate endoderm, it was hypothesized that other cells within the embryo are required for proper ACD of EMS, specifically for specification of the E daughter and/or repression of the MS daughter. To analyze this idea, separated EMS cells were placed next to all remaining embryonic blastomeres, ABa, ABp, or P2, and the descedants of EMS were screened for positive gut differentiation. Only contact with P2 cells induced proper endoderm lineage specification, where the descedants of EMS developed into two populations of cells: one that differentiates into gut and one that fails to do so (Goldstein 1992). Further experiments from Goldstein found that the orientation of EMS division is highly dependent on contact between P2 and EMS since moving P2 (usually in contact with the posterior end of EMS) to other locations next to EMS results in the EMS daughter that is in contact with P2 to assume the endoderm lineage (i.e., moving P2 to the anterior end of EMS will cause the anterior daughter and not the posterior daughter to assume the endoderm fate) (Goldstein 1993). These studies revealed the likelihood that P2 contact or secretion of a diffusible P2 factor polarized EMS and activated endoderm developmental specification program.

4.2.2 A Wnt Pathway Polarizes EMS Division

Following the discovery that P2 contact is required for EMS ACD, forward genetic screens in the Priess, Mello, Schnabel, and Bowerman labs identified mutants that are defective for EMS ACD including those that lack endoderm and have the <u>mo</u>re <u>m</u>esoderm (*mom*) phenotype. Several of such mutants (including *mom-1* through *mom-5*, *pop-1*, and *lit-1*) were mapped and identified as components of a Wnt/β-catenin-related pathway, now known as the WβA pathway (Thorpe et al. 1997; Rocheleau et al. 1997; Mizumoto and Sawa 2007b; Kaletta et al. 1997; Lin et al. 1995). For instance, MOM-2 is a secreted Wnt ligand that initiates and activates Wnt signaling (Rocheleau et al. 1997; Thorpe et al. 1997). In *mom-2* mutants, terminally differentiated embryos lack gut and display an increase in pharyngeal mesodermal cells. Early activity of the *mom-2* gene product was shown by laser ablation experiments. In eight cell wild-type embryos, ablating all blastomeres except the E cell results in the remaining E cell descendants producing gut tissues. However, similar experiments in *mom-2* mutants found that the remaining E cell descendants produce only pharyngeal muscle and lack gut, an indication that both EMS daughter cells adopt an MS-like fate in the absence of MOM-2/Wnt (Thorpe et al. 1997). Thus, MOM-2 ligand was hypothesized to be the polarizing ligand secreted by P2 to induce asymmetric cell division of the EMS mother cell. To determine if P2-derived MOM-2 non-autonomously regulates EMS, different combinations of isolated wild-type or *mom-2* EMS and P2 blastomeres were placed in close contact and the cell fate of the descendant EMS cells was observed (Thorpe et al. 1997). Endoderm failed to develop when an isolated wild-type EMS cell was placed next to an isolated *mom-2* mutant P2 cell. In contrast, the presence of gut tissue was detected when *mom-2* mutant EMS cells contacted wild-type P2 cells. These experiments showed that a Wnt ligand, MOM-2, is non-autonomously required for P2 polarization of EMS during mitosis and further facilitates downstream WβA signaling in E and MS daughters (Thorpe et al. 1997).

4.2.3 WRM-1/β-Catenin and LIT-1/Nemo Kinase Control POP-1 Nuclear Export

The above experiments identified a signal required for EMS polarity, but how does the MOM-2/Wnt ligand polarize the mother cell to control asymmetric fate specification in EMS daughters? MOM-2 polarizing activity results in molecular EMS asymmetry at the time of division producing one signaled daughter cell capable of activating Wnt target genes and one unsignaled daughter cell where Wnt targets are repressed (Thorpe et al. 1997). Initial observations of nuclear POP-1 asymmetry in the two daughters led to the finding that differential WβA pathway activity results in asymmetric regulation of POP-1 nucleo-cytoplasmic distribution in E versus MS resulting in nuclear asymmetry (Maduro et al. 2002; Lin et al. 1998; Lo et al. 2004). Though regulation of TCF levels is apparently not a typical effect of canonical Wnt signal transduction, POP-1 regulation is essential in the worm WβA pathway. POP-1, like other TCF homologs, serves as either transcriptional repressor or activator depending on the availability of corepressors and coactivators (Fig. 4.1) (Sawa and Korswagen 2013). After EMS ACD, POP-1 is increased in the nucleus of the unsignaled MS daughter compared to the signaled E daughter (Lin et al. 1998). The decrease of POP-1 in E nuclei is due to its interaction with WRM-1/β-catenin, which facilitates POP-1 phosphorylation by LIT-1/NEMO kinase and its subsequent nuclear export (Rocheleau et al. 1999; Meneghini et al. 1999; Ishitani et al. 1999; Shin et al. 1999; Lo et al. 2004; Yang et al. 2011). Counterintuitively, POP-1 export is required for activation of Wnt target genes, presumably by lowering excess POP-1 that would otherwise repress gene expression. Indeed, the *pop-1* mutant EMS phenotype is consistent with the idea that WβA represses POP-1: *pop-1* mutants show excess endoderm due to duplication of E fate and loss of MS whereas the Mom phenotype is loss of E and duplication of MS.

4.2.4 Differential POP-1/TCF Regulation by Two β-Catenins

Asymmetric POP-1 nuclear localization in EMS daughters is dependent on the activity of LIT-1/Nemo kinase where LIT-1 phosphorylates POP-1 to facilitate POP-1 export out of the E nucleus (Lo et al. 2004; Yang et al. 2011). In the embryos of *lit-1* mutants, POP-1 displays high symmetrical localization in both E and MS nuclei owning to the lack of POP-1 export that is typically seen in the wild-type E daughter (Rocheleau et al. 1999; Meneghini et al. 1999; Ishitani et al. 1999; Kaletta et al. 1997). *lit-1* mutants, like *wrm-1* mutants, also exhibit a lack of endoderm, a phenotype indicative of conversion of the E cell fate to an MS-like cell fate. Furthermore, immunoprecipitation experiments show that LIT-1 binds to WRM-1 for effective POP-1 phosphorylation, and this binding is likely in the C-terminal domain of POP-1 rather than the typical "beta-catenin binding domain" located in the N-terminus of TCF proteins (Brunner et al. 1997; Molenaar et al. 1996; Yang et al. 2011). The interaction between WRM-1 and LIT-1 promotes LIT-1 phosphorylation at T220, which is in turn essential for LIT-1 kinase activity required for POP-1 phosphorylation (Yang et al. 2015). The binding of POP-1 with the WRM-1/LIT-1 complex and the binding of POP-1 to its transcriptional coactivator SYS-1 (discussed further in the chapter) use distinct binding domains but are mutually exclusive (Yang et al. 2011). Though recent evidence demonstrates that POP-1 can interact with other DNA binding factors to activate transcription in a WβA-independent manner (Murgan and Bertrand 2015; Murgan et al. 2015), the above data show that POP-1 contains distinct domains that bind to either its activator SYS-1 or the effector of its export, WRM-1, to control its transcriptional activity and subsequently the expression of WβA target genes.

4.3 A Genetic Screen of Larval ACDS Identifies SYS-1/β-Catenin

Though POP-1 export was shown to be necessary for WβA signal transduction by the aforementioned studies, several lines of evidence suggested that POP-1 nuclear downregulation might be insufficient for target gene activation and that the pathway was missing a TCF coactivator. These include the realization that loss of POP-1 decreases target gene expression levels in E in addition to derepressing them in MS

(Shetty et al. 2005). Additionally, transcriptional reporters of POP-1 targets require TCF binding sites for expression, whereas they should be dispensable for or decrease expression if POP-1 was an obligate repressor (Maduro et al. 2005). Finally, a conserved feature of TCF proteins is their ability to switch from repressor mode to transcriptional activation upon β-catenin binding, which displaces corepressors such as Groucho (Roose et al. 1998). These data suggest that a transactivator such as β-catenin may be required to activate *C. elegans* POP-1. In addition to WRM-1, which is a weak transcriptional activator specialized for POP-1 nuclear export (Korswagen et al. 2000; Natarajan et al. 2001), the *C. elegans* genome contains two β-catenins recognized by sequence similarity to vertebrate and fly β-catenin, HMP-2 and BAR-1, neither of which is required for *C. elegans* ACD (Sawa and Korswagen 2013; Phillips et al. 2007; Korswagen et al. 2000). However, SYS-1/β-catenin, first found in a genetic screen to identify factors required for ACDs in the somatic gonad (Miskowski et al. 2001; Siegfried and Kimble 2002; Siegfried et al. 2004), proved to be the missing TCF activator. Despite being just ~10% identical to human β-catenin, the conclusion that SYS-1 is a β-catenin that acts as the POP-1 coactivator during ACD is supported by the observations that (1) SYS-1 is required for many worm ACDs similar to POP-1 and WRM-1, (2) as with most TCF–β-catenin interactions, SYS-1 physically interacts with POP-1 via the N-terminal β-catenin binding domain of POP-1 (Kidd et al. 2005), and (3) SYS-1 activates TCF target genes, such as *ceh-22* and *end-1* in worms, but also via the TCF reporter assay TOPLFLASH when coexpressed with POP-1 in human cells (Lam et al. 2006; Kidd et al. 2005; Huang et al. 2007). Finally, (4) SYS-1 expression rescues the defects of loss of a well-established worm β-catenin, BAR-1, which also acts as a POP-1 coactivator (Kidd et al. 2005; Natarajan et al. 2001). The sequence divergence of *C. elegans* SYS-1 therefore concealed its identity as a critical activator of POP-1 transcription until the requirement of SYS-1 for ACD was revealed through the use of forward genetics.

4.3.1 SYS-1 Controls Asymmetric Division of the Somatic Gonadal Precursors

SYS-1, for symmetric sisters, was identified in a mutagenesis screen for defects in asymmetric divisions of the somatic gonad (Miskowski et al. 2001). The two distal tip cells (DTC), which are somatic cells that serve as niches for the germline stem cells and lead to gonadal arm elongation, are formed by ACDs tightly controlled by the WβA pathway (Kimble 1981; Austin and Kimble 1987; Byrd et al. 2014; Chesney et al. 2009; Siegfried et al. 2004; Kidd et al. 2015). The DTCs are large somatic cells that cap the end of both the gonadal arms of the worm and, coupled with cell-specific GFP markers, serve as an easily screened cell type suitable for high-throughput genetic analyses. The development of the DTCs begins during the L1 larval stage where two somatic gonad precursor cells (SGPs), Z1 and Z4, in the gonadal primordium divide asymmetrically to produce a total of 12 cells including the DTCs (Kimble and Hirsh 1979; Kimble and Ward 1988). The asymmetric division of the Z1/Z4 somatic gonadal precursor cells requires SYS-1, POP-1, and the WβA pathway; the distal daughter cells of Z1 and Z4 both assume the signaled cell fate and give rise to the DTCs while the proximal daughters assume the unsignaled fate, including the anchor cells (Fig. 4.2b) (Siegfried and Kimble 2002; Siegfried et al. 2004; Chang et al. 2005; Miskowski et al. 2001; Lam et al. 2006; Phillips et al. 2007). Further experiments in the somatic gonad show that loss of SYS-1 function via mutation or RNAi causes loss of the signaled fate (DTC) and duplicates the unsignaled proximal fate; SYS-1 overexpression results in an increase in DTCs and loss of proximal fates compared to wild-type worms (Kidd et al. 2005). Interestingly, asymmetric SYS-1 expression can be observed in the Z1/Z4 daughters (Phillips et al. 2007). YFP-tagged SYS-1 under the control of the endogenous *sys-1* promoter is elevated in the distal daughter (Z1a, Z4p) compared to the proximal daughters (Z1p and Z4a). Thus, SYS-1 is necessary and sufficient for WβA-dependent target gene expression and is elevated in the WβA-dependent daughter cell that requires its activity.

4.3.2 The SYS-1 Crystal Structure Displays Hallmarks of β-Catenins

SYS-1's β-catenin identity was further confirmed by its crystal structure. β-catenins contain twelve tandemly repeated "armadillo" domains that consist of two or three alpha helices connected by flexible linkers (Liu et al. 2008). SYS-1 shares the hallmark twelve armadillo repeats of canonical β-catenin packing together to form a superhelix similar to human β-catenin (Liu et al. 2008; Huber et al. 1997; Xing et al. 2008; Poy et al. 2001). Further, the crystallized SYS-1/POP-1 complex shows similar interactions to the human β-catenin/TCF complex. SYS-1 contains a positively charged groove, spanning armadillo repeats 5–8, which interacts with POP-1, an interaction anchored by a conserved "charged button" lysine-aspartate salt bridge (Liu et al. 2008). This salt bridge is critical for the complex since mutation of either residue abrogates binding and gives a loss of the signaled fate during worm ACD (Siegfried and Kimble 2002; Liu et al. 2008). Thus, despite significant sequence divergence, SYS-1 retains the functional and structural characteristics of canonical β-catenin. Conversely, the predicted WRM-1 structure shows that, while WRM-1 shares similarity to SYS-1 in the repeat 6–8 region and contains the charged button, in contrast to SYS-1 WRM-1 contains a nearby bulky side chain that prevents WRM-1/POP-1 interactions (Liu et al. 2008).

4.3.3 Reciprocal POP-1 and SYS-1 Asymmetry Is Widespread in C. elegans

Further analysis of SYS-1 function and expression pattern indicates that SYS-1 activates WβA target gene activity in many of the worm ACDs throughout development, including the aforementioned EMS daughter cells, suggesting that SYS-1 may be the sole TCF coactivator involved in WβA-dependent ACD. SYS-1 asymmetry is observed throughout embryogenesis (Bertrand and Hobert 2009a; Huang et al. 2007; Zacharias et al. 2015; Phillips et al. 2007), consistent with broad function promoting WβA-dependent cell fate. Further, SYS-1 depletion causes gutlessness and promotes the posterior fate of various embryonic ACDs (Bertrand and Hobert 2009a, 2010; Huang et al. 2007; Phillips et al. 2007). The SYS-1 localization pattern is WβA dependent as depletion of various Wnt components such as MOM-1/Porcupine, MOM-2/WNT, MOM-5/Frizzled, and Dvl homologs induces a loss of nuclear SYS-1 asymmetry in E and MS daughters (Huang et al. 2007). Thus, given the earlier observation that WβA decreases POP-1 levels, SYS-1 and POP-1 show a reciprocal expression in the nucleus of daughters of an asymmetrically dividing cell. For instance, in the EMS division, the signaled E daughter has elevated SYS-1 but decreased POP-1 compared to the unsignaled MS cell (Lin et al. 1995; Huang et al. 2007; Lo et al. 2004) (Fig. 4.2a). Current models posit that WβA activation of target gene expression is controlled by the ratio of SYS-1 and POP-1. In the unsignaled cell nucleus, the SYS-1:POP-1 ratio is low and most of POP-1 is free of SYS-1 activation and therefore in a transcriptionally repressive state. In the signaled cell, the WβA pathway elevates the SYS-1:POP-1 ratio due to increased SYS-1 levels and decreased nuclear POP-1 due to POP-1 nuclear export (Sawa and Korswagen 2013; Jackson and Eisenmann 2012; Phillips and Kimble 2009). Thus, Wnt regulates asymmetric target gene expression via differential regulation of the two branches of the pathway.

4.4 Hypodermal Stem Cell Divisions Elucidate POP-1/TCF and SYS-1/β-Catenin Regulation

While the ratio model explains several features of the WβA pathway and has withstood many tests, the identity of the cytoplasmic fate determinants actually partitioned by mother cell division has been informed through the study of the *C. elegans* hypodermal tissue known as the seam cells. In addition to EMS and the SGPs, the seam is another tissue that is widely used in the study of the WβA-regulated asymmetric division (Sawa and Korswagen 2013; Eisenmann 2005, 2011; Mizumoto and Sawa 2007b; Herman et al. 1995; Wildwater et al. 2011; Harandi and Ambros 2015). A newly hatched worm possesses 10 seam cells (H0, H1–H2, V1–V6, and T) embedded within the hyp7 syncytium on both the left and right sides of the worm (Fig. 4.2c) (Sulston and Horvitz 1977). With each larval molt, most seam cells divide asymmetrically along the anterior–posterior (A–P) axis to produce a posterior seam cell daughter and an anterior hypodermal cell daughter. In a stem cell-like manner, the anterior hypodermal cell daughter terminally differentiates and fuses with the hyp7 syncytium while the posterior daughter retains the seam cell fate and the ability to divide at each subsequent larval molt. At the end of the L4 larval molt, the adult worm will have 16 seam cells on each side of the worm due to a symmetric cell division at L2. Seam cell fate acquisition is dependent on active WβA signaling where the signaled posterior seam cell daughter exhibits the characteristic high nuclear localization of SYS-1 and lowered POP-1 and the unsignaled hypodermal cell with a lower nuclear SYS-1 localization but higher POP-1 (Banerjee et al. 2010; Gleason and Eisenmann 2010; Mizumoto and Sawa 2007a). The signaled posterior daughter maintains its seam cell fate by upregulating EGL-18 and ELT-6 GATA factors, which are repressed in the unsignaled anterior daughter (Gorrepati et al. 2013, 2015; Gorrepati and Eisenmann 2015).

4.4.1 Seam Cells Show Polarized WβA Pathway Components at Mitosis

Forward and reverse genetic screens have identified additional WβA components, and functional analysis of these players has greatly clarified regulation of the two pathway branches. Loss of negative WβA regulators often renders a seam cell division symmetric, inducing a fate change duplicating the seam cell fate at the expense of the hypodermal cell fate. Depletion of negative regulators, KIN-19 (the homolog of CKIα), the APC homolog APR-1, or the Axin homolog PRY-1, results in a significant increase in seam cells (Banerjee et al. 2010; Gleason and Eisenmann 2010). In the course of WβA-mediated asymmetric seam cell division, multiple cytoplasmic cell fate determinants of the pathway can be observed to asymmetrically localize on the anterior or posterior cortex of the seam mother cells to give rise to two molecularly distinct daughter cells (Mizumoto and Sawa 2007b). Each daughter cell carries different nuclear, cytoplasmic, and cortical localization profiles of the determinants of WβA pathway, as observed by the addition of fluorescent tags (Mizumoto and Sawa 2007a). The asymmetric localization of negative regulators of the pathway such as APR-1 and PRY-1 can be observed at the anterior cortex, while positive regulators such as Frizzled (MOM-5) and Dishevelled proteins (DSH-2 and MIG-5) are distributed to the posterior cortex of the mother seam cell as well as the early embryo (Park et al. 2004; Baldwin et al. 2016; Mizumoto and Sawa 2007a; Park and Priess 2003). The asymmetric cortical localization of negative determinants to the anterior cortex is regulated by Wnt ligands as demonstrated by observing APR-1/APC and PRY-1/Axin localization in mutants lacking Wnt ligand EGL-20 (Mizumoto and Sawa 2007a). In *egl-20* mutants, both APR-1 and PRY-1 no longer asymmetrically localized to the anterior cortex but are radialized on the cortex of the entire cell (Fig. 4.3) (Mizumoto and Sawa 2007a). The finding that a Wnt ligand

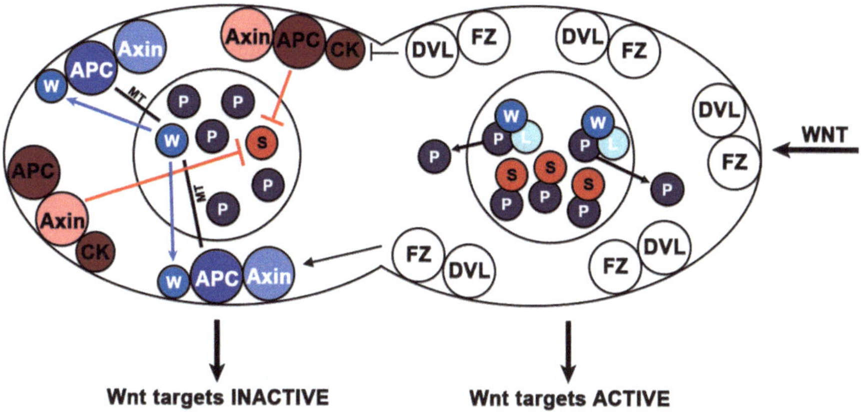

Fig. 4.3 Model of WβA-controlled asymmetric division. Two separate branches of the pathway, the WRM-1-regulating (*blue*) and the SYS-1 regulating (*red*), are balanced by Dishevelled (Dvl) and together generate transcriptional asymmetry in daughter cells following division. The activity of posterior factors including Frizzled and Dishevelled restricts cortical localization or activity of negative factors APC and Axin. In addition, Dishevelled distinguishes between two pools of APC in the unsignaled cell to modulate each downstream branch of the pathway. One pool of APC stabilizes microtubules (MTs) to promote WRM-1/β-catenin (W) export, which results in high nuclear POP-1/TCF (P). The other pool of APC degrades SYS-1/β-catenin in concert with KIN-19/CKIα (CK). Target genes are repressed. In the signaled daughter cell (*right*), neither pool of APC exists so WRM-1 and SYS-1 are both high. POP-1 is low because of nuclear export, mediated by phosphorylation by LIT-1 (L) through association with WRM-1. Target genes are expressed. Modified version of Baldwin et al. (2016)

controls the polarized localization of cell fate-regulating WβA pathway members at mitosis, and likely controls their asymmetric inheritance by daughter cells, suggests that these are the critical cytoplasmic cell fate determinants that make the two daughter cells different at birth.

4.4.2 Distinct Roles for WRM-1/β-Catenin at the Cortex and Nucleus

Interestingly, WRM-1, which shows the aforementioned asymmetric nuclear localization pattern in seam cell daughters, also displays asymmetric cortical localization before division (Takeshita and Sawa 2005; Mizumoto and Sawa 2007a). WRM-1 localizes to the anterior cortex of the mother cell in a similar manner to APR-1 and PRY-1 (Fig. 4.3) (Mizumoto and Sawa 2007a; Takeshita and Sawa 2005). To test the significance of the WRM-1 cortical localization pattern, WRM-1 was uniformly tethered to the cortex of seam cells through the generation of a WRM-1 transgene containing a C-terminal CAAX cortical localization signal (WRM-1::CAAX) (Mizumoto and Sawa 2007a). WRM-1::CAAX reduces the posterior nuclear levels of free WRM-1::GFP, resulting in a low symmetrical localization of WRM-1 in both daughter cells. These studies suggest the existence of two pools of WRM-1: one pool asymmetrically localizes to the anterior cortex of the mother cell and, following cell division, prevents the nuclear localization of the second pool of free WRM-1 in the hypodermal cell daughter. The absence of cortical WRM-1 in the posterior seam cell daughter allows nuclear localization of the free WRM-1 (Fig. 4.3) (Mizumoto and Sawa 2007a).

4.4.3 WRM-1 Asymmetry Is Regulated by Asymmetric Nuclear Export and Differential Microtubule Stability

The cortical localization of WRM-1 is controlled by APR-1/APC, providing both a link to upstream WβA signaling and the canonical Wnt pathway. APR-1 depletion decreases cortical WRM-1, increases nuclear WRM-1, and results in symmetric seam cell divisions that duplicate seam cell fate. Photobleaching experiments during mother cell telophase show that WRM-1 is rapidly removed from the anterior nucleus but is stably present in the posterior nucleus. This nuclear export of WRM-1 is dependent on APR-1, which localizes to the anterior cortex (Takeshita and Sawa 2005; Mizumoto and Sawa 2007a). WRM-1 in turn is required for APR-1 cortical localization, suggesting the formation of a WRM-1/APR-1 complex at the anterior cortex that recruits additional WRM-1 and APR-1 in a feed-forward manner to amplify WRM-1 export from the unsignaled nucleus. A possible mechanism of APR-1-dependent WRM-1 export was provided by the examination of APR-1 regulation of embryonic microtubule dynamics (Sugioka et al. 2011; Sugioka and Sawa 2010). Loss of APR-1 preferentially decreased anterior microtubules during EMS division, suggesting that cortical APR-1 promotes microtubule attachment at the anterior cortex and that APR-1 shares this microtubule-regulating role with its vertebrate homolog, APC (Mimori-Kiyosue et al. 2000; Nakamura et al. 2005; Zumbrunn et al. 2001; Clevers and Nusse 2012; Brocardo et al. 2008; Green and Kaplan 2003; Barth et al. 2008). The importance of asymmetric microtubule organization for nuclear POP-1 and WRM-1 asymmetry was demonstrated by irradiating the anterior centrosome to disrupt microtubule formation. After centrosome irradiation, the nuclear levels of WRM-1 and POP-1 in both daughters are symmetrical, consistent with the idea of microtubule-dependent WRM-1 nuclear export. Furthermore, depletion of several kinesins decreases the level of POP-1 in the anterior daughter, suggesting that nuclear WRM-1 asymmetry is achieved by the active transport of WRM-1 to the cortex in the anterior daughter (Sugioka et al. 2011). Therefore, the emerging model of asymmetric WRM-1 regulation is that APR-1 is restricted to the anterior cortex by upstream WβA signaling, which stabilizes microtubules and promotes WRM-1 transport to the cortex preferentially from the anterior cytoplasm and nucleus. WRM-1 binding to cortical APR-1 stabilizes this localization pattern and increases asymmetric transport. The posterior daughter lacks cortical APR-1 and contains fewer microtubules; thus, the posterior nucleus contains more stable WRM-1, facilitating POP-1 nuclear export (Fig. 4.3, blue factors).

4.4.4 SYS-1 Asymmetry Is Regulated by Members of the Canonical Destruction Complex

In canonical Wnt signaling, the destruction complex formed by Axin, APC, CKIα, and GSK3β targets β-catenin for degradation, and we have recently begun to address whether the destruction complex regulates the transcriptionally active β-catenin in the WβA pathway, SYS-1. Since APR-1/APC, PRY-1/Axin, and KIN-19/CKIα all negatively regulate seam cell fate, we tested the hypothesis that they are required for normal SYS-1 asymmetry. Similar to the embryonic and gonadal divisions described above, SYS-1 is enriched in the signaled (posterior) daughter cell and lost in the unsignaled (anterior) daughter cell (Baldwin and Phillips 2014; Mizumoto and Sawa 2007b) (Fig. 4.2c). After depletion of APR-1 or KIN-19, nuclear SYS-1 levels become symmetrically high in both seam cell daughters, suggesting that negative regulation of β-catenin by the destruction complex is conserved in the WβA pathway (Baldwin and Phillips 2014; Mila et al. 2015). APR-1-dependent negative regulation of SYS-1 is also seen in MS, the unsignaled daughter of EMS, suggesting that destruction complex regulation of SYS-1 may be widespread (Huang et al. 2007). Interestingly, the *pry-1*/Axin mutant phenotype in the seam cells is distinct from that of *apr-1* and *kin-19*; nuclear SYS-1 and APR-1 asymmetry is still seen in most *pry*-1 mutant daughter cell pairs, but their polarity is randomized. Since SYS-1 can be lowered in daughter cells in the absence of PRY-1, it appears that Axin is not required for destruction complex activity. Instead, PRY-1 localizes the destruction complex to the anterior cortex and dictates the correct daughter fate in the correct A–P position. Thus, following cell division, APR-1 and SYS-1 exhibit reciprocal asymmetry in their localization pattern. However, this reciprocal asymmetry can be uncoupled by loss of WRM-1, which decreases cortical APR-1 but has no effect on SYS-1 nuclear asymmetry, or loss of KIN-19, which increases APR-1 on the posterior cortex, but SYS-1 remains high in the posterior nucleus. An explanation for these observations is that mother cells contain two pools of APR-1: one pool negatively regulates WRM-1 (blue APC in Fig. 4.3) and the other negatively regulates SYS-1 (red APC in Fig. 4.3). Loss of WRM-1, for instance, has no effect on SYS-1 asymmetry because only the WRM-1-regulating pool of APR-1 is affected in *wrm-1* mutants. APR-1 entry into the SYS-1-regulating pool appears dependent on KIN-19 because, in the absence of KIN-19, APR-1 can regulate WRM-1 but not SYS-1. By this model, Wnt signaling through PRY-1/Axin not only restricts the localization of the destruction complex but also controls the balance of the two pathway outputs by differentially regulating the two β-catenins that regulate TCF levels and transcriptional activity (Fig. 4.3).

4.4.5 Dishevelled Controls the Balance of the WβA Branches

Since WRM-1 and SYS-1 nuclear asymmetry can be used to analyze the relative function of both pools of APR-1, we next sought to test the function of upstream WβA pathway members in controlling the distribution of these pools during asymmetric seam cell division. Dishevelled (Dvl) in particular seemed a likely suspect due to its role as a gatekeeper for the various Wnt signal transduction pathways including Wnt/β-catenin, Wnt/PCP, and Wnt/Calcium (Sokol 1999; Wallingford and Mitchell 2011; King et al. 2009; Hingwing et al. 2009). Loss of two *C. elegans* Dvl paralogs, *dsh-2* and *mig-5*, causes relatively mild changes to overall seam cell numbers, but gaps and doublets appear in the seam cell A–P distribution suggesting randomized cell fate that resulted in some symmetric divisions (Banerjee et al. 2010; Baldwin et al. 2016). Indeed, APR-1 is mislocalized to both the anterior and posterior cortex of Dvl mutant seam cells, indicating that Dvl restricts the inheritance of negative regulators to the anterior daughter. To analyze the effects of Dvl mutation on APR-1 function, we examined the levels and localization of SYS-1 (to assay the SYS-1-regulating pool of APR-1) and WRM-1 (to assay the WRM-1-regulating pool of APR-1). Consistent with Dishevelled's role as a positive regulator of Wnt/β-catenin signaling, Dvl double mutants show symmetrically decreased nuclear SYS-1 in seam cell daughter nuclei while Dvl overexpression shows increased SYS-1 (Baldwin et al. 2016). However, a positive role regulating SYS-1 levels would suggest that Dvl mutants would have a symmetric loss of seam cell fate, rather than the confused or randomized polarity that is observed. The explanation for this discrepancy proved to be a different Dvl role for WRM-1 regulation compared to SYS-1 regulation. In contrast to decreased SYS-1 levels, Dvl mutants show an approximately 2.5-fold increase in nuclear WRM-1, presumably because APR-1 in both daughters is unable to promote WRM-1 nuclear export. These data suggest that, in the absence of Dvl, nuclear TCF is low and unable to activate transcription because Dvl partitions APR-1 into the WRM-1- and SYS-1-regulating pools in addition to its role of restricting the destruction complex to the anterior cortex. Our current model summarized in Fig. 4.3 is that Wnt restricts Frizzled and Dishevelled to the posterior cortex and APC and Axin to the anterior cortex at mitosis. After cytokinesis, the absence of negative regulation in the posterior daughter leads to the accumulation of WRM-1 and SYS-1, export of excess POP-1, and activation of WβA target genes. The anterior daughter exports WRM-1 from the nucleus via stabilized microtubules and degrades SYS-1 through the function of the destruction complex. Thus, the unsignaled daughter exhibits a low SYS-1:POP-1 ratio and repressed target genes compared to target gene expression in the signaled daughter.

4.4.6 How Is PRY-1 Asymmetric Activity Promoted in Late Seam Cell Divisions?

In the course of our examination of Dvl function in seam cell ACD, we sought to examine whether Dvl regulates the asymmetric PRY-1/Axin localization pattern seen in earlier L1 seam cell divisions (Mizumoto and Sawa 2007a). To our surprise, we found that cortical PRY-1::GFP is symmetric throughout the cell cycle and both daughters are born with high levels of cortical PRY-1 (Baldwin et al. 2016). The finding that PRY-1/Axin is symmetrically localized during L4 seam cell divisions begs the question of how PRY-1 drives asymmetric localization of APR-1/. We propose that PRY-1 is qualitatively altered by Wnt signaling in the L4 division to generate two PRY-1 populations, "active" and "inactive" regulators of APR-1/APC. Active PRY-1 resides at the anterior pole and anchors APR-1/APC; thus, β-catenin negative regulation occurs in this locale. Inactive PRY-1 in the posterior lacks this ability and as such the posterior daughter lacks negative regulation of both β-catenins. This model is an extension of the pattern seen at L1, where PRY-1 is quantitatively different at either pole of the mother cell (Mizumoto and Sawa 2007b). However, it is equally possible that "active" PRY-1 is localized at the posterior pole and functions to prevent APR-1 cortical localization. Identifying the qualitative change that regulates PRY-1 activity will address this. Several interesting possibilities are suggested in the vertebrate literature where ubiquitination, SUMOylation, methylation, and polymerization have all been observed (Song et al. 2014). Perhaps most intriguingly, mammalian Axin is phosphorylated in the absence of Wnt signaling and phospho-Axin displays greater ability to participate in the destruction complex and bind β-catenin (Song et al. 2014; Jho et al. 1999; Yamamoto et al. 1999; Ikeda et al. 1998; Hart et al. 1998; Rubinfeld et al. 1996). Further, due to Axin's posttranslational modifications and intramolecular interactions, two Axin-containing complexes have been proposed: the destruction complex that promotes β-catenin degradation and inhibits Wnt signaling and a second Axin population that catalyzes LRP5/6 phosphorylation by binding GSK3β to elevate Wnt signal transduction (Song et al. 2014; Bilic et al. 2007; Strovel et al. 2000; Leung et al. 2002; Yan et al. 2001). Wnt signaling dephosphorylates Axin and is thought to promote a conformational shape change that promotes association with LRP5/6. Together, these mechanisms suggest that PRY-1 phosphorylation and/or interactions with other Wnt signaling proteins could be responsible for modulating its ability to complex with APR-1/APC.

4.5 WβA Linkage to the Cell Cycle Provides Robustness for Cell Fate Specification

The model depicted in Fig. 4.3 details that polarized localization of destruction complex members in the mother cell just prior to division is likely to be the determining factor that regulates β-catenin nuclear asymmetry for both WRM-1 and SYS-1. However, this model raises a problem that the asymmetrically dividing lineage must solve: mother cells active for Wnt signaling give rise to a signaled daughter that lacks negative SYS-1 regulation, but they also must produce a daughter that can quickly lower its levels of SYS-1. Since SYS-1/β-catenin, at least, is highly expressed at the transcriptional level (Phillips et al. 2007), successive signaled fates would be predicted to have ever-increasing levels of SYS-1 (Fig. 4.4a). In this scenario, the unsignaled daughters inherit a higher and higher level of SYS-1 that would need to be degraded efficiently by the destruction complex, while simultaneously degrading de novo translated SYS-1, to achieve the subthreshold SYS-1 level needed for target gene repression. Eventually, an unsignaled daughter would be unable to degrade its inherited SYS-1 in a timely manner and the cell would assume the signaled fate, resulting in a symmetric division (Fig. 4.4a). That this problem is always overcome in the invariant *C. elegans* lineage and that targets of Wnt signaling are very robust in their asymmetric expression pattern (Fig. 4.4b) (Maduro et al. 2002) suggest that there could be redundant mechanisms limiting the level of SYS-1 in the unsignaled daughter.

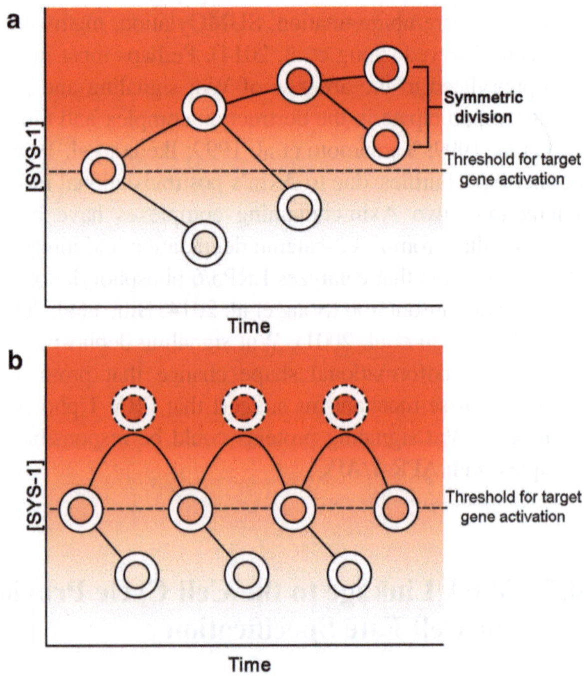

Fig. 4.4 Hypothetical model on SYS-1 robustness. (**a**) SYS-1 level might be expected to inappropriately increase in the signaled daughter following divisions in the absence of negative regulation. When the level of SYS-1 in the unsignaled daughter of successively signaled daughters reaches the threshold (*dashed line*) for gene activation, a change of fate will be observed in the cell resulting in a symmetric cell division. (**b**) Alternatively, following cell division the signaled daughter containing high SYS-1 level (*dashed circles*) is reduced near its threshold to prepare for subsequent cell division. See text for details of this reduction mechanism

4.5.1 Mother Cell SYS-1 Is Cleared by Localization to Mitotic Centrosomes

Though these mechanisms could theoretically include unstable *sys-1* transcripts, or an inherently short SYS-1 protein half-life, we identified a mechanism where SYS-1 protein stability is linked to the cell cycle by localization to the peri-centriolar material (PCM) during cell division (Vora and Phillips 2015). SYS-1 symmetrically localizes to centrosomes specifically during mitosis and is quickly lost after cytokinesis. Fluorescence recovery after photobleach (FRAP) studies shows that centrosomal SYS-1 is rapidly turned over and replaced during mitosis. The SYS-1 centrosomal pattern is accomplished by forming a complex with the centrosomal protein RSA-2 that also accumulates with the PCM; in the absence of RSA-2, centrosomes form but do not contain SYS-1. The RSA-2 loss-of-function background can therefore be used to determine the functional significance of SYS-1 centrosomal localization. Mother cells depleted of RSA-2 (and therefore defective in SYS-1 centrosomal localization) divide to produce daughter cells with higher nuclear SYS-1 levels compared to wild type. Interestingly, this elevation in SYS-1 levels is symmetrical, since both daughter cells show ~50% increase in nuclear SYS-1, suggesting that a symmetric negative regulatory process that requires centrosomal localization is normally responsible for decreasing SYS-1 in both daughters. Both the rapid turnover of centrosomal SYS-1 in wild-type cells and the reliance of centrosomal localization for negative regulation implicate a proteolysis mechanism, such as proteasomal degradation, which is linked to the centrosome. Additionally, proteasome components and proteasomal clients localize to centrosomes in *C. elegans* and other systems (Vora and Phillips 2015, 2016; Fabunmi et al. 2000; Wigley et al. 1999). To test the idea that the proteasome regulates centrosomal SYS-1, we inhibited proteasomal degradation in *C. elegans* embryos and found increased SYS-1 centrosomal levels but decreased SYS-1 turnover in FRAP analyses. These data are consistent with a model where SYS-1 symmetrically localizes to centrosomes during cell division, which leads to SYS-1 clearance from the mother cell before the daughter cells are born. This lowering of inherited SYS-1 allows the unsignaled daughter to properly regulate, via the destruction complex, de novo translated SYS-1 and repress target gene expression (Figs. 4.3 and 4.4b). Wnt signaling inactivates the destruction complex in the signaled daughter cell, leading to elevated de novo SYS-1 levels and activation of target gene expression. Hence, redundant negative regulatory mechanisms for SYS-1/β-catenin during and after division provide robustness for the asymmetric cell fate decision.

4.6 Reiterative Signaling Produces Multiple Cell Fates

The WβA pathway is used repeatedly to activate different targets and specify a new cell type in each iteration, yet the pathway has just one apparent outcome: regulation of Wnt target gene expression. How then does signaling specify differential cell fate in so many distinct cell types? One answer to this question is that Wnt signaling regulation of transcription via SYS-1 and POP-1 is layered with WβA-independent lineage-specific transcription factors to give cell type-specific gene expression. These targets include master regulatory genes, such as *ceh-22*/Nkx 2.1/tinman in the somatic gonad and GATA transcription factor *end-1* in the intestine (Lam et al. 2006; Chesney et al. 2009; Maduro et al. 2002; Shetty et al. 2005), which lock in the fate of the cell or lineage expression. A particularly well-established example is the neural lineage that results in specification of a neuron named AIY (Bertrand and Hobert 2009a, b). Terminal AIY fate is driven by the homeodomain transcription factor TTX-3, which directly regulates dozens of target genes to uniquely define AIY fate (Wenick and Hobert 2004). TTX-3 is initially symmetrically expressed in AIY and its sister cell after mother cell division, but the WβA target gene CEH-10, which is only expressed in AIY, is required for TTX-3 maintenance. Once coexpressed in AIY, CEH-10 and TTX-3 maintain their own expression as well as the AIY-specific pattern of gene expression that drives terminal differentiation. Thus, it can be imagined that WβA transcription factors broadly combine with lineage-specific transcriptional regulators to produce a tapestry of cell-specific gene expression.

4.6.1 Time-Lapse Imaging Identifies a Graded Response to Wnt Signaling

A second mechanism explaining WβA-driven cell type diversity was identified through lineage tracing of cells expressing fluorescently tagged SYS-1 and POP-1 and transcriptional fusions of their target genes (Zacharias et al. 2015; Zacharias and Murray 2016). These time-lapse experiments quantifying gene expression demonstrate that, while sister cells always show asymmetry, their absolute expression level is influenced by the prior expression level in the lineage. For instance, cells with higher levels of SYS-1 give rise to daughters and granddaughters that all have higher SYS-1 (termed "cousin enrichment"), though asymmetry between these daughters is maintained. Similar patterns are seen with POP-1 asymmetry, indicating that variable dosage of the WβA terminal transcription factors in the daughter cells reflects the level present in the mother cell before division. This observation suggests a continuum of Wnt pathway output (for instance, varying from one through ten in terms of WβA pathway activity), rather than a simple binary fate decision ("on-off") that could drive differential gene expression and result in increased cell type diversity.

The graded WβA response seen by Zacharias et al. raises the possibility that this differential effect is due to a Wnt gradient that exists over the A–P length of the embryo. By this model, similar to classic morphogens, cell response would be directly correlated to the distance from ligand source with distinct thresholds that compartmentalize target cell response. In support of this model, the Schnabel lab, in an elegant series of blastomere transplant experiments, showed that the P2 cell and its descendants form a Wnt-secreting polarizing center to direct the cleavage plane of many (and perhaps all) cells of the developing embryo (Bischoff and Schnabel 2006). They demonstrated that the polarizing activity of the P2 lineage persists over time since cells respond to a shift in the position of the Wnt source even after several divisions. Further, promoter activity of three Wnt ligand loci shows a cumulative posterior enrichment of Wnt expression (Zacharias et al. 2015). Thus, a posterior Wnt signal orients cells at a distance during embryogenesis, suggesting that a Wnt gradient across the embryo may be present and could explain the graded response seen by Zacharias et al. However, there is reason to doubt such an attractive model. The Priess lab showed asymmetric POP-1 levels in daughter cells from isolated blastomeres never in contact with the P2 lineage, suggesting that these cells have a Wnt-independent ability to activate WβA (Park and Priess 2003; Park et al. 2004). Additionally, the Murray lab reports many cells with high nuclear SYS-1 in the anterior region of the embryo, despite elevated Wnt expression in the posterior. Indeed, the graded response of cousin enrichment is not a smooth linear distribution from posterior to anterior but a patchwork pattern seen within lineages (Zacharias et al. 2015). Thus, because even nearby Wnt expression fails to increase WβA activity compared to cells located at a distance, it seems unlikely that a gradient of Wnt ligand is the underlying reason for differential output. Instead, the requirement of Frizzled for cousin enrichment suggests inheritance of signal transduction machinery, perhaps in a post-translationally modified form, underlies the ability of certain lineages to respond differentially to the presence or absence of Wnt polarity, rather than a Wnt gradient (Zacharias and Murray 2016).

4.7 Defining the "Default" State

A loss-of-function analysis of the terminal signaling effector in any cell communication pathway is often a useful proxy for the overall function of that pathway, yet POP-1/TCF mutants can show contradictory effects depending on the cell type under analysis. Loss of POP-1/TCF activity gives duplication of either the signaled or unsignaled fate effects depending on the tissue, suggesting that the default state of target gene expression varies in a tissue-specific manner. For instance, loss of POP-1 function in the early embryo duplicates the signaled fate since POP-1 targets become derepressed in MS (Lin et al. 1995; Calvo et al. 2001). Though these targets are expressed at quantitatively lower levels in both E and MS, expression is over the threshold necessary for E fate acquisition (Shetty et al. 2005). In contrast, loss of POP-1 from the somatic gonad duplicates the unsignaled anchor cell fate because the target gene *ceh-22* is expressed at subthreshold levels for fate acquisition (Siegfried and Kimble 2002; Lam et al. 2006). The molecular mechanism responsible for this is likely a target gene-specific differential requirement for POP-1-mediated repression in the unsignaled cell versus a POP-1-mediated activation in the signaled cell. This differential requirement for activation or repression depends on the level of target gene expression due to the effect of basal transcriptional machinery. In cells where the basal transcriptional machinery causes relatively high level of expression, POP-1 derepression via Wnt-induced nuclear export is largely sufficient for gene expression. Conversely, in cells where basal target gene expression level is low, POP-1 derepression must be coupled with POP-1 activation via SYS-1 (Kidd et al. 2005; Shetty et al. 2005). In addition to basal transcription factors, POP-1 target genes are influenced by a second "Helper" DNA sequence that binds to a distinct POP-1 DNA binding domain termed the C-clamp to specifically facilitate transcriptional activation (Bhambhani et al. 2014; Ravindranath and Cadigan 2014). The C-clamp and Helper DNA sequences are dispensable for target gene repression, however. It seems likely that the default transcriptional state of POP-1 targets appears to be influenced by atypical enhancer TCF elements including Helper sites and by the chromatin state that differentially recruits basal transcription factors. It should also be mentioned that, since *pop-1* is an essential gene, tissue-specific analyses are often performed with different loss-of-function approaches. These include *pop-1* RNAi applied at different developmental times or different methods and *pop-1* alleles with mutant β-catenin binding domains, DNA binding domains, or mutant 3′ UTR (Huang et al. 2007; Lin et al. 1995; Siegfried and Kimble 2002; Kidd et al. 2005), none of which may represent a *pop-1* null background. Since POP-1 is a transactivator at intermediate nuclear concentration and a repressor at high levels and these two activities likely result from a distinct use of POP-1 domains, this necessitates a cautious comparison of POP-1 function across tissues. The use of a null *pop-1* allele could remove these confounding variables when examining different tissues. By thus removing the activity of the WβA pathway, the role of the underlying basal transcriptional machinery on target gene expression level can be observed.

4.8 The Global Wnt Gradient Can Be Modified to Produce Mirror Asymmetry Within a Tissue

Consistent with the observation that there exists a gradient of Wnt ligand expression at its peak in posterior tissues and at its lowest in anterior (Harterink and Korswagen 2012; Harterink et al. 2011), most cells in the worm lineage are polarized such that the Wnt-dependent fate is the posterior daughter cell. However, exceptions to this rule include the somatic gonadal precursors (SGPs) and the vulval precursor cells (VPCs) (Figs. 4.2 and 4.5). Though the divisions are still oriented in the A–P dimension, posterior daughters are not equivalent throughout the tissue. For instance, the signaled fate is not exhibited by both posterior cells, as is typical in the WβA pathway. Instead, ACDs in both tissues occur along an internal proximal–distal axis rather than the more common anterior–posterior axis, meaning that distal daughters have equivalent fates within the tissue independent of their anterior or posterior fates. For the SGPs, WβA pathway activity is required

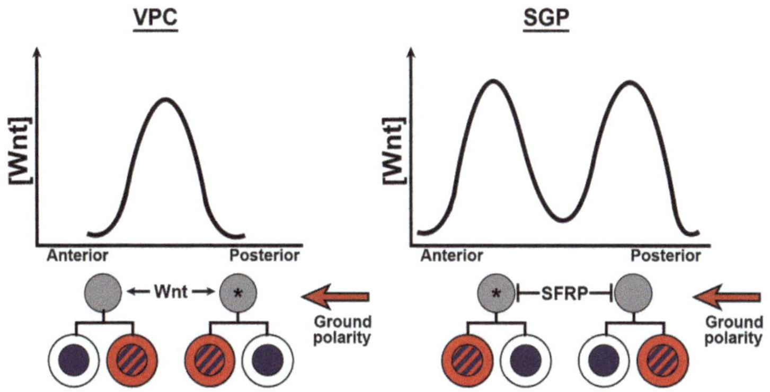

Fig. 4.5 Regulating mirror symmetry in vulval precursor cells (VPCs *left*) and a hypothetical model regulating the somatic gonadal precursors (SGPs *right*). A centrally localized Wnt cue reorients the posterior VPC, while a hypothesized Wnt antagonist locally depletes the central somatic gonad of Wnt to provide a central proximal–distal axis that promotes mirror asymmetry across a tissue. *Asterisks* denote ACDs with refined polarity compared to the global ground polarity orienting most ACDs toward the posterior. *SFRP* secreted Frizzled-related protein. Color scheme as in Fig. 4.1

in the distal daughters but not the proximal daughters (Siegfried and Kimble 2002; Siegfried et al. 2004; Yamamoto et al. 2011; Chang et al. 2005; Jiang and Sternberg 1999; Lam et al. 2006; Hajduskova et al. 2009; Phillips et al. 2007). This pattern is reversed in the vulva, where WβA signaling is active in the proximal daughters (Green et al. 2008; Schmidt et al. 2005; Phillips et al. 2007). While it could be conceived that this pattern is consistent with a single, central Wnt signal or, in the case of the SGPs, a central Wnt antagonist such as secreted Frizzled-related proteins (Harterink et al. 2011), instead evidence from VPC analyses indicates that these tissues as a whole also respond to the overall posterior inducing polarity signal similar to other tissues (termed "ground polarity"). However, posterior vulval cells reorient anteriorly via local Wnt cues based on specific Wnt ligand/receptor combinations [termed "refined polarity" (Harterink et al. 2011; Green et al. 2008)]. Since this Wnt source is posterior to the anterior VPC, it reinforces this cell's polarity, but the local Wnt signal is sufficient to override the long-range posterior Wnt/EGL-20 signaling effect on the posterior VPC. This is demonstrated by examinations of mutants lacking the refined polarity pathway where the posterior divisions reverse their orientation, resulting in a split vulva phenotype due to posterior orientation of both anterior and posterior vulval ACDs (Fig. 4.5 left) (Green et al. 2008; Deshpande et al. 2005; Inoue et al. 2004). In the absence of both the ground and refined pathways, polarity is random, similar to loss of PRY-1/Axin from the seam cells (Green et al. 2008; Baldwin and Phillips 2014). Interestingly, the specific effects of the ground and refined polarity pathways are mediated by different Wnt ligands and receptors, including non-Frizzled receptors such as CAM-1/Ror, a receptor tyrosine kinase, and LIN-18/Ryk, and require Wnt/PCP components in addition to Wnt/β-catenin pathways. This suggests these two polarizing pathways enable differential effects by acting through distinct molecular outputs in the receiving cells that include transcription-independent processes. A similar reorientation could also take place to provide SGP mirror asymmetry, though the geometry would suggest a central depletion of Wnt ligand (Fig. 4.5 right).

4.9 Instructive and Permissive Signaling by Wnt Ligands

Though Wnts act instructively to orient polarity in *C. elegans* EMS and VPC divisions as well as human ES cells (Goldstein et al. 2006; Herman 2002; Habib et al. 2013; Goldstein 1993), the situation appears more complex in the seam cells, where a permissive role has been suggested (Yamamoto et al. 2011). In the absence of multiple redundant Wnt signals, early larval seam cell divisions remain polarized but are randomly oriented, resulting in maintenance of the wild-type number of seam cells. Conversely, loss of Wnt receptors Frizzled and Ror leads to symmetric divisions, suggesting that the receptors act in a Wnt-independent fashion to generate polarity while the Wnt ligand orients that polarity. Further, misexpression of a Wnt ligand in anterior tissues rescues wild-type polarity at similar rates as posterior misexpression (Yamamoto et al. 2011). Together, these data suggest the presence of a possible second, posterior-orienting, polarizing cue that may be intrinsic to the seam cells. One interesting option is that this cue may be mediated by cell contacts to nearby seam cells. The seam cells are in an epithelium such that each seam cell is in tight contact with its anterior and posterior seam cell neighbors (Altun and Hall 2009). Seam cell ablations render neighboring seam cell divisions symmetric, suggesting cell contact may also play a polarity role (Austin and Kenyon 1994). Interestingly, cell adhesion and APC also control the polarity of fly sperm stem cell divisions (Yamashita et al. 2003), suggesting that such a cell-contact polarization mechanism may be conserved. Cell contacts are also important for proper Wnt/PCP-induced cell polarity in vertebrates and flies (Mayor and Theveneau 2014; Bayly and Axelrod 2011; Ehaideb et al. 2014), so the study of multiple Wnt signaling pathways to direct coordinated molecular outputs promises to be a fruitful area of future investigation.

4.10 Concluding Remarks: WβA Consists of Upstream Polarizers and Downstream Effectors

The emerging picture of *C. elegans* WβA-mediated ACD suggests distinct functional differences of the members of the WβA pathway with regard to their effect on polarity. Upstream components of the pathway serve to localize the downstream components that drive asymmetric gene expression, such as APC, WRM-1, LIT-1, SYS-1, and POP-1 (Sugioka et al. 2011; Shetty et al. 2005; Huang et al. 2007; Baldwin et al. 2016; Baldwin and Phillips 2014). Loss of upstream members of the pathway (Wnt, Dvl, Pry-1/Axin) generally leads to randomly asymmetric localization of the downstream proteins (SYS-1, WRM-1, POP-1) and random orientation of daughter cell fate, but polarity is often maintained such that two distinct daughters are still generated. Conversely, loss of downstream components of the pathway, either negative or positive acting factors, leads to symmetric divisions presumably because the output of the asymmetric pathway is unable to be completed in either cell such that the two cells show similar transcriptional profiles. Upstream WβA members thus modulate and orient a cell to give a unified polarity orientation across a tissue or even across the whole organism, but the downstream members are required for the actual differential transcriptional output and are therefore essential for asymmetry. Though important aspects of this pathway remain unknown, such as the nature of the Wnt-independent ability to randomly polarize and the cell type-specific response to a global Wnt gradient, the combination of genetics and in vivo cell observations makes the study of *C. elegans* ACD a powerful system to determine how dividing cells sense a polarizing signal and respond to produce two daughter cells with different transcriptional profiles.

Acknowledgements The authors would like to thank the Phillips lab for helpful comments on the manuscript. This work was supported by the American Cancer Society [grant number RSG-11-140-01-DC], the Roy J. Carver Charitable Trust [grant number 13-4131], and the National Science Foundation [grant number IOS-1456941] to B.T.P.

References

Altun ZF, Hall D (2009) Epithelial system, seam cells. WormAtlas

Austin J, Kenyon C (1994) Cell contact regulates neuroblast formation in the *Caenorhabditis elegans* lateral epidermis. Development 120:313–323

Austin J, Kimble J (1987) *glp-1* is required in the germ line for regulation of the decision between mitosis and meiosis in *C. elegans*. Cell 51:589–599

Axelrod JD, Matsuno K, Artavanis-Tsakonas S, Perrimon N (1996) Interaction between wingless and Notch signaling pathways mediated by Dishevelled. Science 271:1826–1832

Baldwin AT, Phillips BT (2014) The tumor suppressor APC differentially regulates multiple beta-catenins through the function of axin and CKIalpha during *C. elegans* asymmetric stem cell divisions. J Cell Sci 127:2771–2781

Baldwin AT, Clemons AM, Phillips BT (2016) Unique and redundant beta-catenin regulatory roles of two Dishevelled paralogs during *C. elegans* asymmetric cell division. J Cell Sci 129:983–993

Banerjee D, Chen X, Lin SY, Slack FJ (2010) kin-19/casein kinase Ialpha has dual functions in regulating asymmetric division and terminal differentiation in *C. elegans* epidermal stem cells. Cell Cycle 9:4748–4765

Barker N, Hurlstone A, Musisi H, Miles A, Bienz M, Clevers H (2001) The chromatin remodelling factor Brg-1 interacts with beta-catenin to promote target gene activation. EMBO J 20:4935–4943

Barth AI, Caro-Gonzalez HY, Nelson WJ (2008) Role of adenomatous polyposis coli (APC) and microtubules in directional cell migration and neuronal polarization. Semin Cell Dev Biol 19:245–251

Bayly R, Axelrod JD (2011) Pointing in the right direction: new developments in the field of planar cell polarity. Nat Rev Genet 12:385–391

Bertrand V, Hobert O (2009a) Linking asymmetric cell division to the terminal differentiation program of postmitotic neurons in *C. elegans*. Dev Cell 16:563–575

Bertrand V, Hobert O (2009b) Wnt asymmetry and the terminal division of neuronal progenitors. Cell Cycle 8:1973–1974

Bertrand V, Hobert O (2010) Lineage programming: navigating through transient regulatory states via binary decisions. Curr Opin Genet Dev 20:362–368

Bhambhani C, Ravindranath AJ, Mentink RA, Chang MV, Betist MC, Yang YX, Koushika SP, Korswagen HC, Cadigan KM (2014) Distinct DNA binding sites contribute to the TCF transcriptional switch in *C. elegans* and Drosophila. PLoS Genet 10:e1004133

Bilic J, Huang YL, Davidson G, Zimmermann T, Cruciat CM, Bienz M, Niehrs C (2007) Wnt induces LRP6 signalosomes and promotes dishevelled-dependent LRP6 phosphorylation. Science 316:1619–1622

Bischoff M, Schnabel R (2006) A posterior centre establishes and maintains polarity of the *Caenorhabditis elegans* embryo by a Wnt-dependent relay mechanism. PLoS Biol 4:e396

Brocardo M, Lei Y, Tighe A, Taylor SS, Mok MT, Henderson BR (2008) Mitochondrial targeting of adenomatous polyposis coli protein is stimulated by truncating cancer mutations: regulation of Bcl-2 and implications for cell survival. J Biol Chem 283:5950–5959

Brunner E, Peter O, Schweizer L, Basler K (1997) *pangolin* encodes a Lef-1 homologue that acts downstream of Armadillo to transduce the Wingless signal in *Drosophila*. Nature 385:829–833

Byrd DT, Knobel K, Affeldt K, Crittenden SL, Kimble J (2014) A DTC niche plexus surrounds the germline stem cell pool in *Caenorhabditis elegans*. PLoS One 9:e88372

Calvo D, Victor M, Gay F, Sui G, Luke MP-S, Dufourcq P, Wen G, Maduro M, Rothman J, Shi Y (2001) A POP-1 repressor complex restricts inappropriate cell type-specific gene transcription during *Caenorhabditis elegans* embryogenesis. EMBO J 20:7197–7208

Chang W, Lloyd CE, Zarkower D (2005) DSH-2 regulates asymmetric cell division in the early *C. elegans* somatic gonad. Mech Dev 122:781–789

Chesney MA, Lam N, Morgan DE, Phillips BT, Kimble J (2009) *C. elegans* HLH-2/E/Daughterless controls key regulatory cells during gonadogenesis. Dev Biol 331:14–25

Clevers H, Nusse R (2012) Wnt/beta-catenin signaling and disease. Cell 149:1192–1205

Deshpande R, Inoue T, Priess JR, Hill RJ (2005) *lin-17*/Frizzled and *lin-18* regulate POP-1/TCF-1 localization and cell type specification during *C. elegans* vulval development. Dev Biol 278:118–129

Ehaideb SN, Iyengar A, Ueda A, Iacobucci GJ, Cranston C, Bassuk AG, Gubb D, Axelrod JD, Gunawardena S, Wu CF, Manak JR (2014) prickle modulates microtubule polarity and axonal transport to ameliorate seizures in flies. Proc Natl Acad Sci USA 111:11187–11192

Eisenmann D (2005) Wnt signaling. WormBook [Online]. Available: http://www.wormbook.org. Accessed 25 Jun 2005

Eisenmann DM (2011) *C. elegans* seam cells as stem cells: Wnt signaling and casein kinase Ialpha regulate asymmetric cell divisions in an epidermal progenitor cell type. Cell Cycle 10:20–21

Eisenmann DM, Maloof JN, Simske JS, Kenyon C, Kim SK (1998) The β-catenin homolog BAR-1 and LET-60 Ras coordinately regulate the Hox gene *lin-39* during *Caenorhabditis elegans* vulval development. Development 125:3667–3680

Fabunmi RP, Wigley WC, Thomas PJ, DeMartino GN (2000) Activity and regulation of the centrosome-associated proteasome. J Biol Chem 275:409–413

Gao B (2012) Wnt regulation of planar cell polarity (PCP). Curr Top Dev Biol 101:263–295

Gao C, Chen YG (2010) Dishevelled: the hub of Wnt signaling. Cell Signal 22:717–727

Gleason JE, Eisenmann DM (2010) Wnt signaling controls the stem cell-like asymmetric division of the epithelial seam cells during *C. elegans* larval development. Dev Biol 348:58–66

Gleason JE, Szyleyko EA, Eisenmann DM (2006) Multiple redundant Wnt signaling components function in two processes during *C. elegans* vulval development. Dev Biol 298:442–457

Goldstein B (1992) Induction of gut in *Caenorhabditis elegans* embryos. Nature 357:255–257

Goldstein B (1993) Establishment of gut fate in the E lineage of *C. elegans*: the roles of lineage-dependent mechanisms and cell interactions. Development 118:1267–1277

Goldstein B, Hird SN (1996) Specification of the anteroposterior axis in *Caenorhabditis elegans*. Development 122:1467–1474

Goldstein B, Takeshita H, Mizumoto K, Sawa H (2006) Wnt signals can function as positional cues in establishing cell polarity. Dev Cell 10:391–396

Gorrepati L, Eisenmann DM (2015) The *C. elegans* embryonic fate specification factor EGL-18 (GATA) is reutilized downstream of Wnt signaling to maintain a population of larval progenitor cells. Worm 4:e996419

Gorrepati L, Thompson KW, Eisenmann DM (2013) *C. elegans* GATA factors EGL-18 and ELT-6 function downstream of Wnt signaling to maintain the progenitor fate during larval asymmetric divisions of the seam cells. Development 140:2093–2102

Gorrepati L, Krause MW, Chen W, Brodigan TM, Correa-Mendez M, Eisenmann DM (2015) Identification of Wnt pathway target genes regulating the division and differentiation of larval seam cells and vulval precursor cells in *Caenorhabditis elegans*. G3 (Bethesda) 5:1551–1566

Green RA, Kaplan KB (2003) Chromosome instability in colorectal tumor cells is associated with defects in microtubule plus-end attachments caused by a dominant mutation in APC. J Cell Biol 163:949–961

Green JL, Inoue T, Sternberg PW (2008) Opposing Wnt pathways orient cell polarity during organogenesis. Cell 134:646–656

Habib SJ, Chen BC, Tsai FC, Anastassiadis K, Meyer T, Betzig E, Nusse R (2013) A localized Wnt signal orients asymmetric stem cell division in vitro. Science 339:1445–1448

Hajduskova M, Jindra M, Herman MA, Asahina M (2009) The nuclear receptor NHR-25 cooperates with the Wnt/beta-catenin asymmetry pathway to control differentiation of the T seam cell in *C. elegans*. J Cell Sci 122:3051–3060

Harandi OF, Ambros VR (2015) Control of stem cell self-renewal and differentiation by the heterochronic genes and the cellular asymmetry machinery in *Caenorhabditis elegans*. Proc Natl Acad Sci USA 112:E287–E296

Hardin J, King RS (2008) The long and the short of Wnt signaling in *C. elegans*. Curr Opin Genet Dev 18:362–367

Hart MJ, de los Santos R, Albert IN, Rubinfeld B, Polakis P (1998) Downregulation of beta-catenin by human Axin and its association with the APC tumor suppressor, beta-catenin and GSK3 beta. Curr Biol 8:573–581

Harterink M, Korswagen HC (2012) Dissecting the Wnt secretion pathway: key questions on the modification and intracellular trafficking of Wnt proteins. Acta Physiol (Oxf) 204:8–16

Harterink M, Kim DH, Middelkoop TC, Doan TD, van Oudenaarden A, Korswagen HC (2011) Neuroblast migration along the anteroposterior axis of *C. elegans* is controlled by opposing gradients of Wnts and a secreted Frizzled-related protein. Development 138:2915–2924

Hecht A, Vleminckx K, Stemmler MP, van Roy F, Kemler R (2000) The p300/CBP acetyltransferases function as transcriptional coactivators of beta-catenin in vertebrates. EMBO J 19:1839–1850

Herman MA (2002) Control of cell polarity by noncanonical Wnt signaling in *C. elegans*. Semin Cell Dev Biol 13:233–241

Herman MA, Vassilieva LL, Horvitz HR, Shaw JE, Herman RK (1995) The *C. elegans* gene *lin-44*, which controls the polarity of certain asymmetric cell divisions, encodes a Wnt protein and acts cell nonautonomously. Cell 83:101–110

Hingwing K, Lee S, Nykilchuk L, Walston T, Hardin J, Hawkins N (2009) CWN-1 functions with DSH-2 to regulate *C. elegans* asymmetric neuroblast division in a beta-catenin independent Wnt pathway. Dev Biol 328:245–256

Huang S, Shetty P, Robertson SM, Lin R (2007) Binary cell fate specification during *C. elegans* embryogenesis driven by reiterated reciprocal asymmetry of TCF POP-1 and its coactivator β-catenin SYS-1. Development 134:2685–2695

Huber AH, Nelson WJ, Weis WI (1997) Three-dimensional structure of the armadillo repeat region of β-catenin. Cell 90:871–882

Ikeda S, Kishida S, Yamamoto H, Murai H, Koyama S, Kikuchi A (1998) Axin, a negative regulator of the Wnt signaling pathway, forms a complex with GSK-3beta and beta-catenin and promotes GSK-3beta-dependent phosphorylation of beta-catenin. EMBO J 17:1371–1384

Inoue T, Oz HS, Wiland D, Gharib S, Deshpande R, Hill RJ, Katz WS, Sternberg PW (2004) *C. elegans* LIN-18 is a Ryk ortholog and functions in parallel to LIN-17/Frizzled in Wnt signaling. Cell 118:795–806

Ishitani T, Ninomiya-Tsuji J, Nagai S-I, Nishita M, Meneghini M, Barker N, Waterman M, Bowerman B, Clevers H, Shibuya H, Matsumoto K (1999) The TAK1-NLK-MAPK-related pathway antagonizes signalling between β-catenin and transcription factor TCF. Nature 399:798–802

Jackson BM, Eisenmann DM (2012) beta-catenin-dependent Wnt signaling in *C. elegans*: teaching an old dog a new trick. Cold Spring Harb Perspect Biol 4:a007948

Jho E, Lomvardas S, Costantini F (1999) A GSK3beta phosphorylation site in axin modulates interaction with beta-catenin and Tcf-mediated gene expression. Biochem Biophys Res Commun 266:28–35

Jiang LI, Sternberg PW (1999) An HMG1-like protein facilitates Wnt signaling in *Caenorhabditis elegans*. Genes Dev 13:877–889

Kaletta T, Schnabel H, Schnabel R (1997) Binary specification of the embryonic lineage in *Caenorhabditis elegans*. Nature 390:294–298

Kidd AR 3rd, Miskowski JA, Siegfried KR, Sawa H, Kimble J (2005) A β-catenin identified by functional rather than sequence criteria and its role in Wnt/MAPK signaling. Cell 121:761–772

Kidd AR 3rd, Muniz-Medina V, Der CJ, Cox AD, Reiner DJ (2015) The *C. elegans* Chp/Wrch Ortholog CHW-1 Contributes to LIN-18/Ryk and LIN-17/Frizzled signaling in cell polarity. PLoS One 10:e0133226

Kimble J (1981) Alterations in cell lineage following laser ablation of cells in the somatic gonad of *Caenorhabditis elegans*. Dev Biol 87:286–300

Kimble J, Hirsh D (1979) The postembryonic cell lineages of the hermaphrodite and male gonads in *Caenorhabditis elegans*. Dev Biol 70:396–417

Kimble J, Ward S (1988) Germ-line development and fertilization. In: Wood WB (ed) The nematode *Caenorhabditis elegans*. Cold Spring Harbor Laboratory Press, Cold Spring Harbor, NY

King RS, Maiden SL, Hawkins NC, Kidd AR 3rd, Kimble J, Hardin J, Walston TD (2009) The N- or C-terminal domains of DSH-2 can activate the *C. elegans* Wnt/β-catenin asymmetry pathway. Dev Biol 328:234–244

Klein TJ, Mlodzik M (2005) Planar cell polarization: an emerging model points in the right direction. Annu Rev Cell Dev Biol 21:155–176

Korswagen HC, Herman MA, Clevers HC (2000) Distinct β-catenins mediate adhesion and signalling functions in *C. elegans*. Nature 406:527–532

Kramps T, Peter O, Brunner E, Nellen D, Froesch B, Chatterjee S, Murone M, Züllig S, Basler K (2002) WNT/Wingless signaling requires BCL9/Legless-mediated recruitment of Pygopus to the nuclear β-catenin-TCF complex. Cell 109:47–60

Lam N, Chesney MA, Kimble J (2006) Wnt signaling and CEH-22/tinman/Nkx2.5 specify a stem cell niche in *C. elegans*. Curr Biol 16:287–295

Leung JY, Kolligs FT, Wu R, Zhai Y, Kuick R, Hanash S, Cho KR, Fearon ER (2002) Activation of AXIN2 expression by beta-catenin-T cell factor. A feedback repressor pathway regulating Wnt signaling. J Biol Chem 277:21657–21665

Lin R, Thompson S, Priess JR (1995) *pop-1* encodes an HMG box protein required for the specification of a mesoderm precursor in early *C. elegans* embryos. Cell 83:599–609

Lin R, Hill RJ, Priess JR (1998) POP-1 and anterior-posterior fate decision in *C. elegans* embryos. Cell 92:229–239

Lin S, Baye LM, Westfall TA, Slusarski DC (2010) Wnt5b-Ryk pathway provides directional signals to regulate gastrulation movement. J Cell Biol 190:263–278

Liu J, Phillips BT, Amaya MF, Kimble J, Xu W (2008) The *C. elegans* SYS-1 protein is a bona fide β-catenin. Dev Cell 14:751–761

Lo M-C, Gay F, Odom R, Shi Y, Lin R (2004) Phosphorylation by the β-catenin/MAPK complex promotes 14-3-3-mediated nuclear export of TCF/POP-1 in signal-responsive cells in *C. elegans*. Cell 117:95–106

Logan CY, Nusse R (2004) The Wnt signaling pathway in development and disease. Annu Rev Cell Dev Biol 20:781–810

MacDonald BT, He X (2012) Frizzled and LRP5/6 receptors for Wnt/beta-catenin signaling. Cold Spring Harb Perspect Biol 4:1–23

Maduro MF, Lin R, Rothman JH (2002) Dynamics of a developmental switch: recursive intracellular and intranuclear redistribution of *Caenorhabditis elegans* POP-1 parallels Wnt-inhibited transcriptional repression. Dev Biol 248:128–142

Maduro MF, Kasmir JJ, Zhu J, Rothman JH (2005) The Wnt effector POP-1 and the PAL-1/Caudal homeoprotein collaborate with SKN-1 to activate *C. elegans* endoderm development. Dev Biol 285:510–523

Mayor R, Theveneau E (2014) The role of the non-canonical Wnt-planar cell polarity pathway in neural crest migration. Biochem J 457:19–26

Meneghini MD, Ishitani T, Carter JC, Hisamoto N, Ninomiya-Tsuji J, Thorpe CJ, Hamill DR, Matsumoto K, Bowerman B (1999) MAP kinase and Wnt pathways converge to downregulate an HMG-domain repressor in *Caenorhabditis elegans*. Nature 399:793–797

Mila D, Calderon A, Baldwin AT, Moore KM, Watson M, Phillips BT, Putzke AP (2015) Asymmetric Wnt pathway signaling facilitates stem cell-like divisions via the nonreceptor tyrosine kinase FRK-1 in *Caenorhabditis elegans*. Genetics 201:1047–1060

Mimori-Kiyosue Y, Shiina N, Tsukita S (2000) The dynamic behavior of the APC-binding protein EB1 on the distal ends of microtubules. Curr Biol 10:865–868

Miskowski J, Li Y, Kimble J (2001) The *sys-1* gene and sexual dimorphism during gonadogenesis in *Caenorhabditis elegans*. Dev Biol 230:61–73

Mizumoto K, Sawa H (2007a) Cortical β-catenin and APC regulate asymmetric nuclear β-catenin localization during asymmetric cell division in *C. elegans*. Dev Cell 12:287–299

Mizumoto K, Sawa H (2007b) Two βs or not two βs: regulation of asymmetric division by β-catenin. Trends Cell Biol 17:465–473

Molenaar M, van de Wetering M, Oosterwegel M, Peterson-Maduro J, Godsave S, Korinek V, Roose J, Destrée O, Clevers H (1996) XTcf-3 transcription factor mediates β-catenin-induced axis formation in Xenopus embryos. Cell 86:391–399

Munro E, Bowerman B (2009) Cellular symmetry breaking during *Caenorhabditis elegans* development. Cold Spring Harb Perspect Biol 1:a003400

Murgan S, Bertrand V (2015) How targets select activation or repression in response to Wnt. Worm 4:e1086869

Murgan S, Kari W, Rothbacher U, Iche-Torres M, Melenec P, Hobert O, Bertrand V (2015) Atypical transcriptional activation by TCF via a Zic transcription factor in *C. elegans* neuronal precursors. Dev Cell 33:737–745

Nakamura K, Kim S, Ishidate T, Bei Y, Pang K, Shirayama M, Trzepacz C, Brownell DR, Mello CC (2005) Wnt signaling drives WRM-1/β-catenin asymmetries in early *C. elegans* embryos. Genes Dev 19:1749–1754

Natarajan L, Witwer NE, Eisenmann DM (2001) The divergent *Caenorhabditis elegans* β-catenin proteins BAR-1, WRM-1 and HMP-2 make distinct protein interactions but retain functional redundancy *in vivo*. Genetics 159:159–172

Park FD, Priess JR (2003) Establishment of POP-1 asymmetry in early *C. elegans* embryos. Development 130:3547–3556

Park FD, Tenlen JR, Priess JR (2004) *C. elegans* MOM-5/frizzled functions in MOM-2/Wnt-independent cell polarity and is localized asymmetrically prior to cell division. Curr Biol 14:2252–2258

Parker DS, Jemison J, Cadigan KM (2002) Pygopus, a nuclear PHD-finger protein required for Wingless signaling in Drosophila. Development 129:2565–2576

Phillips BT, Kimble J (2009) A new look at TCF and β-catenin through the lens of a divergent *C. elegans* Wnt pathway. Dev Cell 17:27–34

Phillips BT, Kidd AR 3rd, King R, Hardin J, Kimble J (2007) Reciprocal asymmetry of SYS-1/beta-catenin and POP-1/TCF controls asymmetric divisions in *Caenorhabditis elegans*. Proc Natl Acad Sci USA 104:3231–3236

Poy F, Lepourcelet M, Shivdasani RA, Eck MJ (2001) Structure of a human Tcf4-β-catenin complex. Nat Struct Biol 8:1053–1057

Ravindranath AJ, Cadigan KM (2014) Structure-function analysis of the C-clamp of TCF/Pangolin in Wnt/ss-catenin signaling. PLoS One 9:e86180

Rocheleau CE, Downs WD, Lin R, Wittmann C, Bei Y, Cha Y-H, Ali M, Priess JR, Mello CC (1997) Wnt signaling and an APC-related gene specify endoderm in early *C. elegans* embryos. Cell 90:707–716

Rocheleau CE, Yasuda J, Shin TH, Lin R, Sawa H, Okano H, Priess JR, Davis RJ, Mello CC (1999) WRM-1 activates the LIT-1 protein kinase to transduce anterior/posterior polarity signals in *C. elegans*. Cell 97:717–726

Roose J, Clevers H (1999) TCF transcription factors: molecular switches in carcinogenesis. Biochim Biophys Acta 97456:M23–M37

Roose J, Molenaar M, Peterson J, Hurenkamp J, Brantjes H, Moerer P, van de Wetering M, Destree O, Clevers H (1998) The Xenopus Wnt effector XTcf-3 interacts with Groucho-related transcriptional repressors. Nature 395:608–612

Rose L, Gonczy P (2014) Polarity establishment, asymmetric division and segregation of fate determinants in early *C. elegans* embryos. WormBook, pp 1–43

Rubinfeld B, Albert I, Porfiri E, Fiol C, Munemitsu S, Polakis P (1996) Binding of GSK3beta to the APC-beta-catenin complex and regulation of complex assembly. Science 272:1023–1026

Sawa H, Korswagen HC (2013) Wnt signaling in *C. elegans*. WormBook, pp 1–30

Schmid T, Hajnal A (2015) Signal transduction during *C. elegans* vulval development: a NeverEnding story. Curr Opin Genet Dev 32:1–9

Schmidt DJ, Rose DJ, Saxton WM, Strome S (2005) Functional analysis of cytoplasmic dynein heavy chain in *Caenorhabditis elegans* with fast-acting temperature-sensitive mutations. Mol Biol Cell 16:1200–1212

Shetty P, Lo M-C, Robertson SM, Lin R (2005) *C. elegans* TCF protein, POP-1, converts from repressor to activator as a result of Wnt-induced lowering of nuclear levels. Dev Biol 285:584–592

Shin TH, Yasuda J, Rocheleau CE, Lin R, Soto M, Bei Y, Davis RJ, Mello CC (1999) MOM-4, a MAP kinase kinase kinase-related protein, activates WRM-1/LIT- 1 kinase to transduce anterior/posterior polarity signals in *C. elegans*. Mol Cell 4:275–280

Siegfried K, Kimble J (2002) POP-1 controls axis formation during early gonadogenesis in *C. elegans*. Development 129:443–453

Siegfried KR, Kidd AR 3rd, Chesney MA, Kimble J (2004) The sys-1 and sys-3 genes cooperate with Wnt signaling to establish the proximal-distal axis of the *Caenorhabditis elegans* gonad. Genetics 166:171–186

Sokol SY (1999) Wnt signaling and dorso-ventral axis specification in vertebrates. Curr Opin Genet Dev 9:405–410

Sokol SY (2015) Spatial and temporal aspects of Wnt signaling and planar cell polarity during vertebrate embryonic development. Semin Cell Dev Biol 42:78–85

Song X, Wang S, Li L (2014) New insights into the regulation of Axin function in canonical Wnt signaling pathway. Protein Cell 5:186–193

Strovel ET, Wu D, Sussman DJ (2000) Protein phosphatase 2Calpha dephosphorylates axin and activates LEF-1-dependent transcription. J Biol Chem 275:2399–2403

Sugioka K, Sawa H (2010) Regulation of asymmetric positioning of nuclei by Wnt and Src signaling and its roles in POP-1/TCF nuclear asymmetry in *Caenorhabditis elegans*. Genes Cells 15:397–407

Sugioka K, Mizumoto K, Sawa H (2011) Wnt regulates spindle asymmetry to generate asymmetric nuclear beta-catenin in *C. elegans*. Cell 146:942–954

Sulston JE, Horvitz HR (1977) Post-embryonic cell lineages of the nematode, *Caenorhabditis elegans*. Dev Biol 56:110–156

Takemaru KI, Moon RT (2000) The transcriptional coactivator CBP interacts with beta-catenin to activate gene expression. J Cell Biol 149:249–254

Takeshita H, Sawa H (2005) Asymmetric cortical and nuclear localizations of WRM-1/β-catenin during asymmetric cell division in *C. elegans*. Genes Dev 19:1743–1748

Thompson B, Townsley F, Rosin-Arbesfeld R, Musisi H, Bienz M (2002) A new nuclear component of the Wnt signalling pathway. Nat Cell Biol 4:367–373

Thorpe CJ, Schlesinger A, Carter JC, Bowerman B (1997) Wnt signaling polarizes an early *C. elegans* blastomere to distinguish endoderm from mesoderm. Cell 90:695–705

Vora S, Phillips BT (2015) Centrosome-associated degradation limits beta-catenin inheritance by daughter cells after asymmetric division. Curr Biol 25:1005–1016

Vora SM, Phillips BT (2016) The benefits of local depletion: The centrosome as a scaffold for ubiquitin-proteasome-mediated degradation. Cell Cycle 15:2124–2134

Wallingford JB, Mitchell B (2011) Strange as it may seem: the many links between Wnt signaling, planar cell polarity, and cilia. Genes Dev 25:201–213

Wehrli M, Tomlinson A (1998) Independent regulation of anterior/posterior and equatorial/polar polarity in the Drosophila eye; evidence for the involvement of Wnt signaling in the equatorial/polar axis. Development 125:1421–1432

Wenick AS, Hobert O (2004) Genomic *cis*-regulatory architecture and *trans*-acting regulators of a single interneuron-specific gene battery in *C. elegans*. Dev Cell 6:757–770

Wigley WC, Fabunmi RP, Lee MG, Marino CR, Muallem S, DeMartino GN, Thomas PJ (1999) Dynamic association of proteasomal machinery with the centrosome. J Cell Biol 145:481–490

Wildwater M, Sander N, de Vreede G, van den Heuvel S (2011) Cell shape and Wnt signaling redundantly control the division axis of *C. elegans* epithelial stem cells. Development 138:4375–4385

Xing Y, Takemaru K, Liu J, Berndt JD, Zheng JJ, Moon RT, Xu W (2008) Crystal structure of a full-length beta-catenin. Structure 16:478–487

Yamamoto H, Kishida S, Kishida M, Ikeda S, Takada S, Kikuchi A (1999) Phosphorylation of axin, a Wnt signal negative regulator, by glycogen synthase kinase-3beta regulates its stability. J Biol Chem 274:10681–10684

Yamamoto Y, Takeshita H, Sawa H (2011) Multiple Wnts redundantly control polarity orientation in *Caenorhabditis elegans* epithelial stem cells. PLoS Genet 7:e1002308

Yamashita YM, Jones DL, Fuller MT (2003) Orientation of asymmetric stem cell division by the APC tumor suppressor and centrosome. Science 301:1547–1550

Yan D, Wiesmann M, Rohan M, Chan V, Jefferson AB, Guo L, Sakamoto D, Caothien RH, Fuller JH, Reinhard C, Garcia PD, Randazzo FM, Escobedo J, Fantl WJ, Williams LT (2001) Elevated expression of axin2 and hnkd mRNA provides evidence that Wnt/beta -catenin signaling is activated in human colon tumors. Proc Natl Acad Sci USA 98:14973–14978

Yang XD, Huang S, Lo MC, Mizumoto K, Sawa H, Xu W, Robertson S, Lin R (2011) Distinct and mutually inhibitory binding by two divergent {beta}-catenins coordinates TCF levels and activity in *C. elegans*. Development 138:4255–4265

Yang XD, Karhadkar TR, Medina J, Robertson SM, Lin R (2015) beta-Catenin-related protein WRM-1 is a multifunctional regulatory subunit of the LIT-1 MAPK complex. Proc Natl Acad Sci USA 112:E137–E146

Yoda A, Kouike H, Okano H, Sawa H (2005) Components of the transcriptional Mediator complex are required for asymmetric cell division in *C. elegans*. Development 132:1885–1893

Zacharias AL, Murray JI (2016) Combinatorial decoding of the invariant *C. elegans* embryonic lineage in space and time. Genesis 54:182–197

Zacharias AL, Walton T, Preston E, Murray JI (2015) Quantitative differences in nuclear beta-catenin and TCF pattern embryonic cells in *C. elegans*. PLoS Genet 11:e1005585

Zumbrunn J, Kinoshita K, Hyman AA, Nathke IS (2001) Binding of the adenomatous polyposis coli protein to microtubules increases microtubule stability and is regulated by GSK3 beta phosphorylation. Curr Biol 11:44–49

Chapter 5
Asymmetric Cell Division in the One-Cell *C. elegans* Embryo: Multiple Steps to Generate Cell Size Asymmetry

Anne Pacquelet

Abstract The first division of the one-cell *C. elegans* embryo has been a fundamental model in deciphering the mechanisms underlying asymmetric cell division. Polarization of the one-cell zygote is induced by a signal from the sperm centrosome and results in the asymmetric distribution of PAR proteins. Multiple mechanisms then maintain PAR polarity until the end of the first division. Once asymmetrically localized, PAR proteins control several essential aspects of asymmetric division, including the position of the mitotic spindle along the polarity axis. Coordination of the spindle and cytokinetic furrow positions is the next essential step to ensure proper asymmetric division. In this chapter, I review the different mechanisms underlying these successive steps of asymmetric division. Work from the last 30 years has revealed the existence of multiple and redundant regulatory pathways which ensure division robustness. Besides the essential role of PAR proteins, this work also emphasizes the importance of both microtubules and actomyosin throughout the different steps of asymmetric division.

5.1 Introduction

As highlighted in multiple places in this book, asymmetric cell division is essential to generate cell diversity during both development and adult life. It plays a crucial role in stem cell biology, and defects in asymmetric division have been shown to be linked to tumorigenesis. Asymmetric cell division may be controlled by extracellular signals or arise as a consequence of events within the mother cell. This review will focus on the latter case.

A. Pacquelet (✉)
CNRS, UMR6290, Rennes, France

Université de Rennes 1, Institut de Génétique et Développement de Rennes, Rennes, France

CNRS UMR6290—IGDR, 2 avenue du Professeur Léon Bernard, 35043 Rennes Cedex, France
e-mail: anne.pacquelet@univ-rennes1.fr

J.-P. Tassan, J.Z. Kubiak (eds.), *Asymmetric Cell Division in Development, Differentiation and Cancer*, Results and Problems in Cell Differentiation 61,
DOI 10.1007/978-3-319-53150-2_5

An essential feature of intrinsic asymmetric cell division is the asymmetric inheritance of cell fate determinants between the two daughter cells. It may also be accompanied by the asymmetric distribution of cell structures such as centrosomes and the cytokinetic midbody or even by nonrandom sister chromatid segregation as reviewed in (Roubinet and Cabernard 2014; Yadlapalli and Yamashita 2013). In many cases, asymmetric cell division generates two daughter cells that are not only different in fate but also in size. Extreme examples of size asymmetry are observed for instance during mitosis in budding yeast or during female meiotic divisions with the formation of polar bodies. Several steps are essential to ensure faithful asymmetric division: first, a polarity axis must be established; this is followed by the alignment of the mitotic spindle and the asymmetric distribution of cell fate determinants along this polarity axis. In some instances, the mitotic spindle is displaced toward one pole of the mother cell, thereby leading to daughter cell size asymmetry. In this chapter, I will review the mechanisms underlying the first asymmetric division of *C. elegans* embryos including the mechanisms that establish and maintain the asymmetric distribution of the PAR proteins and mechanisms required to generate cell size asymmetry. Regulation of cell fate determinants segregation has been reviewed elsewhere (Rose and Gönczy 2014) and will not be addressed here.

Thanks to their accessibility for microscopy and genetic amenability, early *C. elegans* embryos have proven to be a powerful system to decipher the different steps of asymmetric division. Shortly after fertilization, oocyte meiotic divisions complete, the maternal and paternal pronuclei decondense their chromosomes, both pronuclei replicate their DNA, and the sperm centrosome (which is the unique centrosome provided to the zygote) duplicates (Fig. 5.1a–b). The maternal pronucleus then migrates toward the paternal pronucleus, until they meet in the posterior half of the embryo (Fig. 5.1b). During meiosis completion, the embryonic cortex is very active and undergoes intense ruffling (Fig. 5.1a). When the pronuclei decondense, ruffling gradually stops at the posterior and becomes confined to the anterior (Fig. 5.1b), ultimately forming a deep and transient membrane invagination, called pseudocleavage furrow, which retracts during pronuclear migration (Hird and White 1993).

Following pronuclear meeting, the pronuclei become centered in the embryo and rotate so that the centrosomes align with the antero-posterior axis (Fig. 5.1c). The pronuclear envelopes then break down and the mitotic spindle assembles. The mitotic spindle first forms in the center of the embryo and is next displaced toward the posterior during anaphase (Fig. 5.1d). The cytokinesis furrow then cleaves through the spindle midzone allowing proper DNA segregation and generating a big anterior cell, called AB, and a small posterior cell, called P1 (Fig. 5.1e–f). AB and P1 not only differ in size, they also have different cytoplasmic compositions (e.g., P1 contains P granules and higher concentrations of the cell fate determinants PIE-1 and SKN-1 than does AB), different cell cycle timing, and different spindle orientation during the second division (Fig. 5.1f) (Rose and Gönczy 2014). Notably, P1 then undergoes three more rounds of asymmetric cell divisions which are

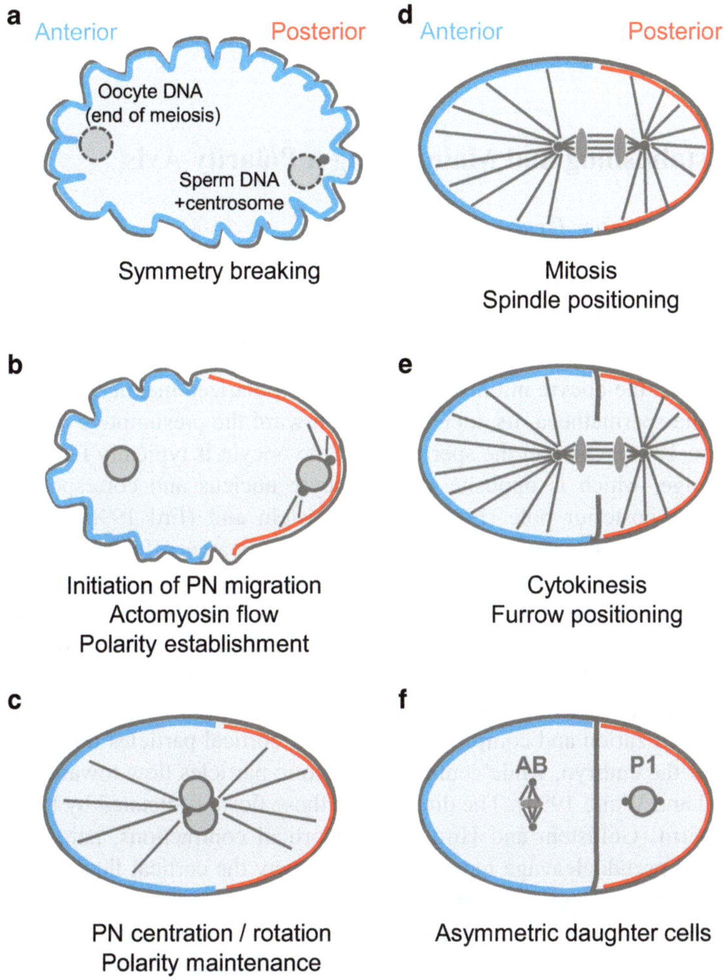

Fig. 5.1 Schematic representation of the different steps leading to asymmetric cell division of the one-cell *C. elegans* embryo. (**a**) After fertilization and completion of meiosis, myosin-dependent cortical ruffling occurs uniformly; PAR polarity proteins are uniformly localized at the cortex. (**b**) The sperm centrosome breaks symmetry and induces a cortical flow of myosin directed toward the anterior. At the beginning of pronuclear migration, cortical ruffling and myosin are restricted to the anterior cortex; polarization of PAR proteins is established. (**c**) After PN migration and meeting, the two pronuclei are centered in the embryo. Polarity is maintained through multiple regulatory mechanisms. (**d**) During anaphase, asymmetric cortical pulling forces displace the spindle toward the posterior of the embryo. (**e**) The cytokinesis furrow is positioned so that cleavage occurs through the spindle midzone. (**f**) This asymmetric cell division produces a big anterior cell (AB) and a small posterior cell (P1). During the second division, AB divides earlier than P1; the spindle in P1 orients along the long axis of the embryo and orthogonally to the spindle in AB

essential to further asymmetrically segregate cell fate determinants that specify the germline and endomesoderm cell fates (Rose and Gönczy 2014).

5.2 Establishing and Maintaining a Polarity Axis

5.2.1 Symmetry Breaking

5.2.1.1 Induction of Polarization and Actomyosin Flow upon Sperm Entry

In *C. elegans*, the oocyte initially forms in a non-polarized manner. Shortly before entering the spermatheca, its nucleus moves toward the presumptive anterior pole (Albertson 1984). Once in the spermatheca, the oocyte is typically fertilized at its leading edge, which is opposite to the oocyte nucleus and corresponds to the presumptive posterior pole (Fig. 5.1a) (Goldstein and Hird 1996). The oocyte nucleus position can occasionally vary without any apparent effect on development, suggesting that it is not responsible for the specification of the antero-posterior axis of the zygote (Albertson 1984). Furthermore, entry of the sperm at the abnormal end of the oocyte (i.e., on the oocyte nucleus side) leads to the reversal of the polarity axis, indicating that the position of the sperm defines the embryo antero-posterior axis (Goldstein and Hird 1996).

After fertilization and completion of meiosis, cortical particles flow toward the anterior of the embryo, while central cytoplasmic particles flow toward the posterior (Hird and White 1993). The direction of these flows is dictated by the position of the sperm (Goldstein and Hird 1996). Cortical contractions, namely cortical ruffling and pseudocleavage ingression, accompany the cortical flow. The embryo initially undergoes cortical ruffling along its whole antero-posterior axis, but as polarization progresses, ruffling becomes restricted to the anterior cortex while the posterior cortex smoothens (Fig. 5.1a–b). Cortical flow and contractions both cease as pronuclei migrate and pseudocleavage regresses (Hird and White 1993). Early experiments showed that cortical flow depends on the presence of intact microfilaments but not of microtubules (Hird and White 1993). Consistently, it was later observed that the flow of cortical particles coincides with a flow of actomyosin foci (Munro et al. 2004). Close to the end of meiosis, shortly before the onset of cortical flow, non-muscle-myosin II (NMY-2) is enriched throughout the cortex and forms a dynamic network of filaments and numerous dense foci. Upon meiosis completion, the sperm centrosome moves close to the posterior cortex and NMY-2 foci rapidly move away from the sperm centrosome. This collective movement of NMY-2 foci toward the anterior generates a foci-rich anterior cap with cortical ruffles and a smooth posterior zone devoid of foci. Actin foci colocalize with and behave similarly to NMY-2 foci (Munro et al. 2004). Furthermore, depletion of the regulatory myosin light chain *mlc-4* reduces myosin-based contractility and leads to a similar reduction of cortical particle and actomyosin foci flows. The flow of cortical particles and actomyosin foci thus appear to be both driven by myosin-based contractility (Munro et al. 2004).

From a mechanical point of view, measurements of cortical tension during polarization show that tension at the anterior cortex is anisotropic, with strongest tension orthogonal to the antero-posterior axis. Experimental data together with modeling of cortical mechanics indicate that cortical viscosity explains this anisotropy in cortical tension and is essential to support long-range flow toward a contracting region of the cortex (Mayer et al. 2010). NMY-2 is required not only for the anterior-directed cortical flow but also for the simultaneous posterior-directed cytoplasmic flow (Shelton et al. 1999). Quantitative analysis of cortical and cytoplasmic flows coupled to computer simulations showed that the hydrodynamic properties of the cytoplasm can explain that the forces generated by NMY-2 at the cortex are transmitted to the central cytoplasmic region where they create a countercurrent flow (Niwayama et al. 2011).

5.2.1.2 Role of the Sperm Centrosome and Microtubules

The position of the sperm corresponds to the position at which cortical flow will initiate. However, induction of flow does not require the presence of the sperm nucleus (Sadler and Shakes 2000) but rather the sperm centrosome. After the completion of meiosis II, the sperm centrosome starts to nucleate microtubules and approaches the cortex (Cowan and Hyman 2004). Laser ablation of the sperm centrosome or depletion of the essential centrosomal components SPD-5 and SPD-2 prevents polarization and NMY-2 cortical flow, indicating that the centrosome or its associated microtubules play a critical role in controlling actomyosin contractility and polarity establishment (Cowan and Hyman 2004; O'Connell et al. 2000; Munro et al. 2004). The centrosome is required only for polarity initiation, as ablating the centrosome after polarity has initiated does not block polarity establishment or maintenance (Cowan and Hyman 2004).

The centrosome–cortex distance decreases at the time when polarity is initiated (Cowan and Hyman 2004). Moreover, increasing the centrosome–cortex distance increases the time required to initiate polarity (Bienkowska and Cowan 2012). Similarly, mutants in which the centrosome remains far from the cortex, such as mutants lacking the aminopeptidase PAM-1 or the deubiquitylation enzymes MATH-33 and USP-47, display polarization defects and decreased cortical flow; these phenotypes are rescued if the centrosome is forced to stay close to the cortex (Lyczak et al. 2006; Fortin et al. 2010; McCloskey and Kemphues 2012). Altogether, these data indicate that the proximity of the sperm centrosome to the cortex is critical for efficient polarization.

Contrary to the loss of centrosome, depletion of microtubules does not strongly impair polarity, suggesting that the centrosome can regulate polarity independently of its ability to nucleate microtubules (Cowan and Hyman 2004). Depletion of tubulin was nevertheless reported to delay polarization (Tsai and Ahringer 2007). Microtubules appear to bias centrosome movements toward the cortex prior to symmetry breaking, possibly explaining some of their effect on polarization (Bienkowska and Cowan 2012). Moreover, microtubules also contribute to polarization by directly controlling the loading of the polarity protein PAR-2 to the posterior cortex, independently of actomyosin flow (Motegi et al. 2011) (for details see Sect. 5.2.2.2).

5.2.1.3 Mechanisms Regulating Actomyosin Flow

Formation of the network of actomyosin foci and cortical contractions requires the small GTPase RHO-1, its GEF ECT-2, and the Rho binding kinase LET-502 (Motegi and Sugimoto 2006; Jenkins et al. 2006; Kumfer et al. 2010). Both RHO-1 and ECT-2 are necessary for MLC-4 phosphorylation and are initially distributed uniformly at the cortex (Motegi and Sugimoto 2006; Jenkins et al. 2006). Concomitantly with the initiation of cortical flow, a local decrease of ECT-2 is observed at the posterior cortex, close to the sperm centrosome. This decrease is not observed when the centrosomal protein SPD-5 is depleted but is not affected by flow inhibition [e.g., upon *rho-1* or *mlc-4(RNAi)*], suggesting that removal of ECT-2 from the posterior cortex is controlled by the sperm centrosome and is one of the early events leading to polarization (Motegi and Sugimoto 2006). The region depleted of ECT-2 then expands toward the anterior. RHO-1 also clears from the posterior cortex, forming larger foci at the anterior cortex (Fig. 5.2a) (Motegi and Sugimoto 2006).

An important question still not resolved is the nature of the centrosomal cue which leads to the removal of ECT-2 from the posterior cortex. The RhoGAP CYK-4 has been proposed to be the cue leading to polarization (Jenkins et al. 2006). It is present in the sperm and at the posterior cortex of fertilized embryos, including embryos depleted of maternal CYK-4, raising the possibility of a paternal contribution. Furthermore, the actomyosin network remains evenly distributed along the antero-posterior axis in embryos depleted of CYK-4 by RNAi (Jenkins et al. 2006). CYK-4 was thus proposed to induce a gradient of actomyosin contractility by locally inhibiting RHO-1 in the vicinity of the sperm centrosome (Jenkins et al. 2006). However, the involvement of CYK-4 in polarity establishment remains controversial. Indeed, although it has a GAP domain, CYK-4 has been shown to bind ECT-2 and activate RHO-1 during cytokinesis in a GAP-dependent manner (Zhang and Glotzer 2015). Moreover, a mutation (E448K) in the GAP domain of CYK-4 that prevents the localization of CYK-4 to the cortex, likely affects its GAP activity, and impairs furrow ingression during cytokinesis does not have a detectable phenotype during polarity establishment (Tse et al. 2012; Canman et al. 2008; Zhang and Glotzer 2015). The hypothesis of Jenkins and Mango implies that CYK-4 inactivates, rather than activates, RHO-1. To support this hypothesis, one would thus need to explain how CYK-4 interaction with ECT-2 is blocked or modified in very early embryos.

Several other proteins are also involved in the formation of the actomyosin network and cortical contractions, albeit in a more subtle manner than ECT-2 and

Fig. 5.2 Main mechanisms involved in polarity establishment (**a**) and maintenance (**b**). Proteins displaced to the anterior upon cortical flow during polarity establishment (in **a**) or enriched at the anterior cortex during polarity maintenance (in **b**) are depicted in *blue*. Posterior proteins are depicted in *red*. PAR proteins which have been shown to directly bind membrane phospholipids are circled in *gray*. *MT* microtubules

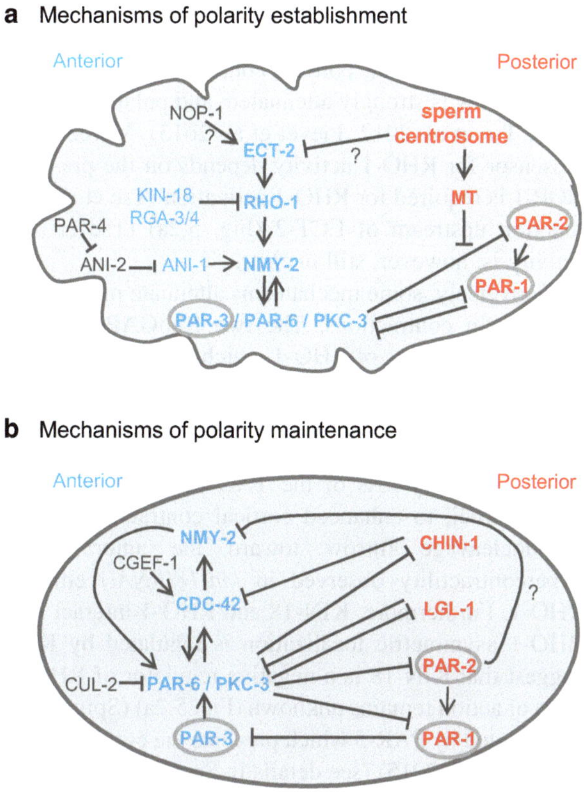

RHO-1. Those include the serine-threonine kinase PAR-4, the scaffolding protein anillin ANI-1, and the serine-rich protein NOP-1 (Fig. 5.2a). PAR-4 is required for pseudocleavage formation and efficient actomyosin cortical flow (Morton et al. 1992; Chartier et al. 2011). In *par-4* mutants, NMY-2::GFP foci form but move slower toward the anterior. As a result, the NMY-2 anterior domain spreads further toward the posterior than in control embryos (Chartier et al. 2011). Interestingly, the anillin ANI-2 accumulates abnormally at the cortex in *par-4* embryos (Chartier et al. 2011; Pacquelet et al. 2015). Moreover, ANI-2 depletion restores the velocity of NMY-2 flow in *par-4* mutants, suggesting that the reduced actomyosin flow observed in *par-4* embryos is due to the excess or mislocalization of ANI-2 (Chartier et al. 2011). ANI-2 is a shorter isoform of anillin which may act as a dominant negative for the classical anillin ANI-1, suggesting that ANI-1 activity may be reduced in *par-4* mutants (Fig. 5.2a) (Chartier et al. 2011; Amini et al. 2014). Consistently, ANI-1, which forms cortical patches at the anterior cortex during polarity establishment, is required for cortical contractility, for pseudocleavage, and for the formation of NMY-2 foci (Maddox et al. 2005; Tse et al. 2012). Understanding precisely how ANI-1 regulates actomyosin activity will require further investigation. The serine-rich protein NOP-1 is also required for proper actomyosin contractions: in *nop-1* embryos, NMY-2 and ANI-1 foci are smaller than normal, cortical contractions and pseudocleavage are not observed, cortical flow is strongly attenuated, and polarization is less pronounced (Rose et al. 1995; Tse et al. 2012; Fievet et al. 2013). Moreover, the cortical recruitment of a biosensor for RHO-1 activity depends on the presence of NOP-1, indicating that NOP-1 is required for RHO-1 activation (Tse et al. 2012). NOP-1 was proposed to function upstream of ECT-2 (Fig. 5.2a) (Tse et al. 2012); its exact molecular activity is, however, still unclear.

Conversely, some mechanisms attenuate myosin activity and limit the extent of actomyosin contractions. The two RhoGAPs RGA-3 and RGA-4 are essential negative regulators of RHO-1 which localize at the anterior cortex during polarity establishment (Fig. 5.2a). In *rga-3/4(RNAi)* embryos, the cortical myosin network appears denser and contracts excessively; moreover, the pseudocleavage furrow forms more anteriorly than in control embryos (Schmutz et al. 2007; Schonegg et al. 2007). Similarly loss of the TAO kinase KIN-18 leads to the accumulation of NMY-2 foci, to enhanced cortical contractions, and to the displacement of the pseudocleavage furrow toward the anterior (Spiga et al. 2013). The hypercontractility observed in *kin-18(RNAi)* embryos requires the activity of RHO-1. Furthermore, KIN-18 and RHO-1 interact in yeast two-hybrid assays and RHO-1 asymmetric localization is regulated by KIN-18. Altogether, these data suggest that KIN-18 is a negative regulator of RHO-1, although its exact mechanism of action remains unknown (Fig. 5.2a) (Spiga et al. 2013). RHO-1 activation is also limited by PAR-5 which prevents the cortical recruitment of CYK-4 (Fig. 5.2a) (Basant et al. 2015) (see details in Sect. 5.3.2).

Finally, systematic RNAi screens also identified several putative new regulators of actomyosin flow (Fievet et al. 2013). Some of these genes (i.e., the ezrin *erm-1*, the septin *unc-59*, the protein phosphatase 1 subunit *gsp-1*, the plastin ortholog *plst-1*, the ARF GTPase *cnt-2*) have homologs known to be functionally linked to the actin cytoskeleton. Others, such as the nuclear pore protein NPP-2 or the CCR4-NOT component NTL-2, have not previously been shown to be linked to actomyosin (Fievet et al. 2013). Further characterization of their function will extend our understanding of how actomyosin flow is regulated to establish polarity.

5.2.2 PAR Polarity Establishment and Maintenance

5.2.2.1 PAR Proteins, Essential Polarity Determinants

The *par* genes were initially identified by Ken Kemphues in a screen performed to isolate genes required for early asymmetric cell division in *C. elegans* embryos ("*par*-titioning defective" genes) (Kemphues et al. 1988). Subsequent work showed that the serine-threonine kinase PAR-1 and the RING domain containing protein PAR-2 localize at the posterior cortex of polarized one-cell embryos (Guo and

Kemphues 1995; Levitan et al. 1994; Boyd et al. 1996). The PDZ domain containing proteins PAR-3 and PAR-6 localize at the anterior cortex together with the serine-threonine kinase PKC-3 (Etemad-Moghadam et al. 1995; Tabuse et al. 1998; Hung and Kemphues 1999). In the absence of PAR-3, PAR-6, or PKC-3, PAR-1 and PAR-2 spread toward the anterior side; the first division is symmetric and gives rise to two daughter cells that divide synchronously and behave like P1 cells with respect to the orientation of the second cleavage (Kemphues et al. 1988; Guo and Kemphues 1995; Etemad-Moghadam et al. 1995; Boyd et al. 1996; Watts et al. 1996; Tabuse et al. 1998). Conversely, embryos without PAR-2 show PAR-3, PAR-6, and PKC-3 spreading toward the posterior and divide symmetrically, forming two daughter cells that divide synchronously and behave like AB cells with respect to the orientation of the second cleavage (Hung and Kemphues 1999; Tabuse et al. 1998; Etemad-Moghadam et al. 1995; Kemphues et al. 1988). Loss of PAR-1 also alters some aspects of the asymmetry of division, in particular cell cycle asynchrony at the two-cell stage and partitioning of cytoplasmic components such as P granules. However, contrary to PAR-2, PAR-1 is not strictly required for the asymmetric localization of anterior PAR proteins (Guo and Kemphues 1995; Hung and Kemphues 1999).

The serine-threonine kinase PAR-4 and the 14.3.3 protein PAR-5 are uniformly localized in one-cell embryos (Watts et al. 2000; Morton et al. 2002). Similar to PAR-1, PAR-4 is required for cell cycle asynchrony at the two-cell stage and for P granule segregation but not for cell size asymmetry (Morton et al. 1992). It also moderately affects the localization of anterior PAR proteins (Hung and Kemphues 1999; Chartier et al. 2011). *par-5* mutants divide symmetrically and display an expanded anterior PAR domain which overlaps with the PAR-2 domain (Morton et al. 2002).

The cortical localization of PAR proteins likely involves the direct binding of PAR-1, PAR-2, and PAR-3 to membrane phosphoinositides (Li et al. 2010a; Motegi et al. 2011). PAR-1 is also recruited to the cortex through a direct interaction with PAR-2 (Motegi et al. 2011). PAR-6, PKC-3, and PAR-3 can be detected as a complex in vivo, and PAR-6 directly interacts with both PKC-3 and PAR-3 (Li et al. 2010b). PAR-3 is required for PAR-6 and PKC-3 cortical localization and partially colocalizes with PAR-6 (Tabuse et al. 1998; Hung and Kemphues 1999; Beers and Kemphues 2006). PAR-6 and PKC-3 are required for each other's localization (Tabuse et al. 1998; Hung and Kemphues 1999). PAR-3 and PKC-3 may stabilize PAR-6 at the cortex by counteracting the effect of CDC-37, a Hsp90 cochaperone which inhibits PAR-6 cortical localization (Beers and Kemphues 2006). While the direct interaction between PAR-6 and PKC-3 is needed to localize PAR-6 at the cortex, the direct interaction between PAR-6 and PAR-3 is required neither for PAR-6 cortical localization nor for its colocalization with PAR-3, suggesting that PAR-3 may also indirectly interact with PAR-6 (Li et al. 2010b). Importantly, only 65% of PAR-6 puncta colocalize with PAR-3, revealing the existence of a second mode of cortical association for PAR-6 (Beers and Kemphues 2006). The recruitment of this second population of PAR-6 requires the small GTPase CDC-42, which is known to interact with PAR-6 (Beers and Kemphues

2006; Gotta et al. 2001). Consistently, both *cdc-42* depletion and mutations disrupting PAR-6/CDC-42 interaction reduce PAR-6 cortical recruitment (Schonegg and Hyman 2006; Aceto et al. 2006).

Analysis of PAR protein dynamics revealed that polarization of one-cell embryos involves two temporally distinct phases, namely polarity establishment and maintenance (Cuenca et al. 2003). During the first phase, which corresponds to the phase of actomyosin cortical flow, PAR-3, PAR-6, and PKC-3 are excluded from the posterior cortex while PAR-2 accumulates at the posterior, thereby establishing polarity. After pronuclear meeting, several redundant mechanisms ensure polarity maintenance. The molecular mechanisms involved in these two phases are detailed below.

5.2.2.2 Establishment of PAR Polarity

After fertilization, PAR-6 and PAR-3 uniformly localize at the cortex while PAR-2 is weakly detected at both the anterior and posterior cortex. Concomitantly with the appearance of the anterior-directed cortical flow, PAR-6/PAR-3 become restricted to the anterior cortex while PAR-2 accumulates at the posterior (Boyd et al. 1996; Etemad-Moghadam et al. 1995; Hung and Kemphues 1999; Cuenca et al. 2003; Munro et al. 2004). Cortical PAR-6::GFP puncta and NMY-2::GFP foci move toward the anterior at identical speed, suggesting that they move within a common cortical flow (Munro et al. 2004). Consistently, NMY-2 is required to polarize PAR proteins (Fig. 5.2a) (Guo and Kemphues 1996; Cuenca et al. 2003). Moreover, reduction of myosin-based contractility upon depletion of the regulatory myosin light chain MLC-4 decreases both PAR-6::GFP and actomyosin flows (Munro et al. 2004). Similarly, PAR-6::GFP is not polarized in the absence of ECT-2 or RHO-1 (Jenkins et al. 2006; Motegi and Sugimoto 2006) (Fig. 5.2a). The anterior PAR domain also spreads toward the posterior in mutants with reduced myosin flow such as *par-4* or *nop-1* (Fig. 5.2a) (Chartier et al. 2011; Fievet et al. 2013). Conversely, excessive contractility in *rga-3/4(RNAi)* or *kin-18(RNAi)* embryos leads to the formation of a smaller PAR-6 domain (Fig. 5.2a) (Schonegg et al. 2007; Spiga et al. 2013). Altogether, these data indicate that actomyosin contractility plays an essential role during polarization by creating a directed cortical flow which localizes PAR-6/PAR-3/PKC-3 to the anterior cortex. PAR-6/PAR-3/PKC-3 in turn exclude PAR-2 from the anterior cortex (Cuenca et al. 2003) (see details in next paragraph). Notably, PAR proteins and CDC-42 also regulate actomyosin flow in a positive feedback loop (Fig. 5.2a). Indeed, depleting either PAR-3 or CDC-42 severely reduces cortical flow and prevents the expansion of the region devoid of ECT-2 (Munro et al. 2004; Motegi and Sugimoto 2006).

Actomyosin cortical flow is clearly essential to establish PAR polarity. However, a second, independent mechanism also appears to contribute to PAR polarity establishment. Indeed, when actomyosin contractility is severely reduced, PAR-2 can still load on the cortex adjacent to the sperm centrosome and exclude PAR-3 from the posterior cortex, albeit less efficiently than in wild-type embryos (Shelton et al. 1999; Zonies et al. 2010). Cortical loading of PAR-2 involves binding of PAR-2 to microtubules nucleated by the sperm centrosome. Preventing the interaction between PAR-2 and microtubules delays symmetry breaking but does not prevent final polarization in embryos with normal cortical flow; by contrast, it completely prevents symmetry breaking in embryos lacking cortical flow (Motegi et al. 2011). The interaction with microtubules protects PAR-2 from being phosphorylated by PKC-3 (Motegi et al. 2011). Phosphorylation of PAR-2 by PKC-3 prevents the binding of PAR-2 to membrane lipids and leads to its cortical exclusion (Hao et al. 2006; Motegi et al. 2011). The interaction between PAR-2 and microtubules therefore allows PAR-2 to be loaded on the cortex adjacent to the sperm centrosome (Fig. 5.2a) (Motegi et al. 2011). Once at the cortex, PAR-2 recruits PAR-1 which in turn phosphorylates PAR-3 to exclude it from the posterior cortex (Fig. 5.2a) (Motegi et al. 2011). Thus, the asymmetric loading of PAR-2 in the vicinity of the sperm centrosome followed by reciprocal phosphorylation events is sufficient to break symmetry and to drive the mutual exclusion of anterior and posterior PAR proteins.

PAR-5 is also required for polarity establishment: in the absence of PAR-5, PAR-6 polarizes less and slower than in wild-type embryos; it also overlaps with the PAR-2 domain (Cuenca et al. 2003; Mikl and Cowan 2014). PAR-5 expression levels are tightly regulated by alternative splicing (Mikl and Cowan 2014). Interestingly, the production of the most translationally active *par-5* isoform increases in *par-2* mutants. This feedback regulation of PAR-2 on PAR-5 expression is likely to contribute to the robustness of polarity establishment (Mikl and Cowan 2014).

5.2.2.3 Maintenance of PAR Polarity

After the anterior and posterior PAR domains have been established, GFP::PAR-6 and GFP::PAR-2 can diffuse across domain boundary, indicating that there is no physical barrier between the two domains and that some other mechanisms are required to prevent the spreading of the anterior and posterior PAR proteins (Goehring et al. 2011). Mechanisms leading to the mutual exclusion between the anterior and posterior PARs are crucial during this maintenance phase (Cuenca et al. 2003). PKC-3 phosphorylates PAR-2 and PAR-1 to exclude them from the anterior cortex (Fig. 5.2b) (Hao et al. 2006; Motegi et al. 2011). PAR-2 plays a critical role to maintain polarity and prevent the spreading of anterior PARs toward the posterior after polarity has been established (Cuenca et al. 2003). The effect of PAR-2 on anterior PAR proteins can be partly explained by the recruitment of PAR-1 which phosphorylates PAR-3 and excludes it from the posterior cortex (Fig. 5.2b) (Motegi et al. 2011). However, *par-1* mutants have milder polarity defects than *par-2* mutants, indicating that PAR-2 also prevents the spreading of anterior PARs independently of PAR-1 (Cuenca et al. 2003; Sailer et al. 2015). The ability of PAR-2 to regulate myosin may be one way by which PAR-2 also contributes to polarity maintenance (Fig. 5.2b). Indeed, in the absence of PAR-2, NMY-2 spreads toward the posterior during the maintenance phase (Munro et al.

2004). Moreover, depletion of MRCK-1 [myotonic dystrophy-related Cdc42 binding kinase, a putative CDC-42 effector and a regulator of MLC-4 (Galli et al. 2011)] leads to the lack of cortical myosin and is sufficient to abolish the expansion of anterior PAR proteins observed in *par-2* mutants (Sailer et al. 2015). The accumulation of NMY-2 on the posterior cortex of *par-2* mutants is not a secondary consequence of the mislocalization of anterior PAR proteins as it is also observed in *par-3;par-2* and *par-2;par-6* embryos (Beatty et al. 2013). The exact mechanism involved in this regulation is, however, unknown.

CDC-42 and its interaction with PAR-6 also have a crucial role to restrict PAR-6 at the anterior cortex during polarity maintenance (Fig. 5.2b) (Gotta et al. 2001; Motegi and Sugimoto 2006; Schonegg and Hyman 2006; Aceto et al. 2006). Likewise, CDC-42 is required to maintain actomyosin at the anterior cortex (Fig. 5.2b) (Motegi and Sugimoto 2006). During the maintenance phase, active CDC-42 localizes at the anterior cortex in a PAR-6-dependent manner (Aceto et al. 2006; Kumfer et al. 2010). CDC-42 activity is promoted by the RhoGEF CGEF-1 and inhibited by the RhoGAP CHIN-1 (Kumfer et al. 2010). CHIN-1 localizes at the posterior cortex and is excluded from the anterior cortex by the anterior PAR proteins (Fig. 5.2b) (Kumfer et al. 2010; Sailer et al. 2015). Depletion of CGEF-1 leads to the lack of cortical myosin during maintenance. Conversely, cortical myosin expands toward the posterior in the absence of CHIN-1 (Kumfer et al. 2010). However, similar to *par-1* mutants, single *chin-1* mutants have only minor PAR-6 localization defects. By contrast, PAR-6 asymmetry is not maintained in double *chin-1;par-1* embryos, indicating that the two pathways act redundantly during maintenance to prevent the spreading of PAR-6 toward the posterior pole (Sailer et al. 2015). These defects are not rescued by *mrck-1(RNAi)*, suggesting that they are not due to defects in myosin flow (Sailer et al. 2015). Rather, PAR-6 accumulation at the posterior cortex in *chin-1;par-1* embryos is due to the simultaneous presence at the posterior cortex of small amounts of PAR-3—due to the lack of PAR-1—and of active CDC-42—due to the lack of CHIN-1 (Fig. 5.2b) (Sailer et al. 2015).

A third pathway contributing to polarity maintenance involves LGL-1, the homolog of the *drosophila* tumor suppressor Lethal Giant Larvae (Fig. 5.2b). LGL-1 localizes at the posterior cortex in a PKC-3-dependent manner: PKC-3 and putative PKC-3 phosphorylation sites in LGL-1 are both required to exclude LGL-1 from the anterior cortex (Beatty et al. 2010; Hoege et al. 2010). *lgl-1* single mutants do not show polarity defects on their own. However, they enhance the lethality and polarity defects of *par-2* thermosensitive mutants grown at permissive temperature (Beatty et al. 2010; Hoege et al. 2010). In particular, PAR-2 and LGL-1 both contribute to restrict the anterior PARs at the anterior cortex during polarity maintenance (Beatty et al. 2010). LGL-1 can interact with PAR-6 and PKC-3 (Hoege et al. 2010). Moreover, *lgl-1* mutants show higher levels of PAR-6 protein, suggesting that LGL-1 may restrict the expansion of the anterior PAR domain by limiting PAR-6 levels (Beatty et al. 2013). However, the exact molecular mechanisms by which LGL-1 contributes to polarity maintenance remain elusive.

The balance between anterior and posterior PAR protein levels also contributes to polarity maintenance. In *par-2* mutants, anterior PAR proteins spread toward the posterior during the maintenance phase leading to polarity defects and embryonic lethality. Interestingly, it was observed that *par-2* lethality can be suppressed by reducing the levels of any of the anterior cortical proteins PAR-6, PAR-3, PKC-3, and CDC-42 (Watts et al. 1996; Gotta et al. 2001; Labbé et al. 2006). Those observations suggested that new regulators of anterior PAR proteins could be identified through their ability to suppress *par-2* lethality. One *par-2* suppressor identified in a genome-wide RNAi screen is the translation repressor NOS-3 (Labbé et al. 2006). NOS-3 was first shown to repress the expression of FEM-3, a protein known for its role in spermatogenesis and which is part of a ubiquitin ligase complex formed by the cullin CUL-2 and the substrate adaptor FEM-1 ($CBC^{FEM\text{-}1}$) (Kraemer et al. 1999; Starostina et al. 2007). It was found that *nos-3* mutants suppress *par-2* polarity defects by decreasing PAR-6 levels. The suppression of *par-2* polarity defects and the regulation of PAR-6 levels observed in *nos-3* mutants require $CBC^{FEM\text{-}1}$ activity. Moreover, PAR-6 interacts with the substrate adaptor FEM-1. Collectively, this suggests that PAR-6 may be a substrate of the ubiquitin ligase $CBC^{FEM\text{-}1}$ (Fig. 5.2b) (Pacquelet et al. 2008). Similarly, *par-2* polarity defects are suppressed and PAR-6 levels decreased in cyclins *cyb-2.1/2* and *cdk-1* mutants. The suppression of *par-2* defects is CUL-2 dependent, suggesting that CYB-2/CDK-1 also regulate PAR-6 through CUL-2. This new regulatory pathway is, however, independent of NOS-3 (Rabilotta et al. 2015).

Finally, intracellular trafficking may also be involved in polarity maintenance. After polarity establishment, dynamin (DYN-1) and early endosomes localize to the anterior cortex in a PAR-dependent manner (Andrews and Ahringer 2007; Nakayama et al. 2009). While DYN-1 does not regulate polarity establishment, it is required to maintain PAR-6, CDC-42, and RHO-1 at the anterior cortex. It also contributes to the enrichment of endocytic vesicles close to the anterior cortex and partially colocalizes with PAR-6 on vesicle-like puncta (Nakayama et al. 2009). PAR-6 also colocalizes with the small GTPase RAB-5, a marker of early endosomes (Nakayama et al. 2009; Hyenne et al. 2012). *rab-5(RNAi)* suppresses *par-2* polarity defects and leads to the formation of a smaller PAR-6 domain (Hyenne et al. 2012). Altogether, these results suggest that intracellular trafficking contributes to polarity. However, whether the defects observed in *dyn-1*- or *rab-5*-depleted embryos are due to a direct role of trafficking on polarity proteins or to indirect effects remains to be investigated.

Importantly, correction of weak polarity defects occurring during polarity establishment has been observed in several instances (Schonegg et al. 2007; Schenk et al. 2010; Spiga et al. 2013; Fievet et al. 2013). Correction can occur early during the maintenance phase (Spiga et al. 2013) or later during division (Schenk et al. 2010). In the latter case, it has been shown that formation of a smaller or larger anterior domain does not significantly affect the position of the mitotic spindle and the cytokinetic furrow. However, during mitosis, the position of the anterior/posterior boundary is shifted to match the position of the furrow. This change in PAR domain boundary is accompanied by a flow of cortical myosin toward the nascent furrow and myosin contractility is required to correct the boundary position (Schenk et al. 2010). This correction mechanism ensures proper segregation of the anterior and posterior PAR proteins in the two daughter cells.

Thus several independent mechanisms appear to contribute to polarity maintenance in the one-cell *C. elegans* embryo. The variety of these multiple and redundant pathways and the ability to correct slight polarity defects are likely to contribute to the robustness of polarity maintenance and asymmetric cell division.

5.3 From Polarity to Cytokinesis Furrow Position

5.3.1 From Polarity to Spindle Position

Once established, anterior–posterior polarity controls mitotic spindle position. The mitotic spindle first assembles at the center of the one-cell embryo in prometaphase but is then displaced toward the posterior during anaphase. Pulling forces exerted from the cortex on astral microtubules play a major role in spindle posterior displacement. As evidenced by spindle severing experiments and centrosome fragmentation assays, those pulling forces are stronger on the posterior side of the embryo (Grill et al. 2001, 2003). A cortical complex composed of the Gα proteins GOA-1 and GPA-16, the activators GPR-1 and GPR-2 (Pins/LGN/AGS3 homologs), and the coiled-coil protein LIN-5 (Mud/Numa homolog) is required to generate pulling forces and displace the spindle toward the posterior (Colombo et al. 2003; Gotta et al. 2003; Nguyen-Ngoc et al. 2007; Park and Rose 2008). When any of those proteins are depleted, the mitotic spindle does not elongate as much as in control embryos and stays in the center of the embryo which consequently divides symmetrically (Lorson et al. 2000; Gotta and Ahringer 2001; Colombo et al. 2003; Gotta et al. 2003). The microtubule minus end-directed motor dynein is also required to generate pulling forces on the spindle and interacts with the Gα/GPR/LIN-5 complex. These results have led to a model according to which dynein is anchored at the cortex by the Gα/GPR/LIN-5 complex and generates pulling forces on astral microtubules as it attempts to move toward microtubule minus end (Pecreaux et al. 2006; Couwenbergs et al. 2007; (Nguyen-Ngoc et al. 2007).

From metaphase on, GPR-1/2 and LIN-5 are enriched at the posterior cortex (Fig. 5.3a). This asymmetric localization is regulated by the PAR proteins and is likely to explain why the pulling forces are stronger on the posterior side (Colombo et al. 2003; Gotta et al. 2003; Park and Rose 2008). PAR proteins regulate GPR-1/2 and LIN-5 localization at least through the Casein Kinase 1 (CNSK-1) and the phosphatidylinositol-4-phosphate PI(4)P5 kinase (PPK-1) (Fig. 5.3a) (Panbianco et al. 2008). During mitosis, GFP::CSNK-1 localizes at the anterior cortex and PPK-1 at the posterior cortex. Both localizations are regulated by the PAR proteins. In the absence of CSNK-1, excessive pulling forces are detected in the embryo and PPK-1, GPR-1/2, and LIN-5 localize uniformly at the cortex (Panbianco et al.

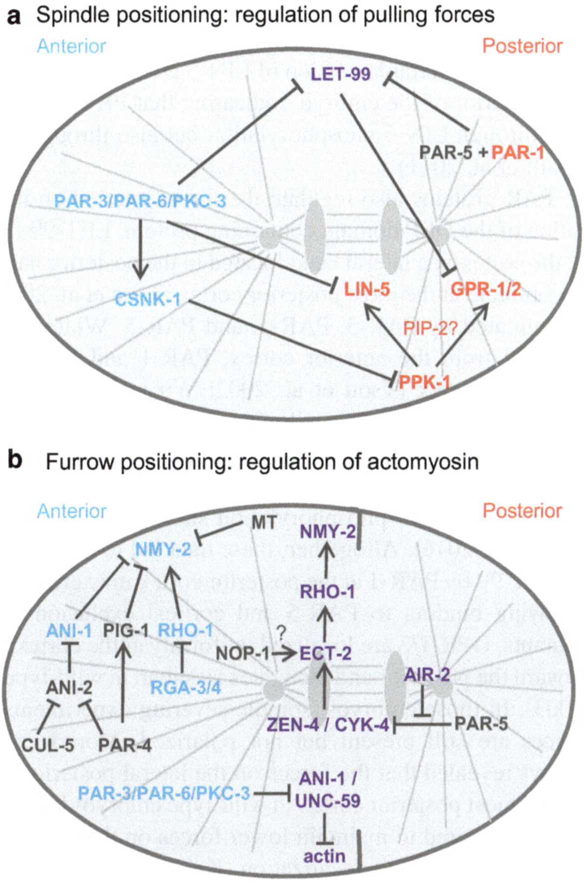

Fig. 5.3 Main mechanisms regulating pulling forces exerted on the mitotic spindle (**a**) and cytokinetic furrow ingression and position (**b**). Proteins enriched at the anterior cortex are depicted in *blue* and proteins enriched at the posterior cortex in *red*. Proteins which localize in a lateral band (in **a**) or in the vicinity of the spindle midzone (in **b**) are depicted in purple. During cytokinesis, actomyosin is enriched and activated in proximity to the spindle midzone while astral microtubules prevent furrow ingression. Several signaling pathways also prevent myosin localization and/or activity at the anterior cortex. This regulation is essential to keep spindle and furrow positions coordinated throughout cytokinesis. *MT* microtubules

2008). The excess of pulling forces is not observed if GPR-1/2 or LIN-5 is depleted at the same time as CSNK-1. On the contrary, loss of PPK-1 leads to weaker pulling forces and reduced accumulation of cortical GPR-1/2. Altogether, these data indicate that CSNK-1 is likely to inhibit the forces exerted on the spindle by preventing the cortical recruitment of PPK-1 and thereby that of GPR-1/2 and LIN-5. As PPK-1 is required for PIP2 synthesis, it was suggested that GPR-1/2 and LIN-5 asymmetry may be due to PIP2 asymmetry (Panbianco et al. 2008). Validation of this hypothesis by direct observation of PIP2 asymmetry has nevertheless not been reported.

LIN-5 is also directly regulated by PKC-3 which phosphorylates LIN-5 at the anterior cortex (Fig. 5.3a) (Galli et al. 2011). Expression of a nonphosphorytable version of LIN-5 leads to increased pulling forces at the anterior cortex. Conversely, a phosphomimetic version of LIN-5 reduces posterior forces. These data support a model in which the phosphorylation of LIN-5 by PKC-3 at the anterior cortex contributes to the asymmetry of pulling forces. However, the nonphosphorylatable version of LIN-5 is able to displace the mitotic spindle toward the posterior of the embryo, indicating that PKC-3 regulates spindle positioning in part through LIN-5 phosphorylation but also through complementary mechanisms (Galli et al. 2011).

PAR proteins also regulate the forces exerted on the spindle through the regulation of the DEP domain containing protein, LET-99 (Fig. 5.3a). LET-99 localizes at the cortex in a lateral band located in the posterior half of the embryo but does not accumulate at the most posterior cortex (Tsou et al. 2002). This localization pattern is regulated by PAR-3, PAR-1, and PAR-5. While PAR-3 is required to exclude LET-99 from the anterior cortex, PAR-1 and PAR-5 exclude it from the most posterior cortex (Tsou et al. 2002; Wu and Rose 2007; Wu et al. 2016). Both PAR-1 and PAR-5 physically interact with LET-99 (Wu and Rose 2007; Wu et al. 2016). PAR-1 kinase activity and predicted LET-99 phosphorylation sites are required to control LET-99 localization (Wu and Rose 2007; Wu et al. 2016). Predicted LET-99 phosphorylation sites are also necessary for binding to PAR-5 (Wu et al. 2016). Altogether, these data led to the hypothesis that phosphorylation of LET-99 by PAR-1 at the posterior cortex may create 14.3.3 binding sites, thereby allowing binding to PAR-5 and cortical exclusion (Wu et al. 2016). In *let-99* mutants, GPR-1/2 are localized uniformly at the cortex; the spindle is not displaced toward the posterior and elongates less than in wild-type embryos (Tsou et al. 2002, 2003). In those embryos, spindle severing experiments show that cortical pulling forces are still present but not polarized. Moreover, centrosome fragmentation assays revealed that the forces on the lateral posterior cortex are lower than forces on the most posterior cortex in wild-type embryos but not in *let-99* mutants. LET-99 is thus required to maintain lower forces on the lateral cortex; this regulation may contribute to spindle localization (Krueger et al. 2010).

Finally, actin also contributes to the regulation of the pulling forces exerted on the spindle. Indeed, acute actin depolymerization increases pulling forces at the anterior pole (Afshar et al. 2010; Berends et al. 2013). These data are consistent with a computer model that proposes that increased cortical stiffness reduces force generation by increasing the rate of force generator detachment. Actin enrichment at the anterior cortex would thus negatively regulate anterior pulling forces by increasing cortex rigidity (Kozlowski et al. 2007).

5.3.2 *Defining Cytokinesis Furrow Position According to Spindle Position*

Classical experiments performed by Ray Rappaport in sand dollar eggs showed that the presence of the mitotic spindle is essential to induce cytokinesis furrowing (Rappaport 1981). Furthermore, experimental displacement of the mitotic spindle changes the site of furrowing (Rappaport 1985). All these experiments emphasized the essential role of the mitotic spindle in inducing and positioning the cleavage furrow. Consistently, in *C. elegans* one-cell embryos, the anaphase spindle is pulled toward the posterior; the cytokinetic furrow then forms in the vicinity of the spindle midzone, giving rise to a big anterior cell and a small posterior cell (Fig. 5.1e–f). In mutants where the mitotic spindle stays in the center of the embryo—e.g., *par* or *goa-1;gpa-16* mutants (Kemphues et al. 1988; Gotta and Ahringer 2001)—the cytokinetic furrow is shifted toward the center of the embryo and cleaves to form two equally sized cells. Spindle severing experiments have shown that two redundant signals, emanating from the central spindle and from astral microtubules, induce cytokinetic furrowing. When a laser is used to asymmetrically cut the spindle and to separate the spindle midzone from one of the asters, the embryo forms two spatially and temporally distinct furrows: the first furrow forms midway of the asters and the second close to the spindle midzone (Bringmann and Hyman 2005). Loss of SPD-1, a protein essential for spindle midzone assembly, prevents formation of this second furrow but not of the furrow located midway of the asters (Bringmann and Hyman 2005). Moreover, loss of SPD-1 does not prevent furrowing in embryos with an intact mitotic spindle but completely inhibits furrow ingression when aster separation is delayed (Verbrugghe and White 2004; Lewellyn et al. 2010). Finally, forcing the mitotic spindle to localize to the posterior pole of the embryo also induces the formation of two furrows, one close to the spindle midzone and one in the anterior region, away from the two spindle asters (Werner et al. 2007). Altogether, these data show the existence of two pathways inducing furrowing, one dependent on the spindle midzone and another independent of the spindle midzone but apparently regulated by the position of spindle asters (Fig. 5.3b).

The centraspindlin complex, which consists of a dimer of the kinesin ZEN-4/MKLP1 and a dimer of CYK-4/MgcRacGAP, localizes to a narrow region in the center of the spindle midzone and weakly at the equatorial cortex (Raich et al. 1998; Jantsch-Plunger et al. 2000; Mishima et al. 2002; Nishimura and Yonemura 2006; Basant et al. 2015). Similar to SPD-1, ZEN-4 and CYK-4 are required for spindle midzone assembly but not for initial furrow ingression in otherwise normal embryos (Raich et al. 1998; Jantsch-Plunger et al. 2000). Spindle severing experiments have shown that both ZEN-4 and CYK-4 are essential for midzone-induced furrowing (Fig. 5.3b) (Bringmann and Hyman 2005). They may also weakly contribute to the ingression of aster-mediated furrowing (Bringmann and Hyman 2005). Simultaneous inhibition of both midzone and aster signaling is expected to completely inhibit furrow ingression. This prediction has allowed the genetic identification of factors involved in aster-mediated furrowing. Those include proteins required for spindle elongation, such as the Gα proteins, GOA-1 and GPA-16, as well as GPR-1/2. While loss of GOA-1/GPA-16 or GPR-1/2 does not affect furrow ingression on its own, it completely blocks ingression when either SPD-1 or centralspindlin is missing (Dechant and Glotzer 2003; Bringmann et al. 2007). Other factors identified and regulating furrowing redundantly with centralspindlin are the myosin regulators NOP-1 and ANI-1 (Tse et al. 2011, 2012).

Furrow ingression requires myosin and its activators, the RHO-1 GTPase and the RHO GEF ECT-2 (Guo and Kemphues 1996; Jantsch-Plunger et al. 2000; Morita et al. 2005). As discussed above, despite its RHO GAP domain, CYK-4 has been recently shown to unconventionally activate RHO (Fig. 5.3b). This activation involves the interaction between the CYK-4 GAP domain and the ECT-2 GEF domain (Zhang and Glotzer 2015). CYK-4 interacts with membrane lipids through its C1 domain; this domain is required for both CYK-4 cortical localization and activity during furrow ingression (Lekomtsev et al. 2012; Zhang and Glotzer 2015). Centralspindlin cortical localization is also regulated by PAR-5: *par-5* embryos display increased CYK-4 and ZEN-4 cortical localization and RHO-1 activity, suggesting that PAR-5 restricts RHO-1 activation by inhibiting centralspindlin cortical localization (Basant et al. 2015). This function of PAR-5 is independent of its role in polarity and is likely to involve the direct binding of PAR-5 to phosphorylated ZEN-4 (Basant et al. 2015). Furthermore, genetic data suggest that the kinase AIR-2/Aurora B promotes centralspindlin cortical localization in the equatorial region through the inhibition of PAR-5 dependent exclusion (Fig. 5.3b) (Basant et al. 2015).

Aster microtubules are likely to regulate furrowing by preventing myosin accumulation and contraction in the vicinity of the asters. In wild-type embryos, prior to cytokinesis, myosin foci are enriched at the anterior and the equatorial cortex. In embryos devoid of organized microtubule arrays or lacking GPR-1/2, myosin accumulates throughout the cortex (Werner et al. 2007; Tse et al. 2011). Furthermore, *par-5(RNAi)* embryos, which have increased RHO-1 activity, show some cortical hypercontractility which is restricted to the equator during anaphase; on the contrary, *par-5(RNAi)* embryos in which the spindle has been forced to stay close to the posterior cortex show widespread contractility, except close to the spindle asters (Basant et al. 2015). Altogether, these data strongly suggest that aster microtubules contribute to pattern myosin localization and may locally inhibit myosin recruitment (Fig. 5.3b). However, the molecular mechanisms involved in this regulation are not yet clarified.

Finally, PAR proteins are also required for robust cytokinesis when activity of an F-actin nucleator, the formin CYK-1, is weakened (Jordan et al. 2016) (Fig. 5.3b). In the absence of either PAR-6 or PAR-2, the anillin ANI-1 and the septin UNC-59 levels at the contractile ring increase while F-actin levels decrease. Depletion of ANI-1 or UNC-59 leads to increased F-actin levels at the contractile ring and suppresses the cytokinesis defects observed in *par-2/6; cyk-1* mutant embryos. Altogether, these data suggest PAR-6 or PAR-2 limit ANI-1 and UNC-59 accumulation in order to maintain high F-actin levels at the contractile ring and allow robust cytokinesis (Jordan et al. 2016).

5.3.3 Regulating Myosin: An Additional Mechanism Required for Precise Furrow Positioning

Although cues originating from the mitotic spindle are crucial for furrow positioning, recent work showed that a myosin-dependent but spindle independent mechanism also controls furrow position during some asymmetric divisions. This mechanism has been described both in *drosophila* and *C. elegans* neuroblasts as well as in *drosophila* germ cells (Cabernard et al. 2010; Ou et al. 2010; Cinalli and Lehmann 2013). In *drosophila* neuroblasts, the mitotic spindle grows asymmetrically during anaphase such that the central spindle is closer to the basal side of the cell. This asymmetric elongation of the spindle is preceded by the asymmetric accumulation of myosin on the basal pole. Genetic experiments were used to uncouple spindle and myosin positions and elegantly showed that both spindle and myosin can induce furrowing (Cabernard et al. 2010).

In neuroblasts, myosin and the mitotic spindle are localized on the same side of the mother cell and both pathways cooperate to robustly position the furrow (Cabernard et al. 2010; Roth et al. 2015). However, the ability of asymmetrically localized myosin to induce furrowing independently of the mitotic spindle raises the question of how cytokinesis is regulated in cells where myosin and spindle localizations are not coincident and may thus lead to furrow induction at different places in the cell. This is of particular significance in the one-cell *C. elegans* embryo as myosin localizes at the anterior cortex during polarization while the mitotic spindle is displaced toward the posterior half of the embryo. Intriguingly, embryos, such as *zyg-9* mutants, in which the spindle localizes very close to the posterior cortex and transverse to the long axis, form two furrows, one close to the spindle and another at the edge of the anterior myosin domain (Werner et al. 2007), suggesting that anterior myosin may indeed also be able to induce furrowing independently of the mitotic spindle in one-cell *C. elegans* embryos.

In wild-type embryos, the amount of myosin localizing at the anterior cortex increases shortly after anaphase onset and then decreases during furrow ingression. These changes in myosin cortical accumulation are tightly regulated by the kinases PAR-4 and PIG-1 and the anillin ANI-1. This regulation is crucial to ensure that central spindle and furrow positions coincide throughout division (Pacquelet et al. 2015). PAR-4 impinges on myosin via two pathways, one anillin-dependent pathway, which also responds to the cullin CUL-5, and an anillin-independent pathway, which involves the kinase PIG-1/MELK (Fig. 5.3b). Upon removal of both PIG-1 and ANI-1, myosin accumulation increases and lasts longer. This excessive accumulation of myosin shifts the cytokinetic furrow toward the anterior. Consequently, furrow and spindle midzone positions do not coincide. In the most extreme cases, this leads to DNA segregation defects (Pacquelet et al. 2015). Importantly, furrow mispositioning is also observed when myosin activity is upregulated by completely independent means, for instance, by depleting the Rho GAPs RGA-3/4 or by expressing a gain-of-function allele of *ect-2* (Pacquelet et al. 2015; Zhang and Glotzer 2015). Thus, increased myosin activity at the anterior cortex is sufficient to shift the cytokinetic furrow toward the anterior. This underscores the importance of regulating myosin to maintain the equilibrium between the two opposite effects of anterior myosin and posterior spindle on furrow positioning; this appears to be crucial to keep the positions of the spindle midzone and of the cytokinetic furrow coordinated throughout cytokinesis.

The exact mechanisms by which PAR-4, PIG-1, and ANI-1 regulate cortical myosin remain to be investigated. Nevertheless, it is intriguing that they appear to act as negative regulators of myosin during cytokinesis (Pacquelet et al. 2015), while PAR-4 and ANI-1 on the contrary positively regulate myosin contractility during polarity establishment (see also Sect. 5.2.1.3) (Morton et al. 1992; Chartier et al. 2011; Maddox et al. 2005). PAR-4 and ANI-1 thus appear to differentially regulate myosin at different steps of asymmetric cell division.

5.4 Concluding Remarks

The one-cell *C. elegans* embryo has proven to be a fruitful model to dissect the multiple steps that lead to asymmetric cell division. Beyond the seminal discovery of PAR proteins, the joint efforts of many research groups have highlighted the existence of numerous and redundant pathways that establish and maintain polarity. Similarly, localization of the force generators that position the mitotic spindle as well as cytokinetic furrow positioning is regulated at multiple levels. In addition, rescue mechanisms can correct initial polarization defects. Altogether, these complex regulatory networks provide considerable robustness to the different steps that lead to asymmetric cell division and ensure that asymmetric cell division can still occur in challenging contexts which may arise from genetic or environmental modifications.

Finally, I would like to emphasize the role of actomyosin which appears as an essential player of asymmetric cell division. While early work had underlined the critical role of myosin during polarity establishment, more recent work shows that myosin also impinges on furrow position during cytokinesis. Moreover, actin also appears to influence pulling forces exerted on the mitotic spindle by affecting cortical rigidity. Regulation of actomyosin localization and activity throughout polarization and mitosis is thus essential to ensure proper asymmetric division.

Acknowledgments The author would like to thank members of the Michaux and Pécréaux labs for discussions and Ken Kemphues for helpful comments on the manuscript. Work in the Michaux lab is supported by grants from the Ligue contre le Cancer (22/29/35/36/72), the Fondation Maladies Rares, the CNRS, and Université de Rennes 1.

References

Aceto D, Beers M, Kemphues KJ (2006) Interaction of PAR-6 with CDC-42 is required for maintenance but not establishment of PAR asymmetry in *C. elegans*. Dev Biol 299:386–397. doi:10.1016/j.ydbio.2006.08.002

Afshar K, Werner ME, Tse YC et al (2010) Regulation of cortical contractility and spindle positioning by the protein phosphatase 6 PPH-6 in one-cell stage *C. elegans* embryos. Dev Camb Engl 137:237–247. doi:10.1242/dev.042754

Albertson DG (1984) Formation of the first cleavage spindle in nematode embryos. Dev Biol 101:61–72

Amini R, Goupil E, Labella S et al (2014) *C. elegans* Anillin proteins regulate intercellular bridge stability and germline syncytial organization. J Cell Biol 206:129–143. doi:10.1083/jcb.201310117

Andrews R, Ahringer J (2007) Asymmetry of early endosome distribution in *C. elegans* embryos. PloS One 2:e493. doi:10.1371/journal.pone.0000493

Basant A, Lekomtsev S, Tse YC et al (2015) Aurora B kinase promotes cytokinesis by inducing centralspindlin oligomers that associate with the plasma membrane. Dev Cell 33:204–215. doi:10.1016/j.devcel.2015.03.015

Beatty A, Morton D, Kemphues K (2010) The *C. elegans* homolog of Drosophila Lethal giant larvae functions redundantly with PAR-2 to maintain polarity in the early embryo. Dev Camb Engl 137:3995–4004. doi:10.1242/dev.056028

Beatty A, Morton DG, Kemphues K (2013) PAR-2, LGL-1 and the CDC-42 GAP CHIN-1 act in distinct pathways to maintain polarity in the *C. elegans* embryo. Dev Camb Engl 140:2005–2014. doi:10.1242/dev.088310

Beers M, Kemphues K (2006) Depletion of the co-chaperone CDC-37 reveals two modes of PAR-6 cortical association in *C. elegans* embryos. Dev Camb Engl 133:3745–3754. doi:10.1242/dev.02544

Berends CWH, Muñoz J, Portegijs V et al (2013) F-actin asymmetry and the endoplasmic reticulum-associated TCC-1 protein contribute to stereotypic spindle movements in the *Caenorhabditis elegans* embryo. Mol Biol Cell 24:2201–2215. doi:10.1091/mbc.E13-02-0076

Bienkowska D, Cowan CR (2012) Centrosomes can initiate a polarity axis from any position within one-cell *C. elegans* embryos. Curr Biol 22:583–589. doi:10.1016/j.cub.2012.01.064

Boyd L, Guo S, Levitan D et al (1996) PAR-2 is asymmetrically distributed and promotes association of P granules and PAR-1 with the cortex in *C. elegans* embryos. Dev Camb Engl 122:3075–3084

Bringmann H, Hyman AA (2005) A cytokinesis furrow is positioned by two consecutive signals. Nature 436:731–734. doi:10.1038/nature03823

Bringmann H, Cowan CR, Kong J, Hyman AA (2007) LET-99, GOA-1/GPA-16, and GPR-1/2 are required for aster-positioned cytokinesis. Curr Biol 17:185–191. doi:10.1016/j.cub.2006.11.070

Cabernard C, Prehoda KE, Doe CQ (2010) A spindle-independent cleavage furrow positioning pathway. Nature 467:91–94. doi:10.1038/nature09334

Canman JC, Lewellyn L, Laband K et al (2008) Inhibition of Rac by the GAP activity of centralspindlin is essential for cytokinesis. Science 322:1543–1546. doi:10.1126/science.1163086

Chartier NT, Salazar Ospina DP, Benkemoun L et al (2011) PAR-4/LKB1 mobilizes nonmuscle myosin through anillin to regulate *C. elegans* embryonic polarization and cytokinesis. Curr Biol 21:259–269. doi:10.1016/j.cub.2011.01.010

Cinalli RM, Lehmann R (2013) A spindle-independent cleavage pathway controls germ cell formation in Drosophila. Nat Cell Biol 15:839–845. doi:10.1038/ncb2761

Colombo K, Grill SW, Kimple RJ et al (2003) Translation of polarity cues into asymmetric spindle positioning in *Caenorhabditis elegans* embryos. Science 300:1957–1961. doi:10.1126/science.1084146

Couwenbergs C, Labbé J-C, Goulding M et al (2007) Heterotrimeric G protein signaling functions with dynein to promote spindle positioning in *C. elegans*. J Cell Biol 179:15–22. doi:10.1083/jcb.200707085

Cowan CR, Hyman AA (2004) Centrosomes direct cell polarity independently of microtubule assembly in *C. elegans* embryos. Nature 431:92–96. doi:10.1038/nature02825

Cuenca AA, Schetter A, Aceto D et al (2003) Polarization of the *C. elegans* zygote proceeds via distinct establishment and maintenance phases. Dev Camb Engl 130:1255–1265

Dechant R, Glotzer M (2003) Centrosome separation and central spindle assembly act in redundant pathways that regulate microtubule density and trigger cleavage furrow formation. Dev Cell 4:333–344

Etemad-Moghadam B, Guo S, Kemphues KJ (1995) Asymmetrically distributed PAR-3 protein contributes to cell polarity and spindle alignment in early *C. elegans* embryos. Cell 83:743–752

Fievet BT, Rodriguez J, Naganathan S et al (2013) Systematic genetic interaction screens uncover cell polarity regulators and functional redundancy. Nat Cell Biol 15:103–112. doi:10.1038/ncb2639

Fortin SM, Marshall SL, Jaeger EC et al (2010) The PAM-1 aminopeptidase regulates centrosome positioning to ensure anterior-posterior axis specification in one-cell *C. elegans* embryos. Dev Biol 344:992–1000. doi:10.1016/j.ydbio.2010.06.016

Galli M, Muñoz J, Portegijs V et al (2011) aPKC phosphorylates NuMA-related LIN-5 to position the mitotic spindle during asymmetric division. Nat Cell Biol 13:1132–1138. doi:10.1038/ncb2315

Goehring NW, Hoege C, Grill SW, Hyman AA (2011) PAR proteins diffuse freely across the anterior-posterior boundary in polarized *C. elegans* embryos. J Cell Biol 193:583–594. doi:10.1083/jcb.201011094

Goldstein B, Hird SN (1996) Specification of the anteroposterior axis in *Caenorhabditis elegans*. Dev Camb Engl 122:1467–1474

Gotta M, Ahringer J (2001) Distinct roles for Galpha and Gbetagamma in regulating spindle position and orientation in *Caenorhabditis elegans* embryos. Nat Cell Biol 3:297–300. doi:10.1038/35060092

Gotta M, Abraham MC, Ahringer J (2001) CDC-42 controls early cell polarity and spindle orientation in *C. elegans*. Curr Biol 11:482–488

Gotta M, Dong Y, Peterson YK et al (2003) Asymmetrically distributed *C. elegans* homologs of AGS3/PINS control spindle position in the early embryo. Curr Biol CB 13:1029–1037

Grill SW, Gönczy P, Stelzer EH, Hyman AA (2001) Polarity controls forces governing asymmetric spindle positioning in the *Caenorhabditis elegans* embryo. Nature 409:630–633. doi:10.1038/35054572

Grill SW, Howard J, Schäffer E et al (2003) The distribution of active force generators controls mitotic spindle position. Science 301:518–521. doi:10.1126/science.1086560

Guo S, Kemphues KJ (1995) par-1, a gene required for establishing polarity in *C. elegans* embryos, encodes a putative Ser/Thr kinase that is asymmetrically distributed. Cell 81:611–620

Guo S, Kemphues KJ (1996) A non-muscle myosin required for embryonic polarity in *Caenorhabditis elegans*. Nature 382:455–458. doi:10.1038/382455a0

Hao Y, Boyd L, Seydoux G (2006) Stabilization of cell polarity by the *C. elegans* RING protein PAR-2. Dev Cell 10:199–208. doi:10.1016/j.devcel.2005.12.015

Hird SN, White JG (1993) Cortical and cytoplasmic flow polarity in early embryonic cells of *Caenorhabditis elegans*. J Cell Biol 121:1343–1355

Hoege C, Constantinescu A-T, Schwager A et al (2010) LGL can partition the cortex of one-cell *Caenorhabditis elegans* embryos into two domains. Curr Biol 20:1296–1303. doi:10.1016/j.cub.2010.05.061

Hung TJ, Kemphues KJ (1999) PAR-6 is a conserved PDZ domain-containing protein that colocalizes with PAR-3 in *Caenorhabditis elegans* embryos. Dev Camb Engl 126:127–135

Hyenne V, Tremblay-Boudreault T, Velmurugan R et al (2012) RAB-5 controls the cortical organization and dynamics of PAR proteins to maintain *C. elegans* early embryonic polarity. PloS One 7:e35286. doi:10.1371/journal.pone.0035286

Jantsch-Plunger V, Gönczy P, Romano A et al (2000) CYK-4: a Rho family gtpase activating protein (GAP) required for central spindle formation and cytokinesis. J Cell Biol 149:1391–1404

Jenkins N, Saam JR, Mango SE (2006) CYK-4/GAP provides a localized cue to initiate anteroposterior polarity upon fertilization. Science 313:1298–1301. doi:10.1126/science.1130291

Jordan SN, Davies T, Zhuravlev Y et al (2016) Cortical PAR polarity proteins promote robust cytokinesis during asymmetric cell division. J Cell Biol 212:39–49. doi:10.1083/jcb.201510063

Kemphues KJ, Priess JR, Morton DG, Cheng NS (1988) Identification of genes required for cytoplasmic localization in early *C. elegans* embryos. Cell 52:311–320

Kozlowski C, Srayko M, Nedelec F (2007) Cortical microtubule contacts position the spindle in *C. elegans* embryos. Cell 129:499–510. doi:10.1016/j.cell.2007.03.027

Kraemer B, Crittenden S, Gallegos M et al (1999) NANOS-3 and FBF proteins physically interact to control the sperm-oocyte switch in *Caenorhabditis elegans*. Curr Biol 9:1009–1018

Krueger LE, Wu J-C, Tsou M-FB, Rose LS (2010) LET-99 inhibits lateral posterior pulling forces during asymmetric spindle elongation in *C. elegans* embryos. J Cell Biol 189:481–495. doi:10.1083/jcb.201001115

Kumfer KT, Cook SJ, Squirrell JM et al (2010) CGEF-1 and CHIN-1 regulate CDC-42 activity during asymmetric division in the *Caenorhabditis elegans* embryo. Mol Biol Cell 21:266–277. doi:10.1091/mbc.E09-01-0060

Labbé J-C, Pacquelet A, Marty T, Gotta M (2006) A genomewide screen for suppressors of par-2 uncovers potential regulators of PAR protein-dependent cell polarity in *Caenorhabditis elegans*. Genetics 174:285–295. doi:10.1534/genetics.106.060517

Lekomtsev S, Su K-C, Pye VE et al (2012) Centralspindlin links the mitotic spindle to the plasma membrane during cytokinesis. Nature 492:276–279. doi:10.1038/nature11773

Levitan DJ, Boyd L, Mello CC et al (1994) par-2, a gene required for blastomere asymmetry in *Caenorhabditis elegans*, encodes zinc-finger and ATP-binding motifs. Proc Natl Acad Sci USA 91:6108–6112

Lewellyn L, Dumont J, Desai A, Oegema K (2010) Analyzing the effects of delaying aster separation on furrow formation during cytokinesis in the *Caenorhabditis elegans* embryo. Mol Biol Cell 21:50–62. doi:10.1091/mbc.E09-01-0089

Li B, Kim H, Beers M, Kemphues K (2010a) Different domains of *C. elegans* PAR-3 are required at different times in development. Dev Biol 344:745–757. doi:10.1016/j.ydbio.2010.05.506

Li J, Kim H, Aceto DG et al (2010b) Binding to PKC-3, but not to PAR-3 or to a conventional PDZ domain ligand, is required for PAR-6 function in *C. elegans*. Dev Biol 340:88–98. doi:10.1016/j.ydbio.2010.01.023

Lorson MA, Horvitz HR, van den Heuvel S (2000) LIN-5 is a novel component of the spindle apparatus required for chromosome segregation and cleavage plane specification in *Caenorhabditis elegans*. J Cell Biol 148:73–86

Lyczak R, Zweier L, Group T et al (2006) The puromycin-sensitive aminopeptidase PAM-1 is required for meiotic exit and anteroposterior polarity in the one-cell *Caenorhabditis elegans* embryo. Dev Camb Engl 133:4281–4292. doi:10.1242/dev.02615

Maddox AS, Habermann B, Desai A, Oegema K (2005) Distinct roles for two *C. elegans* anillins in the gonad and early embryo. Dev Camb Engl 132:2837–2848. doi:10.1242/dev.01828

Mayer M, Depken M, Bois JS et al (2010) Anisotropies in cortical tension reveal the physical basis of polarizing cortical flows. Nature 467:617–621. doi:10.1038/nature09376

McCloskey RJ, Kemphues KJ (2012) Deubiquitylation machinery is required for embryonic polarity in *Caenorhabditis elegans*. PLoS Genet 8:e1003092. doi:10.1371/journal.pgen.1003092

Mikl M, Cowan CR (2014) Alternative 3′ UTR selection controls PAR-5 homeostasis and cell polarity in *C. elegans* embryos. Cell Rep 8:1380–1390. doi:10.1016/j.celrep.2014.08.004

Mishima M, Kaitna S, Glotzer M (2002) Central spindle assembly and cytokinesis require a kinesin-like protein/RhoGAP complex with microtubule bundling activity. Dev Cell 2:41–54

Morita K, Hirono K, Han M (2005) The *Caenorhabditis elegans* ect-2 RhoGEF gene regulates cytokinesis and migration of epidermal P cells. EMBO Rep 6:1163–1168. doi:10.1038/sj.embor.7400533

Morton DG, Roos JM, Kemphues KJ (1992) par-4, a gene required for cytoplasmic localization and determination of specific cell types in *Caenorhabditis elegans* embryogenesis. Genetics 130:771–790

Morton DG, Shakes DC, Nugent S et al (2002) The *Caenorhabditis elegans* par-5 gene encodes a 14-3-3 protein required for cellular asymmetry in the early embryo. Dev Biol 241:47–58. doi:10.1006/dbio.2001.0489

Motegi F, Sugimoto A (2006) Sequential functioning of the ECT-2 RhoGEF, RHO-1 and CDC-42 establishes cell polarity in *Caenorhabditis elegans* embryos. Nat Cell Biol 8:978–985. doi:10.1038/ncb1459

Motegi F, Zonies S, Hao Y et al (2011) Microtubules induce self-organization of polarized PAR domains in *Caenorhabditis elegans* zygotes. Nat Cell Biol 13:1361–1367. doi:10.1038/ncb2354

Munro E, Nance J, Priess JR (2004) Cortical flows powered by asymmetrical contraction transport PAR proteins to establish and maintain anterior-posterior polarity in the early *C. elegans* embryo. Dev Cell 7:413–424. doi:10.1016/j.devcel.2004.08.001

Nakayama Y, Shivas JM, Poole DS et al (2009) Dynamin participates in the maintenance of anterior polarity in the *Caenorhabditis elegans* embryo. Dev Cell 16:889–900. doi:10.1016/j.devcel.2009.04.009

Nguyen-Ngoc T, Afshar K, Gönczy P (2007) Coupling of cortical dynein and G alpha proteins mediates spindle positioning in *Caenorhabditis elegans*. Nat Cell Biol 9:1294–1302. doi:10.1038/ncb1649

Nishimura Y, Yonemura S (2006) Centralspindlin regulates ECT2 and RhoA accumulation at the equatorial cortex during cytokinesis. J Cell Sci 119:104–114. doi:10.1242/jcs.02737

Niwayama R, Shinohara K, Kimura A (2011) Hydrodynamic property of the cytoplasm is sufficient to mediate cytoplasmic streaming in the *Caenorhabditis elegans* embryo. Proc Natl Acad Sci USA 108:11900–11905. doi:10.1073/pnas.1101853108

O'Connell KF, Maxwell KN, White JG (2000) The spd-2 gene is required for polarization of the anteroposterior axis and formation of the sperm asters in the *Caenorhabditis elegans* zygote. Dev Biol 222:55–70. doi:10.1006/dbio.2000.9714

Ou G, Stuurman N, D'Ambrosio M, Vale RD (2010) Polarized myosin produces unequal-size daughters during asymmetric cell division. Science 330:677–680. doi:10.1126/science.1196112

Pacquelet A, Zanin E, Ashiono C, Gotta M (2008) PAR-6 levels are regulated by NOS-3 in a CUL-2 dependent manner in Caenorhabditiselegans. Dev Biol 319:267–272. doi:10.1016/j.ydbio.2008.04.016

Pacquelet A, Uhart P, Tassan J-P, Michaux G (2015) PAR-4 and anillin regulate myosin to coordinate spindle and furrow position during asymmetric division. J Cell Biol 210:1085–1099. doi:10.1083/jcb.201503006

Panbianco C, Weinkove D, Zanin E et al (2008) A casein kinase 1 and PAR proteins regulate asymmetry of a PIP(2) synthesis enzyme for asymmetric spindle positioning. Dev Cell 15:198–208. doi:10.1016/j.devcel.2008.06.002

Park DH, Rose LS (2008) Dynamic localization of LIN-5 and GPR-1/2 to cortical force generation domains during spindle positioning. Dev Biol 315:42–54. doi:10.1016/j.ydbio.2007.11.037

Pecreaux J, Röper J-C, Kruse K et al (2006) Spindle oscillations during asymmetric cell division require a threshold number of active cortical force generators. Curr Biol CB 16:2111–2122. doi:10.1016/j.cub.2006.09.030

Rabilotta A, Desrosiers M, Labbé J-C (2015) CDK-1 and two B-type cyclins promote PAR-6 stabilization during polarization of the early *C. elegans* embryo. PloS One 10:e0117656. doi:10.1371/journal.pone.0117656

Raich WB, Moran AN, Rothman JH, Hardin J (1998) Cytokinesis and midzone microtubule organization in *Caenorhabditis elegans* require the kinesin-like protein ZEN-4. Mol Biol Cell 9:2037–2049

Rappaport R (1981) Cytokinesis:cleavage furrow establishment in cylindrical sand dollar eggs. J Exp Zool 217:365–375

Rappaport R (1985) Repeated furrow formation from a single mitotic apparatus in cylindrical sand dollar eggs. J Exp Zool 234:167–171. doi:10.1002/jez.1402340120

Rose L, Gönczy P (2014) Polarity establishment, asymmetric division and segregation of fate determinants in early *C. elegans* embryos. WormBook 30:1–43. doi:10.1895/wormbook.1.30.2

Rose LS, Lamb ML, Hird SN, Kemphues KJ (1995) Pseudocleavage is dispensable for polarity and development in *C. elegans* embryos. Dev Biol 168:479–489. doi:10.1006/dbio.1995.1096

Roth M, Roubinet C, Iffländer N et al (2015) Asymmetrically dividing Drosophila neuroblasts utilize two spatially and temporally independent cytokinesis pathways. Nat Commun 6:6551. doi:10.1038/ncomms7551

Roubinet C, Cabernard C (2014) Control of asymmetric cell division. Curr Opin Cell Biol 31:84–91. doi:10.1016/j.ceb.2014.09.005

Sadler PL, Shakes DC (2000) Anucleate *Caenorhabditis elegans* sperm can crawl, fertilize oocytes and direct anterior-posterior polarization of the 1-cell embryo. Dev Camb Engl 127:355–366

Sailer A, Anneken A, Li Y et al (2015) Dynamic opposition of clustered proteins stabilizes cortical polarity in the *C. elegans* zygote. Dev Cell 35:131–142. doi:10.1016/j.devcel.2015.09.006

Schenk C, Bringmann H, Hyman AA, Cowan CR (2010) Cortical domain correction repositions the polarity boundary to match the cytokinesis furrow in *C. elegans* embryos. Development 137:1743–1753. doi:10.1242/dev.040436

Schmutz C, Stevens J, Spang A (2007) Functions of the novel RhoGAP proteins RGA-3 and RGA-4 in the germ line and in the early embryo of *C. elegans*. Dev Camb Engl 134:3495–3505. doi:10.1242/dev.000802

Schonegg S, Hyman AA (2006) CDC-42 and RHO-1 coordinate acto-myosin contractility and PAR protein localization during polarity establishment in *C. elegans* embryos. Dev Camb Engl 133:3507–3516. doi:10.1242/dev.02527

Schonegg S, Constantinescu AT, Hoege C, Hyman AA (2007) The Rho GTPase-activating proteins RGA-3 and RGA-4 are required to set the initial size of PAR domains in *Caenorhabditis elegans* one-cell embryos. Proc Natl Acad Sci USA 104:14976–14981. doi:10.1073/pnas.0706941104

Shelton CA, Carter JC, Ellis GC, Bowerman B (1999) The nonmuscle myosin regulatory light chain gene mlc-4 is required for cytokinesis, anterior-posterior polarity, and body morphology during *Caenorhabditis elegans* embryogenesis. J Cell Biol 146:439–451

Spiga FM, Prouteau M, Gotta M (2013) The TAO kinase KIN-18 regulates contractility and establishment of polarity in the *C. elegans* embryo. Dev Biol 373:26–38. doi:10.1016/j.ydbio.2012.10.001

Starostina NG, Lim J, Schvarzstein M et al (2007) A CUL-2 ubiquitin ligase containing three FEM proteins degrades TRA-1 to regulate *C. elegans* sex determination. Dev Cell 13:127–139. doi:10.1016/j.devcel.2007.05.008

Tabuse Y, Izumi Y, Piano F et al (1998) Atypical protein kinase C cooperates with PAR-3 to establish embryonic polarity in *Caenorhabditis elegans*. Dev Camb Engl 125:3607–3614

Tsai M-C, Ahringer J (2007) Microtubules are involved in anterior-posterior axis formation in *C. elegans* embryos. J Cell Biol 179:397–402. doi:10.1083/jcb.200708101

Tse YC, Piekny A, Glotzer M (2011) Anillin promotes astral microtubule-directed cortical myosin polarization. Mol Biol Cell 22:3165–3175. doi:10.1091/mbc.E11-05-0399

Tse YC, Werner M, Longhini KM et al (2012) RhoA activation during polarization and cytokinesis of the early *Caenorhabditis elegans* embryo is differentially dependent on NOP-1 and CYK-4. Mol Biol Cell 23:4020–4031. doi:10.1091/mbc.E12-04-0268

Tsou M-FB, Hayashi A, DeBella LR et al (2002) LET-99 determines spindle position and is asymmetrically enriched in response to PAR polarity cues in *C. elegans* embryos. Dev Camb Engl 129:4469–4481

Tsou M-FB, Hayashi A, Rose LS (2003) LET-99 opposes Galpha/GPR signaling to generate asymmetry for spindle positioning in response to PAR and MES-1/SRC-1 signaling. Dev Camb Engl 130:5717–5730. doi:10.1242/dev.00790

Verbrugghe KJC, White JG (2004) SPD-1 is required for the formation of the spindle midzone but is not essential for the completion of cytokinesis in *C. elegans* embryos. Curr Biol CB 14:1755–1760. doi:10.1016/j.cub.2004.09.055

Watts JL, Etemad-Moghadam B, Guo S et al (1996) par-6, a gene involved in the establishment of asymmetry in early *C. elegans* embryos, mediates the asymmetric localization of PAR-3. Dev Camb Engl 122:3133–3140

Watts JL, Morton DG, Bestman J, Kemphues KJ (2000) The *C. elegans* par-4 gene encodes a putative serine-threonine kinase required for establishing embryonic asymmetry. Dev Camb Engl 127:1467–1475

Werner M, Munro E, Glotzer M (2007) Astral signals spatially bias cortical myosin recruitment to break symmetry and promote cytokinesis. Curr Biol 17:1286–1297. doi:10.1016/j.cub.2007.06.070

Wu J-C, Rose LS (2007) PAR-3 and PAR-1 inhibit LET-99 localization to generate a cortical band important for spindle positioning in *Caenorhabditis elegans* embryos. Mol Biol Cell 18:4470–4482. doi:10.1091/mbc.E07-02-0105

Wu J-C, Espiritu EB, Rose LS (2016) The 14-3-3 protein PAR-5 regulates the asymmetric localization of the LET-99 spindle positioning protein. Dev Biol 412:288–297. doi:10.1016/j.ydbio.2016.02.020

Yadlapalli S, Yamashita YM (2013) DNA asymmetry in stem cells - immortal or mortal? J Cell Sci 126:4069–4076. doi:10.1242/jcs.096024

Zhang D, Glotzer M (2015) The RhoGAP activity of CYK-4/MgcRacGAP functions non-canonically by promoting RhoA activation during cytokinesis. eLife. doi:10.7554/eLife.08898

Zonies S, Motegi F, Hao Y, Seydoux G (2010) Symmetry breaking and polarization of the *C. elegans* zygote by the polarity protein PAR-2. Dev Camb Engl 137:1669–1677. doi:10.1242/dev.045823

Chapter 6
Size Matters: How *C. elegans* Asymmetric Divisions Regulate Apoptosis

Jerome Teuliere and Gian Garriga

Abstract Apoptosis is a form of programmed cell death used by metazoans to eliminate abnormal cells, control cell number, and shape the development of organs. The use of the nematode *Caenorhabditis elegans* as a model for the study of apoptosis has led to important insights into how cells die and how their corpses are removed. Eighty percent of these apoptotic cell deaths occur during nervous system development and in daughters of neuroblasts that divide asymmetrically. Pioneering work defined a conserved apoptosis pathway that is initiated in *C. elegans* by the BH3-only protein EGL-1 and that leads to the activation of the caspase CED-3. While the execution of the apoptotic fate is well understood, much less is known about the mechanisms that specify the apoptotic fate of particular cells. In some cells fated to die, this regulation occurs at the level of the *egl-1* gene transcription, and investigators have identified several lineage-specific transcription factors that both positively and negatively regulate *egl-1*. In this review, we focus on a second set of molecules that appear to influence apoptosis by controlling the position of the cleavage plane in divisions that produce apoptotic cells.

6.1 Daughter Cell Size Asymmetry in Asymmetric Divisions

A striking feature of apoptosis during *C. elegans* development is its association with Asymmetric Cell Divisions (ACDs) that produce daughter cells of different sizes. These divisions produce a larger daughter cell that survives and a smaller daughter cell that dies (Cordes et al. 2006; Ou et al. 2010; Hatzold and Conradt 2008). In mutants that fail to establish this Daughter Cell Size Asymmetry (DCSA), the cell normally fated to die is larger than normal, survives, and sometimes adopts the fate of its sister, leading to the production of extra cells.

J. Teuliere • G. Garriga (✉)
Department of Molecular and Cell Biology, Helen Wills Neuroscience Institute, University of California, Berkeley, CA, USA
e-mail: garriga@berkeley.edu

J.-P. Tassan, J.Z. Kubiak (eds.), *Asymmetric Cell Division in Development, Differentiation and Cancer*, Results and Problems in Cell Differentiation 61,
DOI 10.1007/978-3-319-53150-2_6

DCSA also occurs during divisions that do not generate an apoptotic cell. For instance, size asymmetry is generated during the division of the *C. elegans* zygote to produce the larger anterior AB blastomere and the smaller posterior P1 blastomere (reviewed by Galli and van den Heuvel 2008). Another example is the *Drosophila* embryonic neuroblast, which divides in a stem cell pattern to generate a larger neuroblast and a smaller cell Ganglion Mother Cell (GMC) with more limited developmental potential (reviewed by Wodarz 2005; Egger et al. 2008). Cell size alone might facilitate apoptosis but not cause it, or small cell size may be proapoptotic below a size threshold that is not reached in the *C. elegans* P1 blastomere or in the *Drosophila* GMC. Experiments performed on HeLa cells provide support for the size threshold hypothesis (Kiyomitsu and Cheeseman 2013). Hela cells divide symmetrically and are highly resistant to proapoptotic signals, but inhibiting mitotic spindle centration can induce asymmetric furrow positioning, leading to DCSA. When the smaller daughter of a dividing HeLa cell is below a critical size, its cell cycle is delayed. Occasionally, the small cell displays an apoptosis-like cell death phenotype. A better understanding of the cellular mechanisms that control DCSA is needed to test these hypotheses.

6.2 Mechanisms Regulating Q Neuroblasts Daughter Cell Size Asymmetry

Recent progress in understanding the cellular basis of DCSA has been made by studying the Q neuroblast lineage, which produces three neurons and two apoptotic cells (Sulston and Horvitz 1977) (Fig. 6.1a). The left and right Q neuroblasts are each generated by the division of a lateral epidermal cell shortly before hatching. During the first larval stage (L1), each Q neuroblast delaminates from the epithelium and divides. The anterior (Q.a) and posterior (Q.p) Q daughter cells both give rise to cells that die. The Q.a cell divides first and produces a smaller anterior cell fated to die (Q.aa) and a larger posterior cell (Q.ap) that will differentiate into the AQR (right lineage) or PQR (left lineage) oxygen-sensing neuron. The Q.p cell divides with a reversed polarity, generating a smaller posterior apoptotic cell (Q.pp) and a larger anterior precursor (Q.pa), which will divide to produce the AVM (right lineage) or PVM (left lineage) mechanosensory neuron and the SDQ interneuron.

Unexpectedly, analysis of mitotic spindle position revealed that the Q.a and Q.p cells use different strategies to establish DCSA and specify the apoptotic fate (Ou et al. 2010). The mitotic spindle of Q.p shifts posteriorly, allowing the contractile ring of the cytokinetic furrow to form in a more posterior position (Fig. 6.1b). The *C. elegans* zygote also produces unequally sized daughter cells by asymmetrically positioning the mitotic spindle (Fig. 6.2). Interactions between aster microtubules and the cell cortex mediate both the orientation of the spindle and its movement because pulling forces, driven by dynein, are much stronger at the posterior pole (reviewed by Galli and van den Heuvel 2008). It is unknown whether Q.p and the zygote employ similar strategies to achieve their posterior shifts in spindle position. The Q.a spindle, by contrast, is located

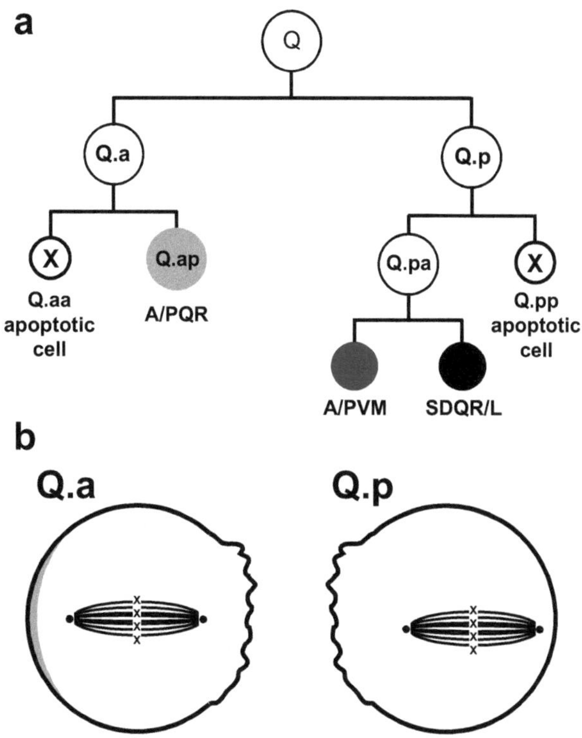

Fig. 6.1 *The C. elegans Q neuroblast lineage*. (**a**) Diagram of the Q lineage divisions that produce the A/PQR, A/PVM, and SDQ neurons and two apoptotic cells. Xs indicate cells that die. (**b**) Cartoon showing the features of the Q.a and Q.p divisions that lead to DCSA. Q.a localizes myosin to its anterior cortex (*gray crescent* on the *left* of the cell) and extends its membrane on the posterior side (wavy membrane on the *right*). Q.p displaces its mitotic spindle posteriorly and extends its membrane on the anterior side (Ou et al. 2010; J.T., unpublished observations)

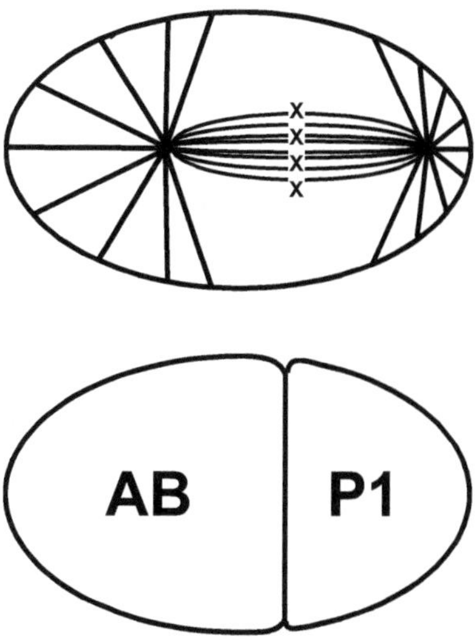

Fig. 6.2 *The asymmetric division of the C. elegans zygote*. The diagram of the zygote at the *top* shows that the position of the spindle at metaphase is shifted toward the posterior (*right*). The two dots at the ends of the spindle represent centrosomes that also nucleate aster microtubules that attach to the cortex. Xs represent chromosomes. The shifted spindle results in the larger AB and smaller P1 blastomeres shown at the *bottom*

centrally. Asymmetry is instead regulated by a novel mechanism that involves the asymmetric accumulation of the non-muscle myosin II at the anterior cortex (Fig. 6.1b), visualized by GFP tagging the myosin heavy chain NMY-2. Chromophore-Assisted Laser Inactivation (CALI) performed to inactivate NMY-2::GFP at the anterior side of Q.a resulted in a more symmetric division. In support of a role for Q.a DCSA in facilitating apoptosis, the anterior daughter Q.aa, which normally dies, often survived and differentiated like its sister Q.ap. CALI performed on myosin at the posterior cortex of Q.p had no effect on Q.p DCSA, suggesting that myosin II polarity is not required for Q.p ACD (Ou et al. 2010).

This novel Q.a DCSA mechanism appears to be evolutionarily conserved as myosin polarity is also involved in generating the smaller basal ganglion mother cells (GMCs) during Drosophila neuroblast stem cell divisions (Fig. 6.3a). Myosin II accumulates asymmetrically at the basal cortex together with the kinesin Pavarotti and anillin, which are components of the centralspindlin protein complex and cytokinetic contractile ring, respectively. Both molecules accumulate on the side that will produce the smaller GMC (Fig. 6.3b) (Cabernard et al. 2010; Connell et al. 2011). Myosin basal polarity in Drosophila neuroblasts may produce an extended contractile ring structure, forming a basal contractile dome. It is unknown if the *C. elegans* anillin homolog ANI-1 and the

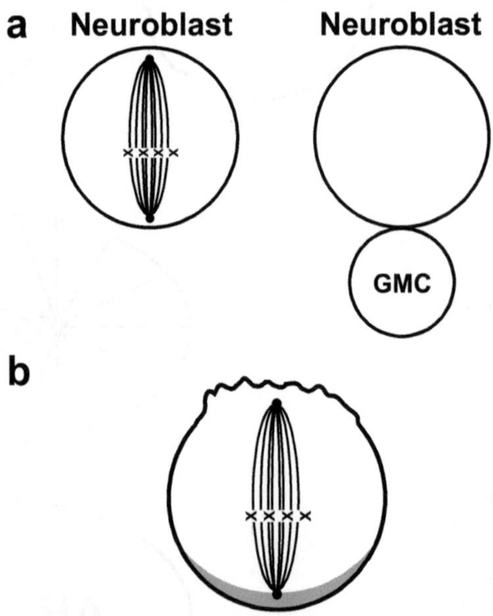

Fig. 6.3 *The asymmetric division of the Drosophila embryonic neuroblast.* (**a**) The neuroblast divides asymmetrically to generate a larger apical neuroblast and a smaller basal Ganglion Mother Cell (GMC). The neuroblast continues to divide in a stem cell pattern to generate additional GMCs. The embryonic GMCs divide once to generate two neurons or glia. (**b**) Neuroblast DCSA is thought to be generated by an asymmetric spindle, the basal localization of anillin, myosin, and the kinesin Pavarotti (crescent at the *bottom* of the neuroblast), and the extension of the apical membrane (*wavy line* at the *top*) (Kaltschmidt et al. 2000; Cabernard et al. 2010; Connell et al. 2011)

Pavarotti homolog ZEN-4 also accumulate at the anterior pole of the Q.a neuroblast with myosin II. Moreover, myosin basal accumulation in *Drosophila* neuroblasts resulted from an apical depletion of anillin, kinesin, and myosin, followed by their partial depletion from the basal tip and accumulation in lateral positions on the basal side of the cell (Cabernard et al. 2010). Unlike the Q.a division, the basal centrosome of *Drosophila* neuroblasts moves toward the basal surface, resulting in an asymmetric spindle (Fig. 6.3a) (Kaltschmidt et al. 2000). It appears that Q.a solely relies on asymmetric myosin to generate DCSA, but both an asymmetric spindle and myosin seem to contribute to DCSA in the Drosophila neuroblast (Cabernard et al. 2010; Connell et al. 2011). A recent report described an asymmetric spindle in the division of murine cortical precursors, but whether asymmetric myosin also contributes to DCSA in these divisions was not addressed (Delaunay et al. 2014).

In *C. elegans* Q.a and Q.p and in *Drosophila* neuroblasts, DCSA is coupled to plasma membrane extension in addition to spindle and cortical polarities, exacerbating the size differences between the daughter cells (Figs. 6.1b and 6.3b) (Ou et al. 2010; Connell et al. 2011). The molecular regulators of membrane extension are not known. In Q.a, as in the *Drosophila* neuroblast, the membrane extends posteriorly and apically, respectively, possibly as a consequence of myosin contraction at the opposite pole. It has been proposed that the directional extension of the Drosophila neuroblast membrane is merely the normal extension that occurs at both poles during anaphase of symmetric divisions, with one of the poles not extending because of cortical myosin contraction (Connell et al. 2011). Whereas this model could explain the posterior membrane extension in the Q.a division, it does not explain the anterior membrane extension observed during the Q.p division.

6.3 The PIG-1/MELK Pathway

6.3.1 The Role of PIG-1 in the Q Neuroblast Lineage

Despite the differences in the mechanisms used by Q.a and Q.p to establish DCSA, the same kinase pathway functions in these cells to regulate both divisions (Chien et al. 2013). Q.a, Q.p, and several other divisions that produce apoptotic cells require the *par-1-like gene-1* (*pig-1*), which encodes the homolog of the vertebrate Maternal Embryonic Leucine zipper Kinase (MELK), a member of the AMP-activated Protein Kinase (AMPK) family that includes the PIG-1 paralog PAR-1 (Cordes et al. 2006; Ou et al. 2010; Hirose and Horvitz 2013). The loss of *pig-1* function results in the loss of myosin asymmetry in Q.a (Ou et al. 2010). Although how PIG-1 regulates Q.p DCSA was not addressed in this study, it seems likely that the regulation of spindle asymmetry is one of its roles. These observations suggest that the distinct mechanisms that impose the asymmetry of the Q.a and Q.p divisions can share common regulators.

The gene *par-4* acts genetically in the same pathway as *pig-1* to establish Q.p DCSA and to regulate cell death by cell extrusion in a caspase-independent cell death pathway (Chien et al. 2012; Denning et al. 2012). The tumor-suppressor Liver Kinase B1 (Lkb1) is the mammalian homolog of PAR-4. Mutations in the *Lkb1* gene are detected in sporadic cancers and can result in Peutz–Jeghers syndrome, an autosomal dominant disorder characterized by benign gastrointestinal tract polyps. PAR-4/Lkb1 and its homologs are master regulators of cell polarity, controlling the polarity of oocytes, asymmetrically dividing cells, epithelia, and neurons (reviewed in Partanen et al. 2013). Mammalian Lkb1 directly activates AMPK family members by phosphorylation of a threonine that is conserved in PIG-1. Although PIG-1 has not been shown to be a direct target of PAR-4, this conserved threonine is essential for PIG-1 activity in DCSA, suggesting that PAR-4 regulates PIG-1 activity in the Q.a and Q.p divisions (Chien et al. 2012) (Fig. 6.4).

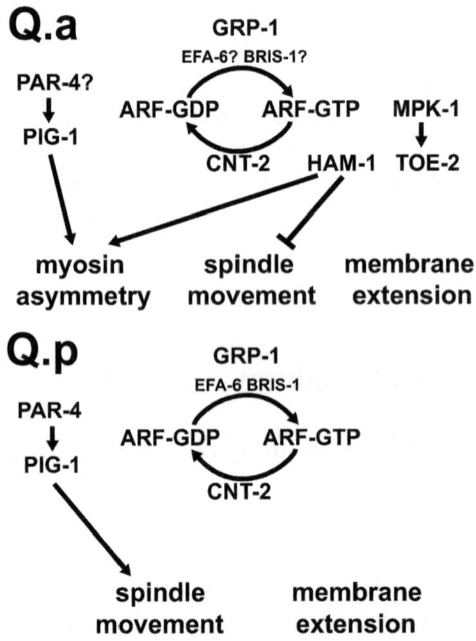

Fig. 6.4 *Regulators that mediate Q.a and Q.p DCSA*. All of the molecules listed are known to function in the Q.a or Q.p divisions, but the functions of PAR-4, EFA-6, and BRIS-1 have not been tested in Q.a. Myosin asymmetry, posterior spindle movement, or membrane extension promote DCSA, and molecules that promote or inhibit these processes are indicated by *arrows* or Ts, respectively. The lack of connections between molecules and the processes that drive DCSA indicates that these processes have not been analyzed in conditions where the function of these molecules is impaired. The lack of a connection between any of the molecules and membrane extension indicates that this process was not analyzed in any of the mutants

6.3.2 The Role of pig-1 in Zygote ACD and Other Systems

PIG-1 also functions in asymmetric divisions that do not produce dying cells. The first indication of a more general role came from RNAi screens in a sensitized genetic background. Although *pig-1* knockdown had little to no effect on embryonic viability, it enhanced the embryonic lethality of a *par-1* temperature-sensitive mutant grown at a permissive temperature, suggesting that the two AMPK-related kinases have overlapping functions (Morton et al. 2012). *pig-1(RNAi)* also significantly enhanced the lethality of a temperature-sensitive *par-2(ts)* mutant raised at a permissive temperature. PAR-1 and the ring-finger protein PAR-2 localize to the posterior of the zygote and exclude the PAR-3/PAR-6/PKC-3 polarity complex, which localizes to the anterior cortex (reviewed by Motegi and Seydoux 2013). Complete loss of *par-1* or *par-3* function, for example, allows PAR-3 protein to extend into the posterior domain normally occupied by PAR-1 and PAR-2. Analysis of PAR-3 distribution revealed that *pig-1 (RNAi)* in the *par-2(ts)* mutant background allowed the PAR-3 protein domain to extend into this posterior domain. The position of the mitotic spindle or furrow, however, was not scored in these experiments (Morton et al. 2012). These and additional observations of Morton et al. (2012) suggest that PIG-1 possesses a function in the zygote that overlaps with that of PAR-1 and PAR-2.

More recently, Pacquelet et al. showed that PIG-1 and PAR-4 play a role in regulating the levels of the myosin II heavy chain NMY-2, and this function controls the position of the zygotic furrow (Pacquelet et al. 2015). They showed that PIG-1 and the Anillin ANI-1 act in parallel and presumably downstream of PAR-4. Working with a temperature-sensitive *par-4* mutant grown at an intermediate temperature, the authors saw a slight anterior shift of the furrow that was enhanced by *ani-1(RNAi)* or a *pig-1* null mutation. Although neither *ani-1(RNAi)* nor the *pig-1* mutation alone produced a phenotype, *ani-1(RNAi)* in the *pig-1* mutant generated a strong anterior shift of the furrow. A surprising observation was that the mitotic spindle position was normal despite these perturbations, indicating that loss or knockdown of these molecules uncoupled the central spindle position from the position of the furrow. Observing furrow ingression over time revealed that the furrow slides forward in *ani-1(RNAi); par-4(ts)* but not in wild-type zygotes.

The other interesting finding of this study is that the levels of myosin at the cortex are regulated by PAR-4, PIG-1, and ANI-1 (Pacquelet et al. 2015). A symmetric network of actomyosin is broken at the site of sperm entry, defining the posterior of the zygote and resulting in the movement of this network toward the anterior (reviewed by Motegi and Seydoux 2013). The levels of NMY-2 peak at the anterior cortex before furrow ingression begins and continue to decline during anaphase in wild-type zygotes. The peak levels of NMY-2 are higher in *ani-1(RNAi); par-4(ts)* and in *ani-1 (RNAi); pig-1(null)* zygotes, and NMY-2 levels declined more slowly in *ani-1(RNAi); par-4(ts)* zygotes and failed to decline in *ani-1(RNAi); pig-1(null)* zygotes. The hypothesis that the high NMY-2 levels are responsible for the furrow positioning defects is supported by the ability of an *nmy-2* mutation to suppress the defect in *ani-1(RNAi); par-4(ts)* zygotes. Pacquelet et al. (2015) also determined the level of myosin activation by staining embryos for phosphorylated myosin light chain. During anaphase in wild-type zygotes, the authors detected no phosphorylated myosin light chain at the cortex. By contrast, phosphorylated myosin light chain accumulated at the furrow and spread toward the anterior in *ani-1(RNAi); pig-1(null)* zygotes. The authors concluded that the ANI-1 and PIG-1 inhibit the accumulation and spread of active myosin in the *C. elegans* zygote.

Although PIG-1 functions in the Q.a, Q.p, and zygotic divisions, the protein appears to play distinct roles in the different divisions. As mentioned earlier, PIG-1 likely controls spindle positioning in Q.p and asymmetric myosin localization in Q.a, and the position of the spindle mediates DCSA in the Q.p and zygotic divisions. Yet, PIG-1 loss did not alter the position of the zygotic spindle. One possible explanation is that the PIG-1 and PAR-1 kinases have overlapping roles in spindle positioning with PAR-1 playing the major role in the zygote and PIG-1 playing the major role in Q.p. This hypothesis is consistent with the enhancement of the *par-1(ts)* embryonic lethality by *pig-1(RNAi)* and with the enhancement of a *pig-1* partial loss-of-function mutant Q.pp apoptotic defect by the *par-1(ts)* allele (Chien et al. 2012; Morton et al. 2012). Another possibility is that PIG-1 activity is regulated by its distribution, which was found to be different in Q.p and in the zygote. In Q.p, PIG-1 fused to GFP localized to centrosomes during mitosis (Chien et al. 2012). In the zygote, endogenous PIG-1 was cytoplasmic and did not concentrate at the centrosomes.

Perhaps the most confusing discrepancy is the role of *pig-1* in Q.a and the zygote. Loss of *pig-1* leads to a loss of asymmetric NMY-2 in Q.a but to an increase in asymmetric NMY-2 in the zygote. One possible explanation is that either PIG-1 or its activity is asymmetric in Q.a but not the zygote. In this model, the effects of *pig-1* loss in Q.a would be to distribute NMY-2 activity symmetrically, whereas in the zygote, it would be to simply reduce the levels of NMY-2 while a second mechanism ensures that it remains at the anterior cortex. Consistent with this hypothesis, the CDC-42 GTPase maintains the anterior localization of the myosin NMY-2 through the anterior PAR protein complex (reviewed by Motegi and Seydoux 2013). Neither CDC-42 nor the anterior PAR complex proteins have been shown to function in the Q-lineage divisions.

Studies of *pig-1* and its homologs in other developmental contexts are consistent with a general function in myosin regulation. For instance, a *pig-1* mutation was identified as an enhancer of the embryonic elongation defects that occur in the plectin mutant *vab-10* (Zahreddine et al. 2010). Elongation relies on cell shape changes in the embryonic hypodermis and the contraction of dorsal–ventral actomyosin cables. VAB-10 is an essential component of the hemidesmosome-like structures that connect and transmit elongation forces between body wall muscles and hypodermis. In *vab-10* mutants, these hemidesmosomes are weaker and are susceptible to rupture if contractile forces are too strong. Perhaps PIG-1 inhibits actomyosin contraction, and its loss leads to increased tension and hemidesmosome rupture, resulting in the enhanced lethality observed in the *vab-10; pig-1* double mutant (Zahreddine et al. 2010).

In *Xenopus laevis* embryos, xMELK (aka pEg3) localizes at the cytokinetic furrow during mitosis and is required to complete cytokinesis (Le Page et al. 2011). Consistent with a role in cell proliferation, several cancers are associated with excessive MELK expression (for an example, see Rhodes et al. 2004). In aggressive basal-like breast cancer cells treated with shRNAs against MELK, proliferation was reduced with many cells failing to achieve cytokinesis and sometimes dividing asymmetrically (Wang et al. 2014). In *Xenopus*, xMELK overexpression also leads to cytokinesis failures and impairs accumulation at the division furrow of the activated small GTPase RhoA, which controls actomyosin contraction (Le Page et al. 2011). One interesting possibility is that both PIG-1 and MELK have a conserved function in inhibiting actomyosin contraction by acting upstream of Rho family GTPases. Additional work will be needed in *C. elegans* to determine if such a function exists and how it could allow PIG-1 to regulate three different types of ACD in the zygote, Q.a, and Q.p.

6.3.3 Transcriptional Regulation of pig-1

Regulators of *pig-1* expression have been identified. SPTF-3, the *C. elegans* homolog of the Sp1 transcription factor, activates *pig-1* expression. SPTF-3 binds to a conserved sequence in the genes *pig-1* and *egl-1* to promote their transcription, providing a rare example of a common regulator of the caspase and DCSA pathways (Hirose and Horvitz 2013). While the paper by Hirose and Horvitz (2013) focuses on the lineage that produces M4 and its apoptotic sister, the authors also showed that SPTF-3 acted through *pig-1* to specify the apoptotic fate of Q.aa. The investigators did not analyze the Q.p division, but because SPTF-3 is expressed broadly and can control *pig-1* expression in Q.a and M4, it seems likely that it controls *pig-1* transcription in the Q.p division as well. The authors did not analyze the role of SPTF-3 in DCSA either, but since *sptf-3* is required for *pig-1* transcription, *sptf-3* is expected to influence DCSA in Q.a.

Another report proposed that the putative DNA-binding protein HAM-1 is also a transcriptional activator of *pig-1* (Feng et al. 2013). HAM-1 was initially shown to function in the specification of the apoptotic and neuronal fates and in DCSA of the HSN/PHB neuroblast lineage (Guenther and Garriga 1996; Frank et al. 2005). HAM-1 contains a divergent Winged-Helix domain, the Stox box, found in the two human proteins Stox1 and Stox2 and in the *Drosophila* protein Knockout (Feng et al. 2013; Leung et al. 2016). The presence of this domain in HAM-1 was initially confusing because antibodies to HAM-1 only detected the protein at the cell cortex: HAM-1 is detected in a subset of embryonic cells and is often localized at the posterior cortex of dividing cells, including the HSN/PHB neuroblast (Guenther and Garriga 1996; Frank et al. 2005). GFP-tagged HAM-1, however, was more recently shown to localize to nuclei as predicted from its sequence (Feng et al. 2013; Leung et al. 2016). In a phenotype that is not understood at the lineage level, the survival of the PLM and ALN neurons requires *ham-1* function, and nuclear localization is essential for this function (Leung et al. 2016). HAM-1 is also necessary for DCSA and the apoptotic fate in the Q.a but not the Q.p division (Feng et al. 2013).

Relying on ModEncode chromatin immunoprecipitation data, Feng et al. (2013) reported that HAM-1 transcriptionally regulated *pig-1* by binding to the same *pig-1* 5′ DNA sequences as SPTF-3. Deleting the *pig-1* 5′ binding region for HAM-1 and SPTF-3 indeed abolished PIG-1 expression. It is, however, unclear if the endogenous regulator of *pig-1* is SPTF-3, HAM-1, or a combination of both transcription factors. Consistent with a role for HAM-1 in *pig-1* transcription, NMY-2 asymmetry in Q.a was lost in a *ham-1* mutant, the same phenotype observed in the *pig-1* mutant. In addition, the expression of a transgenic PIG-1::GFP translational reporter was reduced in *ham-1* mutants in both Q.a and Q.p cells. Why this reduction of *pig-1* expression in Q.p does not induce a Q.p DCSA defect is unclear. Feng et al. (2013) also reported that the *ham-1::gfp* transgene used in their study was expressed broadly and only localized to nuclei, failing to localize to the cortex, a finding that conflicts with several other observations. Although the original studies using anti-HAM-1 antibodies only detected cortical protein, the expression pattern was restricted. In addition, the *ham-1* transgene used by the Hawkins group detected both nuclear and cortical GFP::HAM-1 (Leung et al. 2016). Finally, missense mutations in a conserved sequence near the N-terminus or deletion of a central polyproline-rich sequence disrupt both HAM-1 function and its ability to localize to the cortex (Guenther and Garriga 1996; Leung et al. 2016). It seems likely that HAM-1 has both nuclear and cortical functions, but it is currently unclear whether these functions represent separate or interrelated activities of HAM-1.

6.4 The Arf Pathway

In addition to PIG-1, membrane trafficking mediated by Arf proteins and their regulators has also been implicated in both Q.a and Q.p DCSA (Fig. 6.4). Arfs are small GTPases of the Ras superfamily that mediate coatomer formation, an essential step in the generation of transport vesicles (reviewed by Itzen and Goody 2011). Like other small GTPases, Arfs function as biological switches whose active GTP-bound and inactive GDP-bound states are controlled by guanine nucleotide exchange factors (GEFs) that facilitate GDP release and GTP binding and by GTPase activating proteins (GAPs) that stimulate hydrolysis of GTP to GDP.

Two Arfs, ARF-1 and ARF-6, the AGAP family homolog CNT-2, and three Arf GEFs, the cytohesin GRP-1, EFA-6, and the Brag/IQSEC homolog BRIS-1, are all required to varying extents for Q.a and Q.p DCSA (Singhvi et al. 2011; Teulière et al. 2014). Genetic analysis has been unable to determine whether PIG-1 and the Arfs function in the same or parallel pathways (Singhvi et al. 2011). DCSA requires the CNT-2 GAP and the GRP-1 SEC7 (GEF) activities, indicating that Arf regulation is the main role that these multidomain proteins play in regulating DCSA (Singhvi et al. 2011; Teulière et al. 2014). CNT-2 is required for receptor-mediated endocytosis in the *C. elegans* germline, suggesting that Arf-mediated endocytosis contributes to DCSA. Consistent with this hypothesis, RNAi-mediated knockdown of the endocytosis regulators RAB-5 and DYN-1/Dynamin leads to the production of extra neurons

(Singhvi et al. 2011). One surprising observation was that GFP-tagged GRP-1 localizes to nuclei. Experiments in which the SEC7 domain of GRP-1 was targeted to either the plasma membrane or the nucleus, however, revealed that GRP-1's site of function in DCSA and in specifying the apoptotic fate is at the membrane (Teulière et al. 2014). Why GRP-1 localizes to nuclei is a mystery.

Like PIG-1 and PAR-4, Arfs and Arf regulators also function in the caspase-independent cell death pathway (Denning et al. 2012). Loss of either ARF-1 or GRP-1 function suppresses the cell extrusion-mediated cell death caused by mutations in the caspase gene *ced-3*. Denning et al. (2012) proposed that the cycles defined by ARF-1 and GRP-1 are required for trafficking of the cell adhesion molecules that control cell shedding. We propose that the cell shedding phenotypes of *pig-1*, *par-4*, *grp-1*, and *arf-1* mutants result from cell transformations that occur in the lineages that produce the shed cells: in these mutants, the cells are not shed because like the undead daughters of Q.a or Q.p they survive and adopt the fate of their epithelial sister cells, cells that do not die.

How does Arf activity regulate DCSA? The most obvious hypothesis is that ARF-1 and ARF-6 cycles are necessary for trafficking an essential membrane molecule that mediates Q.a and Q.p polarity. Another possibility is that Arfs could regulate actin dynamics. ARNO, which like GRP-1 belongs to the cytohesin family of Arf GEFs, and Arf6 promote large fan-like lamellipodia in MDCK cells by activating Rac1, an actin regulator (Santy et al. 2005). Acting through Racs or other Rho family GTPases, GRP-1 and ARF-6 could stimulate membrane extension in the Q.a and Q.p divisions and contribute to DCSA.

Consistent with a role for Arfs in *C. elegans* actin dynamics, reduction of *cnt-2* by RNAi was found to suppress *act-2*/actin and enhance *nmy-2*/myosin II heavy chain mutant phenotypes (Fievet et al. 2012). The cortex of embryos treated with *cnt-2(RNAi)* displayed an increase in NMY-2 foci velocity during the establishment of zygotic polarity. How CNT-2 acts on cortical myosin is not known but may require Arfs since the same study also placed ARF-1 among genes that regulate the actomyosin network. These findings suggest a conserved function for the Arf pathway in controlling the cortical actomyosin network of zygotic and neuroblast asymmetric divisions. It will be interesting to determine whether like *pig-1*, the *cnt-2* and Arf mutations perturb the anterior polarity of myosin II in the Q.a division. Together, *pig-1* and Arf pathways may collaborate to coordinate the assembly and contractile behavior of the actomyosin cortical network in neuroblasts. How this control is coupled to asymmetric displacement of the spindle and membrane extension remains to be investigated. It is noteworthy that the genes identified in screens for Q lineage mutants are also regulators of the zygotic division, suggesting that their function is generally required for ACD.

6.5 Genes Regulating DCSA only in the Q.a Division

Certain genes have been found to regulate Q.a but not Q.p DCSA (Fig. 6.4). To date, no Q.p-specific regulator has been identified. As mentioned above, the HAM-1 protein is specifically required for Q.a DCSA. In *ham-1* mutants, Q.a sometimes divides with a reversed DCSA polarity associated with posterior spindle displacement (Feng et al. 2013). If HAM-1 acts on DCSA through *pig-1* activation, it is unclear how it inhibits the wild-type Q.a spindle displacement, since PIG-1 is required for posterior spindle displacement in Q.p. If HAM-1 inhibits spindle displacement independently of *pig-1*, its mechanism of action is unknown.

Another specific regulator of Q.a DCSA is the gene *toe-2* (*target of Erk-2*) (Gurling et al. 2014). Although the TOE-2 protein is required for the execution of the apoptotic fate of both Q.a and Q.p daughter cells, it plays distinct roles in the two divisions. TOE-2 loss alters the DCSA of Q.a but not Q.p. Yet, its loss does alter the fate of Q.pp., which can survive and express A/PVM or SDQ markers, or adopt the fate of its mitotic sister cell to produce extra A/PVM- and SDQ-like neurons.

First identified as a target of MPK-1, the homolog of the MAP kinase Erk, TOE-2 contains docking sites for MPK-1 and a DEP (Dishevelled, EGL-10, and Pleckstrin) domain (Arur et al. 2009). DEP domain containing proteins usually modulate G-protein coupled receptor signaling at the cell surface (Consonni et al. 2014). In a structure–function analysis of TOE-2, Gurling et al. (2014) showed that the Q.pp apoptotic fate but not Q.a DCSA required the TOE-2 DEP domain. By contrast, Q.a DCSA but not the Q.pp. apoptotic fate required the region of the protein containing the MPK-1 docking sites. The deletion removing the docking sites is large and could possibly affect other unknown TOE-2 functions. The finding that MPK-1 regulates the Q.a but not the Q.p division, however, supports the hypothesis that deletion of the TOE-2 MAPK docking sequences is responsible for the lack of mutant protein activity in Q.a DCSA. These findings suggest that TOE-2 has two distinct, separable functions in Q.a and Q.p.

The subcellular localization of the full-length and two mutant proteins discussed above also supports the hypothesis that TOE-2 has two functions (Gurling et al. 2014). During mitosis, TOE-2 localizes to centrosomes and to the cortex, accumulating at the cytoplasmic furrow. In interphase, TOE-2 accumulates in the nucleus. The mutant protein lacking the DEP domain localizes normally, but the mutant protein lacking the docking sites fails to localize to the cortex and furrow. These observations are consistent with a model where cortical TOE-2 under MPK-1 control is required for Q.a DCSA, and either nuclear or centrosomal TOE-2 specifies the Q.pp. apoptotic fate. This model implies that the TOE-2 DEP domain has a nuclear or centrosomal function, which would be unprecedented for a DEP domain containing protein.

Yeast two-hybrid screens identified several TOE-2-interacting proteins (Li et al. 2004; Xin et al. 2009, 2013). One is the *C. elegans* homolog of a family of proteins known as GRAFs. These proteins contain BAR and RhoGEF domains, suggesting that this protein could act with TOE-2 to regulate actin dynamics through the control of Rho family small GTPases activity. Further studies will be necessary to determine if

GRAF homologs or other Rho GTPase regulators are involved in Q.a DCSA and if they function in cooperation with PIG-1 and the Arf pathway.

6.6 The Role of Wnt Signaling in DCSA and Apoptosis

6.6.1 The Model for Wnt Signaling in C. elegans *AP Divisions*

Wnts are secreted glycoproteins that organize a broad range of developmental processes that include cell proliferation, guidance of migrating cells and growth cones, neuronal polarity, and developmental patterning. Wnts play a global role in regulating *C. elegans* asymmetric divisions that are oriented along the anterior–posterior (AP) axis. The model for Wnt function in these divisions posits that Wnts, which are expressed from cells that are usually located posterior to the dividing cell, orient the polarity of the dividing cell to generate two daughter cells with different levels of Wnt signaling activity. The high level of signaling in the posterior cell activates two pathways: a canonical pathway that leads to the accumulation of the β-catenin SYS-1 and a second pathway that leads to the export of the TCF/LEF transcription factor POP-1 from the nucleus. These two pathways culminate in a low SYS-1/POP-1 ratio in the anterior cell where POP-1 represses transcription of target genes and to a high SYS-1/POP-1 ratio in the posterior cell where SYS-1/POP-1 activates transcription of target genes (reviewed in Chap. 4).

One of the yeast two-hybrid screens discussed above also identified the *C. elegans* homolog of the Ran GTPase, RAN-1, as a TOE-2-interacting protein (Li et al. 2004). Because Ran regulates nuclear-cytoplasmic transport (reviewed by Kalab and Heald 2008), one interesting possibility is that this interaction mediates functions carried out by TOE-2 in the Q.p nucleus, perhaps by assisting in the nuclear export of POP-1. This model predicts that the POP-1 asymmetry specifies the apoptotic fate of Q.pp. Consistent with this model, the *toe-2* mutant displays additional posterior-to-anterior fate transformations in the Q lineage (Gurling et al. 2014). Similar transformations are observed in other lineages when genes encoding proteins of the Wnt/POP-1 asymmetry pathway are mutated (reviewed by Sawa and Korswagen 2013).

6.6.2 Wnt Signaling in DCSA

The first *C. elegans* Wnt shown to regulate ACD was *lin-44* (Herman and Horvitz 1994; Herman et al. 1995). Herman and Horvitz (1994) initially showed that loss of *lin-44* function resulted in randomization of ACD polarity. For example, in the AP division of the T epidermal cell, the fates of the T.a and T.p were often reversed. By contrast, mutations in the gene *lin-17,* which encodes a Frizzled family Wnt receptor (Sawa et al. 1996), resulted in symmetric divisions of the T.a and T.p cells, with both cells adopting similar fates that are distinct from the fates of the wild-type T.a or T.p (Sternberg and Horvitz 1988; Herman and Horvitz 1994). Herman and Horvitz (1994) also studied the male B cell and the T.pp. divisions, which appear to produce daughter cells of different sizes. Because the authors used Nomarski optics to follow the divisions and only nuclei can be observed using this approach, they noted that the male B cell divided to produce an anterior cell (B.a) with a larger nucleus than its posterior sister cell (B.p). The T.pp. cell divided with the opposite polarity, producing an anterior nucleus that was smaller than its posterior sister's nucleus. Presumably, nuclear size reflects cell size, although this has not been verified for these divisions. In *lin-44* mutants, the DCSA of both divisions was often reversed. Loss of *lin-17* function, by contrast, resulted in male B daughter cells with similar size nuclei (Sternberg and Horvitz 1988; Herman and Horvitz 1994). The *lin-17 lin-44* double mutant expressed the Lin-17 B-cell phenotype (Herman and Horvitz 1994). These findings are consistent with a model where LIN-44 orients the polarity of DCSA by acting though LIN-17. Why the phenotypes of *lin-44* and *lin-17* mutants are different is unclear. One possibility is LIN-44 orients an undirected AP polarity that requires LIN-17. Another possibility is that other Wnts act through LIN-17 to orient the polarity in the absence of *lin-44* function.

How does LIN-17 signal to establish DCSA? The Herman group identified several pathways that may act downstream of LIN-17 by analyzing male B-cell DCSA in mutants or RNAi-treated animals (Wu and Herman 2006). Focusing on molecules known to function downstream of Frizzled receptors, they identified the *C. elegans* Dishevelled MIG-5 as required for B-cell DCSA. Dishevelleds are cytoplasmic proteins that mediate the effects of Wnt signaling downstream of Frizzled receptors in several signaling pathways, so the identification of a Dishevelled was not unexpected. They also found that mutations in the *pop-1* gene and in the gene encoding the *C. elegans* Nlk homolog LIT-1, which promotes POP-1 nuclear export, produce weak B-cell DCSA phenotypes. It is unclear whether the weak effects result from a partial loss of gene function in the mutant animals or from a minor role of the POP-1 asymmetry pathway in DCSA.

Frizzleds and Dishevelleds also participate in Planar Cell Polarity (PCP), a conserved pathway that organizes the polarity of sheets of cells and shapes the development of tissues that range from the orientation of stereocilia in the mammalian cochlea to the movement of cells during gastrulation (reviewed by Hale and Strutt 2015). PCP can also regulate the polarity of dividing cells. For example, during the divisions of the Drosophila Sensory Organ Precursors (SOPs), which produce chemosensory and mechanosensory organs, the PCP pathway orients the divisions along the axis that is orthogonal to the apical–basal axis (Segalen and Bellaïche 2009). In the absence of PCP components, the divisions are still asymmetric and produce the correct cell types, but the division axes are no longer properly oriented. PCP requires some of the components of the Wnt signaling pathway like Frizzled receptors and Dishevelled molecules but also requires PCP-specific molecules. DCSA of the *C. elegans* male B cell, for example, does not require the PCP-specific Flamingo homolog FMI-1, but does require the Vang Gogh VANG-1 homolog, although the *vang-1(RNAi)* effects are weak (Wu and Herman 2006).

PCP signaling can also activate the Rho GTPase pathway, and knockdown by RNAi of the *C. elegans* RhoA homolog RHO-1 and the Rho kinase Rock homolog LET-502 generates B-cell DCSA phenotypes that are almost as strong as the *lin-17* loss-of-function phenotype (Wu and Herman 2006). Another strong phenotype was observed with RNAi to the myosin light chain gene *mlc-4*, suggesting that Wnt signaling may control myosin contractility through Rho. The severity of the *rho-1*, *let-502*, and *mlc-4* B-cell phenotypes and the lack of an *fmi-1* phenotype or the weak *vang-1* phenotype suggest that the PCP pathway is not the major pathway used by Wnts to regulate RHO-1 activity in B-cell DCSA.

6.6.3 The Role of Wnt Signaling in Specification of the Apoptotic Fate

Although the roles of Wnts have been studied in many A/P divisions, only one study focused on the role of Wnt signaling in divisions that produce apoptotic cells (Bertrand and Hobert 2009). Studying several terminal divisions that produce neurons, the authors showed that disrupting the pathway that leads to POP-1 export resulted in posterior neurons that failed to express their differentiation markers and instead expressed the same markers as their anterior sisters. Two of the divisions studied produced an apoptotic daughter cell. In the division that produces an anterior ASER neuron and a posterior daughter that dies, disrupting POP-1 export led to the production of two neurons that expressed the ASER marker. In the division that produces an anterior daughter that dies and a posterior AIN neuron, disrupting POP-1 export caused both daughter cells to adopt an apoptotic fate. These findings indicate that Wnt signaling does not specify the apoptotic or non-apoptotic fate in a division but rather the fate of the posterior cell. Neither the role of the canonical SYS-1 pathway nor DCSA was explored in this study.

6.7 Daughter Cell Size Asymmetry in the NSM Neuroblast Lineage

6.7.1 The Transcriptional Network Regulating the Apoptotic Fate and Size of the NSM Sister Cell

Most apoptotic deaths in *C. elegans* are the products of embryonic divisions. In contrast to the accessible larval Q neuroblasts, embryonic neuroblasts that produce a daughter cell fated to die have rarely been studied due to the difficulty of identifying and imaging cell divisions in the developing embryo. The most extensively studied embryonic division that produces an apoptotic daughter cell is the division of the NSM neuroblast (NSMnb). The NSMnb divides along a dorsal medial/ventral lateral axis to generate a larger ventral serotonergic NSM neuron and a smaller dorsal cell (NSM sister cell, NSMsc) that is fated to die (Fig. 6.5a). Mutants were isolated on the basis of the survival of the NSMsc and the production of extra NSM-like cells (Ellis and Horvitz 1991; Hatzold and Conradt 2008). The genes identified revealed a complex interplay of transcriptional regulators coordinating cell death specification and smaller size of the NSMsc.

6.7.2 Ectopic CES-1 Activation Leads to DCSA and Apoptosis Defects in the NSM Lineage

Early studies identified the *ces-1* and *ces-2* genes as regulators of apoptosis in the NSMsc (Ellis and Horvitz 1991). Loss-of-function (lf) mutations in *ces-2* or gain-of-function (gf) mutations in *ces-1* lead to survival of the NSMsc and its transformation into a second NSM-like neuron, and genetic interaction studies indicate that *ces-2* is a negative regulator of *ces-1* (Ellis and Horvitz 1991). The genes *ces-2* and *ces-1* encode bZip and Snail transcription factors homologs, respectively, and CES-2 can bind to sequences upstream of the *ces-1* transcriptional start site that are mutated in the gain-of-function alleles (Metzstein et al. 1996; Metzstein and Horvitz 1999). A CES-1::YFP transgenic reporter can only be detected in the NSM of wild-type embryos but is detected in the NSMnb, the NSM, and the NSMsc of a *ces-2* mutant embryo (Hatzold and Conradt 2008). These findings suggest a model where CES-2 represses *ces-1* transcription, which allows the NSMsc to die. One of the targets of CES-1 in the mutants is *egl-1*, the most upstream proapoptotic gene in the apoptotic pathway (Thellmann et al. 2003). In wild-type embryos, the bHLH transcription factors HLH-2 and HLH-3 act together to stimulate transcription of *egl-1* in the NSMsc. Ectopic or excess expression of CES-1 in these cells may block the ability of HLH-2/HLH-3 heterodimers to stimulate *egl-1* transcription. In this model, CES-1 normally plays no role in the apoptotic fate of NSMsc, and only when it is expressed in the lineage of *ces-2(lf)* or *ces-1(gf)* mutants does it have an effect.

Given this model, it was surprising when Hatzhold and Conradt found that the *ces-2 (lf)* and *ces-1(gf)* mutations disrupted the DCSA of the NSMnb to generate daughter cells more equivalent in size (Hatzold and Conradt 2008). Furthermore, many of the divisions were no longer oriented along the appropriate axis. Perhaps excess or ectopic expression of CES-1 alters the coordinated expression of both *egl-1* and DCSA regulators mediated by the bHLH transcription factors. In this model, these transcription factors could function like SPTF-3 in the Q and M4 lineages.

Similar phenotypes were observed in a *dnj-11* mutant (Hatzold and Conradt 2008). DNJ-11 is an ortholog of the mammalian ZRF1/MIDA1/MPP11/DNAJC2 family that function as ribosome-associated chaperones but can also associate with chromatin, competing with the Polycomb-Repressive Complex 1 (PRC) for a mono-ubiquitinated repressive mark on histone 2A (reviewed by Aloia et al. 2015). By removing PRC1 from chromatin, ZRF1 recruits an ubiquitin peptidase that removes the repressive mark

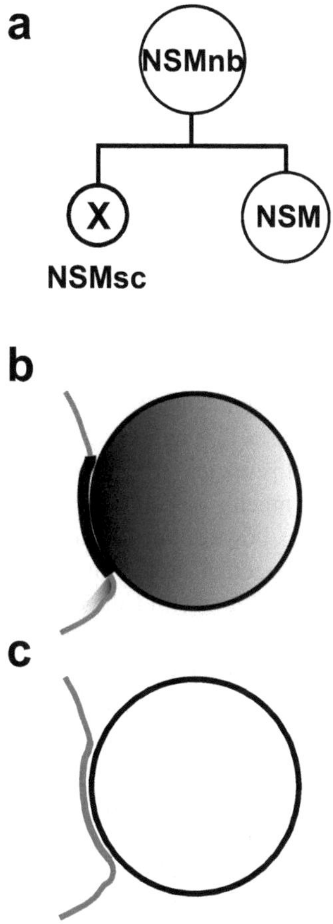

Fig. 6.5 *The NSM neuroblast division.* (**a**) The NSM neuroblast (NSMnb) divides asymmetrically to generate a larger ventral NSM neuron and dorsal sister cell (NSMsc) that will die. (**b**) The gradient in the NSMnb represents the caspase activity gradient, and the *thick line* on the neighboring cell represents the accumulation of the CED-1 receptor to the dorsal side of the NSMnb (Chakraborty et al. 2015). Because CED-1 accumulates on the membrane of phagocytic cells where they contact an apoptotic cell during engulfment, we show CED-1 on the neighboring cell. This localization or the nonautonomous function of CED-1 in the asymmetric distribution of caspase activity in the NSMnb has not been established. (**c**) Both the caspase asymmetry and the asymmetric localization of CED-1 are lost in various apoptosis and engulfment mutants

and leads to the transcription of genes that are involved in differentiation (Richly et al. 2010). This ZRF1 activity is important for neural differentiation (Aloia et al. 2015). Because *ces-2* and *dnj-11* loss leads to similar phenotypes as *ces-1* expression in the NSMnb and its daughter cells, DNJ-11 and CES-2 may act together to inhibit CES-1 expression. If this hypothesis is correct, DNJ-11 would play a repressive function, in contrast to the role its ZRF1 ortholog plays in neuronal development. Alternatively, DNJ-11 could activate another regulatory protein, possibly CES-2 itself, to repress *ces-1*.

6.7.3 Coupling of Cell Cycle and ACD Regulation

More recently, CES-1 has been shown to contribute to NSMnb cell cycle progression (Yan et al. 2013). Screening for suppressors of the extra NSM phenotype generated by the *ces-1(gf)* mutation identified a mutation in *cya-1*, which encodes a Cyclin A homolog. Wild-type animals have two NSMs labeled by a tryptophan hydroxylase (*tph-1*) reporter. Because the NSMscs survive and differentiate like NSMs in *ces-1 (gf)* mutants, these animals have four cells that express the marker. When the *ces-1 (gf); cya-1* double mutants produced three *tph-1*-expressing cells, one was larger than the other two, suggesting that the *cya-1* mutation does not suppress the apoptotic defect but instead blocks the NSMnb division. Further analysis of the division in *cya-1* mutants confirmed this hypothesis and showed that the NSMnb arrested between S phase and mitosis. The penetrance of the *cya-1* NSMnb division defect is enhanced by the *ces-1(gf)* mutation, suggesting that the excess CES-1 in the NSMnb disrupts its ability to divide. CES-1 was found to bind to the *cdc-25.2* gene, and the cell cycle defects caused by excess CES-1 indeed resulted from a reduction of *cdc-25.2* expression. *Cdc-25.2* encodes one of the three *C. elegans* Cdc25 homologs, phosphatases that remove an inhibitory phosphate from Cyclin A.

It is unclear whether the decreased levels of CDC-25.2 mediate any of the NSMnb asymmetry defects caused by the *ces-1(gf)* mutation. In Drosophila neuroblast divisions, however, the maintenance of apical localized Inscuteable and the PAR-3 homolog Bazooka, proteins necessary for the asymmetry of the neuroblast division, requires normal function of the cyclin-dependent kinase *cdc2* (Tio et al. 2001). Reduction of *cdc2* allows the neuroblasts to divide, but as they progress through mitosis, Inscuteable and Bazooka often lose their polarized localization or no longer asymmetrically localize apically. Spindle orientation is also disrupted. Presumably, Cdc2 targets are important for maintaining neuroblast polarity, but those targets have yet to be identified. CDC-25.2 could serve an analogous function in *C. elegans* neuroblasts to maintain polarity. The Xenopus homolog of CDC25B, for example, is a substrate of xMELK (Davezac et al. 2002). It will be interesting to determine whether PIG-1 also phosphorylates CDC-25.2 in *C. elegans* in the Q lineage, as this could shed light on the possible cross talk between DCSA and cell cycle regulators in divisions that generate apoptotic cells.

6.8 Asymmetric Distribution of Caspase Activity in the NSMnb

A recent paper by Barbara Conradt's group showed that caspase activity is asymmetrically distributed in the NSMnb (Chakraborty et al. 2015). The authors found that GFP-tagged CED-3 caspase could be detected in the NSMnb and was distributed equally to both daughter cells with an eventual increase of the CED-3::GFP levels in the NSMsc relative to its surviving NSM sister. They also measured the distribution of caspase activity in the NSMnb by tagging TAC-1 with GFP. TAC-1 is a member of the TACC family of pericentriolar (PCM) proteins and a target of CED-3. Cleavage of TAC-1::GFP by caspase activity results in the release of GFP from the centrosome. At metaphase, the centrosome that is inherited by the NSMsc contains less GFP, and this asymmetry requires *ced-3* function. This finding suggests that the side of the NSMnb that produces the daughter cell that will die already has higher caspase activity prior to cytokinesis (Fig. 6.5b). This active caspase asymmetry in the NSMnb requires the intrinsic apoptosis pathway genes *egl-1*, *ced-9*, and *ced-4*. The *ces-1*, *ces-2*, and *dnj-11* mutations that disrupt NSMsc apoptosis also result in the loss of asymmetric caspase activity, presumably through repression of *egl-1* expression in the NSMnb.

Finally, caspase asymmetry also depends on engulfment regulators that function in surrounding phagocytic cells. Two pathways mediate the clearance of apoptotic cells. The cell surface receptor CED-1 functions with the GULP adaptor CED-6, the ATP-binding cassette transporter CED-7, and the dynamin homolog DYN-1 in one pathway. The CrkII homolog CED-2 and the Rac GEF CED-5/CED-12 regulate the Rho-family GTPase CED-10/Rac in a second pathway (reviewed by Reddien and Horvitz 2004). After the initial discovery of these engulfment pathways, two groups found that the genes also played a role in apoptosis (Hoeppner et al. 2001; Reddien et al. 2001). Loss of *ced-1*, for example, led to a low frequency of apoptotic defects in some lineages. In other cells fated to die, loss of *ced-1* had no effect but enhanced the defect of weak apoptotic mutants.

The CED-1 receptor functions in the phagocytic cell, and during engulfment CED-1::GFP is concentrated at the interface between the phagocytic cell and the apoptotic corpse (Zhou et al. 2001). When a C-terminally truncated CED-1::GFP was analyzed during the division of the NSMnb, it surprisingly localized asymmetrically to the side of the NSMnb that will give rise to the NSMsc, the cell fated to die (Chakraborty et al. 2015) (Fig. 6.5b). Furthermore, loss of the engulfment genes *ced-1*, *ced-2*, or *ced-6* resulted in a loss of the caspase asymmetry in the NSMnb, implicating both pathways in the asymmetric activation of caspases. These observations lead to the intriguing possibility that the role played by CED-1 in apoptosis is to promote the caspase asymmetry in the neuroblast before cell division through a "kiss of death" mechanism involving the future phagocyte of the dying small daughter cell.

Asymmetric localization of CED-1::GFP to the NSMnb requires *ced-3* function, and as discussed above, the asymmetry of CED-3 activity requires *ced-1* function (Chakraborty et al. 2015). These findings suggest that there is a feedback mechanism that ensures CED-3 activity and CED-1 localization asymmetries. Which asymmetry develops first and what initiates the asymmetry are unknown. By analogy to other divisions discussed above, either intracellular cues inherited from the previous division or cell signaling by extracellular cues may initiate these events.

6.9 Future Directions

The molecules discussed in this review provide the tools to identify the mechanisms that mediate DCSA. For example, the identification of PIG-1 targets should provide a framework for thinking about how this kinase controls DCSA. Furthermore, investigating the distribution of the PIG-1 protein or its activity might explain the apparent differences in how it acts in the zygotic, Q.a, and Q.p divisions. It will similarly be important to understand how the Arfs regulate the Q.a and Q.p divisions. If they function in vesicular trafficking of specific cargo, the identification of these molecules should lead to new insights into the specification of DCSA. If they act on actin dynamics, this mechanism would establish an in vivo context for studying the connection between Arfs and Racs.

One mystery is why both Q.a and Q.p DCSA require PIG-1 and Arf function when the two cells divide asymmetrically using distinct mechanisms. Defining their functions more precisely should identify a mechanism or mechanisms that are shared by the two divisions. Studying the functions of HAM-1 and TOE-2, by contrast, should identify the mechanisms that specifically regulate the Q.a division. Why genetic screens have not identified molecules that regulate Q.p but not Q.a DCSA is unclear. One possibility is the screens for extra cells produced from the Q.p division are not saturated. Another possibility is that the Q.p division asymmetry is a default state, and molecules like HAM-1 and TOE-2 modify the Q.p division mechanism to generate the Q.a DCSA. This model is consistent with the requirement for molecules that function in both divisions and for Q.a specific molecules.

We have known that Wnt signaling functions in DCSA since the initial studies identified the Wnt LIN-44 in *C. elegans* ACD, and a single study identified some of the possible downstream signaling molecules (Herman and Horvitz 1994; Herman et al. 1995; Wu and Herman 2006). One interesting possibility is the Frizzled LIN-17 and the Dishevelled MIG-5 control actomyosin contractility through RHO-1 activity. It will be important to establish this connection by identifying the molecules that mediate this connection and defining how they function.

Is the asymmetric caspase activation in the mother cell is a general mechanism used to program cell death in the smaller daughter? Because neither *ced-3* nor *ced-1* mutants display a Q lineage DCSA defect (Cordes et al. 2006, J.T., unpublished observation), the PAR-4/PIG-1 pathway, Arf regulation, HAM-1, and TOE-2 must not require the caspase asymmetry to DCSA. If the Q.a and Q.p cells possess asymmetric caspase activity, does this asymmetry require the DCSA regulators? If these molecules regulate caspase asymmetry, one possible explanation would be that they independently control DCSA and asymmetric caspase activity. Alternatively, these molecules might only position the furrow asymmetrically and the asymmetric furrow regulates caspase activity. If these molecules do not regulate caspase asymmetry, such a finding would suggest that the reason both daughter cells survive in these mutants is because caspase activity, diluted by an increase in cytoplasmic volume, drops below a threshold required for efficient cell killing.

Acknowledgements Work in our laboratory is supported by the National Institutes of Health grant NS32057.

References

Aloia L, Demajo S, Di Croce L (2015) ZRF1: a novel epigenetic regulator of stem cell identity and cancer. Cell Cycle 14:510–515

Arur S, Ohmachi M, Nayak S, Hayes M, Miranda A, Hay A, Golden A, Schedl T (2009) Multiple ERK substrates execute single biological processes in *Caenorhabditis elegans* germ-line development. Proc Natl Acad Sci U S A 106:4776–4781

Bertrand V, Hobert O (2009) Linking asymmetric cell division to the terminal differentiation program of postmitotic neurons in *C. elegans*. Dev Cell 16:563–575

Cabernard C, Prehoda KE, Doe CQ (2010) A spindle-independent cleavage furrow positioning pathway. Nature 467:91–94

Chakraborty S, Lambie EJ, Bindu S, Mikeladze-Dvali T, Conradt B (2015) Engulfment pathways promote programmed cell death by enhancing the unequal segregation of apoptotic potential. Nat Commun 6:10126

Chien S-C, Brinkmann E-M, Teulière J, Garriga G (2013) *Caenorhabditis elegans* PIG-1/MELK acts in a conserved PAR-4/LKB1 polarity to promote asymmetric neuroblast divisions. Genetics 193:897–909

Connell M, Cabernard C, Ricketson D, Doe CQ, Prehoda KE (2011) Asymmetric cortical extension shifts cleavage furrow position in Drosophila neuroblasts. Mol Biol Cell 22:4220–4226

Consonni SV, Maurice MM, Bos JL (2014) DEP domains: structurally similar but functionally different. Nat Rev Mol Cell Biol 15:357–362

Cordes S, Frank CA, Garriga G (2006) The *C. elegans* MELK ortholog PIG-1 regulates cell size asymmetry and daughter cell fate in asymmetric neuroblast divisions. Development 133:2747–2756

Davezac N, Baldin V, Blot J, Ducommun B, Tassan J-P (2002) Human pEg3 kinase associates with and phosphorylates CDC25B phosphatase: a potential role for pEg3 in cell cycle regulation. Oncogene 21:7630–7641

Delaunay D, Cortay V, Patti D, Knoblauch K, Dehay C (2014) Mitotic spindle asymmetry: a Wnt/PCP-regulated mechanism generating asymmetrical division in cortical precursors. Cell Reports 6(2):400–414

Denning DP, Hatch V, Horvitz HR (2012) Programmed elimination of cells by caspase-independent celsl extrusion in *C. elegans*. Nature 488:226–230

Egger B, Chell JM, Brand AH (2008) Insights into neural stem cell biology from flies. Philos Trans R Soc Lond B Biol Sci 363:39–56

Ellis RE, Horvitz HR (1991) Two *C. elegans* genes control the programmed deaths of specific cells in the pharynx. Development 112:591–603

Feng G, Yi P, Yang Y, Chai Y, Tian D, Zhu Z, Liu J, Zhou F, Cheng Z, Wang X et al (2013) Developmental stage-dependent transcriptional regulatory pathways control neuroblast lineage progression. Development 140:3838–3847

Fievet BT, Rodriguez J, Naganathan S, Lee C, Zeiser E, Ishidate T, Shirayama M, Grill S, Ahringer J (2012) Systematic genetic interaction screens uncover cell polarity regulators and functional redundancy. Nat Cell Biol 15:103–112

Frank CA, Hawkins NC, Guenther C, Horvitz HR, Garriga G (2005) *C. elegans* HAM-1 positions the cleavage plane and regulates apoptosis in asymmetric neuroblast divisions. Dev Biol 284:301–310

Galli M, van den Heuvel S (2008) Determination of the cleavage plane in early *C. elegans* embryos. Annu Rev Genet 42:389–411

Guenther C, Garriga G (1996) Asymmetric distribution of the *C. elegans* HAM-1 protein in neuroblasts enables daughter cells to adopt distinct fates. Development 122:3509–3518

Gurling M, Talavera K, Garriga G (2014) The DEP domain-containing protein TOE-2 promotes apoptosis in the Q lineage of *C. elegans* through two distinct mechanisms. Development 141:2724–2734

Hale R, Strutt D (2015) Conservation of planar polarity pathway function across the animal kingdom. Annu Rev Genet 49:529–551

Hatzold J, Conradt B (2008) Control of apoptosis by asymmetric cell division. PLoS Biol 6:e84
Herman MA, Horvitz HR (1994) The *Caenorhabditis elegans* gene lin-44 controls the polarity of asymmetric cell divisions. Development 120:1035–1047
Herman MA, Vassilieva LL, Horvitz HR, Shaw JE, Herman RK (1995) The *C. elegans* gene lin-44, which controls the polarity of certain asymmetric cell divisions, encodes a Wnt protein and acts cell nonautonomously. Cell 83:101–110
Hirose T, Horvitz HR (2013) An Sp1 transcription factor coordinates caspase-dependent and -independent apoptotic pathways. Nature 500:354–358
Hoeppner DJ, Hengartner MO, Schnabel R (2001) Engulfment genes cooperate with ced-3 to promote cell death in *Caenorhabditis elegans*. Nature 412:202–206
Itzen A, Goody RS (2011) GTPases involved in vesicular trafficking: structures and mechanisms. Semin Cell Dev Biol 22:48–56
Kalab P, Heald R (2008) The RanGTP gradient—a GPS for the mitotic spindle. J Cell Sci 121:1577–1586
Kaltschmidt JA, Davidson CM, Brown NH, Brand AH (2000) Rotation and asymmetry of the mitotic spindle direct asymmetric cell division in the developing central nervous system. Nat Cell Biol 2:7–12
Kiyomitsu T, Cheeseman IM (2013) Cortical dynein and asymmetric membrane elongation coordinately position the spindle in anaphase. Cell 154:391–402
Le Page Y, Chartrain I, Badouel C, Tassan J-P (2011) A functional analysis of MELK in cell division reveals a transition in the mode of cytokinesis during Xenopus development. J Cell Sci 124:958–968
Leung A, Hua K, Ramachandran P, Hingwing K, Wu M, Koh PL, Hawkins N (2016) *C. elegans* HAM-1 functions in the nucleus to regulate asymmetric neuroblast division. Dev Biol 410:56–69
Li S, Armstrong CM, Bertin N, Ge H, Milstein S, Boxem M, Vidalain P-O, Han J-DJ, Chesneau A, Hao T et al (2004) A map of the interactome network of the metazoan *C. elegans*. Science 303:540–543
Metzstein MM, Horvitz HR (1999) The *C. elegans* cell death specification gene ces-1 encodes a snail family zinc finger protein. Mol Cell 4:309–319
Metzstein MM, Hengartner MO, Tsung N, Ellis RE, Horvitz HR (1996) Transcriptional regulator of programmed cell death encoded by *Caenorhabditis elegans* gene ces-2. Nature 382:545–547
Morton DG, Hoose WA, Kemphues KJ (2012) A genome-wide RNAi screen for enhancers of par mutants reveals new contributors to early embryonic polarity in *Caenorhabditis elegans*. Genetics 192:929–942
Motegi F, Seydoux G (2013) The PAR network: redundancy and robustness in a symmetry-breaking system. Philos Trans R Soc B 368:20130010
Ou G, Stuurman N, D'Ambrosio M, Vale RD (2010) Polarized myosin produces unequal-size daughters during asymmetric cell division. Science 330:677–680
Pacquelet A, Uhart P, Tassan J-P, Michaux G (2015) PAR-4 and anillin regulate myosin to coordinate spindle and furrow position during asymmetric division. J Cell Biol 210:1085–1099
Partanen JI, Tervonen TA, Klefström J (2013) Breaking the epithelial polarity barrier in cancer: the strange case of LKB1/PAR-4. Philos Trans R Soc Lond B Biol Sci 368:20130111
Reddien PW, Horvitz HR (2004) The engulfment process of programmed cell death in *Caenorhabditis elegans*. Annu Rev Cell Dev Biol 20:193–221
Reddien PW, Cameron S, Horvitz HR (2001) Phagocytosis promotes programmed cell death in *C. elegans*. Nature 412:198–202
Rhodes DR, Yu J, Shanker K (2004) Large-scale meta-analysis of cancer microarray data identifies common transcriptional profiles of neoplastic transformation and progression. Proc Natl Acad Sci U S A 101:9309–9314
Richly H, Rocha-Viegas L, Ribeiro JD, Demajo S, Gundem G, Lopez-Bigas N, Nakagawa T, Rospert S, Ito T, Di Croce L (2010) Transcriptional activation of polycomb-repressed genes by ZRF1. Nature 468:1124–1128

Santy LC, Ravichandran KS, Casanova JE (2005) The DOCK180/Elmo complex couples ARNO-mediated Arf6 activation to the downstream activation of Rac1. Curr Biol 15:1749–1754

Sawa H, Korswagen HC (2013) Wnt signaling in *C. elegans*. In: WormBook (ed) The *C. elegans* research community. WormBook. doi:10.1895/wormbook.1.7.2. http://www.wormbook.org

Sawa H, Lobel L, Horvitz HR (1996) The *Caenorhabditis elegans* gene lin-17, which is required for certain asymmetric cell divisions, encodes a putative seven-transmembrane protein similar to the Drosophila frizzled protein. Genes Dev 10:2189–2197

Segalen M, Bellaïche Y (2009) Cell division orientation and planar cell polarity pathways. Semin Cell Dev Biol 20:972–977

Singhvi A, Teulière J, Talavera K, Cordes S, Ou G, Vale RD, Prasad BC, Clark SG, Garriga G (2011) The Arf GAP CNT-2 regulates the apoptotic fate in *C. elegans* asymmetric neuroblast divisions. Curr Biol 21:948–954

Sternberg PW, Horvitz HR (1988) lin-17 mutations of *Caenorhabditis elegans* disrupt certain asymmetric cell divisions. Dev Biol 130:67–73

Sulston JE, Horvitz HR (1977) Post-embryonic cell lineages of the nematode, *Caenorhabditis elegans*. Dev Biol 56:110–156

Teulière J, Cordes S, Singhvi A, Talavera K, Garriga G (2014) Asymmetric neuroblast divisions producing apoptotic cells require the cytohesin GRP-1 in *Caenorhabditis elegans*. Genetics 198:229–247

Thellmann M, Hatzold J, Conradt B (2003) The Snail-like CES-1 protein of *C. elegans* can block the expression of the BH3-only cell-death activator gene egl-1 by antagonizing the function of bHLH proteins. Development 130:4057–4071

Tio M, Udolph G, Yang X, Chia W (2001) cdc2 links the Drosophila cell cycle and asymmetric division machineries. Nature 409:1063–1067

Wang Y, Lee Y-M, Baitsch L, Huang A, Xiang Y, Tong H, Lako A, Von T, Choi C, Lim E et al (2014) MELK is an oncogenic kinase essential for mitotic progression in basal-like breast cancer cells. Elife 3:e01763

Wodarz A (2005) Molecular control of cell polarity and asymmetric cell division in Drosophila neuroblasts. Curr Opin Cell Biol 17:475–481

Wu M, Herman MA (2006) A novel noncanonical Wnt pathway is involved in the regulation of the asymmetric B cell division in *C. elegans*. Dev Biol 293:316–329

Xin X, Rual J-F, Hirozane-Kishikawa T, Hill DE, Vidal M, Boone C, Thierry-Mieg N (2009) Shifted Transversal Design smart-pooling for high coverage interactome mapping. Genome Res 19:1262–1269

Xin X, Gfeller D, Cheng J, Tonikian R, Sun L, Guo A, Lopez L, Pavlenco A, Akintobi A, Zhang Y et al (2013) SH3 interactome conserves general function over specific form. Mol Syst Biol 9:652

Yan B, Memar N, Gallinger J, Conradt B (2013) Coordination of cell proliferation and cell fate determination by CES-1 snail. PLoS Genet 9:e1003884

Zahreddine H, Zhang H, Diogon M, Nagamatsu Y, Labouesse M (2010) CRT-1/calreticulin and the E3 ligase EEL-1/HUWE1 control hemidesmosome maturation in *C. elegans* development. Curr Biol 20:322–327

Zhou Z, Hartwieg E, Horvitz HR (2001) CED-1 is a transmembrane receptor that mediates cell corpse engulfment in *C. elegans*. Cell 104:43–56

Chapter 7
The Midbody and its Remnant in Cell Polarization and Asymmetric Cell Division

Christian Pohl

Abstract The midbody is a protein-dense assembly that forms during cytokinesis when the actomyosin ring constricts around bundling central spindle microtubules. After its initial description by Walther Flemming in the late nineteenth century and its rediscovery through electron microscopy in the 1960s and 1970s, its ultrastructural organization and the sequential recruitment of its molecular constituents has only been elucidated in the past decade. Recently, it has become clear that the midbody can serve as a polarity cue during asymmetric cell division, cell polarization, and spindle orientation by coordinating cytoskeletal organization, vesicular transport, and localized cortical cues. In this chapter, these newly emerging functions will be discussed as well as asymmetries during midbody formation and their consequences for cellular organization in tissues.

7.1 Formation of the Midbody, Cytokinetic Abscission, and the Midbody Remnant

Using cells isolated from salamander, Walther Flemming (1843–1905) was among the first to observe and correctly describe animal cell mitosis, including many aspects of cytokinesis (Flemming 1874, 1875, 1876, 1965; Paweletz 2001). Remarkably, he also observed a structure only visible late during cell division that he proposed could be similar to the cell plate of plants (Flemming 1891). This structure, a micrometer-sized body which appeared at the site of cytokinetic abscission and as densely stained in his fixed preparations, he coined "*Zwischenkörper*" (midbody, now also "Flemming body") (Flemming 1891; Pohl 2008). Considering his equipment, it is remarkable that he also observed filaments (microtubules) connecting sister cells that became bundled at the abscission site, terminating in the "*Zwischenkörper*." Many of Flemming's observations concerning cytokinesis and the midbody were later confirmed and extended to an ultrastructural level of description through electron microscopy (Buck and Tisdale

C. Pohl (✉)
Buchmann Institute for Molecular Life Sciences, Institute of Biochemistry II, Goethe University Medical School, Max-von-Laue-Strasse 15, 60438 Frankfurt (Main), Germany
e-mail: pohl@em.uni-frankfurt.de

J.-P. Tassan, J.Z. Kubiak (eds.), *Asymmetric Cell Division in Development, Differentiation and Cancer*, Results and Problems in Cell Differentiation 61,
DOI 10.1007/978-3-319-53150-2_7

1962a, b; Robbins and Gonatas 1964; Paweletz 1967; Jones 1969; Mullins and Biesele 1973).

While Flemming's observations—when compared with our current knowledge of midbody formation—are by and large correct, he was of course missing the numerous molecular regulatory mechanisms and the molecular constituents of the midbody which research in the past 15 years has revealed (comprehensively reviewed recently; Chen et al. 2013; Agromayor and Martin-Serrano 2013; Elia et al. 2013; Schiel et al. 2013; Bhutta et al. 2014; Mierzwa and Gerlich 2014; D'Avino et al. 2015; D'Avino and Capalbo 2016). Importantly, it has become clear that the midbody requires furrow constriction to form from the spindle midzone and that it changes its morphology during the progression from final stages of furrow constriction to cytokinetic abscission. Accordingly, midbody components show distinct spatiotemporal dynamics.

The midbody can be subdivided into three domains (Fig. 7.1a): (a) a central midbody core (also called dark zone; Hu et al. 2012), consisting of bundled microtubules, PRC1 (bundling factor), centralspindlin (MKLP1, MgcRacGAP), and the kinesin KIF4; (b) a midbody ring (or bulge), which encircles the core at its midpoint and contains former cleavage furrow proteins (Actin, RhoA, Anillin, Citron kinase), centralspindlin, as well as numerous other factors (ECT2, CEP55, ARF6, Centriolin, Apollon/BRUCE/BIRC6); and (c) the midbody arms (or flanking regions), which contain the chromosomal passenger complex (CPC, with its kinase Aurora B), MKLP2, and other factors, many of which might be degraded during late stages of cytokinesis through ubiquitination (Pohl and Jentsch 2008).

During late stages of cytokinesis, the midbody ring component CEP55 recruits TSG101 and ALIX, a subunit of the endosomal sorting complex required for transport I (ESCRT-I) and an ESCRT-associated factor, respectively (Carlton and Martin-Serrano 2007; Morita et al. 2007). The ESCRT complexes constitute a series of multi-protein assemblies that have been first described to mediate the sorting of ubiquitinated proteins into multi-vesicular bodies (Katzmann et al. 2001; Babst et al. 2002a, b). It has been speculated that recruitment of ESCRTs to the midbody ring could be ubiquitin-dependent (Pohl 2009; Fig. 7.1a, **middle panels**). In accordance with their sequential function and membrane sculpting activity, TSG101 forms two rings adjacent to the midbody ring (Elia et al. 2011), eventually leading to the recruitment of two distal structures composed of ESCRT-III subunits (Christ et al. 2016). Work from several labs has shown that ESCRT-III is required for further ingression of the intercellular bridge distal to the midbody ring (also called secondary ingression site) and for abscission; however, there are different models discussed for the precise mechanism (Bhutta et al. 2014; Fig. 7.1b). Notwithstanding the mechanistic controversy, it seems most likely that ESCRT-III polymerization nucleating at the midbody ring together with recruitment of the ESCRT-III-associated ATPase VPS4 mediates membrane scission (Elia et al. 2013).

Soon after the discovery of ESCRT-III mediating abscission, it has been found that a main target of the abscission checkpoint, which prevents premature abscission through unsegregated chromatids thereby safeguarding the genome (Norden et al. 2006; Mendoza et al. 2009; Steigemann et al. 2009; Fig. 7.1c), is the ESCRT-III subunit CHMP4C (Carlton et al. 2012). Chromatin in the intracellular bridge leads to the activation of the kinase Aurora B, which in turn phosphorylates CHMP4C

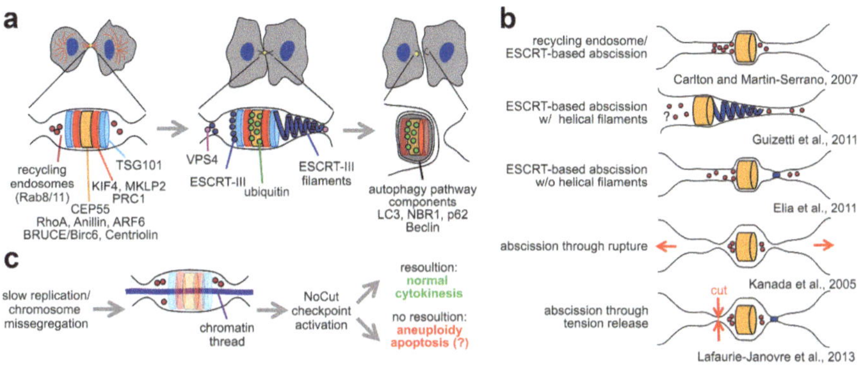

Fig. 7.1 Midbody structure and dynamic changes during abscission. (**a**) Schematic depiction of cell cycle stage (*top*) and spatial organization of the midbody (*bottom*) from late stages of cytokinesis (*left*) to abscission (*center*) and to post-abscission (after midbody remnant uptake). Selected proteins and their localization are listed. For details, see Hu et al. (2012). Notably, it has not been agreed on a "canonical" localization of ESCRT-III yet (see text for details). (**b**) Different models for cytokinetic abscission. *Red arrows* indicate pulling forces (Kanada et al. 2005) or laser cutting (Lafaurie-Janovre et al. 2013). See text for details. (**c**) Mechanism of NoCut checkpoint activation. See text for details

making it unavailable for ESCRT-III-mediated membrane remodeling probably by binding to the chromosomal passenger complex protein Borealin instead (Capalbo et al. 2012). Additionally, a newly identified factor, ANCHR (Abscission/NoCut Checkpoint Regulator), can bind to VPS4 in an Aurora B activity-dependent manner and even seems to form a ternary complex with CHMP4C (Thoresen et al. 2014). Interestingly, it has been shown that in male germ cells, where abscission has to be arrested, the interaction between CEP55 and TSG101/ALIX is blocked by the testis-specific protein TEX14 (Iwamori et al. 2010). Thus, there seem to exist multiple ways to prevent productive ESCRT-III complex assembly, either to prevent aneuploidy during abscission checkpoint signaling or regulated by developmental/differentiation signals.

After abscission, the midbody core and the ring transform into the midbody remnant. Previously, there have been contradicting reports how this might occur—through asymmetric abscission on one side, through rupture of the intercellular bridge, or through sequential abscission on both sides. Recent time-lapse microscopy and ultrastructural analysis strongly suggests that latter seems to occur (at least in cultured cells; Guizetti et al. 2011; Lafaurie-Janovre et al. 2013; Fig. 7.1b), thus transforming the remnant of the midbody into an extracellular structure. At the end of this chapter, potential asymmetries and implications of these asymmetries for cell polarization and fate will be discussed.

Taken together, sequential recruitment of factors during spindle midzone bundling and furrow constriction leads to the formation of the midbody. On its arms, decrease in Aurora B activity and potential protein degradation lead to progression of cytokinesis, eventually leading to the ubiquitination of the midbody ring, the recruitment of ESCRT complex components, membrane deformation, and abscission.

Besides the abovementioned dynamics at the midbody, there are numerous reports showing membrane trafficking to the midbody. Specifically, Rab8/Rab11/Rab35 trans-Golgi and endosomal vesicles are targeted to the midbody by the Sec6/Sec8 or exocyst membrane tethering complex (Horgan et al. 2004; Wilson et al. 2005; Fielding et al. 2005; Gromley et al. 2005; Chen et al. 2006; Kouranti et al. 2006; Pohl and Jentsch 2008; Kaplan and Reiner 2011), which requires RalA and RalB for furrow and midbody localization, respectively (Chen et al. 2006; Cascone et al. 2008). In addition, midbody localization of endosomal vesicles and abscission also depend on the syntaxin Sx16 (Neto et al. 2013) and on the septin Sept9 (Estey et al. 2010; Kim et al. 2011). Polarized trafficking to the midbody and many of the membrane remodeling and targeting factors involved in abscission are also required during the formation of apical membrane domains at the site of the midbody/midbody remnant (see Sect. 7.2 below).

It is important to consider that there is apparently no "standard" way of furrowing, midbody formation, and abscission since each process can follow dramatically different routes depending on the cell type and the developmental or pathophysiological context (i.e., see Sanger et al. 1998; Geddis et al. 2007; Margall-Ducos et al. 2007; Kosodo et al. 2008; Ettinger et al. 2011; Figard et al. 2013). What is summarized above is mainly the presumptive default pathway and only reflects midbody formation and abscission in the common repertoire of cultured (and mostly transformed) mammalian cell lines. Cytokinesis during development or in intact tissues and organs requires maintenance of cell–cell contacts, which imposes spatial constraints (see under Sect. 7.4 below). Therefore, it is crucial that the force-generating molecular machineries of the different phases of cytokinesis—actomyosin during furrowing, microtubule bundling factors during midbody formation, and ESCRT complexes during abscission—are all malleable in terms of modular regulation, cell type-specific accessory factors, etc. On the one hand, this makes cytokinesis a complex process to study since mechanistic details will depend on the particular setting used during investigations. On the other hand, it allows researchers to unravel in which way developmental or pathophysiological constraints entail differential use of ubiquitous molecular machineries.

7.2 The Midbody as a Polarity Cue for Cell Polarization

7.2.1 Luminogenesis

Polarity factors not only control asymmetric cell division but are also crucial for establishing and maintaining apicobasal polarity of epithelia (see chapters by Meraldi and Hinck). If epithelial cells are grown in a medium containing or embedded in extracellular matrix (ECM) components (e.g., collagen, laminin, matrigel), they will form cysts with a lumen (Fig. 7.2). The cyst cell membranes that line the lumen are apically polarized while those contacting the medium are basally polarized. Pioneering work from the lab of the late Alan Hall and of Ben Margolis' lab first demonstrated that the midbody defines the site where apical polarity factors accumulate and where an apical lumen can form *de novo* in cysts (Jaffe et al. 2008; Schlüter et al. 2009). Generally, in apico-basally polarized cells, ingression of the cleavage furrow is asymmetric, from basal to apical (Reinsch and Karsenti 1994; Fleming et al. 2007). If this is preserved in a three-dimensional structure, an apical domain can be maintained at the center of a cyst. Jaffe et al. showed that this requires the function of the GTPase and polarity factor CDC42. CDC42 maintains spindle orientation so that cytokinesis occurs at the apical membrane. Moreover, already during the first division, the apical polarity factor aPKC (part of the PAR3/PAR6/aPKC apical polarity complex) was found to be associated with the midbody at the center of the interface (future lumen) between the daughter cells (Jaffe et al. 2008). Schlüter et al. in turn showed that Rab11-dependent endosomal trafficking along microtubules transports the Crumbs complex (Pocha and Knust 2013) component Crumbs3 also to this site, which assists in the recruitment of aPKC and PAR6. Subsequent work from the Mostov lab then corroborated that polarized trafficking to a membrane domain that forms around the midbody is crucial for *de novo* luminogenesis (Bryant et al. 2010; Fig. 7.2). Bryant et al. find that luminogenesis first requires a polarity inversion: the apical membrane determinant gp135/podocalyxin, which is initially localized to nascent basal membranes of a forming cyst, has to move to the nascent apical site, while β-catenin, which is initially localized to the cell–cell contact, has to move to basal membranes. This is accomplished by podocalyxin being internalized through Rab8/11 recycling endosomes, transporting it to a special site around the midbody which has been coined apical membrane initiation site (AMIS). Subsequently, the AMIS transforms into two domains: the apical membrane proper and nascent tight junctions (containing apical polarity factors like aPKC and PAR3/6). Recently, the Mostov lab (Bryant et al. 2014) has identified that initial polarity inversion is signaled from the extracellular matrix: a RhoA-dependent podocalyxin-containing complex is eventually dissociated by the action of basement membrane-derived signaling to an α2/α3/β1-integrin/FAK/p190ARhoGAP module and by PKCβII. This leads to podocalyxin endocytosis and relocation to the AMIS. How this is coordinated with midbody formation is not clear yet, for instance, whether midbody formation and/or

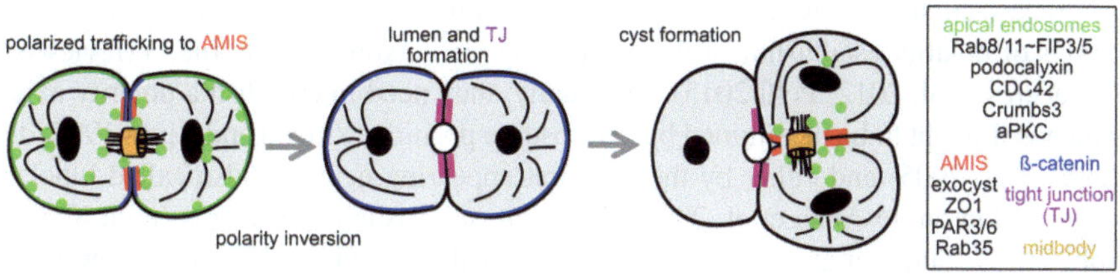

Fig. 7.2 Schematic depiction of *de novo* lumen formation. *Left*: A cell doublet grown in matrix during cytokinesis. Note that factors that later line the apical surface of the nascent lumen (e.g., podocalyxin, Crumbs3) are initially localized to what later becomes a basal surface. Rab8/11/35 vesicles traffic along micortubules to the midbody and the directly adjacent apical membrane initiation site (AMIS). *Middle*: The AMIS transforms into tight junctions (TJ). *Right*: In subsequent divisions, spindle positioning (guided by CDC42) and furrowing have to be coordinated with the existing lumen and tTJs. Box: Representative proteins of apical endosomes (*green*) and the AMIS (*red*) are listed

progression of abscission also triggers re-localization of β-catenin from cell–cell contacts to cell–ECM contacts.

Concerning the regulation of Rab11 vesicle transport to the AMIS, it has been recently shown that the Rab11-binding protein FIP5 is involved in targeting podocalyxin/Crumbs3-positive endosomes to the midbody (Li et al. 2014). Specifically, Li et al. have shown that the kinase GSK-3 phosphorylates FIP5 during metaphase/anaphase, which blocks its apical transport by inhibiting the binding to the sorting nexin SNX18, which has been shown to be involved in luminogenesis previously (Willenborg et al. 2011). This inhibition might act as a timer to allow transport only when the midbody is ready to scaffold or organize the AMIS. In addition to these factors, timely and sequential recruitment of PAR3 and the tight junction component ZO-1 to or around the midbody is also crucial for AMIS specification (Wang et al. 2014). Moreover, besides Rab8 and Rab11, Rab35 is also involved in transport of Cdc42, Crumbs3, and Podocalyxin to the newly forming lumen during cytokinesis (Klinkert et al. 2016). Thus, by coupling division to luminogenesis through potentially redundant mechanisms and by subsequently coupling spindle orientation to the preexisting apical membrane, maintenance of a growing single lumen in a multicellular cyst or tubule can be ensured. Moreover, both the midbody and the extracellular matrix seem to act as symmetry-breaking elements in this case.

It will be interesting to identify how the establishment of an AMIS alters the localization and/or recruitment of ESCRT complexes to the midbody and whether other midbody components show different dynamics when cell division also entails luminogenesis. Furthermore, differences in midbody fate between non-polarized cells, in apico-basally polarized cells in a mononolayer, and in cysts have not been systematically analyzed so far.

7.2.2 *Sprouting and Polarized Divisions in Neuronal Cells*

Neuronal cells are prototypic examples of polarized cells. From an initially non-polarized state, they need to undergo a symmetry-breaking process and several phases of polarity establishment that eventually yield axon and dendrite(s) (reviewed in Cáceres et al. 2012; Pohl 2015). Eventually, they need to establish a polarity axis, which is thought to be determined by centrosome position (Zmuda and Rivas 1998; de Anda et al. 2005) and either by the two first appearing neurites (also called bipolar stage; Calderon de Anda et al. 2008) and/or by stochastic mechanisms (reviewed in Pohl 2015). Surprisingly, Drosophila embryos without centrosomes still show relatively normal development (Basto et al. 2006) and neuronal precursor cells properly polarize (Pollarollo et al. 2011), arguing that additional factors besides centrosomes contribute to neuronal polarity establishment.

Work from the Dotti lab demonstrated that the midbody transforms into the site of initial apical neuronal sprouting (Pollarollo et al. 2011; Fig. 7.3). Specifically, Pollarollo et al. reported that in the Drosophila notum both Rho1 and Aurora A accumulate in the midbody during neural precursor divisions and remain in the remnant.

They also find that phosphatidylinositol-4,5-bisphosphate (PI(4,5)P_2) becomes enriched at the site of the midbody. Rho1/Aurora A accumulation and PI(4,5)P_2 enrichment preceded recruitment of the polarity factor Bazooka/PAR-3, which binds PI(4,5) P_2 and is a main organizer of neuronal polarity. In addition, DE-Cadherin is recruited to this site and it has been later shown that the accumulation of N-Cadherin in mammalian neurons also occurs at the site of initial sprouting (Gärtner et al. 2012). Thus, similar to luminogenesis, the midbody mediates symmetry breaking by recruiting apical polarity factors and other molecules essential for initiating membrane dynamics.

Depending on the developmental context, there seem to be additional signals from surrounding cells that contribute to the asymmetric distribution of polarity factors in *de novo* polarizing neurons: during neural rod formation in zebrafish, neural progenitors that undergo midline crossing divisions accumulate apical polarity markers at the cleavage furrow (Tawk et al. 2007). However, this is not fully dependent on cell cycle/ cell division since apical polarity marker accumulation at the midline also occurs in division-blocked progenitors (Buckley et al. 2013). In this case, polarization is mediated by the centrosome (organizing a mirror-symmetric microtubule array in the cell) and cell–cell as well as cell–ECM contacts. In contrast, midline crossing divisions are followed by luminogenesis for which apical polarity factors and Rab11-mediated polarized trafficking require proper progression of cell division. Unfortunately, it is currently not clear how important midbody position and function as a polarity organizer are for neural rod development. Taken together, this data suggests that the midbody might be dispensable for polarized growth (which is less cell cycle dependent) but may play an important role during luminogenesis (which seems to be cell cycle dependent).

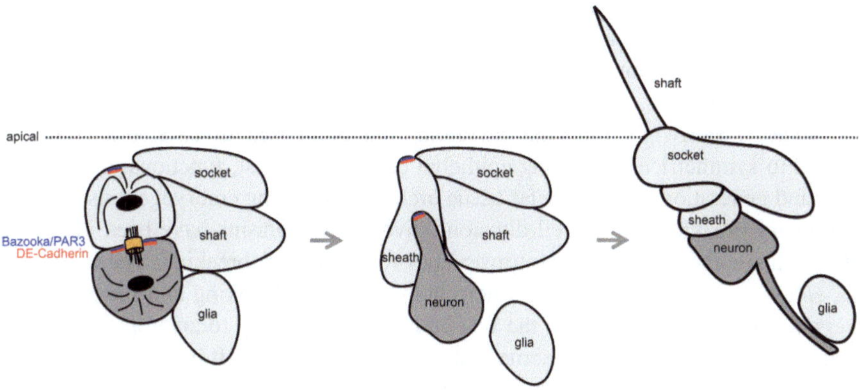

Fig. 7.3 Polarity establishment in Drosophila sensory neurons. During division of IIIb, which gives rise to the neuron and the sheath cell of the sensory organ, PAR3 and DE-cadherin start to accumulate at the membrane around the midbody in the neuron and also at the apical pole of the sheath cell (*left*). Subsequently, cells start to sprout toward the apical membrane (*middle*), eventually leading to a fully differentiated sensory organ

7.3 The Midbody and Spindle Orientation during Asymmetric Cell Division

For proper organization of tissues or organs, cells have to orient a cell division plane according to intrinsic cues and according to cues from their local environment. The major biomechanical mechanism for division plane orientation that inherently integrates both cues is cell shape. Research in experimental embryology during the late nineteenth century (Hofmeister 1863; Pflüger 1884; Driesch 1893; Hertwig 1893) has led to the postulation of what we now call Hertwig's rule. This rule states that a cell will divide along the long axis it had during interphase (Fig. 7.4). Recent work has confirmed this rule quantitatively (Gibson et al. 2011; Minc et al. 2011; Fig. 7.4b) and could also demonstrate that this is due to nuclear centering through microtubules according to the center of mass (Minc et al. 2011) and packing constraints imposed by the local environment (Gibson et al. 2011). However, since cells usually round up during mitosis to escape epithelial confinement (Ramanathan et al. 2015; Sorce et al. 2015), an additional mechanism of shape memory has to operate during division. In cultured cells, shape memory requires cues from the ECM that in turn organize local phospholipid, actin, and microtubule dynamics leading to segregation of cortical components affecting spindle orientation (Théry et al. 2005, 2007; Toyoshima et al. 2007; Toyoshima and Nishida 2007; Fink et al. 2011). In epithelial tissues where cells have to divide planarly polarized, tricellular junctions (showing higher levels of, e.g., E-cadherin, Discs-large/Dlg, or Gliotactin) instead of ECM represent this spatial memory (Bosveld et al. 2016; Fig. 7.4a). These tricelluar junctions recruit cortical force generators during interphase that stay localized at these sites during mitosis to polarize the division plane. Furthermore, it has been shown previously that in planar polarized mammalian epithelia, Cadherin-based cortical cues are crucial for planar oriented divisions (den Elzen et al. 2009). Thus, similar to neuron sprouting, Cadherin-based cortical cues can organize polarized growth although without the apparent requirement for midbody-based assembly of these cues.

Notably, Hertwig's rule and tricellular junction-based division orientation is linked to symmetrically dividing epithelial cells. In contrast, functional diversification and generation of a particular tissue architecture during embryogenesis requires symmetry breaking during cell division. Diverse mechanisms have been identified that utilize either extrinsic or intrinsic cues for symmetry breaking. The nematode *C. elegans* is a model organism where both symmetry breaking and functional diversification already start with the first cell division, leading to anteroposterior axis formation and segregation of soma and germline (Fig. 7.4c). It has been speculated that during the second round of cell divisions, 90° rotation of the nucleus–centrosome complex and, later on, of the spindle in the germline founder cell might require structures that remained from cytokinesis (Hyman and White 1987; Hyman 1989, Waddle et al. 1994, Keating and White 1998; Skop and White 1998). Recently, it was shown that rotation in the germline founder cell indeed requires the midbody remnant from the first embryonic division (Singh and Pohl 2014a; Fig. 7.4c, right panels). This midbody remnant becomes invariably displaced ventrally and serves as platform for the

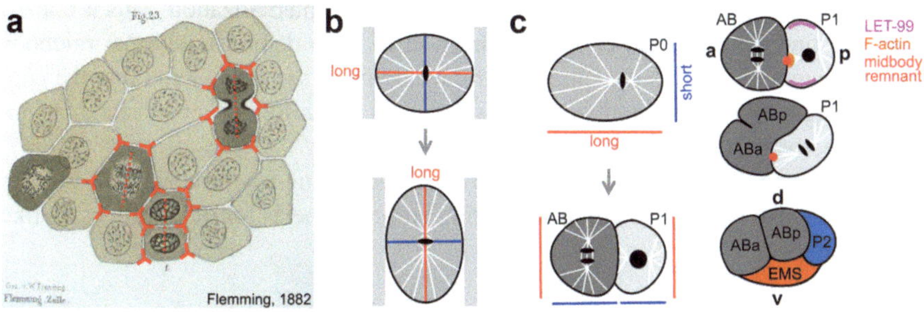

Fig. 7.4 Hertwig's rule and midbody-based spindle orientation in *C. elegans*. (**a**) Walther Flemming's drawing of an epithelium with dividing cells from his 1882 book on cell division (Flemming 1882). *Red dashed lines* indicate the long axis of dividing cells. Red triskelions depict tricellular junctions of dividing cells. (**b**) Reorientation of the division axis according to Hertwig's rule. See also experiments in (Minc et al. 2011). (**c**) *Left*: The first two rounds of cell division in the *C. elegans* embryo. The first division (*top*) follows Hertwig's rule, while during the second round of divisions (*bottom*) the posterior cell P1 violates this rule. *Right*: Schematic depiction of midbody remnant-based spindle rotation during asymmetric division of the germline founder cell P1. See text for details

formation of a specific cortical domain that will direct spindle orientation during asymmetric division of the germline founder cell P1 into P2 and the endomesodermal founder cell EMS. Thus, the midbody remnant of the first embryonic division functions as a cue during symmetry breaking in the subsequent division. Developmentally, this is an important mechanism since it leads to the formation of the dorsoventral axis by placing endomesoderm ventrally and ectoderm dorsally. Moreover, the use of the midbody remnant as a polarity cue allows cell division to occur along the short axis of the cell, thus violating Hertwig's rule.

Mechanistically, a combination of two previously proposed models can explain the interplay between extrinsic (midbody-derived) and intrinsic (anteroposterior polarity-derived) cues necessary for this violation (Singh and Pohl 2014b). The older model—coined "cortical-site" model—argues that a special site in the P1 cortex together with the spatial constraints of the ellipsoid egg will mediate nucleus–centrosome complex/spindle rotation (Hyman and White 1987; Hyman 1989, Waddle et al. 1994, Keating and White 1998; Skop and White 1998). The "cortical site" model has been disputed by the "LET-99" model, proposing that lateral localization of the conserved polarity factor LET-99 to the prospective dorsal and ventral sides of the P1 cell would be sufficient to mediate nucleus–centrosome complex/spindle rotation (Tsou et al. 2002, 2003). Due to LET-99's function reducing cortical pulling forces (Rose and Kemphues 1998; Tsou et al. 2002), its bilateral localization pattern is thought to bias cortical pulling forces along the anteroposterior axis, thereby leading to rotation. Singh and Pohl reconcile these opposing views by arguing that LET-99 might indeed bias spindle rotation (Singh and Pohl 2014b); however, they also demonstrate that the midbody remnant is crucial for tethering the spindle to the cortex during rotation (Singh and Pohl 2014a). Tethering occurs through formation of a dense actin cortical structure around the midbody remnant, which forms asymmetrically requiring proper anteroposterior polarity but not the midbody core or midbody remnant internalization. This is consistent with findings from Ou et al. and Green et al. who demonstrated that midbody remnant internalization is dispensable for development (see below) and that the contractile ring-derived outer midbody ring is sufficient to scaffold the abscission machinery, respectively (Green et al. 2013; Ou et al. 2014).

Thus, the midbody can serve as polarizing cue during asymmetric cell division in vivo and also seems to be required during embryogenesis for other processes like luminogenesis.

7.4 The Furrow and Midbody in Polarized Epithelia

Once polarized tissues have formed during development, it is essential to maintain this architecture during tissue growth. However, each cell division represents a transient lesion since cells often round up during division, have to form a contractile ring that has to ingress, and establish a new cell–cell contact simultaneously. This "lesion" has to be resolved so that tissue polarity and organization are not harmed.

Often, furrowing is directional in order to preserve tissue architecture. It can proceed from basal to apical (e.g., in many differentiating epithelia; Kosodo et al. 2004; Dubreuil et al. 2007; see below), proceed from apical to basal (e.g., in the early *C. elegans* embryo; Carvalho et al. 2009; Pohl and Bao 2010), or show random asymmetry (e.g., in the *C. elegans* one-cell embryo; Maddox et al. 2007). In the latter case, although Anillin and Septins are required for random asymmetry during furrowing in the *C. elegans* one-cell embryo (Maddox et al. 2007), such random asymmetries have not been observed so far in other systems yet. Recent analyses suggest that random asymmetry during furrowing is most likely due to cortical chirality, which manifests as simultaneous rotational and translational actomyosin flows (Singh and Pohl 2014a; Naganathan et al. 2014). Chiral cortical contractile actomyosin flow will lead to the formation of an asymmetrically organized contractile ring (with higher levels of non-muscle myosin II and Septins on one side of the contractile ring) that will progressively constrict from one random point on the embryo's equator. Due to restriction of actomyosin to apical surfaces (once cells have established apicobasal polarity at the 4-cell embryo stage), cortical chirality can also mediate left–right asymmetric furrowing (Schonegg et al. 2014). Cortical chirality is also used to asymmetrically position the midbody remnant during dorsoventral axis formation (see above; Singh and Pohl 2014a). Since we currently have a limited understanding of what establishes intrinsic cellular chirality (Pohl 2015) and since chiral behaviors seem to be cell type specific, it is possible that random asymmetries during furrowing might actually be used to drive various directional processes at different stages of development. Furthermore, these chiral dynamics during cytokinesis would in principle also allow to generate an intrinsically asymmetric midbody remnant.

Using Drosophila development as a model, several recent studies have identified mechanisms that show how directional asymmetry during furrowing and asymmetric midbody positioning contribute to maintenance of tissue architecture during growth

(Founounou et al. 2013; Guillot and Lecuit 2013; Herszterg et al. 2013; Morais-de-Sa and Sunkel 2013; reviewed in Herszterg et al. 2014). In the case of the Drosophila embryonic ectoderm and the follicular epithelium, a symmetrically organized contractile ring undergoes asymmetric furrowing from basal to apical by adherens junction-mediated interactions with the apical part of the furrow through the Cadherin–Catenin complex (Guillot and Lecuit 2013; Morais-de-Sa and Sunkel 2013). Moreover, apically localized E-Cadherin is sufficient to recruit the midbody to the apical side (Morais-de-Sa and Sunkel 2013). This suggests that apical junction components break symmetry during contractile ring closure. However, more similar to embryonic cytokinesis in *C. elegans*, asymmetric contractile rings that lead to directional furrowing are also found in Drosophila, e.g., in the pupal notum (Herszterg et al. 2013). In addition, if cells in an epithelium round up during mitosis and become apically extruded like in the Drosophila pupal wing or follicular epithelium, a transition to symmetric furrowing depending on the degree of extrusion will lead to apical midbody positioning at the end of cytokinesis (Herszterg et al. 2013; Morais-de-Sa and Sunkel 2013).

Remarkably, similar to midbody-mediated spindle orientation in *C. elegans*, an F-actin-rich structure is also observed around the midbody in Drosophila epithelia, which seems to be crucial to force the neighboring cells—that invaded this territory during furrowing—away to form the new cell–cell contact and apical junctions between daughter cells (Herszterg et al. 2013). Importantly, this mechanism of midbody-based *de novo* cell–cell contact establishment also maintains apicobasal polarity by providing directionality for apical junction biogenesis (Morais-de-Sa and Sunkel 2013). It seems very plausible to speculate that this is brought about by similar mechanisms that lead to luminogenesis and sprouting, namely Rab8/11/35-dependent polarized trafficking (see above).

7.5 Asymmetric Midbody Remnant Inheritance and Regulation of Cell Fate

Until now, there is no direct proof that midbodies affect cell fate directly. To give direct proof is difficult for several reasons: (1) It has become clear that the fate of postmitotic midbodies varies depending on the cell type and cellular context (Ettinger et al. 2011; Chen et al. 2013; Singh and Pohl 2014b); (2) the precise stage at which midbodies need to be present to alter cell fate remains elusive; and (3) cells that are completely devoid of midbodies cannot be generated since this would mean to block progression of cytokinesis. Notwithstanding these issues, evidence has accumulated that midbody inheritance seems to be connected to cell fate in several models.

In the *C. elegans* embryo, the first midbody is asymmetrically inherited and is required for proper cell placement and cell fate inductions (Singh and Pohl 2014a). Asymmetric inheritance seems to depend on cortical tension; cells with low cortical tension are more likely to internalize midbodies, probably due to the fact that midbody internalization seems to mainly occur by engulfment/phagocytosis (Singh and Pohl

2014b; Crowell et al. 2014; Fig. 7.5a), which is facilitated when the cortex is more deformable. However, although the presence of the midbody is required for proper patterning, its asymmetric inheritance is not (Green et al. 2013; Ou et al. 2014). Therefore, cell fate regulation through midbodies in *C. elegans* seems most likely to involve nascent midbody remnants organizing asymmetric cortical domains and thereby cell and tissue patterning. Hence, it is also plausible that midbodies in *C. elegans* might also be involved in processes discussed above like luminogenesis, polarization of the cytoskeleton, and maintenance of a polarized tissue architecture.

In *Drosophila*, asymmetric midbody inheritance has been investigated for germline stem cell divisions (Salzmann et al. 2014; Fig. 7.5b). Here, midbody inheritance correlates with asymmetric centrosome inheritance: while during asymmetric male germline stem cell divisions the stem cell inherits the mother centriole and the differentiating blast cell inherits the midbody remnant, female stem cell divisions give rise to a stem cell that inherits the midbody remnant and a blast cell that inherits the mother centrosome. Thus, independently of cell fate, the cell inheriting the daughter centrosome also inherits the midbody remnant. At this stage, it is not possible to compare this mechanism to asymmetric midbody inheritance in *C. elegans* since centrosome asymmetries have not been investigated so far. Moreover, it has been argued that instead of modulating cell fate, midbody remnants in Drosophila might rather regulate cell behaviors such as cell division timing (Salzmann et al. 2014). Importantly, these findings strongly argue against a general inheritance mechanism for midbody remnants and instead suggest cell type-specific regulation of midbody inheritance.

In mammalian cell cultures, midbody remnants also seem to be inherited asymmetrically being retained by the cell with the mother centrosome (although not being absolute but ranging from ~75% to ~90%; Kuo et al. 2011; Fig. 7.5c). Moreover, midbody remnants were found to associate with several stem cell compartments in vivo like basal layers of mouse testes seminiferous tubules, ventricular progenitor cells in the mouse brain, mouse skeletal muscle progenitors, and the bulge of human hair follicles. Also, in vitro, midbodies accumulated when cells were dedifferentiated into induced pluripotent stem cells. Remarkably, in stem cells, midbody accumulation seems to positively affect reprogramming efficiency, while for cancer cells, midbody accumulation results in enhanced tumorigenicity (Kuo et al. 2011). Notably, midbody accumulation was shown to correlate with the level of autophagy activity. Midbody clearance by autophagy requires ubiquitylation of the midbody component MKLP1 and its recognition by the ubiquitin-binding autophagy receptor p62 and—probably redundantly—the midbody component CEP55 interacting with the autophagy receptor NBR1 (Pohl and Jentsch 2008; Kuo et al. 2011; Isakson et al. 2013). However, it seems that midbody degradation by conventional macroautophagy is not the conventional way of disposal; rather, autophagy-assisted phagocytosis seems to be the predominant pathway—at least in cultured cells (Crowell et al. 2014). Thus, the "standard" route of midbody degradation seems to involve recruitment of autophagy pathway components and subsequent engulfment/phagocytosis whereby the engulfed midbody remnant is then routed to lysosomes. Again, as stated above, this "standard" route does not always apply since neuronal stem cells for instance also seem to actively extrude the midbody

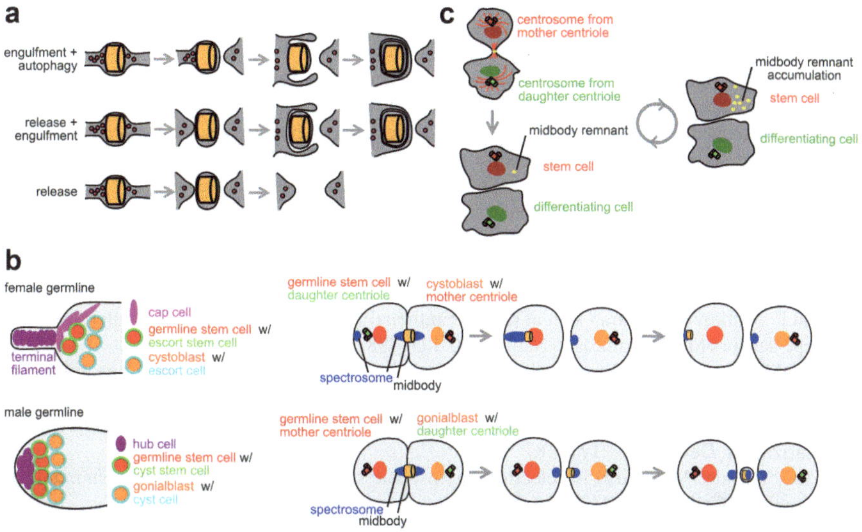

Fig. 7.5 Fate of the midbody and inheritance mechanisms in asymmetrically dividing stem cell. (**a**) Different fates for the midbody that seem to depend on the cellular context. In some cases, midbodies seem to be engulfed after abscission on one side (*top*), while in tissue culture, abscission seems to occur on both sides and midbodies are either engulfed afterward (*middle*) or released (*bottom*). (**b**) Asymmetric stem cell divisions in the Drosophila female (*top*) and male (*bottom*) germlines. *Left*: Schematic depictions of the very proximal part of the respective germline. *Right*: Cartoons illustrating the inheritance of centrosomes and midbodies. Note the contribution of the spectrosome (*blue*) to midbody inheritance during female germline stem cell division. The spectrosome is a germline-specific organelle rich in cytoskeletal and plasma membrane-associated proteins. It is the precursor to the fusome that later connects germline cells in their cysts (not shown). (**c**) Asymmetric inheritance of centrioles and midbodies during asymmetric stem cell divisions in mammalian systems

remnant rather than degrading it intracellularly, thereby even using it as a vehicle for stemness factor hitchhiking (Dubreuil et al. 2007).

Based on this data, it has been speculated that midbody remnants could represent signaling entities due to receptor clustering or accumulation of specific transmembrane proteins (Chen et al. 2013). This would explain the opposing fates observed for midbody remnants in different stem cell niches in vivo and also functions affecting proliferation and (de-)differentiation. To disentangle apparent controversial results, it will be necessary to study midbody fate cell type-specifically and to identify if and through which mechanisms midbody remnant components engage in midbody-cell signaling.

Acknowledgements Research in the laboratory of CP is funded by the Deutsche Forschungsgemeinschaft (EXC 115, FOR 1756, SFB 1177) and the LOEWE Research Cluster Ubiquitin Networks. CP's research concerning developmental functions of the midbody was supported by a European Union Framework Program 7 fellowship (Marie Curie Actions Project 326632).

References

Agromayor M, Martin-Serrano J (2013) Knowing when to cut and run: mechanisms that control cytokinetic abscission. Trends Cell Biol 23:433–441

Babst M, Katzmann DJ, Estepa-Sabal EJ et al (2002a) Escrt-III: an endosome-associated hetero-oligomeric protein complex required for mvb sorting. Dev Cell 3:271–282

Babst M, Katzmann DJ, Snyder WB et al (2002b) Endosome-associated complex, ESCRT-II, recruits transport machinery for protein sorting at the multivesicular body. Dev Cell 3:283–289

Basto R, Lau J, Vinogradova T et al (2006) Flies without centrioles. Cell 125:1375–1386

Bhutta MS, McInerny CJ, Gould GW (2014) ESCRT function in cytokinesis: location, dynamics and regulation by mitotic kinases. Int J Mol Sci 15:21723–21739

Bosveld F, Markova O, Guirao B et al (2016) Epithelial tricellular junctions act as interphase cell shape sensors to orient mitosis. Nature 530:495–498

Bryant DM, Datta A, Rodríguez-Fraticelli AE et al (2010) A molecular network for de novo generation of the apical surface and lumen. Nat Cell Biol 12:1035–1045

Bryant DM, Roignot J, Datta A et al (2014) A molecular switch for the orientation of epithelial cell polarization. Dev Cell 31:171–187

Buck RC, Tidsale JM (1962a) An electron microscopic study of the cleavage furrow in mammalian cells. J Cell Biol 13:117–125

Buck RC, Tisdale JM (1962b) The fine structure of the mid-body of the rat erythroblast. J Cell Biol 13:109–115

Buckley CE, Ren X, Ward LC et al (2013) Mirror-symmetric microtubule assembly and cell interactions drive lumen formation in the zebrafish neural rod. EMBO J 32:30–44

Cáceres A, Ye B, Dotti CG (2012) Neuronal polarity: demarcation, growth and commitment. Curr Opin Cell Biol 24:547–553

Calderon de Anda F, Gärtner A, Tsai LH et al (2008) Pyramidal neuron polarity axis is defined at the bipolar stage. J Cell Sci 121:178–185

Capalbo L, Montembault E, Takeda T et al (2012) The chromosomal passenger complex controls the function of endosomal sorting complex required for transport-III Snf7 proteins during cytokinesis. Open Biol 2:120070

Carlton JG, Martin-Serrano J (2007) Parallels between cytokinesis and retroviral budding: a role for the ESCRT machinery. Science 316:1908–1912

Carlton JG, Caballe A, Agromayor M et al (2012) ESCRT-III governs the Aurora B-mediated abscission checkpoint through CHMP4C. Science 336:220–225

Carvalho A, Desai A, Oegema K (2009) Structural memory in the contractile ring makes the duration of cytokinesis independent of cell size. Cell 137:926–937

Cascone I, Selimoglu R, Ozdemir C et al (2008) Distinct roles of RalA and RalB in the progression of cytokinesis are supported by distinct RalGEFs. EMBO J 27:2375–2387

Chen XW, Inoue M, Hsu SC et al (2006) RalA-exocyst-dependent recycling endosome trafficking is required for the completion of cytokinesis. J Biol Chem 281:38609–38616

Chen CT, Ettinger AW, Huttner WB et al (2013) Resurrecting remnants: the lives of post-mitotic midbodies. Trends Cell Biol 23:118–128

Christ L, Wenzel EM, Liestøl K et al (2016) ALIX and ESCRT-I/II function as parallel ESCRT-III recruiters in cytokinetic abscission. J Cell Biol 212:499–513

Crowell EF, Gaffuri AL, Gayraud-Morel B et al (2014) Engulfment of the midbody remnant after cytokinesis in mammalian cells. J Cell Sci 127:3840–3851

D'Avino PP, Capalbo L (2016) Regulation of midbody formation and function by mitotic kinases. Semin Cell Dev Biol pii:S1084–9521(16)30018–0

D'Avino PP, Giansanti MG, Petronczki M (2015) Cytokinesis in animal cells. Cold Spring Harb Perspect Biol 7:a015834

de Anda FC, Pollarolo G, Da Silva JS et al (2005) Centrosome localization determines neuronal polarity. Nature 436:704–708

den Elzen N, Buttery CV, Maddugoda MP et al (2009) Cadherin adhesion receptors orient the mitotic spindle during symmetric cell division in mammalian epithelia. Mol Biol Cell 20:3740–3750

Driesch HAE (1893) Entwicklungsmechanische Studien. Zeitschrift für Wissenschaftliche Zoologie 55:1–62

Dubreuil V, Marzesco AM, Corbeil D et al (2007) Midbody and primary cilium of neural progenitors release extracellular membrane particles enriched in the stem cell marker prominin-1. J Cell Biol 176:483–495

Elia N, Sougrat R, Spurlin TA et al (2011) Dynamics of endosomal sorting complex required for transport (ESCRT) machinery during cytokinesis and its role in abscission. Proc Natl Acad Sci U S A 108:4846–4851

Elia N, Ott C, Lippincott-Schwartz J (2013) Incisive imaging and computation for cellular mysteries: lessons from abscission. Cell 155:1220–1231

Estey MP, Di Ciano-Oliveira C, Froese CD et al (2010) Distinct roles of septins in cytokinesis: SEPT9 mediates midbody abscission. J Cell Biol 191:741–749

Ettinger AW, Wilsch-Bräuninger M, Marzesco AM et al (2011) Proliferating versus differentiating stem and cancer cells exhibit distinct midbody-release behaviour. Nat Commun 2:503

Fielding AB, Schonteich E, Matheson J et al (2005) Rab11-FIP3 and FIP4 interact with Arf6 and the exocyst to control membrane traffic in cytokinesis. EMBO J 24:3389–3399

Figard L, Xu H, Garcia HG et al (2013) The plasma membrane flattens out to fuel cell-surface growth during Drosophila cellularization. Dev Cell 27:648–655

Fink J, Carpi N, Betz T et al (2011) External forces control mitotic spindle positioning. Nat Cell Biol 13:771–778

Fleming ES, Zajac M, Moschenross DM et al (2007) Planar spindle orientation and asymmetric cytokinesis in the mouse small intestine. J Histochem Cytochem 55:1173–1180

Flemming W (1874) Über die ersten Entwicklungserscheinungen am Ei der Teichmuschel. Arch Mikrosk Anat 10:257–292

Flemming W (1875) Studien in der Entwicklungsgeschichte der Najaden. Sitzungsber Kaiserl Akad Wiss 71:81–212

Flemming W (1876) Beobachtungen über die Beschaffenheit des Zellkerns. Arch Mikrosk Anat 13:693–717

Flemming W (1882) Zellsubstanz, Kern und Zelltheilung. Vogel, Leipzig

Flemming W (1891) Neue Beiträge zur Kenntnis der Zelle. Arch Mikrosk Anat 37:685–751

Flemming W (1965) Contributions to the knowledge of the cell and its vital processes. J Cell Biol 25:3–69

Founounou N, Loyer N, Le Borgne R (2013) Septins regulate the contractility of the actomyosin ring to enable adherens junction remodeling during cytokinesis of epithelial cells. Dev Cell 24:242–255

Gärtner A, Fornasiero EF, Munck S et al (2012) N-cadherin specifies first asymmetry in developing neurons. EMBO J 31:1893–1903

Geddis AE, Fox NE, Tkachenko E et al (2007) Endomitotic megakaryocytes that form a bipolar spindle exhibit cleavage furrow ingression followed by furrow regression. Cell Cycle 6:455–460

Gibson WT, Veldhuis JH, Rubinstein B et al (2011) Control of the mitotic cleavage plane by local epithelial topology. Cell 144:427–438

Green RA, Mayers JR, Wang S et al (2013) The midbody ring scaffolds the abscission machinery in the absence of midbody microtubules. J Cell Biol 203:505–520

Gromley A, Yeaman C, Rosa J et al (2005) Centriolin anchoring of exocyst and SNARE complexes at the midbody is required for secretory-vesicle-mediated abscission. Cell 123:75–87

Guillot C, Lecuit T (2013) Adhesion disengagement uncouples intrinsic and extrinsic forces to drive cytokinesis in epithelial tissues. Dev Cell 24:227–241

Guizetti J, Schermelleh L, Mäntler J et al (2011) Cortical constriction during abscission involves helices of ESCRT-III-dependent filaments. Science 331:1616–1620

Herszterg S, Leibfried A, Bosveld F et al (2013) Interplay between the dividing cell and its neighbors regulates adherens junction formation during cytokinesis in epithelial tissue. Dev Cell 24:256–270

Herszterg S, Pinheiro D, Bellaïche Y (2014) A multicellular view of cytokinesis in epithelial tissue. Trends Cell Biol 24:285–293

Hertwig OWA (1893) Über den Werth der ersten Furchungszellen für die Organbildung des Embryos. Experimentelle Studien am Frosch und Tritonei. Archiv für mikroscopische Anatomie 42:662–807

Hofmeister FWB (1863) Zusatze und Berichtigungen zu den 1851 veröffentlichen Untersuchungengen der Entwicklung höherer Kryptogamen. Jahrbucher für Wissenschaft und Botanik 3:259–293

Horgan CP, Walsh M, Zurawski TH et al (2004) Rab11-FIP3 localises to a Rab11-positive pericentrosomal compartment during interphase and to the cleavage furrow during cytokinesis. Biochem Biophys Res Commun 319:83–94

Hu CK, Coughlin M, Mitchison TJ (2012) Midbody assembly and its regulation during cytokinesis. Mol Biol Cell 23:1024–1034

Hyman AA (1989) Centrosome movement in the early divisions of *Caenorhabditis elegans*: a cortical site determining centrosome position. J Cell Biol 109:1185–1193

Hyman AA, White JG (1987) Determination of cell division axes in the early embryogenesis of *Caenorhabditis elegans*. J Cell Biol 105:2123–2135

Isakson P, Lystad AH, Breen K et al (2013) TRAF6 mediates ubiquitination of KIF23/MKLP1 and is required for midbody ring degradation by selective autophagy. Autophagy 9:1955–1964

Iwamori T, Iwamori N, Ma L et al (2010) TEX14 interacts with CEP55 to block cell abscission. Mol Cell Biol 30:2280–2292

Jaffe AB, Kaji N, Durgan J et al (2008) Cdc42 controls spindle orientation to position the apical surface during epithelial morphogenesis. J Cell Biol 183:625–633

Jones OP (1969) Elimination of midbodies from mitotic erythroblasts and their contribution to fetal blood plasma. J Natl Cancer Inst 42:753–759

Kanada M, Nagasaki A, Uyeda TQ (2005) Adhesion-dependent and contractile ring-independent equatorial furrowing during cytokinesis in mammalian cells. Mol Biol Cell 16:3865–3872

Kaplan A, Reiner O (2011) Linking cytoplasmic dynein and transport of Rab8 vesicles to the midbody during cytokinesis by the doublecortin domain-containing 5 protein. J Cell Sci 124:3989–4000

Katzmann DJ, Babst M, Emr SD (2001) Ubiquitin-dependent sorting into the multivesicular body pathway requires the function of a conserved endosomal protein sorting complex, ESCRT-I. Cell 106:145–155

Keating HH, White JG (1998) Centrosome dynamics in early embryos of *Caenorhabditis elegans*. J Cell Sci 111:3027–3033

Kim MS, Froese CD, Estey MP et al (2011) SEPT9 occupies the terminal positions in septin octamers and mediates polymerization-dependent functions in abscission. J Cell Biol 195:815–826

Klinkert K, Rocancourt M, Houdusse A et al (2016) Rab35 GTPase couples cell division with initiation of epithelial apico-basal polarity and lumen opening. Nat Commun 7:11166

Kosodo Y, Röper K, Haubensak W et al (2004) Asymmetric distribution of the apical plasma membrane during neurogenic divisions of mammalian neuroepithelial cells. EMBO J 23:2314–2324

Kosodo Y, Toida K, Dubreuil V et al (2008) Cytokinesis of neuroepithelial cells can divide their basal process before anaphase. EMBO J 27:3151–3163

Kouranti I, Sachse M, Arouche N et al (2006) Rab35 regulates an endocytic recycling pathway essential for the terminal steps of cytokinesis. Curr Biol 16:1719–1725

Kuo TC, Chen CT, Baron D et al (2011) Midbody accumulation through evasion of autophagy contributes to cellular reprogramming and tumorigenicity. Nat Cell Biol 13:1214–1223

Lafaurie-Janvore J, Maiuri P, Wang I et al (2013) ESCRT-III assembly and cytokinetic abscission are induced by tension release in the intercellular bridge. Science 339:1625–1629

Li D, Mangan A, Cicchini L et al (2014) FIP5 phosphorylation during mitosis regulates apical trafficking and lumenogenesis. EMBO Rep 15:428–437

Maddox AS, Lewellyn L, Desai A et al (2007) Anillin and the septins promote asymmetric ingression of the cytokinetic furrow. Dev Cell 12:827–835
Margall-Ducos G, Celton-Morizur S, Couton D et al (2007) Liver tetraploidization is controlled by a new process of incomplete cytokinesis. J Cell Sci 120:3633–3639
Mendoza M, Norden C, Durrer K et al (2009) A mechanism for chromosome segregation sensing by the NoCut checkpoint. Nat Cell Biol 11:477–483
Mierzwa B, Gerlich DW (2014) Cytokinetic abscission: molecular mechanisms and temporal control. Dev Cell 31:525–538
Minc N, Burgess D, Chang F (2011) Influence of cell geometry on division-plane positioning. Cell 144:414–426
Morais-de-Sa E, Sunkel C (2013) Adherens junctions determine the apical position of the midbody during follicular epithelial cell division. EMBO Rep 14:696–703
Morita E, Sandrin V, Chung HY et al (2007) Human ESCRT and ALIX proteins interact with proteins of the midbody and function in cytokinesis. EMBO J 26:4215–4227
Mullins JM, Biesele JJ (1973) Cytokinetic activities in a human cell line: the midbody and intercellular bridge. Tissue Cell 5:47–61
Naganathan SR, Fürthauer S, Nishikawa M et al (2014) Active torque generation by the actomyosin cell cortex drives left-right symmetry breaking. Elife 3:e04165
Neto H, Kaupisch A, Collins LL et al (2013) Syntaxin 16 is a master recruitment factor for cytokinesis. Mol Biol Cell 24:3663–3674
Norden C, Mendoza M, Dobbelaere J et al (2006) The NoCut pathway links completion of cytokinesis to spindle midzone function to prevent chromosome breakage. Cell 125:85–98
Ou G, Gentili C, Gönczy P (2014) Stereotyped distribution of midbody remnants in early *C. elegans* embryos requires cell death genes and is dispensable for development. Cell Res 24:251–253
Paweletz N (1967) On the function of the "Flemming body" during division of animal cells. Naturwissenschaften 54:533–535
Paweletz N (2001) Walther Flemming: pioneer of mitosis research. Nat Rev Mol Cell Biol 2:72–75
Pflüger EFW (1884) Ueber die Einwirkung der Schwerkraft und anderer Bedingungen auf die Richtung der Zelltheilung. Pflugers Arch 34:607–616
Pocha SM, Knust E (2013) Complexities of Crumbs function and regulation in tissue morphogenesis. Curr Biol 23:289–293
Pohl C (2008) Coordination of late stages of cytokinesis by the inhibitor of apoptosis protein BRUCE. Dissertation, Ludwig-Maximilians.-Universität München. https://edoc.ub.uni-muenchen.de/8848/
Pohl C (2009) Dual control of cytokinesis by the ubiquitin and autophagy pathways. Autophagy 5:561–562
Pohl C (2015) Cytoskeletal symmetry breaking and chirality: From reconstituted systems to animal development. Symmetry 7:2062–2107
Pohl C, Bao Z (2010) Chiral forces organize left-right patterning in *C. elegans* by uncoupling midline and anteroposterior axis. Dev Cell 19:402–412
Pohl C, Jentsch S (2008) Final stages of cytokinesis and midbody ring formation are controlled by BRUCE. Cell 132:832–845
Pollarolo G, Schulz JG, Munck S et al (2011) Cytokinesis remnants define first neuronal asymmetry in vivo. Nat Neurosci 14:1525–1533
Ramanathan SP, Helenius J, Stewart MP et al (2015) Cdk1-dependent mitotic enrichment of cortical myosin II promotes cell rounding against confinement. Nat Cell Biol 17:148–159
Reinsch S, Karsenti E (1994) Orientation of spindle axis and distribution of plasma membrane proteins during cell division in polarized MDCKII cells. J Cell Biol 126:1509–1526
Robbins E, Gonatas NK (1964) The ultrastructure of a mammalian cell during the mitotic cycle. J Cell Biol 21:429–463
Rose LS, Kemphues K (1998) The let-99 gene is required for proper spindle orientation during cleavage of the *C. elegans* embryo. Development 125:1337–1346
Salzmann V, Chen C, Chiang CY et al (2014) Centrosome-dependent asymmetric inheritance of the midbody ring in Drosophila germline stem cell division. Mol Biol Cell 25:267–275

Sanger JM, Dome JS, Sanger JW (1998) Unusual cleavage furrows in vertebrate tissue culture cells: insights into the mechanisms of cytokinesis. Cell Motil Cytoskeleton 39:95–106
Schiel JA, Childs C, Prekeris R (2013) Endocytic transport and cytokinesis: from regulation of the cytoskeleton to midbody inheritance. Trends Cell Biol 23:319–327
Schlüter MA, Pfarr CS, Pieczynski J et al (2009) Trafficking of Crumbs3 during cytokinesis is crucial for lumen formation. Mol Biol Cell 20:4652–4663
Schonegg S, Hyman AA, Wood WB (2014) Timing and mechanism of the initial cue establishing handed left–right asymmetry in *Caenorhabditis elegans* embryos. Genesis 52:572–580
Singh D, Pohl C (2014a) Coupling of rotational cortical flow, asymmetric midbody positioning, and spindle rotation mediates dorsoventral axis formation in *C. elegans*. Dev Cell 28:253–267
Singh D, Pohl C (2014b) A function for the midbody remnant in embryonic patterning. Commun Integr Biol 7:e28533
Skop AR, White JG (1998) The dynactin complex is required for cleavage plane specification in early *Caenorhabditis elegans* embryos. Curr Biol 8:1110–1116
Sorce B, Escobedo C, Toyoda Y et al (2015) Mitotic cells contract actomyosin cortex and generate pressure to round against or escape epithelial confinement. Nat Commun 6:8872
Steigemann P, Wurzenberger C, Schmitz MH et al (2009) Aurora B-mediated abscission checkpoint protects against tetraploidization. Cell 136:473–484
Tawk M, Araya C, Lyons DA et al (2007) A mirror-symmetric cell division that orchestrates neuroepithelial morphogenesis. Nature 446:797–800
Théry M, Racine V, Pépin A et al (2005) The extracellular matrix guides the orientation of the cell division axis. Nat Cell Biol 7:947–953
Théry M, Jiménez-Dalmaroni A, Racine V et al (2007) Experimental and theoretical study of mitotic spindle orientation. Nature 447:493–496
Thoresen SB, Campsteijn C, Vietri M et al (2014) ANCHR mediates Aurora-B-dependent abscission checkpoint control through retention of VPS4. Nat Cell Biol 16:550–560
Toyoshima F, Nishida E (2007) Integrin-mediated adhesion orients the spindle parallel to the substratum in an EB1- and myosin X-dependent manner. EMBO J 26:1487–1498
Toyoshima F, Matsumura S, Morimoto H et al (2007) PtdIns(3,4,5)P3 regulates spindle orientation in adherent cells. Dev Cell 13:796–811
Tsou MF, Hayashi A, DeBella LR et al (2002) LET-99 determines spindle position and is asymmetrically enriched in response to PAR polarity cues in *C. elegans* embryos. Development 129:4469–4481
Tsou MF, Ku W, Hayashi A et al (2003) PAR-dependent and geometry-dependent mechanisms of spindle positioning. J Cell Biol 160:845–855
Waddle JA, Cooper JA, Waterston RH (1994) Transient localized accumulation of actin in *Caenorhabditis elegans* blastomeres with oriented asymmetric divisions. Development 120:2317–2328
Wang T, Yanger K, Stanger BZ et al (2014) Cytokinesis defines a spatial landmark for hepatocyte polarization and apical lumen formation. J Cell Sci 127:2483–2492
Willenborg C, Jing J, Wu C et al (2011) Interaction between FIP5 and SNX18 regulates epithelial lumen formation. J Cell Biol 195:71–86
Wilson GM, Fielding AB, Simon GC et al (2005) The FIP3-Rab11 protein complex regulates recycling endosome targeting to the cleavage furrow during late cytokinesis. Mol Biol Cell 16:849–860
Zmuda JF, Rivas RJ (1998) The Golgi apparatus and the centrosome are localized to the sites of newly emerging axons in cerebellar granule neurons in vitro. Cell Motil Cytoskeleton 41:18–38

Chapter 8
Drosophila melanogaster Neuroblasts: A Model for Asymmetric Stem Cell Divisions

Emmanuel Gallaud, Tri Pham, and Clemens Cabernard

Abstract Asymmetric cell division (ACD) is a fundamental mechanism to generate cell diversity, giving rise to daughter cells with different developmental potentials. ACD is manifested in the asymmetric segregation of proteins or mRNAs, when the two daughter cells differ in size or are endowed with different potentials to differentiate into a particular cell type (Horvitz and Herskowitz, Cell 68:237–255, 1992). *Drosophila* neuroblasts, the neural stem cells of the developing fly brain, are an ideal system to study ACD since this system encompasses all of these characteristics. Neuroblasts are intrinsically polarized cells, utilizing polarity cues to orient the mitotic spindle, segregate cell fate determinants asymmetrically, and regulate spindle geometry and physical asymmetry. The neuroblast system has contributed significantly to the elucidation of the basic molecular mechanisms underlying ACD. Recent findings also highlight its usefulness to study basic aspects of stem cell biology and tumor formation. In this review, we will focus on what has been learned about the basic mechanisms underlying ACD in fly neuroblasts.

8.1 Introduction

The central nervous system of the fruit fly *Drosophila melanogaster* originates from neural stem cells, called neuroblasts (Nbs). Neuroblasts have been used extensively to study fundamental aspects of cytokinesis, cell polarity, and spindle orientation (Knoblich 2010; Cabernard 2012; Lu and Johnston 2013). Recently, this cell type has also been used to model tumor formation in vivo and basic stem cell biology (Homem and Knoblich 2012; Gonzalez 2013). Here, we will review what has been

E. Gallaud
Biozentrum, University of Basel, Klingelbergstrasse 50-70, 4056 Basel, Switzerland

T. Pham
Biozentrum, University of Basel, Klingelbergstrasse 50-70, 4056 Basel, Switzerland
Department of Biology, University of Washington, Seattle, WA 98195, USA

C. Cabernard (✉)
Department of Biology, University of Washington, Seattle, WA 98195, USA
e-mail: ccabern@uw.edu

J.-P. Tassan, J.Z. Kubiak (eds.), *Asymmetric Cell Division in Development, Differentiation and Cancer*, Results and Problems in Cell Differentiation 61,
DOI 10.1007/978-3-319-53150-2_8

learned about the molecular cell biology of asymmetric cell division (ACD) using fly neuroblasts as a model system. Due to space constraints, we will not go into temporal identity and more detailed aspects of fly neurogenesis but refer the interested reader to excellent reviews that have been covering these subjects recently (Reichert 2011; Homem and Knoblich 2012; Kang and Reichert 2014).

Neuroblasts are specified during embryonic stages 9–11 through Notch/Delta signaling. This lateral inhibition mechanism refines the expression of proneural genes to individual cells in the embryonic neuroectoderm, the presumptive neuroblasts (reviewed in Huang et al. 2014). In addition to Notch signaling, the segment polarity gene *wingless* (Wnt in vertebrates) has also been shown to have "proneural capacity." Loss of *wingless* results in the absence or duplication of identified neuroblasts (Chu-LaGraff and Doe 1993). Upon specification of neuroblasts from a proneural field, they undergo an epithelial-to-mesenchymal transition, delaminating from the neuroepithelium in the ventrolateral region of the *Drosophila* embryo (Hartenstein et al. 1994) (Fig. 8.1a–d).

Delaminated neuroblasts undergo a series of asymmetric cell divisions, generating a self-renewed neuroblast and a differentiating ganglion mother cell (GMC). GMCs will divide once more, giving rise to glia, neurons, or both. Embryonic neuroblasts will form 10–20 primary neurons, generating all the neurons of the *Drosophila* larva but only 10% of the adult fly (Homem and Knoblich 2012; Kang and Reichert 2014). Toward the end of neurogenesis, most neuroblasts in the central brain and thoracic regions exit the cell cycle and enter a quiescent state with the exception of four neuroblasts,

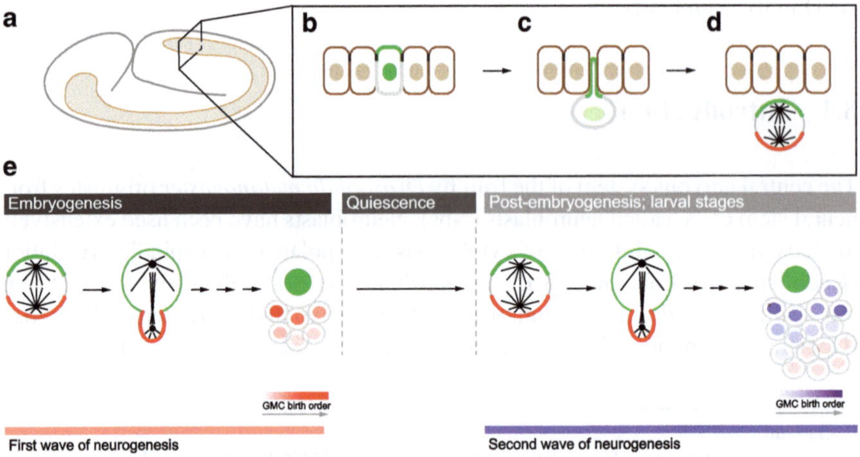

Fig. 8.1 (**a**) Neuroblast behavior during embryonic and larval development. Neuroblasts start to delaminate from the neuroectoderm shortly after gastrulation. (**b, c**) Neuroblasts, specified through lateral inhibition, undergo dramatic cell shape changes due to cytoskeletal changes and extrude from the neuroepithelium, inheriting apical–basal polarity from this tissue. (**d**) Subsequently, neuroblasts start to divide asymmetrically, always oriented in relation to the overlying neuroectoderm. (**e**) During embryogenesis, the first wave of neurogenesis generates first the larval central nervous system. With the exception of mushroom body neuroblasts, all other neuroblasts enter quiescence before the second wave of neurogenesis generates the adult central nervous system. The youngest GMCs are usually in closest proximity to the neuroblast

giving rise to the mushroom body lineage. Entry into quiescence is regulated by cell intrinsic factors such as Hox genes and temporal identity factors but also low levels of nuclear Prospero (Tsuji et al. 2008; Lai et al. 2014) (Fig. 8.1e). In the abdominal regions of the embryo, however, most neuroblasts are eliminated through programmed cell death after completing their neuronal lineages (Campos-Ortega and Hartenstein 1985; Truman and Bate 1988; White et al. 1994).

Nbs start to exit quiescence and reenter mitosis during the late 1st instar stage, approximately 8–10 h after larval hatching. This second wave of neurogenesis is responsible for the formation of 90% of the neurons in the adult CNS. Exit from quiescence is controlled by cell extrinsic factors such as Insulin-Like protein Dilp and the Glycoprotein Anachronism (Ana). These proteins are both secreted from the glia niche (Ebens et al. 1993; Chell and Brand 2010; Sousa-Nunes et al. 2011) (Fig. 8.1e). Neurogenesis continues throughout all larval and into pupal stages, at which point Nbs exit the cell cycle and disappear (reviewed in Knoblich 2010; Homem and Knoblich 2012). This cell cycle exit has recently been shown to be dependent on changes in energy metabolism, induced by the steroid hormone ecdysone, causing a reduction in neuroblast cell size and ultimately terminal differentiation (Homem et al. 2014) (Fig. 8.1e).

Another source of neuroblasts is the optic lobe of the larval brain where symmetrically dividing neuroepithelial cells are gradually converted to neuroblasts in response to a wave of proneural gene expression (Hofbauer and Campos-Ortega 1990; Egger et al. 2007, 2010, 2011; Yasugi et al. 2010; Brand and Livesey 2011).

Asymmetric cell division follows a pattern of four subsequent steps: (1) establishment of an intrinsic polarity axis by localizing polarity proteins to the apical neuroblast cortex, (2) orientation of the mitotic spindle with the apical axis, (3) asymmetric localization of cell fate determinants to the basal cell cortex, and (4) biased segregation of both apical and basal proteins between sibling cells (reviewed in Knoblich 2010; Homem and Knoblich 2012). In the subsequent paragraphs, we will discuss how intrinsic polarity is established and how the mitotic spindle is using these polarity cues to properly align with the intrinsic polarity axis, a requirement for the correct segregation of cell fate determinants. We will further review centrosome asymmetry, spindle geometry, and how this could influence sibling cell size asymmetry.

8.2 Neuroblast Types and Lineages

Two distinct types of larval neuroblasts have been described so far, which differ slightly in their asymmetric division mode. The central brain and ventral nerve cord predominantly contain type I neuroblasts, generating non-self-renewing secondary progenitors called ganglion mother cells (GMCs) and self-renewing neuroblasts. Type I neuroblasts—about 90 per brain lobe—are found in all brain regions, but the dorsoanterior lateral region [DAL; nomenclature according to Pereanu and Hartenstein (2006)] contains only type I neuroblasts. Eight neuroblasts located in the dorsoposterior part of the brain are of type II origin. Type I and type II neuroblasts are morphologically identical and express the bHLH transcription factor

Deadpan (Dpn). However, type I neuroblasts express the bHLH transcription factor Asense (Ase), whereas type II neuroblasts are Asense negative (type II neuroblasts are thus also called Posterior Asense-negative (PAN) neuroblasts) (Bowman et al. 2008; Bello et al. 2008; Boone and Doe 2008). Furthermore, Type II neuroblasts specifically express Pointed P1, an isoform of the Ets (E26 transformation-specific) transcription factor Pointed (Zhu et al. 2011). In addition, type I neuroblasts have cytoplasmic Prospero, which is lacking in type II neuroblasts.

Type II/PAN neuroblasts divide also asymmetrically, but in contrast to Type I neuroblasts, they form an intermediate progenitor (INP), also called trans-amplifying GMC, which subsequently divides asymmetrically to form a GMC and another INP (or TA-GMC). INPs divide several times, generating between 6 and 12 neurons. Due to the occurrence of INPs, type II neuroblast lineages contain many more cells than type I lineages (Bowman et al. 2008; Bello et al. 2008; Boone and Doe 2008). Neuroblast and INP temporal axes contribute to an increase in neural diversity in type II lineages (Bayraktar and Doe 2013).

The central brain also contains eight mushroom body neuroblasts (4 per brain lobe), giving rise exclusively to the neurons and glial cells of the mushroom body. These neuroblasts do not enter quiescence and belong also to type I neuroblasts (Ito and Hotta 1992; Ito et al. 1997). Similarly, based on their division mode, optic lobe neuroblasts can be considered type I neuroblasts although their origin differs from central brain neuroblasts (Egger et al. 2008) (Fig. 8.2).

8.3 Drosophila Neuroblasts as a Model System to Study Asymmetric Cell Division

8.3.1 Establishment of Neuroblast Polarity; Localizing the Par Complex to the Apical Neuroblast Cortex

Drosophila neuroblasts are intrinsically polarized. The polarity axis is established during early prophase with the formation of the apical Par complex. This apical core polarity complex is composed of atypical Protein Kinase C (aPKC; PKCζ, PKCλ in vertebrates), Partitioning defective 6 (Par-6; Par-6 in vertebrates), and Bazooka/Partitioning defective 3 (Baz; Par-3 in vertebrates). By metaphase, this Par complex forms a tight crescent on the apical neuroblast cortex. The Par complex is required for the correct localization of basal cell fate determinants (see below). In addition, it is necessary for the establishment of the Insc/Pins/Gαi/Mud complex, which is essential for the correct orientation of the mitotic spindle (see also below) (Fig. 8.3).

The origin of this intrinsic polarity axis is not completely understood, but localization of the Par complex is most likely inherited during embryonic delamination from the polarized neuroectoderm (Homem and Knoblich 2012). Embryonic neuroblasts can undergo a number of asymmetric divisions, maintaining the position of the apical Par complex in relation to the overlying neuroepithelium. Neuroblast dissociation

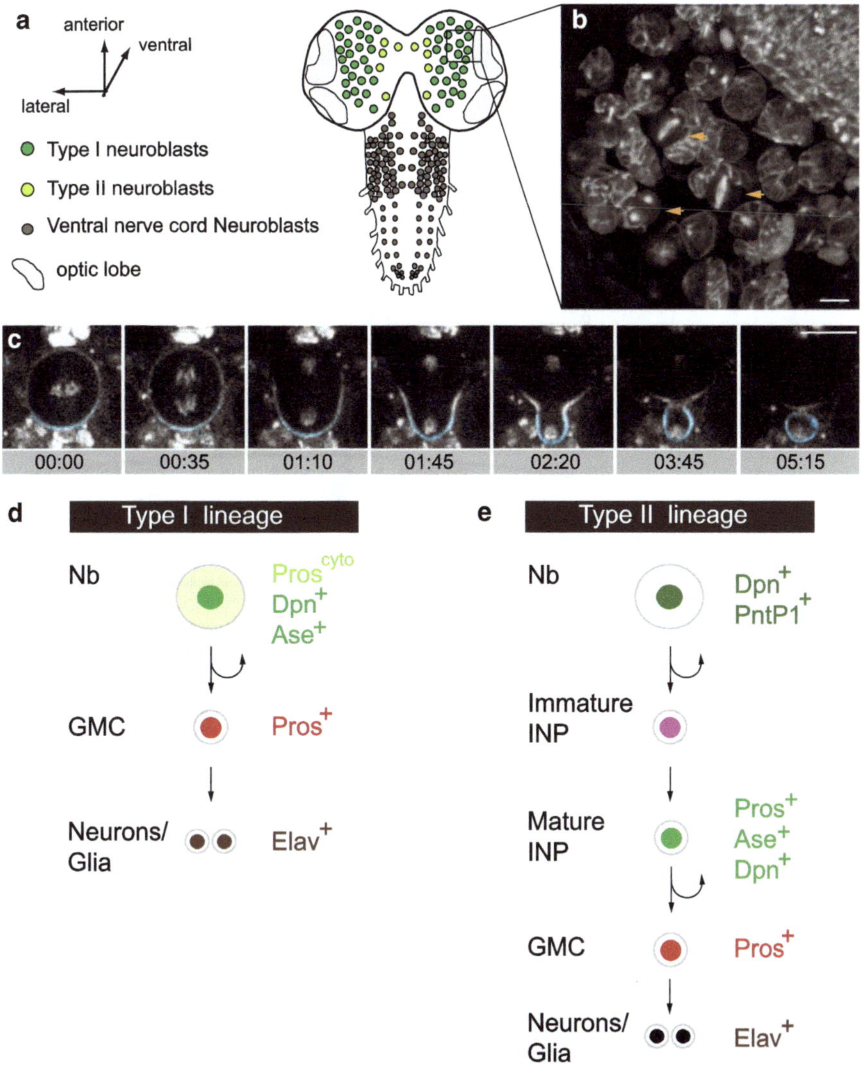

Fig. 8.2 (**a**) Neuroblast types and their lineages in the larval brain. Third instar larval central nervous system contains different neuroblast populations. (**b**) Larval neuroblasts are close to the brain surface, facilitating live imaging experiments; mitotic neuroblasts are highlighted with *orange arrowheads*. (**c**) Image sequence of a central brain neuroblast (Chromosomes and Myosin are shown in *white*.) The basal cell fate determinant Miranda is labeled in *cyan*). Central brain neuroblasts contain both type I (**d**) and type II (**e**) neuroblasts, which differ in their molecular identity and lineage progression. Scale bars: 10μm

experiments have shown that the establishment of intrinsic polarity is cell autonomous, but nonautonomous cues are necessary to maintain the division axis in the embryo (Siegrist and Doe 2006). Recently, it was found that Trapped in Endoderm 1 (Tre1), a

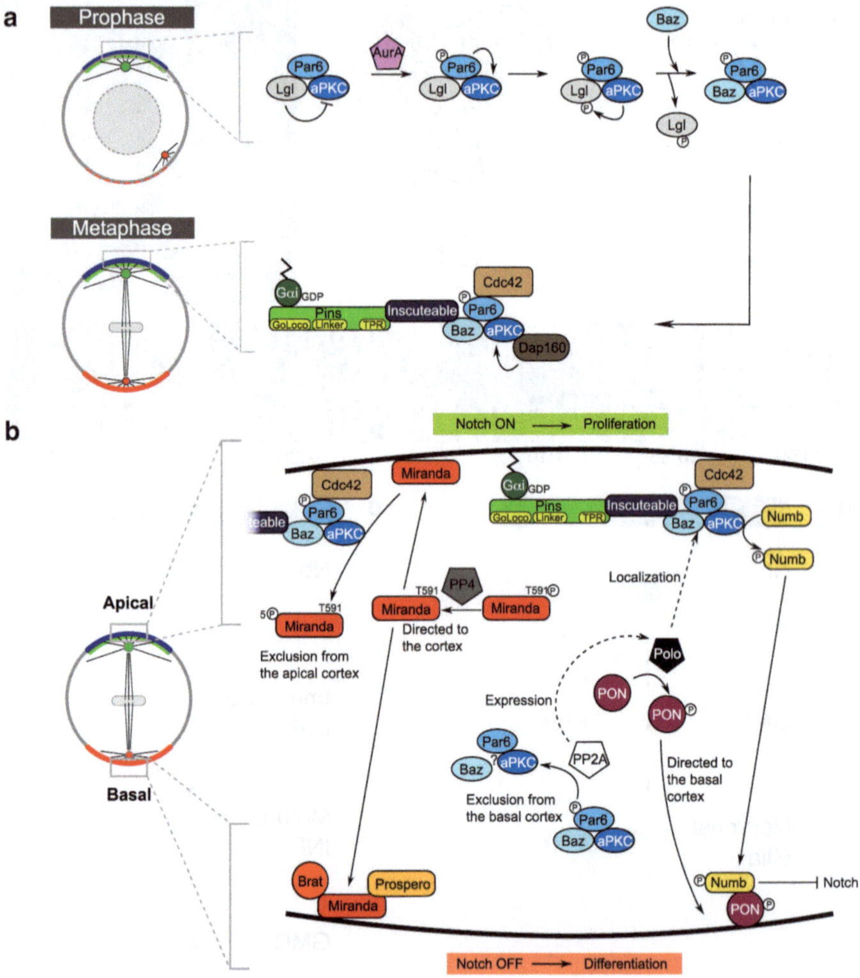

Fig. 8.3 (**a**) Apico-basal polarity in dividing neuroblasts. Polarity establishment starts in prophase but extends into metaphase. aPKC is activated through the phosphorylation of Par-6 via AurA and phosphorylates Lgl. Phosphorylated Lgl is excluded from the complex and gets replaced by Bazooka. Insc binds both Baz and Pins, providing a link between apical polarity proteins and the spindle orientation machinery. The GTP-bound form of Gαi binds the Goloco motif of Pins and is required for its localization. Note that the apical localization of the Par complex is also promoted by Cdc42, binding to Par-6, and the activation of aPKC by Dap160. (**b**) The segregation of basal determinants is the result of phosphorylation and dephosphorylation. The kinase Polo participates in the apical localization of the Par complex and phosphorylates PON to be targeted to the basal cortex. PP2A dephosphorylates Par-6 (antagonizing AurA activity) and excludes the Par complex from the basal cortex. Furthermore, PP2A also regulates Polo expression. Miranda, a cargo protein for Prospero (type I Nb only) and Brat, is regulated by PP4 and aPKC; PP4 dephosphorylates Miranda on the T591, which brings Miranda to the neuroblast cortex. On the apical side, aPKC phosphorylates Miranda's N-terminus, leading to its exclusion from the apical cortex, restricting it to the basal cortex. aPKC also phosphorylates Numb, which is required for Numb's basal localization. Numb segregates into the GMC and will inhibit Notch signaling to induce differentiation. Notch activity remains high in the neuroblast and promotes proliferation. The Miranda cargo proteins Brat and Pros also segregate into the GMC, inducing a differentiation program

rhodopsin family orphan G protein coupled Receptor (GPCR), regulates the relative orientation of cortical polarity in embryonic Nbs with regard to the overlying epithelium (Yoshiura et al. 2012). In wild-type embryos, the Par complex is facing the overlying ectoderm; this stereotypic orientation is lost in *tre1* mutants, showing random orientation of the apical Par complex. Tre1 mediates proper polarization in relation to the overlying neuroectoderm through Goα, facilitating the GDP–GTP exchange. This study led to a model, proposing that an unidentified extrinsic cue from the neuroectoderm activates Goα on the apical side, facing the neuroepithelium. Activated Goα subsequently captures Partner of Inscuteable (Pins; LGN/AGS3 in vertebrates) (Yoshiura et al. 2012). Pins has been shown to bind to Inscuteable (Insc; Insc in vertebrates), providing a physical connection to the apical Par complex (Schober et al. 1999; Wodarz et al. 1999; Yu et al. 2000). Thus, through Pins and Insc, the position of the apical complex can be refined. However, the temporal dynamics of this process are currently unclear.

In embryonic and larval neuroblasts, the localization of the apical Par complex is regulated by interdependent physical interactions. Par-6 and Baz are PDZ containing proteins, localizing to the apical neuroblast cortex. Par-6 can bind Baz in vitro and its localization depends on Baz. Similarly, the apical localization of Baz is lost in Par-6 mutants (Petronczki and Knoblich 2001). *aPKC* mutants retain normal apical Baz localization but compromise Par6 localization (Rolls et al. 2003). This apical complex, however, is subject to change and in interphase is composed of aPKC-Lethal giant larvae (L(2)gl)-Par6. Upon entry into mitosis, AuroraA kinase phosphorylates Par6, thereby activating aPKC. Activated aPKC subsequently phosphorylates L(2)GL, reducing its affinity for aPKC and Par6, releasing it from the complex and allowing Baz to enter (Wirtz-Peitz et al. 2008) (Fig. 8.2a). Live cell imaging from sensory organ precursor cells (SOPs) has shown that by the end of prophase, Baz relocalized to the posterior–lateral cortex as Lgl was released from this side.

This still leaves the question of how Par-6 and aPKC are recruited to the apical cortex. Par-6 binds the RhoGTPase Cdc42, and this interaction is mediated by a short motif in Par-6 termed the "semi-CRIB." This domain ensures that Par-6 selectively binds to activated Cdc42. Binding of Cdc42 to Par-6 in turn activates the PDZ domain of Par-6. Since Cdc42 is lipid modified, associating with the membrane when activated, this mechanism can establish a connection between the apical neuroblast membrane and Par-6. In mitotic neuroblasts, Cdc42 is localized uniformly around the cortex with slight apical enrichment throughout mitosis (Atwood et al. 2007). In *cdc42* loss-of-function mutants, both aPKC and Par-6 are lost from the apical cortex and delocalized into the cytoplasm. Baz, however, shows normal apical localization, placing Baz still upstream of Cdc42. This is consistent with *baz* mutants, which show compromised Cdc42 localization (Atwood et al. 2007). These studies indicate that Baz seems to be the most upstream component of the apical polarity complex, although Par complex components depend on each other for their correct apical localization and activity (Prehoda 2009). For instance, in vertebrate cells, it was shown that PAR-3 is a substrate of PKCζ/PKCλ (Izumi et al. 1998).

Recent studies further identified other factors required for the correct localizing of apical Par complex components. For instance, Dynamin-associated protein

160 (Dap160; related to mammalian intersectin) has recently been shown to be required to regulate aPKC's activity and apical localization (Chabu and Doe 2008), and the phosphatase PP2A is required to exclude aPKC from the basal cortex (Chabu and Doe 2009; Krahn et al. 2009; Ogawa et al. 2009). Similarly, the evolutionary conserved Prefoldin complex, required for the correct folding of proteins (Vainberg et al. 1998), is also implicated in the correct neuroblasts' polarity establishment (Zhang et al. 2016b). Taken together, the core machinery required for the establishment and maintenance of the apical Par complex has been identified, but molecular details are still emerging (Fig. 8.3a, b).

8.3.2 Localization of Basal Cell Fate Determinants

The Par complex is required for the correct localization of cell fate determinants on the basal neuroblast cortex. The adaptor protein Miranda (Mira) (Shen et al. 1997), the transcription factor Prospero (Pros; Prox1 in vertebrates) (Doe et al. 1991; Matsuzaki et al. 1992; Spana and Doe 1995; Hirata et al. 1995; Knoblich et al. 1995), the translational inhibitor Brain tumor (Brat; Trim2, Trim3, Trim32 in vertebrates) (Lee et al. 2006c; Bello et al. 2006; Bowman et al. 2008), the Notch signaling inhibitor Numb (Numb and Numb-like in vertebrates) (Rhyu et al. 1994; Knoblich et al. 1995), and its binding partner, Partner of Numb (Pon) (Lu et al. 1998), all form a tight crescent on the basal neuroblast cortex by metaphase and segregate asymmetrically into the GMC in telophase (Homem and Knoblich 2012) (Fig. 8.3b). Miranda acts as an adaptor protein, binding both Prospero and Brat (Shen et al. 1997, 1998; Ikeshima-Kataoka et al. 1997; Schuldt et al. 1998), and *mira* mutants fail to correctly specify neuronal fate due to mislocalization of Prospero and Brat (Lee et al. 2006c). Genetic and molecular analysis revealed specific localization and interaction domains in Mira, necessary for Prospero and Brat binding but also for correct basal localization. For instance, Miranda's N-terminus is necessary and sufficient for its basal localization. A central cargo domain is necessary for binding to Staufen, Prospero protein, but also *prospero* mRNA. Finally, Mira's C-terminal domain is important for Miranda's timely degradation and cargo release from the GMC cortex (Shen et al. 1998; Fuerstenberg et al. 1998). In wild-type neuroblasts, Miranda is localized on the basal neuroblast cortex, which becomes part of the newly formed GMC. Shortly after cytokinesis, Miranda is released from the GMC cortex, also allowing Prospero protein to detach from the cortex. Subsequently, Pros enters the newly formed GMC nucleus where it activates a number of target genes, necessary for differentiation and repressing self-renewal (Choksi et al. 2006). Mutations in Mira's C-terminus prevent the timely release of both Mira and Prospero from the cortex so that Prospero fails to enter the GMC nucleus (Ikeshima-Kataoka et al. 1997).

The mechanism for basal Miranda localization has been controversial for quite some time. For instance, early reports have shown that Miranda is localized on the apical neuroblast cortex in interphase, requiring Mira's central domain (Shen et al. 1998; Schuldt et al. 1998; Matsuzaki et al. 1998; Fuerstenberg et al. 1998). How does Miranda relocalize to the basal cortex? It has been proposed that the motor protein Non-muscle Myosin II and the unconventional Myosin VI (Jaguar, Jar in Drosophila) are targeting Miranda to the basal cortex by active transport (Petritsch et al. 2003; Barros et al. 2003). However, recent Florescence Recovery after Photobleaching (FRAP) experiments led to the conclusion that Miranda is reaching the basal cortex by passive diffusion throughout the cell rather than Myosin VI-driven cortical transport (Erben et al. 2008). Similarly, since the Rho kinase inhibitor that was used to probe the function of Myosin in this process was found to be non-specific—it also inhibits the apically localized protein Kinase aPKC—this model was later discarded. Molecular analysis showed that aPKC phosphorylates several residues in Miranda's cortical localization domain, thereby excluding it from the apical neuroblast cortex and confining it to the basal cell cortex (Atwood and Prehoda 2009). Thus, the current model proposes that Miranda is confined to the basal cortex through aPKC-mediated apical exclusion (Prehoda 2009).

Protein dephosphorylation—either directly or indirectly—is another important event for localization of Miranda. As mentioned above, Protein phosphatase 2A (PP2A) has been shown to act antagonistically to AurA to suppress aPKC signaling (Chabu and Doe 2009; Krahn et al. 2009; Ogawa et al. 2009) on the basal cortex. The protein phosphatase 4 (PP4) has also been shown to be required for correct Miranda phosphorylation. Larval neuroblasts lacking PP4's regulatory subunit [called Falafel (Flfl)] exhibit defective Mira localization whereas localization of Numb remains unaffected (Sousa-Nunes et al. 2009). PP4 is in a complex with Phosphotyrosyl phosphatase activator (PTPA), promoting Miranda's cortical association by dephosphorylating the amino acid T591. For instance, non-phosphorylatable Miranda T591A is associated primarily with the cell cortex. This step acts before aPKC since PTPA-mediated cortical localization of Mira during early mitotic stages is not related to the phosphorylation of Mira on the five aPKC sites, which displace Miranda from the cortex. Subsequently, aPKC phosphorylates Miranda to exclude it from the apical cortex, restricting its localization basally (Zhang et al. 2016a). Thus, this two-step model proposes that cortical Miranda localization, preceding aPKC-mediated apical exclusion, is mediated by a dephosphorylation step, bringing cytoplasmic Miranda to the cell cortex. The kinase phosphorylating Miranda on T591, preceding dephosphorylation, is currently unknown.

The basal cell fate determinant Numb also requires aPKC for its correct asymmetric localization (Smith et al. 2007). As mentioned above, the Par-6-aPKC-L(2)gl complex in interphase cannot phosphorylate Numb until AurorA-mediated phosphorylation of Par-6 induces a change in the complex, replacing L(2)gl with Par-3. Par-3 can bind both aPKC and Numb, bringing the substrate to the kinase, initiating phosphorylation of Numb by aPKC (Wirtz-Peitz et al. 2008). Thus, proteins that are directly downstream of the Par complex are often aPKC substrates, and phosphorylation is both necessary and sufficient for cortical displacement and concomitant removal from the Par domain (Bailey and Prehoda 2015).

Numb has also been shown to be in a complex with Pon in vivo and although Pon is required for correct Numb localization, it is not absolutely essential; basal Numb localization is delayed in *pon* mutants (Lu et al. 1998). FRAP experiments have shown that Pon exchanges very dynamically with the cytoplasm and is not recruited basally via cortical flow, further supporting a more dynamic model for basal cell fate determinant localization (Mayer et al. 2005). Similarly, Numb contains positively charged amino acids at its N-terminus, enabling recruitment to the plasma membrane. aPKC phosphorylation sites, necessary for Numb's asymmetric localization, are right next to these membrane targeting sites, suggesting that aPKC-mediated phosphorylation neutralizes positive charges, thereby inhibiting Numb's membrane association (Knoblich et al. 1997). Recently, it was also shown that the mitotic kinase Polo is directly phosphorylating Pon, which is required for the correct localization of Numb (Wang et al. 2007). However, since *polo* mutants also affect the localization of aPKC, inputs from both aPKC and Polo might be required for the correct localization of Numb.

8.3.3 Orienting the Mitotic Spindle in Relation to the Apical–Basal Polarity Axis

As described in the previous part, *Drosophila* neuroblasts establish an apicobasal polarity axis shortly before mitotic entry and maintain it throughout cell division. During the process of asymmetric cell division, the orientation of the spindle is crucial for the segregation of cell fate determinants into the GMC (Cabernard and Doe 2009) (see also below) and therefore for neuroblast homeostasis and neural differentiation. The link between cell polarity and spindle orientation has been extensively studied in neuroblasts and the concept that emerged suggests that spindle orientation is the result of a dual involvement of cell polarity and centrosome asymmetry (Morin and Bellaïche 2011).

The key players for spindle orientation are Inscuteable (Insc; mInsc in mice), Partner of Inscuteable (Pins; LGN/AGS3 in vertebrates), the heterotrimeric G-protein alpha subunit called Gαi, and Mushroom body defect (Mud; NuMA in vertebrates). Insc colocalizes with the apical Par complex and directly binds to Bazooka, necessary for its recruitment to the apical cortex of embryonic neuroblasts (Schober et al. 1999; Wodarz et al. 1999). Insc also directly binds to Pins, and Pins is required for apical Insc localization (Yu et al. 2000; Schaefer et al. 2000). Thus, apical Insc localization is regulated by both Baz and Pins; Pins is thus required for both polarity establishment but also for spindle orientation since the mitotic spindle fails to correctly orient along the apical–basal polarity axis in *pins* mutants (Yu et al. 2000; Schaefer et al. 2000). Pins has also been shown to bind to GDP-bound Gαi (but not GTP-bound) playing an important role in cell polarity and spindle orientation, independently from its function in signaling (Schaefer et al. 2000). Pins contains tetratricopeptide repeats (TPRs) and three GoLoco motifs, all of which can bind GDP-bound Gαi. Recent biochemical evidence suggests that Pins is in a closed state and is recruited to the apical cortex upon binding of Gαi to the first GoLoco motif. This binding induces a conformational change, allowing for further Gαi binding but also recruits Mud to the apical neuroblast cortex (Nipper et al. 2007) (Fig. 8.4).

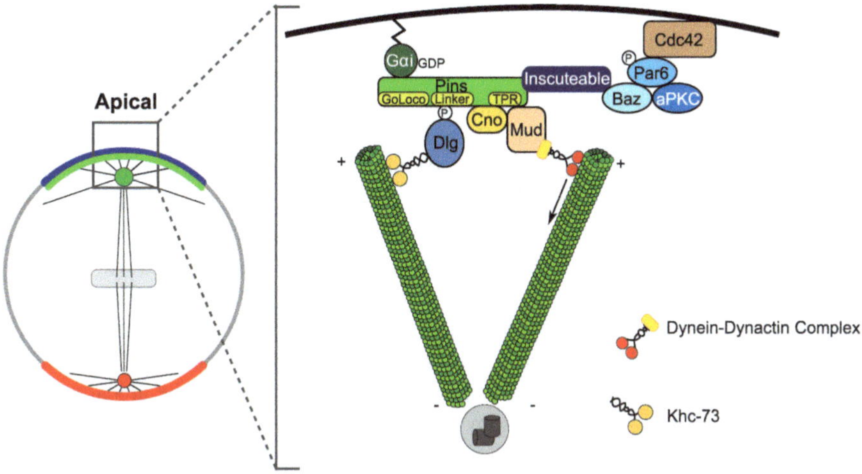

Fig. 8.4 Neuroblast polarity controls mitotic spindle orientation. Pins is restricted to the apical cortex through the binding of Gαi–GDP to its Goloco motif and through the binding of Insc, connecting it to the apical Par complex. Pins serves as a molecular platform for two nonredundant pathways that are involved in spindle orientation. First, AurA phosphorylates Pins in the Linker domain, and this phosphorylation is required for Dlg to bind Pins. The Kinesin Khc-73 can bind to Dlg, capturing astral microtubules and providing a static link between the mitotic spindle and the apical cortex. With the support of Cno, the coiled-coil protein Mud binds the TPR motifs of Pins and in turn associates with the Dynein/Dynactin Complex. The (−) end-directed activity of the Dynein/Dynactin complex exerts pulling forces on the mitotic spindle to regulate its orientation

Mud is an important component for spindle orientation and acts downstream of Insc/Pins/Gαi. Mud binds Pins at its N-terminal TPR domain, creating a Gαi–Pins (TPR)–Mud complex at the apical cortex (Bowman et al. 2006; Izumi et al. 2006; Siller et al. 2006). In contrast to *pins* or *Gαi, mud* mutants compromise spindle orientation without perturbing apical–basal polarity. The coiled-coil protein Mud is localized to the apical cortex, to both centrosomes and weakly to the basal cortex (Siller et al. 2006). Mud's localization is dependent on Pins, the cell cycle regulator AuroraA (AurA; AuroraA in vertebrates), and a centrosomal complex consisting of Anastral spindle 2 (Ana2; STIL in vertebrates) and the dynein light chain protein Cut up (Ctp) (Siller et al. 2006; Wang et al. 2006, 2011). Not surprisingly, mutants of these genes also perturb spindle orientation. How Mud localizes to these different subcellular sites is not entirely clear, but FRAP experiments indicated that Mud is traveling from the cortex to the spindle pole, suggesting that Mud may be a cargo protein of the Dynein complex (Wang et al. 2011). Additional studies have shown that the PDZ protein Canoe (Cno; Afadin/AF-6 in vertebrates) promotes formation of the cortical Pins–Mud complex through several small GTPases (e.g., Ran) and is important for proper spindle orientation (Speicher et al. 2008; Wee et al. 2011).

Mud alone has no enzymatic or motor activity and cannot pull on microtubules on its own. This function is performed by Dynein, a microtubule-based motor, which is necessary to generate the force for correct spindle alignment. Work in flies, *C. elegans*, and other systems showed that spindle positioning requires dynamic attachment of astral microtubules to polarized cortical domains, and the generation of forces exerted on these cortically attached astral microtubules aligns the mitotic spindle (reviewed in Lu and Johnston 2013). For instance, in *Xenopus* it was shown that the Mud orthologue NuMA binds the Dynein Dynactin Complex (DDC), providing a link between cell polarity and spindle poles (Merdes et al. 1996). In agreement with this finding is the observation that lack of Lissencephaly-1 (Lis-1; Dynactin), an activator of Dynein microtubule motor activity, impairs spindle movement and orientation (Siller and Doe 2008). Furthermore, mutants that compromise centrosomes or ablate astral microtubules also show spindle orientation defects (Basto et al. 2006; Siller and Doe 2008; Cabernard and Doe 2009; Wang et al. 2011). Since the Dynein–Dynactin complex moves toward microtubule's (MT's) minus ends, pulling forces are generated on the mitotic spindle that are necessary for its orientation in relation to the apical polarity complex (reviewed in Lu and Johnston 2013).

A second pathway for spindle orientation also involves Gαi and Pins. In *insc* mutant larval neuroblasts, the apical Par complex is mislocalized, but polarized localization of Gαi, Pins, and Disc Large1 (Dlg1; Dlg in vertebrates) occurs at metaphase, independent from cell polarity (Siegrist and Doe 2005). This polarization is induced through astral microtubule plus ends and the kinesin Khc-73, binding to Dlg, reconstituting correct spindle orientation. These findings are consistent with a recent study showing that interphase microtubule asters are partially able to maintain spindle orientation over consecutive divisions by maintaining polarity memory (Januschke and Gonzalez 2010). Molecular insight into the Pins–Dlg–Khc-73 pathway has been elucidated mostly in S2 cells by using the homophilic adhesion protein Echinoid fused to various polarity or spindle orientation proteins (Johnston et al. 2009). With this elegant approach, Johnston and collaborators could induce polarity in S2 cells and test diverse proteins for their involvement in spindle orientation. The authors showed that Aurora A kinase phosphorylates the Linker domain of Pins, allowing direct binding with Dlg, which then associates with the kinesin motor protein Khc-73. The plus-end binding capacity of Khc-73 is thought to provide an MT capture or attachment mechanism. Based on this data, it has been hypothesized that Pins(Linker)–Dlg–Khc-73 is providing a static link between astral microtubules and the apical cortex while Pins (TPR)–Mud–DDC would induce the forces required for a proper centrosome anchoring. These two pathways seem to work in parallel and are only partially redundant for efficient spindle alignment. However, the function of Khc-73 has not yet been assessed with loss-of-function mutants in neuroblasts (Fig. 8.4).

8.3.4 Centrosome Asymmetry and Spindle Orientation Memory

Neuroblast polarity is important to control spindle orientation. Cell polarity has also been shown to influence centrosome asymmetry. In this part, we will review the role of centrosomes and their asymmetry in the context of asymmetric neuroblast division and overall CNS development.

After neuroblasts delaminate from the ectoderm, mitotic spindles undergo a 90° rotation, ensuring that the first neuroblast division occurs perpendicularly to the overlying ectoderm (Kaltschmidt et al. 2000). Interestingly, this 90° rotation of the mitotic spindle only occurs in the first neuroblast division because in successive mitoses, centrosomes are more or less aligned perpendicularly with the overlying ectoderm. The reason for this is that one centrosome remains active during interphase and remains closely associated with the apical cortex (Rebollo et al. 2009). Similarly, larval neuroblasts also maintain an active interphase centrosome, closely associated with the apical neuroblast cortex. The basal centrosome, however, downregulates microtubule activity after centrioles separate following cytokinesis. When neuroblasts enter mitosis again, the basal centrosome acquires microtubule organizing activity, amassing pericentriolar matrix proteins and forming a microtubule aster. Shortly before nuclear envelope breakdown, both centrosomes are roughly positioned at opposite poles on the apical and basal cortex, respectively (Rebollo et al. 2007; Rusan and Peifer 2007). Thus, in interphase, centrosomes display a highly asymmetric behavior, based on differential microtubule organizing (MTOC) activity (Fig. 8.5a). This asymmetry is thought to be necessary to maintain the division axis from one cell cycle to another. Asymmetric interphase MTOC activity is not random but appears to correlate with centrosome age. For instance, it was found that the older centrosome is inactivated during interphase and inherited by the GMC, whereas the centrosome containing the younger centriole maintains MTOC activity in interphase and is inherited by the neuroblast (Conduit and Raff 2010; Januschke et al. 2011) (Fig. 8.5a, b). Centrosome asymmetry and biased centrosome segregation have also been observed in other *Drosophila* stem cells such as male and female germline stem cells (Yamashita et al. 2007; Salzmann et al. 2014) or radial glial progenitors in the mouse neocortex (Wang et al. 2009). In these stem cell models, centrosome asymmetry is associated with nonrandom sister chromatid segregation, midbody inheritance, or proliferative ability (Yadlapalli and Yamashita 2013).

The molecular mechanisms of centrosome asymmetry are not entirely clear, but progress has been made within the last 5 years. The active apical interphase centrosome contains pericentriolar matrix proteins (PCM) such as Centrosomin (Cnn; CDK5Rap2 in vertebrates) and γ-Tubulin, necessary for the formation of a microtubule aster. The basal, inactive centrosome is devoid of PCM markers and MTOC activity (Rebollo et al. 2007; Rusan and Peifer 2007; Conduit and Raff 2010; Lerit and Rusan 2013; Singh et al. 2014). Several proteins were identified regulating MTOC activity maintenance on the apical centrosome and basal centrosome, respectively, and it was found that the protein kinase Polo (orthologue of PLK1) is one of the main players. Polo is enriched on the apical centrosome during interphase and gets downregulated on the mother centriole-containing basal centrosome shortly after centrosomes separate (Rusan and Peifer 2007; Singh et al. 2014). Recently, super resolution microscopy revealed that Polo is localized to both the centriole and the PCM on the apical active interphase centrosome. However, the inactive interphase basal centrosome contains only very little Polo, localized to the centriole (Ramdas Nair et al. 2016) (Fig. 8.5b). Polo is required to maintain an active MTOC in interphase

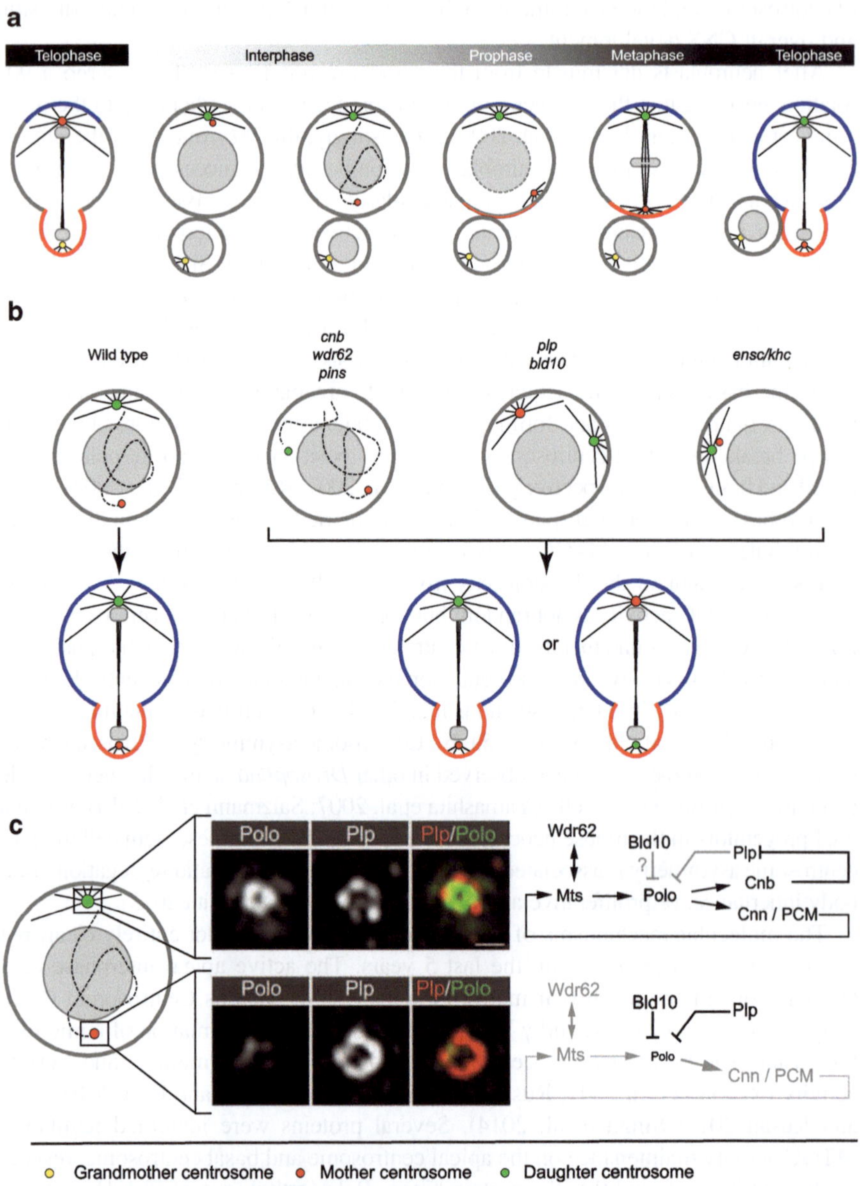

Fig. 8.5 Centrosome asymmetry in interphase neuroblasts. (**a**) Centrosomes split in early interphase neuroblasts and establish centrosome asymmetry based on centriolar age. The daughter centriole-containing centrosome (*green*) maintains a robust microtubule organizing center (MTOC), manifested in the maintenance of a robust microtubule aster. The mother centriole-containing centrosome (*red*) downregulates its MTOC activity and loses its connection with the apical cortex. At mitotic entry, the mother centrosome (*red*) has reached the basal pole of the neuroblast and matures. As a result, the centrosomes are approximately positioned along the polarity axis. This form of centrosome asymmetry leads to biased centrosome inheritance; the youngest centrosome (*red* in the first cell cycle, *green* in the second) is always retained by the self-renewed neuroblast and the older centrosome (*yellow* in the first cell cycle and *red* in the second) segregates with the GMC. (**b**) Several mutants are involved in maintaining the MTOC activity

since *polo* mutants or pharmacological inhibition of Polo result in downregulation of MTOC activity (Januschke and Gonzalez 2010; Januschke et al. 2013; Singh et al. 2014). Recently, we reported that interphase microtubules are necessary to recruit Polo to the apical centrosome and that this stabilization is mediated by the microtubule-associated protein (MAP) Wdr62 (MCPH2/Wdr62 in vertebrates) (Ramdas Nair et al. 2016). Polo is a mitotic kinase and phosphorylates both PCM proteins (Conduit et al. 2014) but also the daughter centriole-specific protein Centrobin (Cnb). Cnb is necessary and sufficient for MTOC activity; *cnb* loss of function disrupts MTOC activity, and ectopic localization to the basal centrosome induces two robust asters throughout interphase (Januschke et al. 2013). Thus, MTOC maintenance is described as a positive feedback loop: Polo generates an active MTOC through phosphorylation of Cnb and Cnn, activating microtubule growth, which is required to maintain the recruitment of Polo to the active apical MTOC (Ramdas Nair et al. 2016). Other molecules, necessary for the stability of microtubules and MTOC maintenance, have recently been identified such as the MAP Ensconsin (Ensc) (Gallaud et al. 2014). Interestingly, in *pins* mutant neuroblasts, interphase MTOC activity is also compromised, providing a link between neuroblast polarity and centrosome polarity (Rebollo et al. 2007). However, it is currently unclear how Pins is influencing centrosome asymmetry (Fig. 8.5b, c).

To establish centrosome asymmetry, MTOC activity has to be downregulated on the basal centrosome. It has been described that centrosomes, destined to be inherited by the GMC, downregulate MTOC activity in early interphase after centrosomes split from each other (Rebollo et al. 2007; Rusan and Peifer 2007; Conduit and Raff 2010). Although the molecular mechanism is not fully understood, it has been reported that Bld10 (Cep135 in vertebrates) and Pericentrin-like protein (Plp) are required for Polo and PCM protein shedding; in *bld10* and *plp* mutants, both centrosomes remain active throughout interphase, resulting in perturbed centrosome positioning, transient spindle orientation defects, and compromised centrosome segregation (Lerit and Rusan 2013; Singh et al. 2014). Plp is asymmetrically localized with a slight enrichment on the basal centrosome where it blocks Polo and therefore early maturation. Moreover,

Fig. 8.5 (continued) asymmetry in interphase; neuroblasts lacking *cnb*, *wdr62*, or *pins* display a complete loss of MTOC activity on both interphase centrosomes. *Plp* and *bld10* mutant neuroblasts maintain two active MTOCs. In *ensc/khc* mutant neuroblasts, centrosomes fail to split and are mispositioned when the cell enters mitosis. In all these mutants, centrosome inheritance is perturbed (albeit the penetrance may differ). (**c**) Centrosome asymmetry is largely controlled by the mitotic kinase Polo. Polo is localized to both the centriole and the pericentriolar matrix on the apical centrosome. The basal centrosome contains weak levels of centriolar Polo. Polo phosphorylates both Cnb and PCM protein such as Cnn. Cnb is only present on the younger centrosome where it gets phosphorylated by Polo and potentially inhibits Pericentrin-like Protein. Plp is required to downregulate Polo but only fulfills this function on the basal centrosome that is devoid of Cnb. This regulation loop maintains Polo level high on the apical daughter centriole-containing centrosome, explaining the presence of the microtubule aster. The MAP Wdr62 is required to maintain high Polo levels on the apical centrosome by promoting microtubule stability. Microtubules recruit Polo to the centrosome through an unknown mechanism. Bld10, equally present on both centrosomes, is also involved in the downregulation of Polo on the basal centrosome

when Cnb is ectopically localized to the basal centrosome, Plp becomes symmetric. This result suggests that Polo, through the activation of Cnb, downregulates its own inhibitor (Lerit and Rusan 2013). Bld10 is localized to both centrioles, and it is currently unclear how this centriolar protein is involved in the downregulation of Polo. Bld10 localization is not perturbed upon *cnb* or *plp* loss of function or ectopic expression of Cnb. Also Cnb and Plp asymmetries are maintained in *bld10* mutant neuroblasts (Singh et al. 2014) (Fig. 8.5b, c).

Defects in centrosome asymmetry perturb centrosome positioning and spindle orientation in early metaphase. However, asymmetric cell division is normal since metaphase spindles realign themselves with the intrinsic polarity axis shortly before anaphase. The molecular nature of this realignment is unclear (Januschke et al. 2013; Singh et al. 2014; Ramdas Nair et al. 2016). Nevertheless, despite this correction mechanism, defects in centrosome asymmetry, centrosome separation, and positioning compromise biased centrosome segregation. For instance, in *bld10*, *wdr62, cnb*, and *plp, but* also *ensc* and *khc* (kinesin-1 heavy chain) mutants, neuroblasts can also inherit the older mother centriole-containing centrosome (Januschke et al. 2013; Lerit and Rusan 2013; Gallaud et al. 2014; Singh et al. 2014; Nair et al. 2016) (Fig. 8.5b). This perturbed centrosome segregation has no consequence for neuroblast polarity and asymmetric cell division, at least in the first few divisions following centrosome missegregation. Thus, the function of biased centrosome segregation is currently unclear since no defects associated with centrosome missegregation have emerged. However, long-term defects and subtle changes in brain development, cell fate, and behavior have not been carefully investigated so far.

8.3.5 Asymmetric Segregation of Cell Fate Determinants

Polarity establishment and spindle orientation is orchestrated in a defined temporal sequence, ensuring that when the neuroblast enters anaphase, basal cell fate determinants can be correctly distributed between the two sibling cells. Thus, upon entry into anaphase, this intrinsic polarity axis has to be in place and the mitotic spindle needs to be correctly oriented to ensure correct segregation of basal cell fate determinants. In wild-type neuroblasts, Miranda (Mira) transports the cell fate determinant Pros and Brat as a cargo into the GMC. Pros enters the nucleus of the GMC where it induces and represses various cell type-specific target genes to trigger neuronal differentiation (see also above and reviewed in Knoblich 2010; Homem and Knoblich 2012). In addition, the basal cell fate determinant Numb represses Notch signaling in the differentiating GMC, further inducing differentiation and inhibiting self-renewal (Lee et al. 2006a; Wang et al. 2006). Thus, mutants compromising cell polarity and the correct localization and segregation of basal cell fate determinants (e.g., Brat, Pros, Mira) compromise differentiation and neuroblast homeostasis (Lee et al. 2006a, b, c; Bowman et al. 2008). Similarly, mutants affecting spindle orientation without perturbing cell polarity, such as *cnn, asl, DSas4*, or *mud*, perturb the asymmetric segregation of both apical and basal cell fate determinants (Siller et al. 2006; Basto et al. 2006; Cabernard and Doe 2009). For instance, neuroblasts containing mitotic spindles that are 90° misaligned in relation to the neuroblast's intrinsic apical–basal axis divide symmetrically by size, generating two functional neuroblasts (Cabernard and Doe 2009). Under such circumstances, both apical polarity proteins and basal cell fate determinants are equally distributed between the two resulting neuroblasts. However, in several instances, it was found that basal cell fate determinants can still segregate asymmetrically. This is due to defects in cleavage furrow positioning, which we will discuss in more detail below. Nevertheless, these results indicate that inheritance of basal cell fate determinants alone is not sufficient to drive differentiation, but the correct ratio of apical versus basal cell fate determinants ultimately determines the resulting cell fate (Cabernard and Doe 2009). Whereas wild-type third instar larval brains contain ~100 neuroblasts/brain lobe, larval brains derived from spindle orientation mutants (e.g., *mud, cnn*, or *Sas4*) show a significant increase in the neural stem cell pool (Cabernard and Doe 2009). Thus, spindle orientation is necessary for the accurate segregation of cell fate determinants and normal brain development.

Classical transplantation experiments also demonstrated that grafting of overproliferative brain tissue—due to mutations in polarity or cell fate determinant genes (pins, miranda, prospero or numb)—into the abdomen of wild-type adult host flies leads to tumor formation. These induced tumors could be immortalized, possessed metastatic potential and displayed two hallmarks of cancer: centrosome alterations and genome instability (Caussinus and Gonzalez 2005).

8.3.6 Spindle Asymmetry and Physical Sibling Cell Size Asymmetry

Drosophila neuroblasts undergo molecular but also physical asymmetric cell divisions. For instance, neuroblasts divide to give rise to a self-renewed neuroblast and a differentiating GMC; the neuroblast is about 2 times larger than the GMC. In this section, we will discuss how neuroblast intrinsic polarity affects this physical asymmetry and we will summarize work addressing how spindle asymmetry is generated but also how it contributes to the resulting sibling cell size asymmetry.

8.3.6.1 Molecular Determinants of Spindle Asymmetry

Spindle asymmetry in neuroblasts has first been investigated by Kaltschmidt and coworkers (Kaltschmidt et al. 2000), measuring spindle arm length, which is a measure of the distance from the centrosome to the mid-zone of the central spindle, a region of overlapping microtubules (MTs). Live cell imaging experiments with spindle and cell membrane markers revealed that in embryonic neuroblasts the spindle is symmetric throughout metaphase but becomes asymmetric at the onset of anaphase, where the microtubules elongate on the apical side and shrink on the basal side of the cell. The apical spindle arm was found to be 30% longer than the basal spindle arm at anaphase (Kaltschmidt et al. 2000). In a complementary study, spindle geometry was quantified in fixed embryonic neuroblasts using γ-tubulin and DNA as reference for the spindle poles and the metaphase plate and revealed that relative to the cell diameter at metaphase, the apical spindle arm is approximately 8% longer than the basal arm. In addition, it was also found that the entire spindle is slightly shifted toward the basal cortex, resulting in a larger gap between the apical centrosome and the apical cortex and a smaller gab between the basal centrosome and the basal cortex. Thus, the authors concluded that spindle asymmetry has already been established in metaphase. In agreement with Kaltschmidt and colleagues, Fuse et al. observed that the apical spindle arm is much longer than the basal spindle arm at telophase (Fuse et al. 2003). Recently, we showed in larval neuroblast that Survivin, a component of the Chromosomal Passenger Complex (CPC), is localizing to the central spindle and that this structure of overlapping microtubules is slightly shifted apically in relation to the ingressing anaphase cleavage furrow. It is only when the furrow ring approaches its closure stage (i.e., late telophase) that Survivin aligns with the furrow site and then fuses together to form the midbody (Roth et al. 2015). Thus, all these studies agree that the mitotic spindle is asymmetric in anaphase and that the distance between the cell membrane and the apical centrosome/microtubule aster is larger than on the basal cortex. These results suggest that the inherent asymmetry in the mitotic spindle is an important factor for the generation of physical sibling cell size asymmetry. However, since the central spindle only overlaps late with the cleavage furrow, spindle geometry might not be the only determinant for the generation of sibling cell size asymmetry.

Mitotic spindle geometry is controlled by two parallel signaling pathways: (1) Baz/aPKC/Insc and (2) Pins/Gαi (Cai et al. 2003; Fuse et al. 2003; Izumi et al. 2004). These two apical pathways act redundantly, preventing basal astral microtubule formation, and suppress spindle growth, generating unequal spindle arm length and hence spindle asymmetry. Perturbing any component in these complexes results in the formation of symmetric spindles and equal-size daughter cells (Cai et al. 2003; Fuse et al. 2003; Yu et al. 2003, 2005; Izumi et al. 2004). Spindle geometry is largely controlled by the activity of Gβγ, a subunit of the heterotrimeric G proteins. In wild-type Nbs, Gβγ binds to excess GDP–Gαi at the apical cortex and inactivates it, which could allow astral MTs and the apical half of the mitotic spindle to grow. On the basal cortex, limited GDP–Gαi allows free Gβγ to suppress spindle growth. Thus, the level of free Gβγ regulated by components of the apical complex governs spindle geometry and daughter cell size in *Drosophila* Nbs (Fuse et al. 2003; Yu et al. 2003) (Fig. 8.6).

Apart from Gβγ's contribution to spindle asymmetry and daughter cell size, it has also been reported that Dlg/Scrib/Lgl are important in regulating cortical polarity, mitotic spindle asymmetry, and cell size asymmetry (Albertson and Doe 2003). It has been shown that neuroblasts mutant for *dlg, scrib*, and *lgl*—or mutant combination thereof—undergo (1) symmetric cell division or (2) divide asymmetrically but with inverted cell polarity. For instance, inverted polarity is observed in 22% of *dlg* mutant neuroblasts, lacking also one copy of *lgl* and *scrib*. In these neuroblasts, the larger

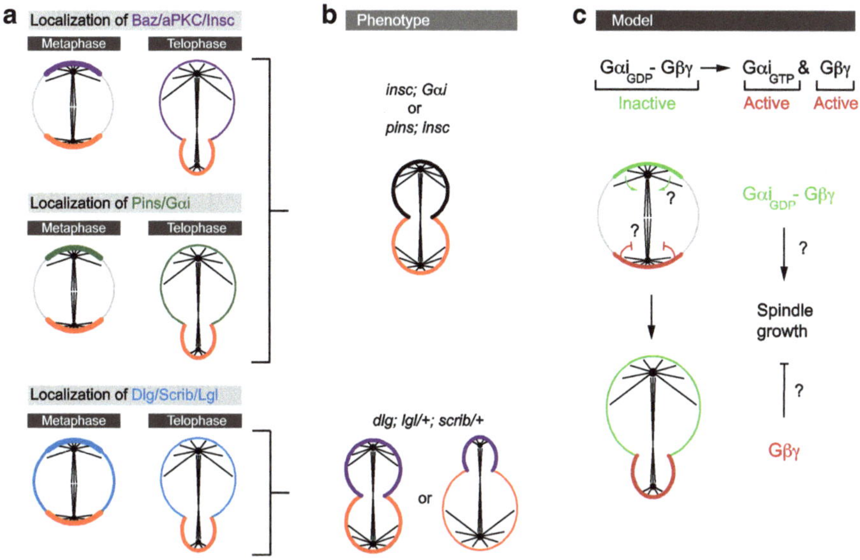

Fig. 8.6 Polarity cues influence spindle geometry. (**a**) Polarity proteins implicated in spindle asymmetry either localize to the apical cortex or are localized uniformly cortical with apical enrichment. (**b**) Mutant combinations of the implicated polarity complexes result in symmetric spindles and physical symmetric cell divisions. Depending on the mutation, basal cell fate determinants such as Miranda are still properly localized and segregate asymmetrically. In some instances, spindles can also become inverted, causing neuroblasts to divide with inverted polarity. (**c**) The molecular mechanism underlying spindle asymmetry involved the heterotrimeric G protein complex. The Gαi:GDP–Gβγ complex is inactive and becomes activated upon exchange of GDP with GTP. This causes the dissociation of Gβγ from Gαi. Whereas inactive Gαi:GDP–Gβγ, localized on the apical cortex, does not inhibit apical spindle growth, active Gβγ molecules inhibit spindle growth on the basal cortex. How cortical G proteins can inhibit spindle growth is unclear and could involve a mediator between the spindle and the neuroblast cortex

daughter cell has GMC fate, as it is Miranda positive and has a longer spindle arm with large astral MTs attached. In contrast, the small daughter cell contains cortical aPKC and has a shorter spindle arm with shorter astral MTs. It is not known how Gβγ is distributed in these inverted polarity Nbs and whether disruption of the apical complex through perturbation of these components redistributes Gβγ in a way that leads to this phenotype (Albertson and Doe 2003) (Fig. 8.6).

Previously, it was found by immunostaining that in wild-type neuroblasts the apical centrosome is slightly larger than the basal centrosome (Spana and Doe 1995). To test whether centrosome asymmetry has any influences on the spindle asymmetry, Cai et al. took advantage of the fact that 10% of Nbs in *sna/esg/wor* triple mutant embryos divide with reversed centrosome polarity. However, they found that switching centrosome position does not change spindle asymmetry since spindle arm length did only correlate with proximity to the Pins crescent but not with centrosome size (Cai et al. 2001). These results suggest that cortical polarity controls spindle asymmetry. This is consistent with recent findings, showing that whenever the mitotic spindle is 90° misaligned in relation to the apical–basal polarity axis, or when neuroblast polarity is compromised, neuroblasts can divide symmetrically by size (Siller et al. 2006; Cabernard and Doe 2009). Similarly, astral microtubules do not seem to influence spindle arm length, since neuroblasts defective for *asterless* (*asl*; Cep152 in humans) (Varmark et al. 2007) or *sas4*, which fail to form functional centrosomes and lack astral microtubules, still form asymmetric spindles (Giansanti et al. 2001; Basto et al. 2006). Recently, it was shown that in *Drosophila* sensory organ precursor (SOP) cells, spindle asymmetry is generated by *Klp10A* and *Patronin*, but whether the same mechanism also applies to fly neuroblasts is currently unknown (Derivery et al. 2015).

Taken together, current data suggest that spindle asymmetry in *Drosophila* Nbs is governed by neuroblast intrinsic polarity cues. How these cues directly influence differential spindle growth and spindle asymmetry is incompletely understood (Fig. 8.6).

8.3.6.2 Polarity and Cleavage Furrow Positioning

In most metazoan cells, cleavage furrow positioning is controlled by cues from the mitotic spindle (White and Glotzer 2012; D'Avino et al. 2015). Thus, spindle asymmetry would provide a plausible mechanism for the positioning of the cleavage furrow, which could explain the observed sibling cell size asymmetry; wild-type neuroblasts position the cleavage furrow closer to the basal cortex, thereby generating a small GMC. Classical spindle manipulation experiments provided ample evidence that spindle-dependent cues lead to the activation of the small GTPase RhoA, causing an actomyosin contractile ring to assemble at the correct position (reviewed in White and Glotzer 2012; D'Avino et al. 2015). However, it has been observed that cleavage furrow markers such as non-muscle Myosin II (Myosin, hereafter), Anillin, or Pavarotti (Pav; MKLP1 in vertebrates) show an asymmetric distribution in early anaphase, before clear inherent spindle asymmetry becomes apparent, supporting the notion that spindle-independent cues could be important for basal cleavage furrow positioning. For instance, Myosin is localized almost uniformly around the neuroblast cortex before metaphase. Upon entry into anaphase, Myosin first disappears from the apical cortex and subsequently from the basal cortex before enriching at the basally shifted cleavage furrow. Apical Myosin clearing is not regulated by spindle cues, since chemical spindle ablation experiments did not compromise apical Myosin relocalization. Furthermore, if the mitotic spindle is rotated 90° in relation to the neuroblast intrinsic polarity axis, Myosin still becomes asymmetrically localized but accumulates on the basal cortex, forming a first cleavage furrow. Subsequently, a second cleavage furrow forms based on the position of the central spindle (Cabernard et al. 2010). These results strongly suggested the existence of a spindle-independent cleavage furrow positioning pathway. Genetic analysis revealed that the polarity proteins Pins and Dlg are required for asymmetric Myosin localization. For instance, in *dlg;;pins* double mutants, Myosin fails to localize in an asymmetric fashion, clearing from both cell poles at the same time (Cabernard et al. 2010). The molecular mechanisms of this polarity-dependent cleavage furrow positioning pathway are still

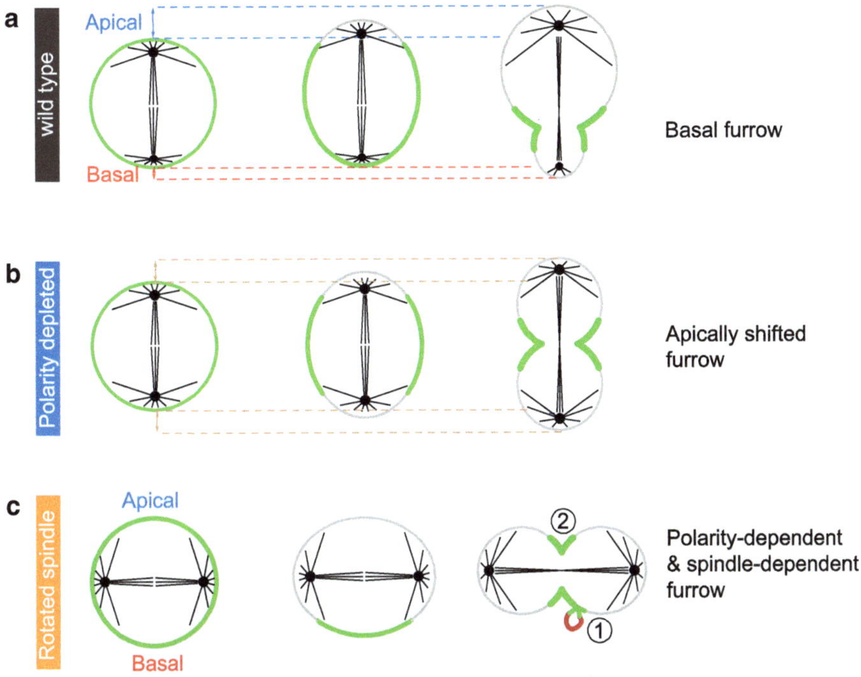

Fig. 8.7 Asymmetric Myosin localization as a mechanism to generate physical sibling cell size asymmetry. (**a**) Myosin is localized all around the metaphase neuroblast cortex. Upon entry into anaphase, Myosin first disappears from the apical cortex and subsequently from the basal cortex, accumulating at the forming cleavage furrow. Myosin's disappearance from the apical cortex could induce apical expansion and unequal cortical growth. (**b**) Asymmetric Myosin localization is regulated by neuroblast polarity cues. *Dlg, pins*, or *dlg;;pins* double mutants all show compromised asymmetric Myosin localization (with variable penetrance). Since these mutants also affect spindle asymmetry, the corresponding neuroblast divisions are often symmetric by size. (**c**) Asymmetric Myosin localization is independent of the mitotic spindle since mutants that misalign the mitotic spindle in relation to the intrinsic polarity axis (e.g., *mud*) show asymmetric Myosin localization (Myosin still disappears from the apical neuroblast cortex) but divide symmetric by size. In this case, two cleavage furrows appear: (1) a first furrow on the basal cortex, resulting in the formation of a Mira-containing (*red*) cortical lobe, and (2) a second furrow centered on the central spindle

being elucidated, but it strongly suggests that polarity cues influence the distribution of Myosin and possibly also other cortical proteins (Fig. 8.7). Recent results suggested that spindle-independent furrow positioning cues are not confined to fly neuroblasts but have also been observed in *C. elegans* (Ou et al. 2010; Pacquelet et al. 2015; Jordan et al. 2016), in fly germ cell formation (Cinalli and Lehmann 2013), or in human cells (Sedzinski et al. 2011). However, although it has been shown that spindle cues and polarity cues together specify the correct position of the cleavage furrow in wild-type neuroblasts (Roth et al. 2015), the exact mechanism and regulatory pathways remain to be elucidated.

Asymmetric Myosin localization could regulate cortical tension and cleavage furrow positioning. For instance, in wild-type Nbs the apical cortex expands much more than the basal cortex during anaphase, effectively shifting the cleavage furrow toward the small daughter cell. This apical cortical expansion correlates with the lack of apical Myosin. In Colcemid-treated *rod* (*rod* removes the spindle assembly checkpoint, allowing neuroblasts to enter anaphase) mutant Nbs, the apical cortex expands to the same extent as observed in wild-type Nbs because Myosin still clears from the apical cortex. However, the basal cortex does not expand much because Myosin prevents basal expansion. Loss of Myosin asymmetry was also found to result in symmetric extension and equal-size daughter cells (Connell et al. 2011). Thus, these findings suggest that cortical expansion is inversely proportional to cortical Myosin levels (Fig. 8.7).

8.4 Conclusions and Outlook

Drosophila neuroblasts have served as an excellent model to elucidate the basic molecular cell biology underlying asymmetric cell division. Much has been learned about the molecular mechanisms involved in polarity establishment, spindle orientation, and cell fate determinant segregation, but unanswered questions abound. For instance, how polarity cues regulate spindle morphology and asymmetry is incompletely understood. Similarly, how spindle asymmetry and actomyosin dynamics are regulated and coordinated to ensure the generation of physical asymmetry remains to be seen. Also, in the future it will be instrumental to elucidate how developmental signals are integrated with neuroblast intrinsic polarity and cell cycle cues to ensure correct asymmetric cell division. Novel genetic approaches such as CRISPR/Cas9 combined with innovative new imaging methods provide new tools, necessary to elucidate many of these open questions in the future.

References

Albertson R, Doe CQ (2003) Dlg, Scrib and Lgl regulate neuroblast cell size and mitotic spindle asymmetry. Nat Cell Biol 5:166–170. doi:10.1038/ncb922

Atwood SX, Prehoda KE (2009) aPKC phosphorylates Miranda to polarize fate determinants during neuroblast asymmetric cell division. Curr Biol 19:723–729. doi:10.1016/j.cub.2009.03.056

Atwood SX, Chabu C, Penkert RR et al (2007) Cdc42 acts downstream of Bazooka to regulate neuroblast polarity through Par-6-aPKC. J Cell Sci 120:3200–3206. doi:10.1242/jcs.014902

Bailey MJ, Prehoda KE (2015) Establishment of par-polarized cortical domains via phosphoregulated membrane motifs. Dev Cell 35:199–210. doi:10.1016/j.devcel.2015.09.016

Barros CS, Phelps CB, Brand AH (2003) Drosophila nonmuscle myosin II promotes the asymmetric segregation of cell fate determinants by cortical exclusion rather than active transport. Dev Cell 5:829–840

Basto R, Lau J, Vinogradova T et al (2006) Flies without centrioles. Cell 125:1375–1386. doi:10.1016/j.cell.2006.05.025

Bayraktar OA, Doe CQ (2013) Combinatorial temporal patterning in progenitors expands neural diversity. Nature 498:449–455. doi:10.1038/nature12266

Bello B, Reichert H, Hirth F (2006) The brain tumor gene negatively regulates neural progenitor cell proliferation in the larval central brain of Drosophila. Development 133:2639–2648. doi:10.1242/dev.02429

Bello BC, Izergina N, Caussinus E, Reichert H (2008) Amplification of neural stem cell proliferation by intermediate progenitor cells in Drosophila brain development. Neural Dev 3:5. doi:10.1186/1749-8104-3-5

Boone JQ, Doe CQ (2008) Identification of Drosophila type II neuroblast lineages containing transit amplifying ganglion mother cells. Dev Neurobiol 68:1185–1195. doi:10.1002/dneu.20648

Bowman SK, Neumüller RA, Novatchkova M et al (2006) The Drosophila NuMA Homolog Mud regulates spindle orientation in asymmetric cell division. Dev Cell 10:731–742. doi:10.1016/j.devcel.2006.05.005

Bowman SK, Rolland V, Betschinger J, Kinsey KA (2008) The tumor suppressors Brat and Numb regulate transit-amplifying neuroblast lineages in Drosophila. Dev Cell 14:535–546. doi:10.1016/j.devcel.2008.03.004

Brand AH, Livesey FJ (2011) Neural stem cell biology in vertebrates and invertebrates: more alike than different? Neuron 70:719–729. doi:10.1016/j.neuron.2011.05.016

Cabernard C (2012) Cytokinesis in *Drosophila melanogaster*. Cytoskeleton (Hoboken) 69:791–809. doi:10.1002/cm.21060

Cabernard C, Doe CQ (2009) Apical/basal spindle orientation is required for neuroblast homeostasis and neuronal differentiation in Drosophila. Dev Cell 17:134–141. doi:10.1016/j.devcel.2009.06.009

Cabernard C, Prehoda KE, Doe CQ (2010) A spindle-independent cleavage furrow positioning pathway. Nature 467:91–94. doi:10.1038/nature09334

Cai Y, Chia W, Yang X (2001) A family of snail-related zinc finger proteins regulates two distinct and parallel mechanisms that mediate Drosophila neuroblast asymmetric divisions. EMBO J 20:1704–1714. doi:10.1093/emboj/20.7.1704

Cai Y, Yu F, Lin S et al (2003) Apical complex genes control mitotic spindle geometry and relative size of daughter cells in Drosophila neuroblast and pI asymmetric divisions. Cell 112:51–62

Campos-Ortega JA, Hartenstein V (1985) The embryonic development of *Drosophila melanogaster*. Springer, Berlin

Caussinus E. Gonzalez C (2005) Induction of tumor growth by altered stem-cell asymmetric division in *Drosophila melanogaster*. Nat Genet 37:1125–1129. doi:10.1038/ng1632

Chabu C, Doe CQ (2008) Dap160/intersectin binds and activates aPKC to regulate cell polarity and cell cycle progression. Development 135:2739–2746. doi:10.1242/dev.024059

Chabu C, Doe CQ (2009) Twins/PP2A regulates aPKC to control neuroblast cell polarity and self-renewal. Dev Biol 330:399–405. doi:10.1016/j.ydbio.2009.04.014

Chell JM, Brand AH (2010) Nutrition-responsive glia control exit of neural stem cells from quiescence. Cell 143:1161–1173. doi:10.1016/j.cell.2010.12.007

Choksi SP, Southall TD, Bossing T et al (2006) Prospero acts as a binary switch between self-renewal and differentiation in Drosophila neural stem cells. Dev Cell 11:775–789. doi:10.1016/j.devcel.2006.09.015

Chu-LaGraff Q, Doe CQ (1993) Neuroblast specification and formation regulated by wingless in the Drosophila CNS. Science 261:1594–1597. doi:10.1126/science.8372355

Cinalli RM, Lehmann R (2013) A spindle-independent cleavage pathway controls germ cell formation in Drosophila. Nat Cell Biol 15:839–845. doi:10.1038/ncb2761

Conduit PT, Raff JW (2010) Cnn dynamics drive centrosome size asymmetry to ensure daughter centriole retention in Drosophila neuroblasts. Curr Biol 20:2187–2192. doi:10.1016/j.cub.2010.11.055

Conduit PT, Feng Z, Richens JH et al (2014) The centrosome-specific phosphorylation of Cnn by Polo/Plk1 drives Cnn scaffold assembly and centrosome maturation. Dev Cell. doi:10.1016/j.devcel.2014.02.013

Connell M, Cabernard C, Ricketson D et al (2011) Asymmetric cortical extension shifts cleavage furrow position in Drosophila neuroblasts. Mol Biol Cell 22:4220–4226. doi:10.1091/mbc.E11-02-0173

D'Avino PP, Giansanti MG, Petronczki M (2015) Cytokinesis in animal cells. Cold Spring Harb Perspect Biol 7:a015834. doi:10.1101/cshperspect.a015834

Derivery E, Seum C, Daeden A et al (2015) Polarized endosome dynamics by spindle asymmetry during asymmetric cell division. Nature 528:280–285. doi:10.1038/nature16443

Doe CQ, Chu-LaGraff Q, Wright DM, Scott MP (1991) The prospero gene specifies cell fates in the Drosophila central nervous system. Cell. doi:10.1016/0092-8674(91)90463-9

Ebens AJ, Garren H, Cheyette BNR, Zipursky SL (1993) The Drosophila anachronism locus: a glycoprotein secreted by glia inhibits neuroblast proliferation. Cell 74:15–27. doi:10.1016/0092-8674(93)90291-W

Egger B, Boone JQ, Stevens NR et al (2007) Regulation of spindle orientation and neural stem cell fate in the Drosophila optic lobe. Neural Dev 2:1. doi:10.1186/1749-8104-2-1

Egger B, Chell JM, Brand AH (2008) Insights into neural stem cell biology from flies. Philos Trans R Soc Lond B Biol Sci 363(1489):39–56

Egger B, Gold KS, Brand AH (2010) Notch regulates the switch from symmetric to asymmetric neural stem cell division in the Drosophila optic lobe. Development 137:2981–2987. doi:10.1242/dev.051250

Egger B, Gold KS, Brand AH (2011) Regulating the balance between symmetric and asymmetric stem cell division in the developing brain. Fly (Austin) 5:237–241

Erben V, Waldhuber M, Langer D et al (2008) Asymmetric localization of the adaptor protein Miranda in neuroblasts is achieved by diffusion and sequential interaction of Myosin II and VI. J Cell Sci 121:1403–1414. doi:10.1242/jcs.020024

Fuerstenberg S, Peng CY, Alvarez-Ortiz P et al (1998) Identification of Miranda protein domains regulating asymmetric cortical localization, cargo binding, and cortical release. Mol Cell Neurosci 12:325–339. doi:10.1006/mcne.1998.0724

Fuse N, Hisata K, Katzen AL, Matsuzaki F (2003) Heterotrimeric G proteins regulate daughter cell size asymmetry in Drosophila neuroblast divisions. Curr Biol 13:947–954. doi:10.1016/S0960-9822(03)00334-8

Gallaud E, Caous R, Pascal A et al (2014) Ensconsin/Map7 promotes microtubule growth and centrosome separation in Drosophila neural stem cells. J Cell Biol 204:1111–1121. doi:10.1083/jcb.201311094

Giansanti MG, Gatti M, Bonaccorsi S (2001) The role of centrosomes and astral microtubules during asymmetric division of Drosophila neuroblasts. Development 128:1137–1145

Gonzalez C (2013) *Drosophila melanogaster*: a model and a tool to investigate malignancy and identify new therapeutics. Nat Rev Cancer 13:172–183. doi:10.1038/nrc3461

Hartenstein V, Younossi-Hartenstein A, Lekven A (1994) Delamination and division in the Drosophila neurectoderm: spatiotemporal pattern, cytoskeletal dynamics, and common control by neurogenic and segment polarity genes. Dev Biol 165:480–499. doi:10.1006/dbio.1994.1269

Hirata J, Nakagoshi H, Nabeshima Y, Matsuzaki F (1995) Asymmetric segregation of the homeodomain protein Prospero during Drosophila development. Nature 377:627–630. doi:10.1038/377627a0

Hofbauer A, Campos-Ortega JA (1990) Proliferation pattern and early differentiation of the optic lobes in *Drosophila melanogaster*. Roux's Arch Dev Biol 198:264–274

Homem CCF, Knoblich JA (2012) Drosophila neuroblasts: a model for stem cell biology. Development 139:4297–4310. doi:10.1242/dev.080515

Homem CCF, Steinmann V, Burkard TR et al (2014) Ecdysone and mediator change energy metabolism to terminate proliferation in Drosophila neural stem cells. Cell 158:874–888. doi:10.1016/j.cell.2014.06.024

Huang C, Chan JA, Schuurmans C (2014) Proneural bHLH genes in development and disease. Curr Top Dev Biol 110:75–127. doi:10.1016/B978-0-12-405943-6.00002-6

Ikeshima-Kataoka H, Skeath JB, Nabeshima Y et al (1997) Miranda directs Prospero to a daughter cell during Drosophila asymmetric divisions. Nature 390:625–629. doi:10.1038/37641

Ito K, Hotta Y (1992) Proliferation pattern of postembryonic neuroblasts in the brain of Drosophila-melanogaster. Dev Biol 149:134–148

Ito K, Awano W, Suzuki K et al (1997) The Drosophila mushroom body is a quadruple structure of clonal units each of which contains a virtually identical set of neurones and glial cells. Development 124:761–771

Izumi Y, Hirose T, Tamai Y et al (1998) An atypical PKC directly associates and colocalizes at the epithelial tight junction with ASIP, a mammalian homologue of *Caenorhabditis elegans* polarity protein PAR-3. J Cell Biol 143:95–106

Izumi Y, Ohta N, Itoh-Furuya A et al (2004) Differential functions of G protein and Baz-aPKC signaling pathways in Drosophila neuroblast asymmetric division. J Cell Biol 164:729–738. doi:10.1083/jcb.200309162

Izumi Y, Ohta N, Hisata K et al (2006) Drosophila Pins-binding protein Mud regulates spindle-polarity coupling and centrosome organization. Nat Cell Biol 8:586–593. doi:10.1038/ncb1409

Januschke J, Gonzalez C (2010) The interphase microtubule aster is a determinant of asymmetric division orientation in Drosophila neuroblasts. J Cell Biol 188:693–706. doi:10.1083/jcb.200905024

Januschke J, Llamazares S, Reina J, Gonzalez C (2011) Drosophila neuroblasts retain the daughter centrosome. Nat Commun 2:243. doi:10.1038/ncomms1245

Januschke J, Reina J, Llamazares S et al (2013) Centrobin controls mother–daughter centriole asymmetry in Drosophila neuroblasts. Nat Cell Biol 15:241–248. doi:10.1038/ncb2671

Johnston CA, Hirono K, Prehoda KE, Doe CQ (2009) Identification of an Aurora-A/PinsLINKER/Dlg spindle orientation pathway using induced cell polarity in S2 cells. Cell 138:1150–1163. doi:10.1016/j.cell.2009.07.041

Jordan SN, Davies T, Zhuravlev Y et al (2016) Cortical PAR polarity proteins promote robust cytokinesis during asymmetric cell division. J Cell Biol 212:39–49. doi:10.1083/jcb.201510063

Kaltschmidt JA, Davidson CM, Brown NH, Brand AH (2000) Rotation and asymmetry of the mitotic spindle direct asymmetric cell division in the developing central nervous system. Nat Cell Biol 2:7–12. doi:10.1038/71323

Kang KH, Reichert H (2014) Control of neural stem cell self-renewal and differentiation in Drosophila. Cell Tissue Res 359:33–45. doi:10.1007/s00441-014-1914-9

Knoblich JA (2010) Asymmetric cell division: recent developments and their implications for tumour biology. Nat Rev Mol Cell Biol 11:849–860. doi:10.1038/nrm3010

Knoblich JA, Jan LY, Jan YN (1995) Asymmetric segregation of Numb and Prospero during cell division. Nature 377:624–627. doi:10.1038/377624a0

Knoblich JA, Jan LY, Jan YN (1997) The N terminus of the Drosophila Numb protein directs membrane association and actin-dependent asymmetric localization. Proc Natl Acad Sci U S A 94:13005–13010. doi:10.1016/0092-8674(92)90468-R

Krahn MP, Egger-Adam D, Wodarz A (2009) PP2A antagonizes phosphorylation of Bazooka by PAR-1 to control apical-basal polarity in dividing embryonic neuroblasts. Dev Cell 16:901–908. doi:10.1016/j.devcel.2009.04.011

Lai S-L, Doe CQ, Brand A (2014) Transient nuclear Prospero induces neural progenitor quiescence. eLife 3:e03363. doi:10.7554/eLife.03363

Lee C-Y, Andersen RO, Cabernard C et al (2006a) Drosophila Aurora-A kinase inhibits neuroblast self-renewal by regulating aPKC/Numb cortical polarity and spindle orientation. Genes Dev 20:3464–3474. doi:10.1101/gad.1489406

Lee C-Y, Robinson KJ, Doe CQ (2006b) Lgl, Pins and aPKC regulate neuroblast self-renewal versus differentiation. Nature 439:594–598. doi:10.1038/nature04299

Lee C-Y, Wilkinson BD, Siegrist SE et al (2006c) Brat is a Miranda cargo protein that promotes neuronal differentiation and inhibits neuroblast self-renewal. Dev Cell 10:441–449. doi:10.1016/j.devcel.2006.01.017

Lerit DA, Rusan NM (2013) PLP inhibits the activity of interphase centrosomes to ensure their proper segregation in stem cells. J Cell Biol 202:1013–1022. doi:10.1083/jcb.201303141

Lu MS, Johnston CA (2013) Molecular pathways regulating mitotic spindle orientation in animal cells. Development 140:1843–1856. doi:10.1242/dev.087627

Lu BW, Rothenberg M, Jan LY, Jan YN (1998) Partner of numb colocalizes with numb during mitosis and directs numb asymmetric localization in Drosophila neural and muscle progenitors. Cell 95:225–235

Matsuzaki F, Koizumi K, Hama C et al (1992) Cloning of the Drosophila prospero gene and its expression in ganglion mother cells. Biochem Biophys Res Commun 182:1326–1332

Matsuzaki F, Ohshiro T, Ikeshima-Kataoka H, Izumi H (1998) Miranda localizes Staufen and Prospero asymmetrically in mitotic neuroblasts and epithelial cells in early Drosophila embryogenesis. Development 125:4089–4098

Mayer B, Emery G, Berdnik D et al (2005) Quantitative analysis of protein dynamics during asymmetric cell division. Curr Biol 15:1847–1854. doi:10.1016/j.cub.2005.08.067

Merdes A, Ramyar K, Vechio JD, Cleveland DW (1996) A complex of NuMA and cytoplasmic dynein is essential for mitotic spindle assembly. Cell 87:447–458

Morin X, Bellaïche Y (2011) Mitotic spindle orientation in asymmetric and symmetric cell divisions during animal development. Dev Cell 21:102–119. doi:10.1016/j.devcel.2011.06.012

Nair AR, Singh P, Garcia DS, Rodriguez-Crespo D (2016) The microcephaly-associated protein Wdr62/CG7337 is required to maintain centrosome asymmetry in Drosophila neuroblasts. Cell Rep 14:1100–1113. doi:10.1016/j.celrep.2015.12.097

Nipper RW, Siller KH, Smith NR et al (2007) Galphai generates multiple Pins activation states to link cortical polarity and spindle orientation in Drosophila neuroblasts. Proc Natl Acad Sci U S A 104:14306–14311. doi:10.1073/pnas.0701812104

Ogawa H, Ohta N, Moon W, Matsuzaki F (2009) Protein phosphatase 2A negatively regulates aPKC signaling by modulating phosphorylation of Par-6 in Drosophila neuroblast asymmetric divisions. J Cell Sci 122:3242–3249. doi:10.1242/jcs.050955

Ou G, Stuurman N, D'Ambrosio M, Vale RD (2010) Polarized myosin produces unequal-size daughters during asymmetric cell division. Science 330:677–680. doi:10.1126/science.1196112

Pacquelet A, Uhart P, Tassan J-P, Michaux G (2015) PAR-4 and anillin regulate myosin to coordinate spindle and furrow position during asymmetric division. J Cell Biol 210:1085–1099. doi:10.1083/jcb.201503006

Pereanu W, Hartenstein V (2006) Neural lineages of the Drosophila brain: a three-dimensional digital atlas of the pattern of lineage location and projection at the late larval stage. J Neurosci 26:5534–5553. doi:10.1523/JNEUROSCI.4708-05.2006

Petritsch C, Tavosanis G, Turck CW et al (2003) The Drosophila myosin VI Jaguar is required for basal protein targeting and correct spindle orientation in mitotic neuroblasts. Dev Cell. doi:10.1016/S1534-5807(03)00020-0

Petronczki M, Knoblich JA (2001) DmPAR-6 directs epithelial polarity and asymmetric cell division of neuroblasts in Drosophila. Nat Cell Biol 3:43–49. doi:10.1038/35050550

Prehoda KE (2009) Polarization of Drosophila neuroblasts during asymmetric division. Cold Spring Harb Perspect Biol 1:a001388. doi:10.1101/cshperspect.a001388

Ramdas Nair A, Singh P, Salvador Garcia D et al (2016) The microcephaly-associated protein Wdr62/CG7337 is required to maintain centrosome asymmetry in Drosophila neuroblasts. Cell Rep 14:1100–1113. doi:10.1016/j.celrep.2015.12.097

Rebollo E, Sampaio P, Januschke J et al (2007) Functionally unequal centrosomes drive spindle orientation in asymmetrically dividing Drosophila neural stem cells. Dev Cell 12:467–474. doi:10.1016/j.devcel.2007.01.021

Rebollo E, Roldán M, Gonzalez C (2009) Spindle alignment is achieved without rotation after the first cell cycle in Drosophila embryonic neuroblasts. Development 136:3393–3397. doi:10.1242/dev.041822

Reichert H (2011) Drosophila neural stem cells: cell cycle control of self-renewal, differentiation, and termination in brain development. Cell Cycle Dev 53:529–546. doi:10.1007/978-3-642-19065-0_21

Rhyu MS, Jan LY, Jan YN (1994) Asymmetric distribution of numb protein during division of the sensory organ precursor cell confers distinct fates to daughter cells. Cell 76:477–491
Rolls MM, Albertson R, Shih H-P et al (2003) Drosophila aPKC regulates cell polarity and cell proliferation in neuroblasts and epithelia. J Cell Biol 163:1089–1098. doi:10.1083/jcb.200306079
Roth M, Roubinet C, Iffländer N et al (2015) Asymmetrically dividing Drosophila neuroblasts utilize two spatially and temporally independent cytokinesis pathways. Nat Commun 6:6551. doi:10.1038/ncomms7551
Rusan NM, Peifer M (2007) A role for a novel centrosome cycle in asymmetric cell division. J Cell Biol 177:13–20. doi:10.1083/jcb.200612140
Salzmann V, Chen C, Chiang CYA et al (2014) Centrosome-dependent asymmetric inheritance of the midbody ring in Drosophila germline stem cell division. 25:267–275. doi:10.1091/mbc.E13-09-0541
Schaefer M, Shevchenko A, Shevchenko A, Knoblich JA (2000) A protein complex containing Inscuteable and the Gα-binding protein Pins orients asymmetric cell divisions in Drosophila. Curr Biol 10:353–362. doi:10.1016/S0960-9822(00)00401-2
Schober M, Schaefer M, Knoblich JA (1999) Bazooka recruits Inscuteable to orient asymmetric cell divisions in Drosophila neuroblasts. Nature 402:548–551. doi:10.1038/990135
Schuldt AJ, Adams J, Davidson CM et al (1998) Miranda mediates asymmetric protein and RNA localization in the developing nervous system. Genes Dev 12:1847–1857
Sedzinski J, Biro M, Oswald A et al (2011) Polar actomyosin contractility destabilizes the position of the cytokinetic furrow. Nature 476:462–466. doi:10.1038/nature10286
Shen CP, Jan LY, Jan YN (1997) Miranda is required for the asymmetric localization of Prospero during mitosis in Drosophila. Cell 90:449–458
Shen CP, Knoblich JA, Chan YM et al (1998) Miranda as a multidomain adapter linking apically localized Inscuteable and basally localized Staufen and Prospero during asymmetric cell division in Drosophila. Genes Dev 12:1837–1846
Siegrist SE, Doe CQ (2005) Microtubule-induced Pins/Galphai cortical polarity in Drosophila neuroblasts. Cell 123:1323–1335. doi:10.1016/j.cell.2005.09.043
Siegrist SE, Doe CQ (2006) Extrinsic cues orient the cell division axis in Drosophila embryonic neuroblasts. Development 133:529–536. doi:10.1242/dev.02211
Siller KH, Doe CQ (2008) Lis1/dynactin regulates metaphase spindle orientation in Drosophila neuroblasts. Dev Biol 319:1–9. doi:10.1016/j.ydbio.2008.03.018
Siller KH, Cabernard C, Doe CQ (2006) The NuMA-related Mud protein binds Pins and regulates spindle orientation in Drosophila neuroblasts. Nat Cell Biol 8:594–600. doi:10.1038/ncb1412
Singh P, Ramdas Nair A, Cabernard C (2014) The centriolar protein Bld10/Cep135 is required to establish centrosome asymmetry in Drosophila neuroblasts. Curr Biol 24:1548–1555. doi:10.1016/j.cub.2014.05.050
Smith CA, Lau KM, Rahmani Z et al (2007) aPKC-mediated phosphorylation regulates asymmetric membrane localization of the cell fate determinant Numb. EMBO J 26:468–480. doi:10.1038/sj.emboj.7601495
Sousa-Nunes R, Chia W, Somers WG (2009) Protein phosphatase 4 mediates localization of the Miranda complex during Drosophila neuroblast asymmetric divisions. Genes Dev 23:359–372. doi:10.1101/gad.1723609
Sousa-Nunes R, Yee LL, Gould AP (2011) Fat cells reactivate quiescent neuroblasts via TOR and glial insulin relays in Drosophila. Nature 471:508–512. doi:10.1038/nature09867
Spana EP, Doe CQ (1995) The Prospero transcription factor is asymmetrically localized to the cell cortex during neuroblast mitosis in Drosophila. Development 121:3187–3195
Speicher S, Fischer A, Knoblich J, Carmena A (2008) The PDZ protein Canoe regulates the asymmetric division of Drosophila neuroblasts and muscle progenitors. Curr Biol 18:831–837. doi:10.1016/j.cub.2008.04.072
Truman JW, Bate M (1988) Spatial and temporal patterns of neurogenesis in the central nervous system of *Drosophila melanogaster*. Dev Biol 125:145–157. doi:10.1016/0012-1606(88)90067-X
Tsuji T, Hasegawa E, Isshiki T (2008) Neuroblast entry into quiescence is regulated intrinsically by the combined action of spatial Hox proteins and temporal identity factors. Development 135:3859–3869. doi:10.1242/dev.025189

Vainberg IE, Lewis SA, Rommelaere H et al (1998) Prefoldin, a chaperone that delivers unfolded proteins to cytosolic chaperonin. Cell 93:863–873

Varmark H, Llamazares S, Rebollo E et al (2007) Asterless is a centriolar protein required for centrosome function and embryo development in Drosophila. Curr Biol 17:1735–1745. doi:10.1016/j.cub.2007.09.031

Wang H, Somers GW, Bashirullah A et al (2006) Aurora-A acts as a tumor suppressor and regulates self-renewal of Drosophila neuroblasts. Genes Dev 20:3453–3463. doi:10.1101/gad.1487506

Wang H, Ouyang Y, Somers WG et al (2007) Polo inhibits progenitor self-renewal and regulates Numb asymmetry by phosphorylating Pon. Nature 449:96–100. doi:10.1038/nature06056

Wang X, Tsai J-W, Imai JH et al (2009) Asymmetric centrosome inheritance maintains neural progenitors in the neocortex. Nature 461:947–955. doi:10.1038/nature08435

Wang C, Li S, Januschke J et al (2011) An ana2/ctp/mud complex regulates spindle orientation in Drosophila neuroblasts. Dev Cell 21:520–533. doi:10.1016/j.devcel.2011.08.002

Wee B, Johnston CA, Prehoda KE, Doe CQ (2011) Canoe binds RanGTP to promote PinsTPR/Mud-mediated spindle orientation. J Cell Biol 195:369–376. doi:10.1083/jcb.201102130

White EA, Glotzer M (2012) Centralspindlin: at the heart of cytokinesis. Cytoskeleton (Hoboken) 69:882–892. doi:10.1002/cm.21065

White K, Grether ME, Abrams JM, Young L (1994) Genetic control of programmed cell death in Drosophila. Science 264:677–683. doi:10.1126/science.8171319

Wirtz-Peitz F, Nishimura T, Knoblich JA (2008) Linking cell cycle to asymmetric division: Aurora-A phosphorylates the Par complex to regulate Numb localization. Cell 135:161–173. doi:10.1016/j.cell.2008.07.049

Wodarz A, Ramrath A, Kuchinke U, Knust E (1999) Bazooka provides an apical cue for Inscuteable localization in Drosophila neuroblasts. Nature 402:544–547. doi:10.1038/990128

Yadlapalli S, Yamashita YM (2013) Chromosome-specific nonrandom sister chromatid segregation during stem-cell division. Nature 498:251–254. doi:10.1038/nature12106

Yamashita YM, Mahowald AP, Perlin JR, Fuller MT (2007) Asymmetric inheritance of mother versus daughter centrosome in stem cell division. Science 315:518–521. doi:10.1126/science.1134910

Yasugi T, Sugie A, Umetsu D, Tabata T (2010) Coordinated sequential action of EGFR and Notch signaling pathways regulates proneural wave progression in the Drosophila optic lobe. Development 137:3193–3203. doi:10.1242/dev.048058

Yoshiura S, Ohta N, Matsuzaki F (2012) Tre1 GPCR signaling orients stem cell divisions in the Drosophila central nervous system. Dev Cell 22:79–91. doi:10.1016/j.devcel.2011.10.027

Yu F, Morin X, Cai Y et al (2000) Analysis of partner of inscuteable, a novel player of Drosophila asymmetric divisions, reveals two distinct steps in inscuteable apical localization. Cell 100:399–409. doi:10.1016/S0092-8674(00)80676-5

Yu F, Cai Y, Kaushik R et al (2003) Distinct roles of Galphai and Gbeta13F subunits of the heterotrimeric G protein complex in the mediation of Drosophila neuroblast asymmetric divisions. J Cell Biol 162:623–633. doi:10.1083/jcb.200303174

Yu F, Wang H, Qian H et al (2005) Locomotion defects, together with Pins, regulates heterotrimeric G-protein signaling during Drosophila neuroblast asymmetric divisions. Genes Dev 19:1341–1353. doi:10.1101/gad.1295505

Zhang F, Huang ZX, Bao H et al (2016a) Phosphotyrosyl phosphatase activator facilitates localization of Miranda through dephosphorylation in dividing neuroblasts. Development 143:35–44. doi:10.1242/dev.127233

Zhang Y, Rai M, Wang C et al (2016b) Prefoldin and Pins synergistically regulate asymmetric division and suppress dedifferentiation. Sci Rep 6:23735. doi:10.1038/srep23735

Zhu S, Barshow S, Wildonger J et al (2011) Ets transcription factor pointed promotes the generation of intermediate neural progenitors in Drosophila larval brains. Proc Natl Acad Sci U S A. doi:10.1073/pnas.1118595109

Chapter 9
Asymmetric Divisions in Oogenesis

Szczepan M. Bilinski, Jacek Z. Kubiak, and Malgorzata Kloc

Abstract In the majority of animals, the oocyte/egg is structurally, molecularly, and functionally asymmetric. Such asymmetry is a prerequisite for a flawless fertilization and faithful segregation of maternal determinants during subsequent embryonic development. The oocyte asymmetry develops during oogenesis and must be maintained during consecutive and obligatorily asymmetric oogonial divisions, which depending on the species lead to the formation of either oocyte alone or oocyte and nurse cell complex. In the following chapter, we summarize current knowledge on the asymmetric oogonial divisions in invertebrate (insects) and vertebrate (*Xenopus*) species.

9.1 Asymmetric Oogonial Divisions in the Ovaries of the Fruit Fly, *Drosophila*, and Other Insects with Meroistic Ovaries

Insect ovaries are paired organs composed of several discrete units, termed ovarioles (see Büning 1994 for a review). As a rule, the ovarioles consist of three well-defined regions: the terminal filament, the germarium, and the vitellarium. The terminal filament is a stack of flat, disk-shaped somatic cells, oriented perpendicular to the long axis of ovariole. The germarium contains dividing and/or differentiating

S.M. Bilinski (✉)
Institute of Zoology, Jagiellonian University, Gronostajowa 9, 30-387 Krakow, Poland
e-mail: szczepan.bilinski@uj.edu.pl

J.Z. Kubiak
CNRS UMR 6290, Institute of Genetics and Development of Rennes, Cell Cycle Group, IFR 140 GFAS, Faculty of Medicine, University of Rennes 1, 35 043 Rennes, France

M. Kloc (✉)
The Houston Methodist Research Institute, Houston, TX, USA

Department of Surgery, The Houston Methodist Hospital, 6550 Fannin Street, Houston, TX 77030, USA
e-mail: mkloc@houstonmethodist.org

J.-P. Tassan, J.Z. Kubiak (eds.), *Asymmetric Cell Division in Development, Differentiation and Cancer*, Results and Problems in Cell Differentiation 61,
DOI 10.1007/978-3-319-53150-2_9

oogonial cells, whereas the vitellarium consists of growing ovarian follicles (reviewed in Büning 1994).

In the majority of insect species, oogenesis starts with the formation of germline cysts composed of several interconnected sibling cells. Although the processes underlying the cyst formation are similar in all studied insect species, the fate of interconnected cyst cells is remarkably different in panoistic versus meroistic ovaries. In the panoistic ovaries, the germline cysts split into equivalent cells that become functional oocytes (Pritsch and Büning 1989; Gottanka and Büning 1990; for details, see Sect. 9.3). In the meroistic ovaries, characteristic of *Drosophila* and other derived insects, the cyst cells remain joined; one of them differentiates into the oocyte and remaining cells form the nurse cells (e.g., Büning 1993, 1994; Kubrakiewicz 1997; Bilinski et al. 1998; Jaglarz 1998; Michalik et al. 2013). The main function of nurse cells is the synthesis and subsequent transport of the macromolecules and organelles to the oocyte cytoplasm, while the oocyte nucleus, as a rule, remains transcriptionally quiescent.

Processes leading to the formation of female germline cysts are best described and understood for the fruit fly, *Drosophila melanogaster* (King 1970; De Cuevas et al. 1997; Pepling et al. 1999). The germaria of this species contain 2–3 germline stem cells (GSCs) that divide asymmetrically to produce new GSCs and progenitor germ cells, termed cystoblasts (see below). Subsequently, each cystoblast divides synchronously four times that leads to the formation of the cyst, composed of 16 sibling germline cells, the cystocytes (Fig. 9.1a). Because all consecutive divisions of the cystoblast and descendant cystocytes are followed by incomplete cytokinesis, the arising cyst cells are interconnected by narrow cytoplasmic strands, termed the intercellular bridges or ring canals (King 1970; Ong and Tan 2010). Initially, all the cystocytes of the cyst are morphologically almost identical. They differ mainly in the number of intercellular bridges connecting given cell to its neighbors (Fig. 9.1a). The central part of each cyst contains two cells with 4 bridges and 2 cells with three bridges. In the peripheral cyst regions, cells with two or only single bridge reside. Classical histological studies showed that the intercellular bridges connecting the cells of the developing/dividing cyst contain specialized cytoplasm, referred to as a fusome, and that all the fusomes of a given cyst merge (fuse) with each other to form a polyfusome, a large ramified structure that extends through all the cyst bridges. The fusome contains elements (membranes) of endoplasmic reticulum (ER) and several membrane cytoskeletal or adapter proteins such as α- and β-spectrin, ankyrin, and the adducin-like protein Hts (Hu-li tai shao; Yue and Spradling 1992; Lin et al. 1994; Lin and Spradling 1995; De Cuevas et al. 1997; Snapp et al. 2004; Petrella et al. 2007; Lighthouse et al. 2008). The ER membranes of the fusome form an intricate continuous system shared by all the cystocytes of the cyst. Interestingly, this system falls apart before the onset of oocyte differentiation and entry into meiosis (Snapp et al. 2004). The fusome/polyfusome mediates essential steps during cyst development and oocyte differentiation (reviewed in McKearin 1997). It is well documented that in *Drosophila* the fusomes anchor mitotic spindles of dividing cystocytes ensuring the proper orientation of the consecutive division planes (King 1970; Lin and Spradling 1995; Deng and Lin 1997;

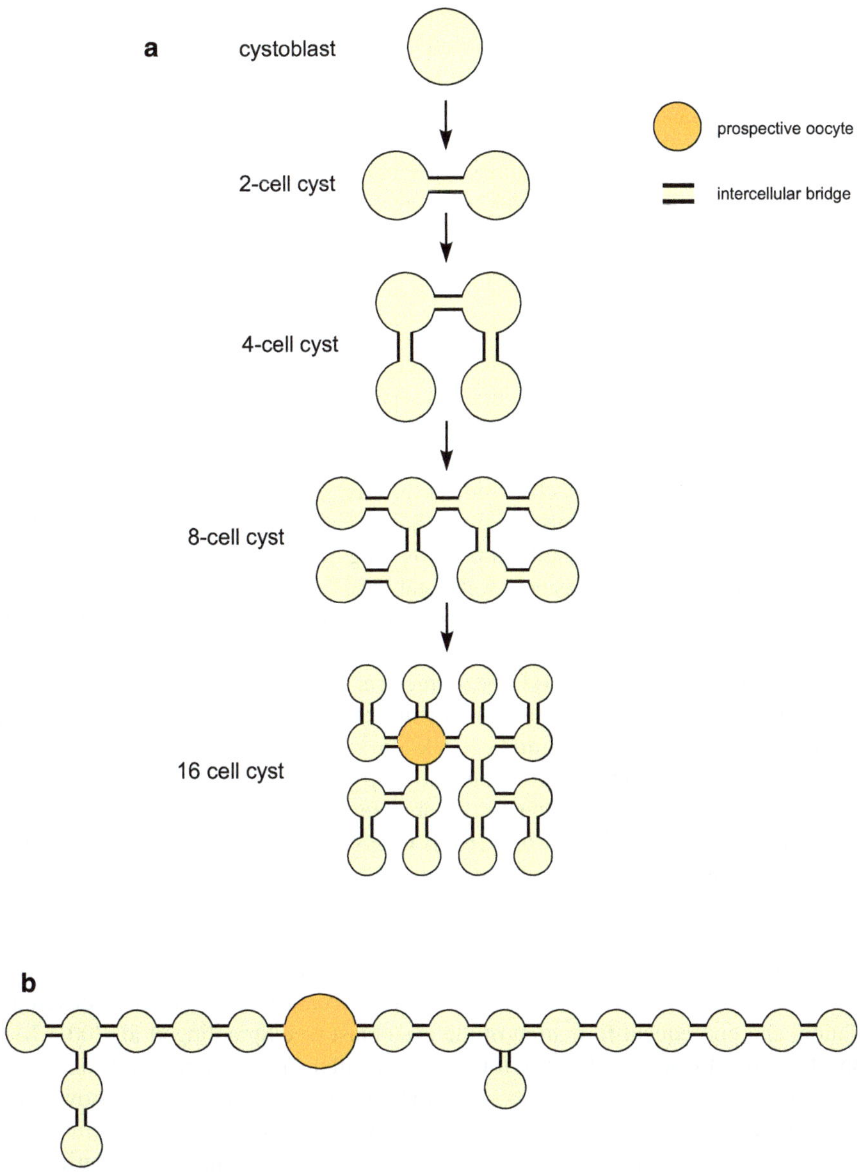

Fig. 9.1 Female germline cysts in meroistic ovaries. (**a**) Successive stages of formation of 16-cell cyst in *Drosophila melanogaster*. (**b**) Arrangement of sibling cells within linear cyst in neuropteran, *Euroleon nostras* (Kubrakiewicz et al. 1998)

McKearin 1997). This in turn leads to a specific "branched" geometry of the cyst (Storto and King 1989; McKearin 1997). Somewhat later, but still within the germarium, the spectraplakin, large cytoskeletal linker protein encoded by *short stop* (*shot*), associates with the fusome (Röper and Brown 2004). This association is required for organizing a polarized microtubule-based transport system (Röper and Brown 2004), indispensable for asymmetric localization of specific transcripts into one of the two cystocytes with four bridges (Deng and Lin 1997; De Cuevas and Spradling 1998), which subsequently differentiates into the oocyte. Other 15 cyst cells become highly polyploid and transcriptionally active nurse cells. Differentiated germline cysts (oocyte–nurse cell complexes) associate with somatic (prefollicular) cells that form epithelium on the cyst surface. The resulting functional units, i.e., ovarian follicles, bud off from the germarium and move to the vitellarium where they develop further, accumulate yolk proteins, and become surrounded by egg envelopes.

The female GSCs of *Drosophila* (similar to germline and non-germline stem cells of other species) reside in specialized microenvironments, termed the stem cell niches (reviewed in Spradling et al. 2001; Xie and Li 2007; Pearson et al. 2009; Losick et al. 2011). The *Drosophila* female germline niche consists of three types of somatic cells: terminal filament cells, cap cells, and escort cells (Kirilly et al. 2011; Losick et al. 2011). The cap cells reside at the base of the terminal filament and are connected to GSCs by adherens junctions (Song et al. 2002). These junctions anchor the GSCs in place and prevent them from moving away from the niche (Song et al. 2002; Losick et al. 2011). It is now well established that short-range signaling from the cap and terminal filament cells controls maintenance and long-term self-renewal of the GSCs and that Jak/Stat-to-Bmp signaling cascade is implicated in these processes (Spradling et al. 2001; López-Onieva et al. 2008; Wang et al. 2008; Losick et al. 2011). Briefly, the terminal filament cells produce cytokine encoded by *unpaired* (*upd*); its secretion stimulates Jak/Stat signaling in cap cells and synthesis of Bmp ligand encoded by *decapentaplegic* (*dpp*) (López-Onieva et al. 2008; Wang et al. 2008). Activation of Bmp receptors prevents transcription of the master differentiation gene *bag-of marbles* (*bam*) in GSCs repressing their differentiation inside the niche. The divisions of GSCs are asymmetrical not because the daughter cells are different but because only one of them inherits a contact (and adherens junctions) with the cap cell and niche microenvironment. This cell remains a stem cell; another daughter cell is drawn away from the niche and its repressive microenvironment. This, in turn, leads to the de-repression (activation) of *bam*, differentiation, and consequent formation of the cystoblast (see Spradling et al. 2001; Xie and Li 2007; Losick et al. 2011 for further details). Interestingly, both daughter cells—the GSC and the cystoblast—contain large intracellular organelle known as a spectrosome. The spectrosome contains α-spectrin and Hts protein and is believed to be a precursor of the fusome (Lin et al. 1994; Lin and Spradling 1995). During the mitotic division of the cystoblast, the spectrosome segregates asymmetrically to one of the descendant cells (Lin and Spradling 1995; de Cuevas and Spradling 1998). The cell that inherits the spectrosome will become, after additional three rounds of mitotic divisions, the oocyte. Above results indicate that during initial stages of *Drosophila* oogenesis, two mitotic divisions are asymmetrical, i.e., the division of the GSC and the division of the cystoblast.

Similar processes have been described in dozens of closer and more distant relatives of the fruit fly, including flies, beetles, wasps, butterflies, lacewings, hemipterans (bugs, aphids, coccoids, etc.), earwigs, and lice (see Büning 1994; Klag and Bilinski 1994; Bilinski 1998; Bilinski et al. 1998; Jaglarz 1992, 1998 for further details). In this review, we describe these processes in four unorthodox systems: earwigs, aphids, lacewings, and rove beetles.

The ovarian follicles of earwigs (order Dermaptera) are morphologically simple and comprise only two germline cells: an oocyte and a single polyploid nurse cell (Fig. 9.2a; Zinsmeister and Zinsmeister 1976; Tworzydlo and Bilinski 2008; Tworzydlo et al. 2010a). Despite similar morphology, the processes leading to the formation of oocyte–nurse cell units are apparently different in basal versus derived species. In *Anisolabis maritima* and *Labidura riparia* (belonging to the basal families), the cystoblasts divide 3 times generating eight-cell cysts (Yamauchi and Yoshitake 1982). In each cyst, four cells (with two or three intercellular bridges) differentiate into the oocytes, and remaining four (with only one bridge) become the nurse cells. Such diversified cysts secondarily split into four oocyte–nurse cell units (Yamauchi and Yoshitake 1982; Tworzydlo et al. 2010a). In advanced taxa (derived families: Forficulidae, Arixenidae, Chelisochidae, and Spongiphoridae), the cystoblast divides only once, and thus two-cell germline cysts form without intermediate eight-cell stage (Tworzydlo and Bilinski 2008; Tworzydlo et al. 2010a). Detailed ultrastructural studies showed that the niche of the forficuloid earwig, *Opisthocosmia silvestris*, is different from that of the fruit fly and comprises two types of somatic cells only, the terminal filament cells and the escort cells (Fig. 9.2d; Tworzydlo et al. 2010b). *Drosophila*-type cap cells, i.e., somatic cells residing in direct contact with GSCs and connected with them via adherens junctions, are absent. Interestingly, between the somatic cells of the *Opisthocosmia* niche two unusual structures implicated in intracellular communication, argosome-like vesicles and cytoneme-like extensions, were found (Fig. 9.2d). This implies that the signaling between niche cells might be mediated by noncanonical plasma membrane specializations (Tworzydlo et al. 2010b).

In aphids, all germline cells of a given ovariole belong to a single, relatively large cyst (Büning 1985; Pyka-Fosciak and Szklarzewicz 2008; Michalik et al. 2013). As a rule, the cysts of derived aphids (Aphididae, Drepanosiphidae) are composed of 32 sister cells (cystocytes), suggesting that in these families the initial cell of the cyst, the cystoblast, divides synchronously 5 times (Büning 1985). In the basal family Adelgidae, the situation is even more complicated. Here, synchronous mitotic divisions of the cystoblast are followed by the supernumerary divisions of individual cystocytes, leading to the formation of “irregular” cysts composed of 50–92 germline cells (Szklarzewicz et al. 2000). In all studied aphids, the cyst cells are interconnected via simple intercellular bridges that are devoid of fusomes/polyfusomes (Büning 1985; Pyka-Fosciak and Szklarzewicz 2008). In contrast to *Drosophila*, in each germline cyst about a half of the cystocytes (usually 16 in derived species, more than 20 in basal ones) differentiate into the oocytes (Fig. 9.2b, c) and migrate to the vitellarium (Szklarzewicz et al. 2000; Michalik et al. 2013). The remaining cystocytes are retained in the germarium, become the

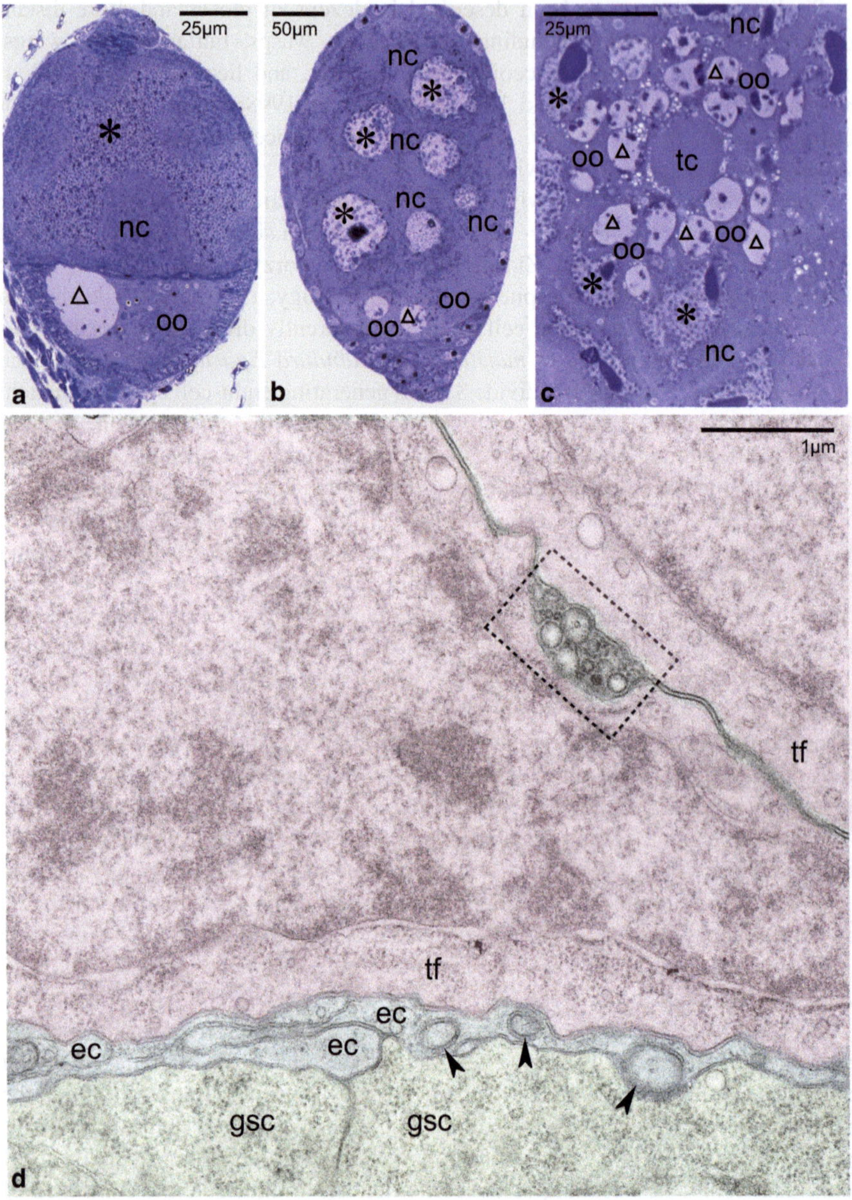

Fig. 9.2 Germline cysts in dermapterans and aphids. (**a**) Ovarian follicle of dermapteran, *Doru lineare*; note that the follicle consists of an oocyte (oo) and single nurse cell (nc). Semithin section stained with *methylene blue*. Courtesy of dr. W. Tworzydlo (Institute of Zoology, Jagiellonian University). (**b**) Differentiated germline cyst of aphid, *Stomaphis quercus*; note that the cyst contains more than one oocyte (oo). Semithin section stained with *methylene blue*. Courtesy of dr. T. Szklarzewicz (Institute of Zoology, Jagiellonian University). (**c**) Transverse section through young ovariole of aphid, *Prociphilus fraxini*; note nurse cells (nc), trophic cord (tc) connecting growing oocyte with the trophic chamber, and several small (early) oocytes (oo) present around trophic cord. Semithin section stained with *methylene blue*. Courtesy of dr. (**a**) Michalik (Institute of Zoology, Jagiellonian University). Nurse cell nuclei (*asterisks*); oocyte nuclei (*triangles*). (**d**)

nurse cells, and form a trophic chamber. The contact between the growing oocytes and the distant tropharium is ensured by elongated cytoplasmic extensions (Fig. 9.2c) referred to as trophic cords (see Büning 1993, 1994 for a review). Because the morphology and functioning of the female GSC niche in aphids have not been analyzed, the mechanisms responsible for the determination of multiple oocytes within each germline cyst are unknown.

The organization of female germline cysts of lacewings (order Neuroptera) substantially differs from that described for *Drosophila*. In neuropterans, the number of sister cells per cyst is, as a rule, variable, not species-specific, and ranges from 12 to 24 (Rousset 1978; Kubrakiewicz 1997; Kubrakiewicz et al. 1998). Three-dimensional reconstruction of serial sections showed that the arrangement of cells within such irregular cysts is almost linear (Fig. 9.1b) rather than branched (Kubrakiewicz 1997). Rarely, when branchings are present their localization sites are apparently random, and the oocyte develops always from the cystocyte located within the central (linear) part of the cyst, i.e., from the cell connected only to two neighbors (Fig. 9.1b). The cyst cells are connected via wide intercellular bridges filled with the fusomes (Kubrakiewicz 1997). The fusomes of the main "linear" part of the cyst merge to form characteristic linear (not branched) polyfusome (Kubrakiewicz 1997). Above observations are in line with the idea that branched (extended through all the intercellular bridges of the cyst) polyfusome is indispensable for the formation of highly branched germline cysts.

One of the most spectacular examples of asymmetric oogonial divisions is the formation of oocyte and nurse cells (trophocytes) in telotrophic ovary of the rove beetle *Creophilus maxillosus*. The nuclei of *Creophilus* oocytes contain multiple nucleoli and multitude of copies (equivalent of 14–40C) of amplified ribosomal DNA (rDNA), which manifests itself morphologically as a huge extrachromosomal DNA body (Kloc 1976, 1980; Kloc et al. 1995; Matuszewski and Kloc 1976). The amplification of rDNA starts in the nuclei of chordoblasts (equivalents of *Drosophila* cystoblasts and the precursors of pro-oocytes and eventually the oocytes) during larval/pupal transition, continues in interphase nuclei of the pro-oocytes of successive generations, and culminates in the formation of giant extrachromosomal DNA body in the oocyte nucleus (Kloc 1976, 1980; Kloc et al. 1995; Matuszewski et al. 1985, 1999; Matuszewski and Kloc 1976). The chordoblasts are located atop somatic prefollicular cells and are connected by horizontally oriented intercellular bridges. The presence of bridges between the chordoblasts suggests that they are descendants of a single progenitor germline stem cell. Interestingly, although chordoblasts are interconnected, they divide and amplify rDNA asynchronously

Fig. 9.2 (continued) EM micrograph of female GSC niche in dermapteran, *Opisthocosmia silvestris*; note the germline stem cells (gsc, *light green*), terminal filament cells (tf, *pink*), processes of escort cells (ec, *light blue*), argosome-like vesicles (*boxed*), and cross-sectioned cytoneme-like extensions (*arrowheads*). Courtesy of dr. W. Tworzydlo (Institute of Zoology, Jagiellonian University)

(Fig. 9.3; Matuszewski et al. 1985, 1999). Each chordoblast undergoes a series of consecutive asymmetric divisions producing a linear cyst (cluster/cord/chain) of sister oogonial cells (chordocytes) connected by vertically oriented intercellular bridges (Fig. 9.3). In each cyst, the most basal/posterior cell (former chordoblast) becomes the oocyte and the remaining chordocytes become the nurse cells (trophocytes). The extrachromosomal DNA body does not disperse during chordoblast/pro-oocyte division and always segregates to the posterior pole of the cell, which makes it a magnificent marker of divisional asymmetry. Detailed studies of divisions within linear cysts showed that during consecutive mitoses the segregation of extrachromosomal DNA toward the posterior pole of chordoblast is imperfect and small amounts of extrachromosomal DNA segregate apically/anteriorly ending up in the nuclei of future nurse cells (Fig. 9.3; Matuszewski et al. 1985, 1999). These studies also showed that each linear cyst not only has a very distinct postero–anterior gradient of extrachromosomal DNA content but also a gradient of mitotic phases; the most posterior cell (chordoblast, i.e., future oocyte), which contacts somatic prefollicular cells, has the largest quantity of extrachromosomal DNA and is the most advanced in mitosis (Fig. 9.3; Matuszewski et al. 1985, 1999). This suggests that somatic cells are the source of signals for posterior segregation of amplified DNA and timing of mitotic entry (Fig. 9.3). Although the identity of these signals is unknown, the postero–anterior gradient of mitotic phases and of amplified rDNA quantity indicates that signaling molecules have postero–anterior concentration gradient and are able to diffuse through the intercellular bridges connecting cells within each linear cyst (Fig. 9.3). It is very possible that similar to *Drosophila* the communication between somatic and germ cells in *Creophilus* involves Jak/Stat and Bmp pathways (see Sect. 9.1 for details). In *Creophilus*, the basal/posterior, abutting the somatic cells, part of chordoblast contains a prominent spectrin-rich germ cell-specific structure called the spectrosome (Matuszewski et al. 1999). It is known that in *Drosophila* the spectrosome contains α- and β-spectrin, adducin, and bam proteins (Wawersik and Van Doren 2005) and by anchoring the pole of mitotic spindle plays a crucial role in asymmetric division of germ cells (Deng and Lin 1997). Thus, by analogy to *Drosophila*, it is very plausible that also in *Creophilus* the chrodoblast's spectrosome anchors the posterior pole of the spindle and enforces directionality/plane of cell division. The studies of cell divisions during *Drosophila* oogenesis showed that the spectrosome present in the cystoblast (precursor of the oocyte) gives rise to the fusome and eventually to the polyfusome, which spans the intercellular bridges connecting 16 cystocytes, and by anchoring spindles' poles imposes polarity of consecutive cell divisions (Huynh 2000). In *Creophilus*, the intercellular bridges connecting chordocytes also contain fusomes (Matuszewski et al. 1999), but because chordocytes are arranged linearly there is no polyfusome. This implies that in *Creophilus* each fusome individually enforces postero–anterior polarity of chordocyte divisions.

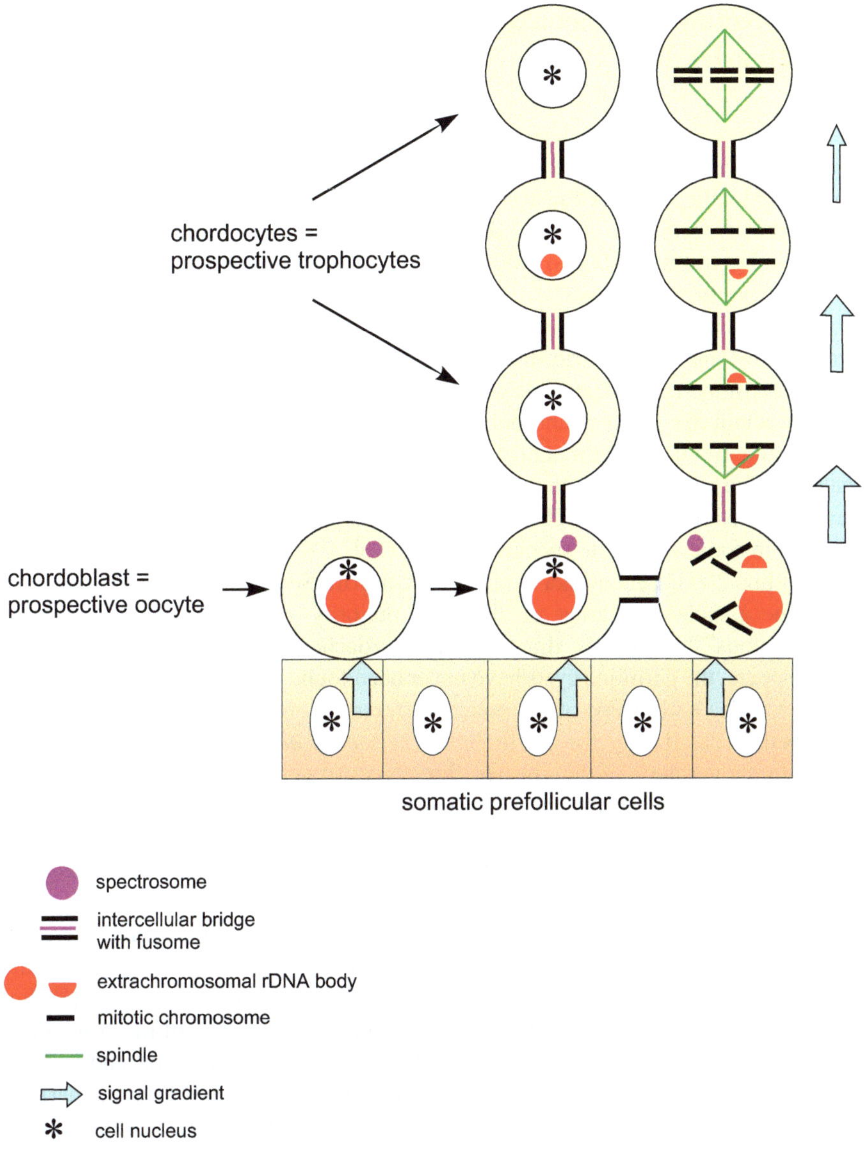

Fig. 9.3 Female germline cyst in telotrophic ovary of *Creophilus maxillosus*. The chordoblasts (equivalent of *Drosophila* cystoblasts) are located atop somatic prefollicular cells. Chordoblasts are connected by horizontal intercellular bridges, amplify rDNA, and form extrachromosomal rDNA bodies. Each chordoblast divides several times resulting in the cord of interconnected cells (chordocytes). The intercellular bridges between chordocytes contain fusomes. In each cyst, the most basal cell (former chordoblast) becomes the oocyte while the remaining chordocytes become the nurse cells (trophocytes). The chordocytes contain low amount of amplified rDNA, which derives from imperfect segregation of rDNA body into the basal pole of the dividing chordoblast. There is gradient of mitotic phases within each cyst (cord); the most basal cell is always most advanced in the mitotic cycle. This and the fact that only the most basal cell amplifies rDNA and

9.2 Asymmetric Oogonial Divisions in Insect Panoistic Ovaries

Panoistic ovaries are characterized by the lack of nurse cells or trophocytes (compare Sects. 9.1 and 9.2), and consequently, the oocyte nucleus is highly transcriptionally active. Ultrastructural analyses revealed that such organization might result from two clearly different developmental processes: secondary division of germline cysts into individual oocytes and differentiation of the oocytes directly from the cystoblasts.

In some insects, e.g., stoneflies (Plecoptera) and thrips (Thysanoptera), the synchronous and asymmetric divisions of the cystoblast lead to the formation of typical germline cysts. The cysts cells are interconnected by intercellular bridges, which contain the fusomes and polyfusomes (Pritsch and Büning 1989; Gottanka and Büning 1990; Tsutsumi et al. 1995). During subsequent developmental stages, the germline cysts split into equivalent cells that enter meiotic prophase and become functional oocytes. This process is fundamentally similar to that described in detail in some vertebrate species (Pepling et al. 1999; Kloc et al. 2004a; Marlow and Mullins 2008; Lechowska et al. 2012; compare Sect. 4).

In the panoistic ovaries of the vast majority of the basal insects, e.g., "apterygotous" bristle-tails, dragonflies, cockroaches, stick insects, crickets, and grasshoppers, the germline cysts have never been found (see Büning 1993, 1994 for a review). The early steps of the oocyte differentiation in such ovaries have been described in detail only for the firebrat, *Thermobia domestica* (Tworzydlo et al. 2014). In the germaria of this species, three types of germline cells are present: the GSCs, the cystoblasts, and the early meiotic oocytes (Fig. 9.4). The GSCs are present along the anterior apex of the germarium. They are separated from each other and from the basement lamina covering the germarium by elongated processes of large somatic cells, termed the apical somatic cells (Fig. 9.4; Tworzydlo et al. 2014). The GSCs of the firebrat divide asymmetrically generating cells with clearly different developmental potentials, i.e., the new GSCs and the cystoblasts. The mitotic spindles of dividing GSCs are always oriented perpendicular to the germarium surface. Such orientation ensures that one of the descendant cells (new GSC) remains in contact with the apical somatic cells and/or their processes, whereas the other (the cystoblast) shifts to the center of the germarium. These findings suggest that the apical somatic cells of the firebrat are the equivalents of *Drosophila* cap cells (Tworzydlo et al. 2014). Analysis of serial sections unambiguously showed that in *Thermobia*, the cystoblasts do not divide mitotically;

Fig. 9.3 (continued) segregates it basally indicate the presence of a gradient of unknown substance emanating from the somatic prefollicular cells, which induces mitosis and amplification (Matuszewski et al. 1985)

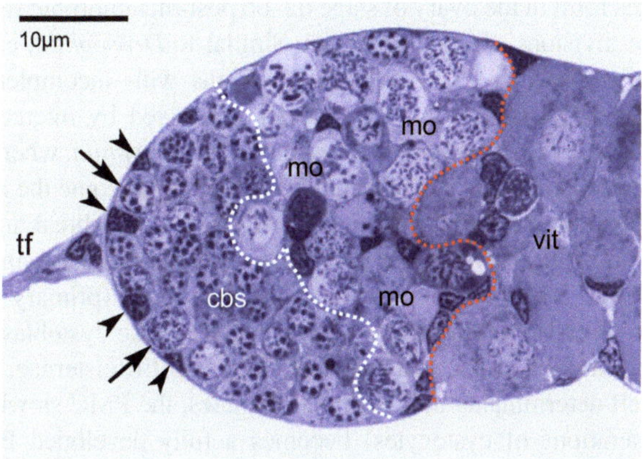

Fig. 9.4 Organization of the germarium in *Thermobia domestica*. Note the apical somatic cells (*arrowheads*), germline stem cells (*arrows*), and terminal filament (tf); *red dotted line* demarcates the border between germarium and vitellarium (vit); *white dotted line* demarcates the border between apical (filled with cystoblasts, cbs) and basal (filled with meiotic oocytes, mo) regions of the germarium. Semithin section stained with *methylene blue*. Courtesy of dr. W. Tworzydlo (Institute of Zoology, Jagiellonian University)

instead, they "directly" enter the meiotic prophase and start to differentiate. Ultimately, these cells become fertilizable oocytes/eggs. This implies that in some basal insect lineages the syncytial phase of oogenesis has been eliminated (lost) during evolution.

9.3 Asymmetric Oogonial Divisions in *Xenopus laevis* Ovary

Although majority of vertebrates have polar/asymmetric oocytes (de Smedt et al. 2000; Gupta et al. 2010; Kloc et al. 2004b, 2008, 2012; Lechowska et al. 2012; Marlow and Mullins 2008; Pepling et al. 2007; Rodler and Sinowatz 2013; Zelazowska et al. 2007), a detailed description of asymmetric oogonial divisions is available only for one vertebrate species–the African clawed frog (toad) *Xenopus laevis* (Kloc et al. 2004a). *Xenopus* oogonia (cystoblasts and cystocytes) and oocytes are polar and asymmetrical along animal/vegetal axis. Their vegetal hemisphere contains spherical cytoplasmic organelle called the Balbiani body (mitochondrial cloud)–an excellent marker of animal/vegetal asymmetry (Kloc et al. 2004a, b). The Balbiani body is an aggregate of thousands of mitochondria, germ cell determinants (mitochondrial cement/germ plasm/germinal granules), localized coding and non-coding RNAs, and Golgi and endoplasmic reticulum (ER) cisternae–all arranged around centrally located pair of centrioles (Kloc et al. 1998, 2001, 2002, 2004a).

Xenopus oocytes form in the ovary of stage 62–66 post-metamorphic froglets through the asymmetric divisions of the cystoblasts. Similar to *Drosophila*, each cystoblast undergoes four consecutive asymmetrical divisions with incomplete cytokinesis giving rise to the cyst (nest) of 16 cystocytes connected by intercellular bridges (Figs. 9.5, 9.6; Kloc et al. 2004a). However, unlike in *Drosophila*, where only one out of 16 cystocytes becomes the oocyte (other 15 cystocytes become the nurse cells), in *Xenopus* all 16 cystocytes, after partitioning of the cyst and breaking cytoplasmic bridges by the invading prefollicular cells, enter meiotic prophase and become the oocytes (Kloc et al. 2004a). The rudiment of Balbiani body (primary mitochondrial cloud, PMC) is already present in the vegetal region of the cystoblast. Through an addition of more mitochondria, expansion of Golgi and ER cisternae, and organization of germ cell determinants into germinal granules, the PMC slowly (throughout successive generations of cystocytes) becomes a fully developed Balbiani body. During consecutive asymmetric divisions of the cystoblast and cystocytes, the PMC/Balbiani body always segregates to the vegetal pole of the cell and remains in close vicinity of intercellular bridge. All sister cells within the cyst contain PMC/Balbiani bodies of comparable size, which implies that either during each division the PMC/Balbiani body is partitioned equally between sister cells or that original PMC/Balbiani body remains in parental cell while the daughter cell generates de novo its own PMC/Balbiani body. Similar to *Drosophila*, the intercellular bridges in *Xenopus* cyst contain actin and kelch protein-rich ring canal rims and spectrin and Hts protein-rich fusomes, which are interconnected into the polyfusome (Kloc et al. 2004a). This suggests that in *Xenopus*, similar to *Drosophila*, the polyfusome anchors the poles of mitotic spindles and enforces divisional asymmetry. Recently, Sidova et al. (2015) using qPCR tomography discovered a set of microRNAs asymmetrically localized in *Xenopus* oocytes. They showed that miR-16c, miR-18b, miR-363-3p, miR-20b, miR-93a, and miR-5102-5p have animal hemisphere localization and miR-19b, miR-221, miR-148b, miR-25, miR-22, and miR-100 are localized in the vegetal hemisphere (Sidova et al. 2015). Although at present there is no experimental proof, the authors suggest that these miRNAs may, in addition to vegetally localized mRNAs and proteins, play a role in asymmetric divisions and cell fate determination in *Xenopus* embryogenesis (Sidova et al. 2015). Because authors studied these miRNAs in fully grown oocytes where the asymmetry is already firmly set, it will be very interesting to see if these miRNAs have also asymmetrical distribution during oogenesis and how they segregate and what function they play during asymmetrical oogonial divisions in developing *Xenopus* ovary.

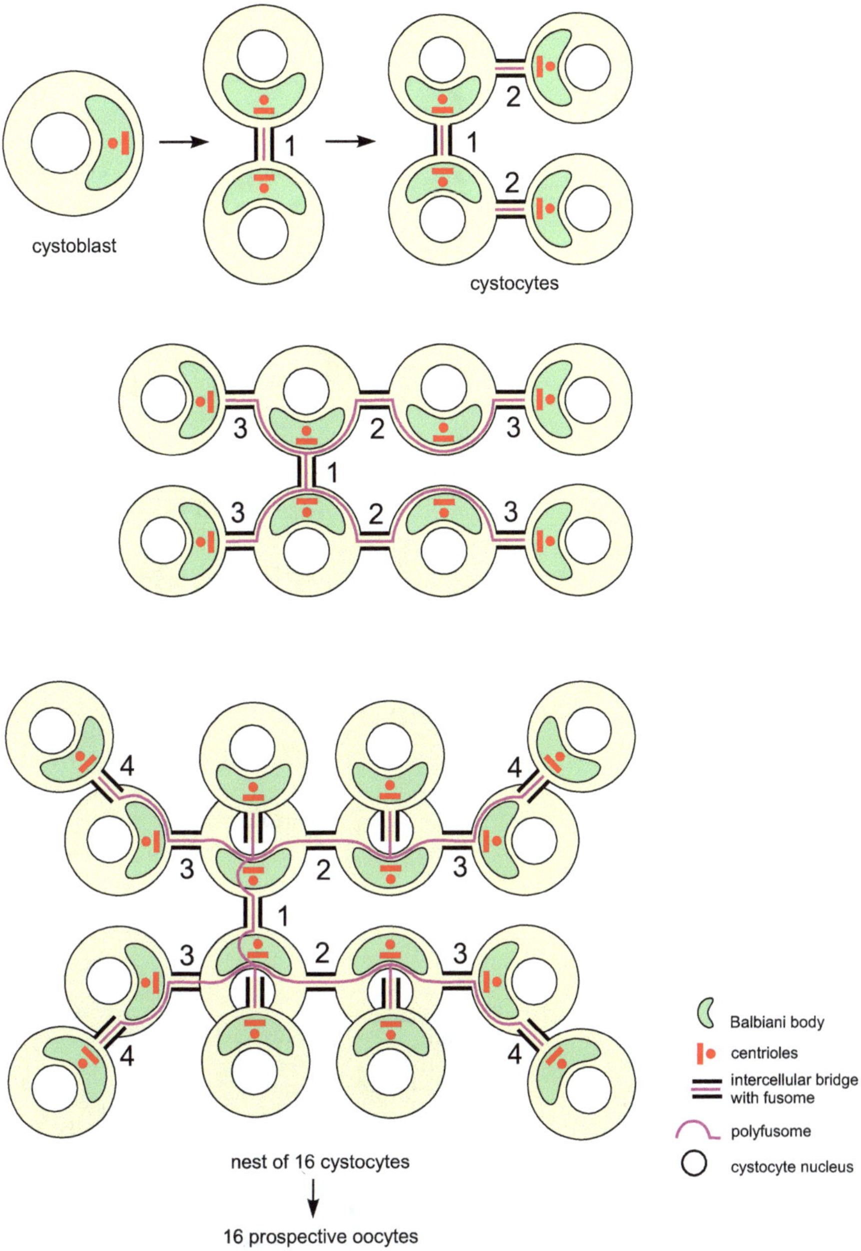

Fig. 9.5 Formation of female germline cyst in *Xenopus laevis* ovary. The cystoblast divides 4 times giving rise to 16 cystocytes connected by intercellular bridges with fusomes interconnected into polyfusome. Eventually each cystocyte will transform into the oocyte. The cystocytes contain Balbiani body, which form through the accumulation of mitochondria, ER, Golgi, and germinal granules around the pair of centrioles (Kloc et al. 2004a)

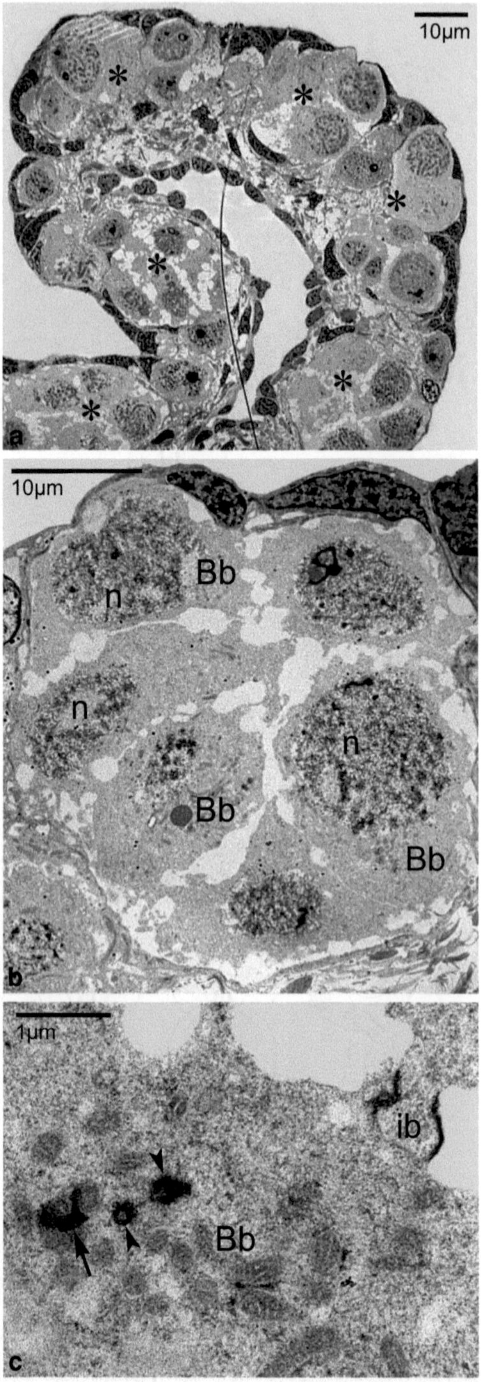

Fig. 9.6 Morphology of *Xenopus* germline cyst. (**a**) EM micrograph of the fragment of *Xenopus* froglet ovary showing several germline cysts (nests), *asterisks*. (**b**) EM micrograph of a single cyst in *Xenopus* froglet ovary; cystocyte nuclei (*n*). Note the Balbiani bodies (Bb) in the vegetal hemisphere of the cystocytes. (**c**) EM micrograph of intercellular bridge (ib) connecting 2 cystocytes. Note the Balbiani body (Bb), centrioles (*arrowheads*), and mitochondrial cement (*arrow*) located in the vicinity of the bridge (Kloc et al. 2004a)

Acknowledgements We are grateful to Ms. E. Kisiel for excellent drawings and help with the figures.

References

Bilinski SM (1998) Introductory remarks. Folia Histochem Cytobiol 36:143–145
Bilinski SM, Büning J, Simiczyjew B (1998) The ovaries of mecoptera: basic similarities and one exception to the rule. Folia Cytochem Cytobiol 36:189–195
Büning J (1985) Morphology, ultrastructure, and germ cell cluster formation in ovarioles of aphids. J Morphol 186:209–221
Büning J (1993) Germ cell cluster formation in insect ovaries. Int J Insect Morphol Embryol 22:237–253
Büning J (1994) The insect ovary. Ultrastructure, previtellogenic growth and evolution. Chapman and Hall, London
De Cuevas M, Spradling AC (1998) Morphogenesis of the *Drosophila* fusome and its implications for oocyte specification. Development 125:2781–2789
De Cuevas M, Lilly MA, Spradling AC (1997) Germline cyst formation in *Drosophila*. Annu Rev Genet 31:405–428
de Smedt V, Szöllösi D, Kloc M (2000) The balbiani body: asymmetry in the mammalian oocyte. Genesis 26:208–212
Deng W, Lin H (1997) Spectrosomes and fusomes anchor mitotic spindles during asymmetric germ cell divisions and facilitate the formation of a polarized microtubule array for oocyte specification in *Drosophila*. Dev Biol 189:79–94
Gottanka J, Büning J (1990) Oocytes develop from interconnected cystocytes in the ppanoistic ovary of *Nemoura* sp. (Pictet) (Plecoptera: Nemouridae). Int J Insect Morphol Embryol 19:219–225
Gupta T, Marlow FL, Ferriola D, Mackiewicz K, Dapprich J, Monos D, Mullins MC (2010) Microtubule actin crosslinking factor 1 regulates the Balbiani body and animal–vegetal polarity of the zebrafish oocyte. PLoS Genet. doi:10.1371/journal.pgen.1001073
Huynh JR (2000) Fusome as a cell-cell communication channel of *Drosophila* ovarian cyst. In: Huynh JR (ed) Madame curie bioscience database. Landes Bioscience, Austin. http://www.ncbi.nlm.nih.gov/books/NBK6300
Jaglarz M (1992) Peculiarities of the organization of egg chambers in carabid ground beetles and their phylogenetic implications. Tissue Cell 24:397–409
Jaglarz MK (1998) The number that counts. Phylogenetic implications of the number of nurse cells in ovarian follicles of Coleoptera–Adephaga. Folia Histochem Cytobiol 36:167–178
King RC (1970) Ovarian development in *Drosophila melanogaster*. Academic Press, New York
Kirilly D, Wang S, Xie T (2011) Self-maintained escort cells form a germline stem cell differentiation niche. Development 138:5087–5097
Klag J, Bilinski S (1994) Germ cell cluster formation and oogenesis in the hymenopteran *Coleocentrotus soldanskii*. Tissue Cell 26:699–706
Kloc M (1976) Extrachromosomal DNA and RNA-synthesis in oocytes of *Creophilus maxillosus* (Staphylinidae, Coleoptera, Polyphaga). Experientia 32:375–377
Kloc M (1980) Extrachromosomal DNA and its activity in RNA synthesis in oogonia and oocytes in the pupal ovary of *Creophilus maxillosus* (Staphylinidae, Coleoptera-Polyphaga). Eur J Cell Biol 21:328–334
Kloc M, Matuszewski B, Nurkowska J (1995) Ribosomal gene amplification in the oocytes of *Creophilus maxillosus* (Staphylinidae, Coleoptera-Polyphaga)–an insect with telotrophic ovaries. Folia Histochem Cytobiol 33:267–276

Kloc M, Larabell C, Chan AP, Etkin LD (1998) Contribution of METRO pathway localized molecules to the organization of the germ cell lineage. Mech Dev 75:81–93
Kloc M, Bilinski S, Chan AP, Allen LH, Zearfoss NR, Etkin LD (2001) RNA localization and germ cell determination in *Xenopus*. Int Rev Cytol 203:63–91
Kloc M, Dougherty MT, Bilinski S, Chan AP, Brey E, King ML, Patrick CW Jr, Etkin LD (2002) Three-dimensional ultrastructural analysis of RNA distribution within germinal granules of *Xenopus*. Dev Biol 241:79–93
Kloc M, Bilinski S, Dougherty MT, Brey EM, Etkin LD (2004a) Formation, architecture and polarity of female germline cyst in *Xenopus*. Dev Biol 266:43–61
Kloc M, Bilinski S, Etkin LD (2004b) The Balbiani body and germ cell determinants: 150 years later. Curr Top Dev Biol 59:1–36
Kloc M, Jaglarz M, Dougherty M, Stewart MD, Nel-Themaat L, Bilinski S (2008) Mouse early oocytes are transiently polar: three-dimensional and ultrastructural analysis. Exp Cell Res 314:3245–3254
Kloc M, Ghobrial RM, Borsuk E, Kubiak JZ (2012) Polarity and asymmetry during mouse oogenesis and oocyte maturation. Results Probl Cell Differ 55:23–44
Kubrakiewicz J (1997) Germ cell cluster organization in polytrophic ovaries of Neuroptera. Tissue Cell 29:221–228
Kubrakiewicz J, Jedrzejowska I, Bilinski SM (1998) Neuropteroidea–different ovary structure in related groups. Folia Histochem Cytobiol 36:179–187
Lechowska A, Bilinski SM, Rasweiler JJ 4th, Cretekos CJ, Behringer RR, Kloc M (2012) Early oogenesis in the short-tailed fruit bat *Carollia perspicillata*: transient germ cell cysts and noncanonical intercellular bridges. Genesis 50:18–27
Lighthouse DV, Buszczak M, Spradling AC (2008) New components of the *Drosophila* fusome suggest it plays novel roles in signaling and transport. Dev Biol 317:59–71
Lin H, Spradling AC (1995) Fusome asymmetry and oocyte determination in *Drosophila*. Dev Genet 16:6–12
Lin H, Yue L, Spradling AC (1994) The *Drosophila* fusome, a germline-specific organelle, contains membrane skeletal proteins and functions in cyst formation. Development 120:947–956
López-Onieva L, Fernández-Miñán A, Gonzáles-Reyes A (2008) Jak/Stat signaling in niche support cells regulates dpp transcription to control germline stem cell maintenance in the *Drosophila* ovary. Development 135:533–540
Losick VP, Morris LX, Fox DT, Spradling AC (2011) *Drosophila* stem cell niches: a decade of discovery suggests a unified view of stem cell regulation. Dev Cell 21:159–171
Marlow FL, Mullins MC (2008) Bucky ball functions in Balbiani body assembly and animal–vegetal polarity in the oocyte and follicle cell layer in zebrafish. Dev Biol 321:40–50
Matuszewski B, Kloc M (1976) Gene amplification in oocytes of the rove beetle *Creophilus maxillosus* (Staphylinidae, Coleoptera-Polyphaga). Experientia 32:34–36
Matuszewski B, Ciechomski K, Nurkowska J, Kloc M (1985) The linear clusters of oogonial cells in the development of telotrophic ovarioles in polyphage Coleoptera. Dev Genes Evol 194:462–469
Matuszewski B, Ciechomski K, Kloc M (1999) Extrachromosomal rDNA and polarity of pro-oocytes during ovary development in *Creophilus maxillosus* (Coleoptera, Staphylinidae). Folia Histochem Cytobiol 37:179–190
McKearin D (1997) The *Drosophila* fusome, organelle biogenesis and germ cell differentiation: if you build it.... BioEssays 19:147–152
Michalik A, Szklarzewicz T, Wegierek P, Wieczorek K (2013) The ovaries of aphids (Hemiptera, Sternorrhyncha, Aphidoidea): morphology and phylogenetic implications. Invertebr Biol 132:226–240
Ong SK, Tan C (2010) Germline cyst formation and incomplete cytokinesis during *Drosophila melanogaster* oogenesis. Dev Biol 337:84–98

Pearson J, López-Onieva L, Rojas-Rios P, Gonzáles-Ryes A (2009) Recent advances in *Drosophila* stem cell biology. Int J Dev Biol 53:1329–1339
Pepling ME, de Cuevas M, Spradling AC (1999) Germline cysts: a conserved phase of germ cell development? Trends Cell Biol 9:257–262
Pepling ME, Wilhelm JE, O'Hara AL, Gephardt GW, Spradling AC (2007) Mouse oocytes within germ cell cysts and primordial follicles contain a Balbiani body. Proc Natl Acad Sci USA 104:187–192
Petrella NL, Smith-Leiker T, Cooley L (2007) The Ovhts polyprotein is cleaved to produce fusome and ring canal proteins required for *Drosophila* oogenesis. Development 134:703–712
Pritsch M, Büning J (1989) Germ cell cluster in the panoistic ovary of Thysanoptera (Insecta). Zoomorphology 108:309–313
Pyka-Fosciak G, Szklarzewicz T (2008) Germ cell cluster formation and ovariole structure in viviparous and oviparous generations of the aphid *Stomaphis quercus*. Int J Dev Biol 52:259–265
Rodler D, Sinowatz F (2013) Expression of intermediate filaments in the Balbiani body and ovarian follicular wall of the Japanese quail (*Coturnix japonica*). Cells Tissues Organs 197:298–311
Röper K, Brown NH (2004) A spectraplakin is enriched on the fusome and organizes microtubules during oocyte specification. Curr Biol 14:99–110
Rousset A (1978) La formation des cysts dans l'ovariole de *Chrysopa perla* (L.) (Neuroptera: Chrysopida). Int J Insect Morphol Embryol 7:45–57
Sidova M, Sindelka R, Castoldi M, Benes V, Kubista M (2015) Intracellular microRNA profiles form in the *Xenopus laevis* oocyte that may contribute to asymmetric cell division. Sci Rep 5:11157
Snapp EL, Iida T, Frescas D, Lippincot-Schwartz J, Lilly MA (2004) The fusome mediates intercellular endoplasmic reticulum connectivity in *Drosophila* ovarian cysts. Mol Biol Cell 15:4512–4521
Song X, Zhu CH, Doan C, Xie T (2002) Germline stem cells anchored by adherens junctions in the *Drosophila* ovary niches. Science 296:1855–1857
Spradling AC, Drummond-Barbosa D, Kai T (2001) Stem cells find their niche. Nature 414:98–104
Storto PD, King RC (1989) The role of polyfusomes in generating branched chains of cystocytes during *Drosophila* oogenesis. Dev Genet 10:70–86
Szklarzewicz T, Wnek A, Bilinski SM (2000) Structure of ovarioles in *Adelges laricis*, a representative of the primitive family Adelgidae. Acta Zool (Stockholm) 81:307–313
Tsutsumi T, Matsuzaki M, Haga K (1995) Formation of germ cell cluster in tubuliferan thrips (Thysanoptera). Int J Insect Morphol Embryol 24:287–296
Tworzydlo W, Bilinski SM (2008) Structure of ovaries and oogenesis in dermapterans. I. Origin and functioning of the ovarian follicles. Arthropod Struct Dev 37:310–320
Tworzydlo W, Bilinski SM, Kočárek P, Haas F (2010a) Ovaries and germline cysts and their evolution in Dermaptera (Insecta). Arthropod Struct Dev 39:360–368
Tworzydlo W, Kloc M, Bilinski SM (2010b) Female germline stem cell niches of earwigs are structurally simple and different from those of *Drosophila*. J Morphol 271:634–640
Tworzydlo W, Kisiel E, Jankowska W, Bilinski SM (2014) Morphology and ultrastructure of the germarium in panoistic ovarioles of a basal "apterygotous" insect *Thermobia domestica*. Zoology 117:200–206
Wang L, Li Z, Cai Y (2008) The JAK/STAT pathway positively regulates DPP signaling in the *Drosophila* germline stem cell niche. J Cell Biol 180:721–728
Wawersik M, Van Doren M (2005) Nanos is required for formation of the spectrosome, a germ cell-specific organelle. Dev Dyn 234:22–27
Xie T, Li L (2007) Stem cells and their niche: an inseparable relationship. Development 134:2001–2006

Yamauchi H, Yoshitake N (1982) Origin and differentiation of the oocyte-nurse cell complex in the germarium of the earwig, *Anisolabis maritima* Borelli (Dermaptera: Labiduridae). Int J Insect Morphol Embryol 11:293–305

Yue L, Spradling AC (1992) *hu-li tai shao*, a gene required for ring canal formation during *Drosophila* oogenesis, encodes a homolog of adducing. Genes Dev 6:2443–2454

Zelazowska M, Kilarski W, Bilinski SM, Podder DD, Kloc M (2007) Balbiani cytoplasm in oocytes of a primitive fish, the sturgeon *Acipenser gueldenstaedtii*, and its potential homology to the Balbiani body (mitochondrial cloud) of *Xenopus laevis* oocytes. Cell Tissue Res 329:137–145

Zinsmeister PP, Zinsmeister DD (1976) Ultrastructural aspects of the earwig ovary. J Insect Physiol 22:1057–1060

Chapter 10
Asymmetric Localization and Distribution of Factors Determining Cell Fate During Early Development of *Xenopus laevis*

Radek Sindelka, Monika Sidova, Pavel Abaffy, and Mikael Kubista

Abstract Asymmetric division is a property of eukaryotic cells that is fundamental to the formation of higher life forms. Despite its importance, the mechanism behind it remains elusive. Asymmetry in the cell is induced by polarization of cell fate determinants that become unevenly distributed among progeny cells. So far dozens of determinants have been identified. *Xenopus laevis* is an ideal system to study asymmetric cell division during early development, because of the huge size of its oocytes and early-stage blastomeres. Here, we present the current knowledge about localization and distribution of cell fate determinants along the three body axes: animal–vegetal, dorsal–ventral, and left–right. Uneven distribution of cell fate determinants during early development specifies the formation of the embryonic body plan.

10.1 Introduction

Asymmetric localization of cell fate determinants is a key biological strategy that produces unequal progeny cells after division. Asymmetric cell division has been observed in many biological systems including differentiation and stem cell proliferation (Blanpain et al. 2004; Knoblich 2008), cancer progression (Knoblich 2010; Shahriyari and Komarova 2013), and during embryogenesis (Pereira and Yamashita 2011; Schatten and Sun 2010). From the perspective of early embryogenesis, formation of cellular polarity leads to asymmetrical cell division and subsequent determination of the main body axes and the developmental plan. Localization of

R. Sindelka • M. Sidova • P. Abaffy
Laboratory of Gene Expression, Institute of Biotechnology, Academy of Sciences of the Czech Republic–Biocev, Prumyslova 595, 252 50 Vestec, Czech Republic

M. Kubista (✉)
Laboratory of Gene Expression, Institute of Biotechnology, Academy of Sciences of the Czech Republic–Biocev, Prumyslova 595, 252 50 Vestec, Czech Republic
TATAA Biocenter AB, Odinsgatan 28, 411 03 Göteborg, Sweden
e-mail: mikael.kubista@tataa.com

J.-P. Tassan, J.Z. Kubiak (eds.), *Asymmetric Cell Division in Development, Differentiation and Cancer*, Results and Problems in Cell Differentiation 61,
DOI 10.1007/978-3-319-53150-2_10

maternal determinants in defined regions of the cytoplasm inside the oocyte is a highly conserved mechanism that leads to unequal distribution of these maternal biomolecules across blastomeres during the first cell divisions. Specific presence and concentration gradients of these determinants in blastomeres control the formation of body axes and determine the developmental plan. Any deviation leads to abnormal development and even premature termination. Here, we summarize our current view of and hypothesis about the formation of the body plan and how it depends on the asymmetric localization of biomolecules.

10.2 The Oocyte and its Animal–Vegetal Axis

In comparison to the mammalian oocyte, the *Xenopus laevis* oocyte is huge. It is over 1 mm in diameter, and it is prepared to form a body plan as evidenced by different pigmentation of its hemispheres (exception are oocytes from albino frogs missing pigmentation). The dark hemisphere is called animal and contains granules of melanin-rich melanosomes. The light hemisphere is called vegetal and harbors the majority of yolk proteins (Danilchik and Gerhart 1987). This asymmetry is established already during oogenesis (Dumont 1972; Danilchik and Gerhart 1987; Kloc et al. 2001), and it specifies the first body axis: animal–vegetal (A–V) (Fig. 10.1). Asymmetrical localization along the A–V axis is observed also for other cell components. The nucleus of the mature oocyte (stage VI), for example, is offset toward the animal hemisphere being located roughly 1/3 from the animal pole (Gurdon 1968; Jullien et al. 2011). The Golgi apparatus, certain components of the endoplasmic reticulum, and mitochondria are more abundant in the vegetal hemisphere (Kloc et al. 2001).

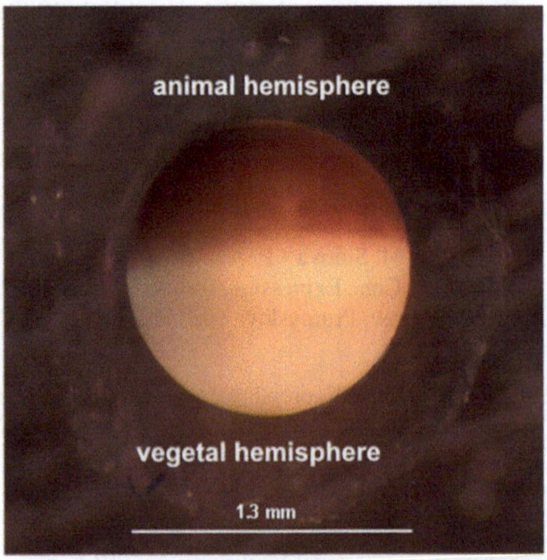

Fig. 10.1 *Xenopus laevis* oocyte

Different types of oocytes can be found among animal species that differ in yolk concentration and compartmentalization. The *Xenopus* oocyte belongs to the group of mesolecithal oocytes with medium concentration of yolk that is stored primarily in the vegetal hemisphere. In addition to yolk and cellular organelles, the mature oocyte contains proteins and RNAs that support early development. These are synthesized during oogenesis when only the maternal chromosomes are present (Heasman 2006). The amount of proteins and RNAs in an individual *Xenopus laevis* oocyte is about 4 μg, which is enormous compared to any other cell making it ideal for temporal (Smits et al. 2014; Sun et al. 2014) and spatial (Sindelka et al. 2008) studies. An important aspect for expression analysis of the developing *Xenopus laevis* embryo is the absence of *de novo* transcription for most of the zygotic genes until the mid-blastula transition (MBT) that occurs after 12 cell divisions. The few known exceptions include components of the Nodal family gene regulatory network for mesendoderm induction (Skirkanich et al. 2011). Notably, while overall transcription is silent/reduced, translation is active (Etkin and Balcells 1985; Newport and Kirschner 1982).

The first developmental axis formed during oogenesis is the animal–vegetal (Deshler et al. 1997; King et al. 2005). External signs of asymmetry appear during the middle and late phases of oogenesis (stage III—IV), when pigment granules accumulate in the animal hemisphere and yolk localizes to the vegetal half of the oocyte (Danilchik and Gerhart 1987). This internal reorganization is associated with the localization of cell fate determinants, which are thought to be specific mRNAs and proteins. Localization of cell fate determinants to the vegetal region of the oocyte has been studied for many years, and today, two main mechanisms are known and some 300 localized transcripts have been identified (Cuykendall and Houston 2010; Houston 2013).

One mechanism is the early transportation pathway of biomolecules, known as *METRO* (messenger transport organizer), which are associated with the mitochondrial cloud (Kloc et al. 1998). The mitochondrial cloud forms during early oogenesis (stage I) in the vegetal region close to the germinal vesicle (Kloc et al. 2004). It is rich in mitochondria and contains also endoplasmic reticulum (Heasman et al. 1984). Newly transcribed RNAs are transferred from the germinal vesicle to the cytoplasm. Their regulatory sequences are bound by RNA-binding proteins and transcripts are transported to the mitochondrial cloud (Snedden et al. 2013). Later during oogenesis, the mitochondrial cloud fragments and its parts translocate to the vegetal cortex (Heasman et al. 1984). Several independent functional sequences within the 3′ UTR of RNAs have been identified, some of which have been related to the mitochondrial cloud localization, while other to the binding to the vegetal cortex (Kloc et al. 1993; Zhou and King 1996a, b). The METRO pathway is important for the localization of germ plasm determinants such as *nanos1*, *dazl* (deleted in azoospermia-like), *ddx25* (deadsouth), *pgat* (primordial germ cell-associated transcript, also called *xpat*), and *wnt11* (wingless-type member 11) (Kloc et al. 2002; Zhou and King 2004). The METRO pathway is also used for localization of functional RNAs. The *Xlsirt* untranslated RNA is important for the anchoring of other RNAs at the cortex of vegetal hemisphere (Zearfoss et al. 2003;

Kloc and Etkin 1994). Localization of RNAs by the METRO pathway is cytoskeletal independent (Kloc and Etkin 1995; Kloc et al. 1996), but the anchoring of METRO-localized mRNAs to the vegetal cortex depends to some degree on cytokeratin and actin filaments (Kloc and Etkin 1995). The gradient of specific mRNAs within the cell created by the METRO pathway is very steep with the vast majority of the transcripts localized to the vegetal pole (Sindelka et al. 2010).

The other localization pathway is called the *late transporting pathway* or just the late pathway. It is active in the middle and late phases of oogenesis (stages III and IV) and uses a transport mechanism that relies on microtubules (Yisraeli et al. 1990). The transcripts organized by the late pathway are not bound to the mitochondrial cloud but accumulate in the vegetal ooplasm (Deshler et al. 1997; Gautreau et al. 1997). The late pathway transports transcripts coding for the maternal transcription factor VegT and for Gdf1 (growth differentiation factor 1, also called Vg1 and member of the TGF-beta family), which are critical for germ layer specification (Messit et al. 2008; White and Heasman 2008; Gagnon et al. 2013). Localization sequences of these transcripts contain the elements UUUCU and UUCAC in the 3′ UTR untranslated region (Bubunenko et al. 2002; Kwon et al. 2002; Lewis et al. 2004). The transport mechanism is thought to depend on the formation of specific vesicles from the endoplasmic reticulum, which are transported to the vegetal hemisphere (Deshler et al. 1997; Etkin 1997; Kwon et al. 2002). The gradient formed by the late pathway is based on the microfilament network and is smooth compared to the steep gradient created by the METRO pathway (Sindelka et al. 2010).

While there are two known pathways to localize mRNAs to the vegetal hemisphere of the oocyte, no active localization mechanism is known that would localize RNAs to the animal hemisphere. However, passive diffusion of RNAs produced in the nucleus, which is off center in the oocyte, results by default in spontaneous enrichment of transcripts in the animal hemisphere (Sidova et al. 2015).

Another important mechanism in the formation of the developmental plan is posttranscriptional regulation. Many studies indicate roles for noncoding RNAs in the regulation of translation during development. Most recently, also microRNAs (miRNA) were found to be polarized within the oocyte, suggesting that they too may be fate determinants (Sidova et al. 2015). The miRNAs were found in two different distributions along the A–V axis. Some miRNAs were more abundant in the animal hemisphere corresponding to the spontaneous animal localization found for mRNAs, while other miRNAs showed vegetal localization similar to the gradients observed for mRNAs organized by the late pathway. No miRNA has yet been found to have the extreme vegetal polarization observed for the mRNAs organized by METRO. The mechanism behind miRNA localization is currently unknown.

The localization of RNAs along the A–V axis established in the oocyte is preserved after fertilization. As the fertilized oocyte divides the mRNA, gradients along the A–V axis remain until at least the blastula stage (Sindelka et al. 2008).

Later in development three germ layers are formed (Fig. 10.2). The ectoderm is formed from cells derived from the animal region of the oocyte and develops into

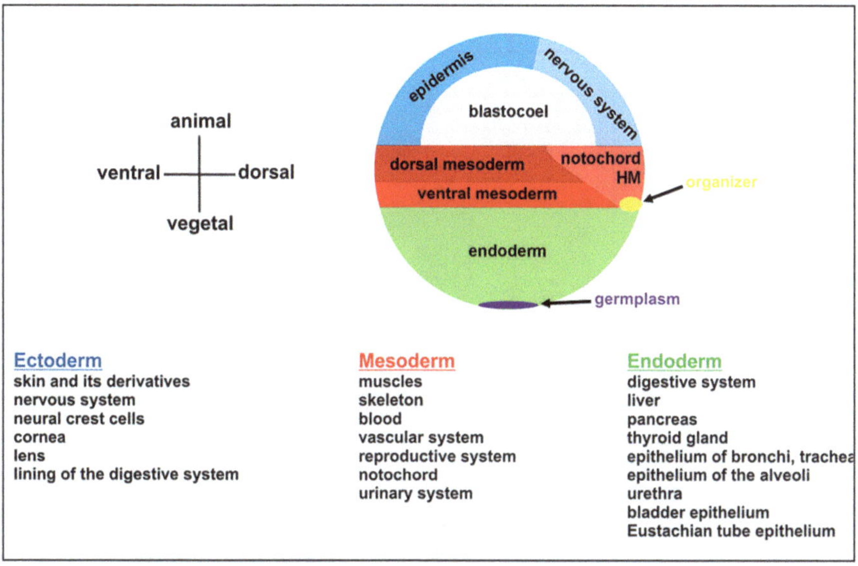

Fig. 10.2 Formation of main germ layers (adapted from Gilbert 2010)

epidermis and neural tissue. Cells from the vegetal region differentiate into the endodermal cell types and form internal organs including the gastrointestinal system. Cells derived from the vegetal pole become germ cells forming gametes. The equatorial region of the oocyte differentiates into mesodermal cells that develop into tissues such as the notochord and muscles. Mesoderm formation is regulated by the activity of vegetally localized transcription factors Vegt and Gdf1 in synergy with Wnt signaling connected to dorsal localization of β-catenin (Zhang and King 1996; Agius et al. 2000). The oocyte A–V axis reflects the ectoderm–endoderm axis in development.

10.3 The Left–Right and Dorsal–Ventral Body Axes

Three main body axes can be distinguished in animals: cranial–caudal, left–right, and dorsal–ventral (back–belly). *Xenopus laevis* is a great model to study the establishment of those body axes. The first embryonic cleavage after fertilization bisects the point of sperm entry, which is always in the animal pole, and the groove elongates toward the vegetal pole within 90 min postfertilization. This division establishes the left–right (L–R) asymmetry at 2-cell stage with one blastomere predestined to develop into the left side of the embryo and the other blastomere into the right side. The left–right asymmetry manifests during gastrulation, when *de novo* embryonic transcription is initiated. In a traditional view, the second cleavage is perpendicular to the first and creates the dorsal (back)–ventral (front) (D–V)

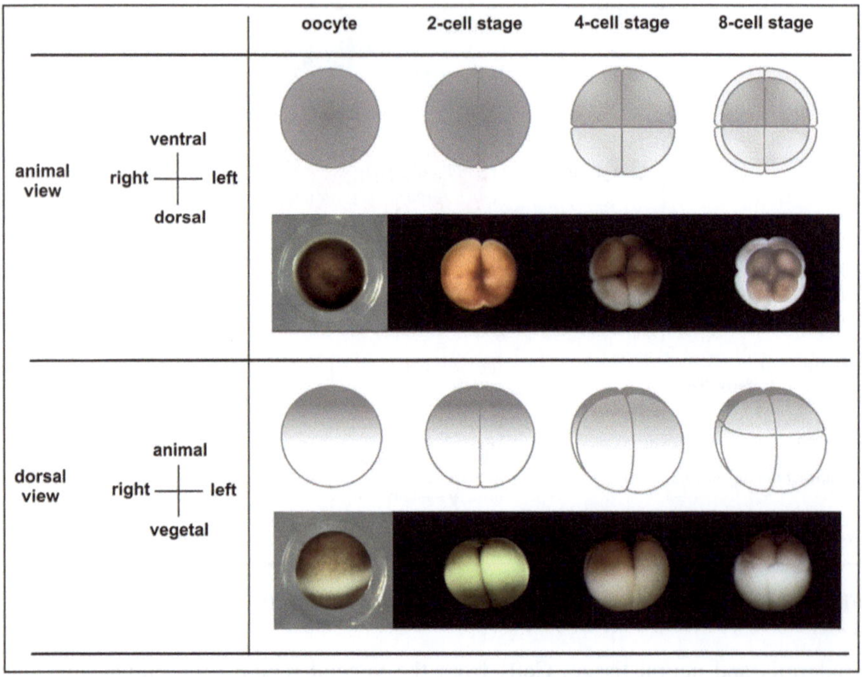

Fig. 10.3 Formation of body plan and determination of the three main body axes

polarity. This cleavage results in different shading and also different size of the dorsal and ventral blastomeres in the 4-cell stage embryo. The third cleavage is equatorial and separates the embryo into four animal and four vegetal blastomeres (Fig. 10.3). An alternative body plan with a shifted dorsal–ventral axis and a specified cranial–caudal axis has also been proposed in the literature (Lane and Sheets 2006), but the original plan with the dorsal–ventral separation at the 4-cell stage is currently the preferred view.

10.4 Cortical Rotation and Dorsal–Ventral Asymmetry

The sperm enters through the animal pole of the oocyte and induces extensive internal movement called cortical rotation (Gerhart et al. 1989). The point of sperm entry also specifies the future ventral side of the embryo. The cortical rotation happens roughly 20 min after sperm entry, when the outer layer of the cortex rotates about 30° relative to the inner cytoplasm (Vincent and Gerhard 1987; Denegre and Danilchik 1993) (Fig. 10.4). The cortical rotation requires the microtubule network of the vegetal cortex (Schroeder and Gard 1992; Vincent et al. 1987). The cortical rotation introduces additional asymmetry inside the fertilized oocyte that leads to

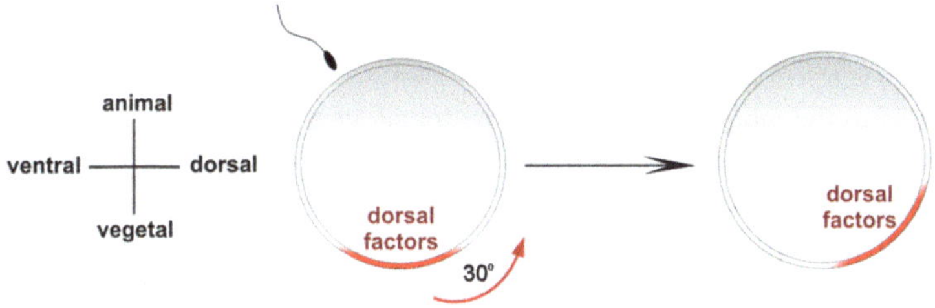

Fig. 10.4 Scheme of cortical rotation and accumulation of dorsalizing factors (adapted from Gilbert 2010)

the dorsal–ventral axis formation during development. Determinants of the dorsal fate accumulate at the dorsal side of the embryo, which is opposite to the site of sperm entry (Weaver and Kimelman 2004; White and Heasman 2008). Some amphibian species such as *Rana* show localized color change called the gray crescent as a consequence of cortical rotation (in *Xenopus* embryos, the gray crescent is not visible). Several maternal proteins have been identified as the dorsal determinants, the majority of them being members of the Wnt pathway. One is Dvl (Dishevelled) (Miller et al. 1999), which inactivates Gsk3β (glycogen synthase kinase 3 beta) in the dorsal part resulting in the blockage of β-catenin (Ctnnb1) degradation (Heasman et al. 2000; Hyatt et al. 1996). Accumulated β-catenin locally activates zygotic expression of *xnr3* (*Xenopus* nodal related 3) and of *siamois* in the dorsal part. These genes determine the dorsal phenotype of the embryo (Darras et al. 1997; Marikawa et al. 1997).

Dorsal and ventral blastomeres can be distinguished in the 4-cell stage embryo as a consequence of the cortical rotation that leads to the differences in shading and size (Sive et al. 2000; Blum et al. 2014). The ventral blastomeres are somewhat larger and darker than the dorsal ones (Fig. 10.3). Several studies have demonstrated differences in the cellular content between the dorsal and ventral blastomeres. For example, Dvl and β-catenin proteins are enriched in the dorsal blastomeres (Rowning et al. 1997; Miller et al. 1999). One study has reported that ventral blastomeres contain more *wnt8b* mRNA than the dorsal (Pandur et al. 2002), but this has been contradicted by recent studies that found no difference between the dorsal and ventral blastomeres at the mRNA level (Flachsova et al. 2013). These authors collected single blastomeres from 8-, 16-, and 32-cell stage embryos and measured, using qRT-PCR, the expression of 41 maternal transcripts. Distinct differences between the animal and vegetal blastomeres were found based on the asymmetry along the A–V axis that developed already in the oocyte. But none of the studied genes showed differential expression among blastomeres that could be related to asymmetry along the D–V axis. This finding is supported by recent whole transcriptome analysis that found no differences at the mRNA level between the dorsal and ventral blastomeres (De Domenico et al. 2015). However, a recent large-scale protein analysis reported dozens of candidates that might be asymmetrically distributed along the D–V axis (Lombard-Banek et al. 2016).

10.5 Left–Right Asymmetry

In contrast to the animal–vegetal and the dorsal–ventral axes, the left–right axis is in some aspects asymmetric, while in other aspects it retains certain symmetry. One of the characteristics of chordates is the bilateral symmetry, which means that there is one plane along which they can be cut into two mirror-image halves that are compositionally the same. However, uneven distribution of internal organs will eventually break the left–right mirror symmetry.

The embryo obtains its left and right halves at the first cell division after fertilization (Fig. 10.3). *Xenopus laevis* has been used as a model to test if components of the H+/K+-ATPase (Levin et al. 2002) and of the H+-V-ATPase ion pumps (Adams et al. 2006) are asymmetrically distributed in 2- and 4-cell stage embryos. Asymmetric ion production leading to different membrane potentials in the left and right halves has been suggested to distribute the left–right determinants. The right side blastomeres produce more ATPase pumps, which results in a local negative potential. This induces transport of small charged molecules through the gap channels between the blastomeres (Levin et al. 2002; Vandenberg and Levin 2010). Serotonin that accumulates in the left and ventral blastomeres during development has been suggested to be a determinant of left–right asymmetry (Fukumoto et al. 2005a). Two serotonin transporters have been found expressed during the formation of the left–right body axis (Fukumoto et al. 2005b). Another candidate behind the left–right asymmetry is protein 14-3-3E, which has been reported differentially expressed between the left and right blastomeres in 2-cell embryos (Bunney et al. 2003). The Nodal pathway, which is an important cascade for left–right asymmetry, is activated in the neurula stage. Three genes, Nodal, its inhibitor Lefty, and Pitx2, are predominantly expressed in the left side of the embryo. The Nodal cascade is formed by leftward flow of extracellular fluids via ciliary activity (Blum et al. 2014). A few transcripts, including *14-3-3E,* have been reported asymmetrically distributed between the left and right blastomeres at the 2- and 4-cell stage embryos based on in situ hybridization (Bunney et al. 2003). *14-3-3E* was not among the 41 maternal transcripts whose distribution was studied using high-throughput RT-qPCR (Flachsova et al. 2013). This study showed lack of asymmetry between the left and right blastomeres in the distribution of members of the Wnt pathway, early cell fate determinants, and several transcription factors. Recent whole transcriptome analysis of single blastomeres by next-generation sequencing also reported the absence of left–right asymmetry at the mRNA level (De Domenico et al. 2015).

10.6 Asymmetrical Localization of Small Molecules

Recently, the metabolites were identified as a novel group of cell fate determinants (Onjiko et al. 2015). A new method combining capillary electrophoresis (CE), microsampling electrospray ionization (ESI), and mass spectrometry (MS) was introduced to quantify the amounts of metabolites in individual *Xenopus laevis* blastomeres. 40 metabolites were measured in each blastomeres revealing differences between animal–vegetal and dorsal–ventral sides in 16-cell embryos (Onjiko et al. 2015). Studies using 8-cell embryos to assess the left–right asymmetry identified ten metabolites with asymmetrical distribution levels (Onjiko et al. 2016).

10.7 Other Models of Asymmetric Cell Division During Development

Asymmetric distributions of fate determinants are studied in detail also in the fruit fly *Drosophila melanogaster* and in the nematode *Caenorhabditis elegans* (Knoblich 2010; Gönczy and Rose 2005; Cowan and Hyman 2004). Several conserved pathways have been identified among those species, and their roles during development have been elucidated. Amphibian and fish species including *Xenopus laevis* are also popular models for studies of asymmetric cell division during development.

Mammalian asymmetric cell division during early developmental has been found to be more complicated, and interpretation of mammalian experimental data is often controversial. Introduction of new methods for gene expression profiling such as qRT-PCR and next-generation sequencing (NGS) has made detailed transcriptional studies possible even for single cells from developing mouse embryos (reviewed in Saiz et al. 2015). Large-scale proteome analysis of mouse single blastomeres is, however, not yet feasible.

10.8 Future Prospects

Precise control of biomolecule localization and distribution is required for proper development of the organism. Localization and function of many RNAs and proteins has been described in several organisms, but the mechanisms behind the minutiously controlled developmental process remain puzzling. Recently introduced methods for large-scale profiling such as next-generation sequencing, deep quantitative proteomics, and metabolomics have initiated a new era of global biomolecule quantification and localization studies. Next challenge will be the analysis of the huge amounts of data these methods are generating and assembling it into localization pathways that control early development.

Acknowledgements This study was supported by Ministry of Youth, Education and Sports of the Czech Republic AV0Z50520701 and grant **LH15074**; BIOCEV CZ.1.05/1.1.00/02.0109 from the ERDF and by the Czech Science Foundation GA CR—**GA16-07500S**.

References

Adams DS, Robinson KR, Fukumoto T, Yuan S, Albertson RC, Yelick P, Kuo L, McSweeney M, Levin M (2006) Early, H+-V+ATPase-dependent proton flux is necessary for consistent left-right patterning of non-mammalian vertebrates. Development 133:1657–1671

Agius E, Oelgeschläger M, Wessely O, Kemp C, De Robertis EM (2000) Endodermal nodal-related signals and mesoderm induction in *Xenopus*. Development 127(6):1173–1178

Blanpain C, Lowry WE, Geoghegan A, Polak L, Fuchs E (2004) Self-renewal, multipotency, and the existence of two cell populations within an epithelial stem cell niche. Cell 118:635–648

Blum M, Schweickert A, Vick P, Wright CV, Danilchik MV (2014) Symmetry breakage in the vertebrate embryo: when does it happen and how does it work? Dev Biol 393:109–123

Bubunenko M, Kress TL, Vempati UD, Mowry KL, King ML (2002) A consensus RNA signal that directs germ layer determinants to the vegetal cortex of Xenopus oocytes. Dev Biol 248:82–92

Bunney TD, De Boer AH, Levin M (2003) Fusicoccin signaling reveals 14-3-3 protein function as a novel step in left-right patterning during amphibian embryogenesis. Development 130:4847–4858

Cowan CR, Hyman AA (2004) Centrosomes direct cell polarity independently of microtubule assembly in *C. elegans* embryos. Nature 431(7004):92–96

Cuykendall TN, Houston DW (2010) Identification of germ plasm-associated transcripts by microarray analysis of Xenopus vegetal cortex RNA. Dev Dyn 239:1838–1848

Danilchik MV, Gerhadt JC (1987) Differentiation of the animal–vegetal axis in *Xenopus laevis* oocytes. I. Polarized intracellular translocation of platelets establishes the yolk gradient. Dev Biol 122:101–112

Darras S, Maikawa Y, Elinson RP, Lemaire P (1997) Animal and vegetal pole cells of early Xenopus embryos respond differently to maternal dorsal determinants: implications for the patterning of the organiser. Development 124:4275–4286

De Domenico E, Owens ND, Grant IM, Gomes-Faria R, Gilchrist MJ (2015) Molecular asymmetry in the 8-cell stage Xenopus tropicalis embryo described by single blastomere transcript sequencing. Dev Biol 408(2):252–268

Denegre JM, Danilchik MV (1993) Deep cytoplasmic rearrangements in axis-respecified Xenopus embryos. Dev Biol 160:157–164

Deshler JO, Highett J, Schnapp BJ (1997) Localization of Xenopus Vg1 mRNA by vera protein and the endoplasmatic reticulum. Science 276:1128–1131

Dumont JN (1972) Oogenesis in *Xenopus laevis* (Daudin) I. Stages of oocyte development in laboratory maintained animals. J Morphol 136:153–179

Etkin LD (1997) A new face from the endoplasmic reticulum: RNA localization. Science 276:1092–1093

Etkin LD, Balcells S (1985) Transformed Xenopus embryos as a transient expression system to analyze gene expression at the midblastula transition. Dev Biol 108:173–178

Flachsova M, Sindelka R, Kubista M (2013) Single blastomere expression profiling of *Xenopus laevis* embryos of 8 to 32-cells reveals developmental asymmetry. Sci Rep 3:2278

Fukumoto T, Kema IP, Levin M (2005a) Serotonin signaling is a very early step in patterning of the left-right axis in chick and frog embryos. Curr Biol 15:794–803

Fukumoto T, Blakely R, Levin M (2005b) Serotonin transporter function is an early step in left-right patterning in chick and frog embryos. Dev Neurosci 27:349–363

Gagnon JA, Kreiling JA, Powrie EA, Wood TR, Mowry KL (2013) Directional transport is mediated by a Dynein-dependent step in an RNA localization pathway. PLoS Biol 11(4): e1001551

Gautreau D, Cote CA, Mowry KL (1997) Two copies of a subelement from the Vg1 RNA localization sequence are sufficient to direct vegetal localization in Xenopus oocytes. Development 124:5013–5020

Gerhart J, Danilchik M, Doniach T, Roberts S, Rowning B, Stewart R (1989) Cortical rotation of the Xenopus egg: consequences for the anteroposterior pattern of embryonic dorsal development. Development 107:37–51

Gilbert SF (2010) Developmental biology, 9th edn. Sinauer Associates, Sunderland, MA

Gönczy P, Rose LS (2005) Asymmetric cell division and axis formation in the embryo. In: WormBook (ed) The *C. elegans* research community. WormBook, Pasadena, CA. doi:10.1895/wormbook.1.30.1

Gurdon JB (1968) Changes in somatic cell nuclei inserted into growing and maturing amphibian oocytes. J Embryol Exp Morphol 20:401–414

Heasman J (2006) Maternal determinants of embryonic cell fate. Semin Cell Dev Biol 17:93–98

Heasman J, Quarmby J, Wylie CC (1984) The mitochondrial cloud of Xenopus ocytes: the source of germinal granule material. Dev Biol 105:458–469

Heasman J, Kofron M, Wylie C (2000) Beta-catenin signaling activity dissected in the early Xenopus embryo: a novel antisense approach. Dev Biol 222:124–134

Houston DW (2013) Regulation of cell polarity and RNA localization in vertebrate oocyte. Int Rev Cell Mol Biol 306:127–185

Hyatt BA, Lohr JL, Yost HJ (1996) Initiation of vertebrate left-right axis formation by maternal Vg1. Nature 384:62–65

Jullien J, Pasque V, Halley-Stott RP, Miyamoto K, Gurdon JB (2011) Mechanism of nuclear reprogramming by eggs and oocytes: a deterministic process? Nat Rev Mol Cell Biol 12:453–459

King ML, Messitt TJ, Mowry KL (2005) Putting RNAs in the right place at the right time: RNA localization in the frog oocyte. Biol Cell 97:19–33

Kloc M, Etkin LD (1994) Delocalization of Vg1 mRNA from the vegetal cortex in Xenopus oocytes after destruction of Xlsirt RNA. Science 265:1101–1103

Kloc M, Etkin LD (1995) Two distinct pathways for the localization of RNAs at the vegetal cortex in Xenopus oocytes. Development 121:287–297

Kloc M, Spohr G, Etkin LD (1993) Translocation of repetitive RNA sequences with the germ plasm in Xenopus oocytes. Science 262:1712–1714

Kloc M, Larabell C, Etkin LD (1996) Elaboration of the messenger transport organizer pathway for localization of RNA to the vegetal cortex of Xenopus oocytes. Dev Biol. 180(1):119–30

Kloc M, Larabell C, Chan AP, Etkin LD (1998) Contribution of METRO pathway localized molecules to the organization of the germ cell lineage. Mech Dev 75(1–2):81–93

Kloc M, Bilinski S, Chan AP, Allen LH, Zearfoss NR, Etkin LD (2001) RNA localization and germ cell determination in Xenopus. Int Rev Cytol 203:63–91

Kloc M, Dougherty MT, Bilinski S, Chan AP, Brey E, King ML, Patrick CW Jr, Etkin LD (2002) Three-dimensional ultrastructural analysis of RNA distribution within germinal granules of Xenopus. Dev Biol 241:79–93

Kloc M, Bilinski S, Dougherty MT, Brey EM, Etkin LD (2004) Formation, architecture and polarity of female germline cyst in Xenopus. Dev Biol 266(1):43–61

Knoblich JA (2008) Mechanisms of asymmetric stem cell division. Cell 132:583–597

Knoblich JA (2010) Asymmetric cell division: recent developments and their implications for tumour biology. Nat Rev Mol Cell Biol 11:849–597

Lane MC, Sheets MD (2006) Heading in a new direction: implications of the revised fate map for understanding *Xenopus laevis* development. Dev Biol 296(1):12–28

Levin M, Thorlin T, Robinson KR, Nogi T, Mercola M (2002) Asymmetries in H+/K+-ATPase and cell membrane potentials comprise a very early step in left-right patterning. Cell 111:77–89

Lewis RA, Kress TL, Cote CA, Gautreau D, Rokop ME, Mowry KL (2004) Conserved and clustered RNA recognition sequences are a critical feature of signals directing RNA localization in Xenopus oocytes. Mech Dev 121:101–109

Lombard-Banek C, Moody SA, Nemes P (2016) Single-cell mass spectrometry for discovery proteomics: quantifying translational cell heterogeneity in the 16-cell frog (Xenopus) embryo. Angew Chem Int Ed Engl 55(7):2454–2458

Marikawa Y, Li Y, Elinson RP (1997) Dorsal determinants in the Xenopus egg are firmly associated with the vegetal cortex and behave like activators of the Wnt pathway. Dev Biol 191:69–79

Messitt TJ, Gagnon JA, Kreiling JA, Pratt CA, Yoon YJ, Mowry KL (2008) Multiple kinesin motors coordinate cytoplasmic RNA transport on a subpopulation of microtubules in Xenopus oocytes. Dev Cell 15(3):426–436

Miller JR, Rowning BA, Larabell CA, Yang-Snyder JA, Bates RL, Moon RT (1999) Establishment of the dorsal-ventral axis in Xenopus embryos coincides with dorsal enrichment of dishevelled that is dependent on cortical rotation. J Cell Biol 146:427–437

Newport J, Kirschner M (1982) A major developmental transition in early Xenopus embryos: I. characterization and timing of cellular changes at the midblastula stage. Cell 30:675–686

Onjiko RM, Moody SA, Nemes P (2015) Single-cell mass spectrometry reveals small molecules that affect cell fates in the 16-cell embryo. Proc Natl Acad Sci USA 112(21):6545–6550

Onjiko RM, Morris SE, Moody SA, Nemes P (2016) Single-cell mass spectrometry with multisolvent extraction identifies metabolic differences between left and right blastomeres in the 8-cell frog (Xenopus) embryo. Analyst 141(12):3648–3656

Pandur PD, Sullivan SA, Moody SA (2002) Multiple maternal influences on dorsal-ventral fate of Xenopus animal blastomeres. Dev Dyn 225:581–587

Pereira G, Yamashita YM (2011) Fly meets yeast: checking the correct orientation of cell division. Trends Cell Biol 21:526–533

Rowning BA, Wells J, Wu M, Gerhart JC, Moon RT, Larabell CA (1997) Microtubule-mediated transport of organelles and localization of beta-catenin to the future dorsal side of Xenopus eggs. Proc Natl Acad Sci USA 94:1224–1229

Saiz N, Plusa B, Hadjantonakis AK (2015 Dec) Single cells get together: high-resolution approaches to study the dynamics of early mouse development. Semin Cell Dev Biol 47–48:92–100

Schatten H, Sun QY (2010) The role of centrosomes in fertilization, cell division and establishment of asymmetry during embryo development. Semin Cell Dev Biol 21:174–184

Schroeder MM, Gard DL (1992) Organization and regulation of cortical microtubules during the first cell cycle of Xenopus eggs. Development 114:699–709

Shahriyari L, Komarova NL (2013) Symmetric vs. asymmetric stem cell divisions: an adaptation against cancer? PLoS One 8:e76195

Sidova M, Sindelka R, Castoldi M, Benes V, Kubista M (2015) Intracellular microRNA profiles form in the *Xenopus laevis* oocyte that may contribute to asymmetric cell division. Sci Rep 5:11157

Sindelka R, Jonak J, Hands R, Bustin SA, Kubista M (2008) Intracellular expression profiles measured by real-time PCR tomography in the *Xenopus laevis* oocyte. Nucleic Acids Res 36:387–392

Sindelka R, Sidova M, Svec D, Kubista M (2010) Spatial expression profiles in the *Xenopus laevis* oocytes measured with qPCR tomography. Methods 51(1):87–91

Sive HL, Grainger MR and Harland MR (2000) Early development of *Xenopus laevis*—a laboratory manual. Cold Spring Harbor Laboratory Press, Chap 2, Fig. 2.1

Skirkanich J, Luxardi G, Yang J, Kodjabachian L, Klein PS (2011) An essential role for transcription before the MBT in *Xenopus laevis*. Dev Biol 357(2):478–491

Smits AH, Lindeboom RG, Perino M, van Heeringen SJ, Veenstra GJ, Vermeulen M (2014) Global absolute quantification reveals tight regulation of protein expression in single Xenopus eggs. Nucleic Acids Res 42:9880–9891

Snedden DD, Bertke MM, Vernon D, Huber PW (2013) RNA localization in Xenopus oocytes uses a core group of trans-acting factors irrespective of destination. RNA 19:889–895

Sun L, Bertke MM, Champion MM, Zhu G, Huber PW, Dovichi NJ (2014) Quantitative proteomics of *Xenopus laevis* embryos: expression kinetics of nearly 4000 proteins during early development. Sci Rep 4:4365

Vandenberg LN, Levin M (2010) Far from solved: a perspective on what we know about early machanisms of left-right asymmetry. Dev Dyn 239:3131–3146

Vincent JP, Gerhart JC (1987) Subcortical rotation in Xenopus eggs: an early step in embryonic axis specification. Dev Biol 123:526–539

Vincent JP, Scharf SR, Gerhart JC (1987) Subcortical rotation in Xenopus eggs: a preliminary study of its machanochemical basis. Cell Motil Cytoskeleton 8:143–154

Weaver C, Kimelman D (2004) Move it or lose it: axis specification in Xenopus. Development 131:3491–3499

White JA, Heasman J (2008) Maternal control of pattern formation in *Xenopus laevis*. J Exp Zool B Mol Dev Evol 310:73–84

Yisraeli JK, Sokol S, Melton DA (1990) A two-step model for the localization of maternal mRNA in Xenopus oocytes: involvement of microtubules and microfilaments in the translocation and anchoring of Vg1 mRNA. Development 108:289–298

Zearfoss NR, Chan AP, Kloc M, Allen LH, Etkin LD (2003) Identification of new Xlsirt family members in the *Xenopus laevis* oocyte. Mech Dev 120(4):503–509

Zhang J, King ML (1996) Xenopus VegT RNA is localized to the vegetal cortex during oogenesis and encodes a novel T-box transcription factor involved in mesodermal patterning. Development 122:4119–4129

Zhou Y, King ML (1996a) Localization of Xcat-2 RNA, a putative germ plasm component, to the mitochondrial cloud in Xenopus stage I oocyte. Development 122:2947–2953

Zhou Y, King ML (1996b) RNA transport to the vegetal cortex of Xenopus oocytes. Dev Biol 179:173–183

Zhou Y, King ML (2004) Sending RNAs into the future: RNA localization and germ cell fate. IUBMB Life 56:19–27

Kwon S, Abramson T, Munro TP, CM J, Kohrmann M, Schnapp BJ (2002) UUCAC- and vera-dependent localization of VegT TNA in Xenopus oocytes. Curr Biol 12:558–564

Chapter 11
Asymmetries in Cell Division, Cell Size, and Furrowing in the *Xenopus laevis* Embryo

Jean-Pierre Tassan, Martin Wühr, Guillaume Hatte, and Jacek Kubiak

Abstract Asymmetric cell divisions produce two daughter cells with distinct fate. During embryogenesis, this mechanism is fundamental to build tissues and organs because it generates cell diversity. In adults, it remains crucial to maintain stem cells. The enthusiasm for asymmetric cell division is not only motivated by the beauty of the mechanism and the fundamental questions it raises, but has also very pragmatic reasons. Indeed, misregulation of asymmetric cell divisions is believed to have dramatic consequences potentially leading to pathogenesis such as cancers. In diverse model organisms, asymmetric cell divisions result in two daughter cells, which differ not only by their fate but also in size. This is the case for the early *Xenopus laevis* embryo, in which the two first embryonic divisions are perpendicular to each other and generate two pairs of blastomeres, which usually differ in size: one pair of blastomeres is smaller than the other. Small blastomeres will produce embryonic dorsal structures, whereas the larger pair will evolve into ventral structures. Here, we present a speculative model on the origin of the asymmetry of this cell division in the Xenopus embryo. We also discuss the apparently coincident asymmetric distribution of cell fate determinants and cell-size asymmetry of the 4-cell stage embryo. Finally, we discuss the asymmetric furrowing during epithelial cell cytokinesis occurring later during *Xenopus laevis* embryo development.

11.1 Introduction

The cell-size asymmetry and the asymmetric distribution of cell fate determinants have been shown to be coupled in diverse model organisms. However, it is not clear to what extent the two asymmetries are really linked and/or co-dependent.

J.-P. Tassan (✉) • G. Hatte • J. Kubiak
CNRS UMR 6290, Rennes, France
Université de Rennes 1, Institut de Génétique et Développement de Rennes, Rennes, France
e-mail: jean-pierre.tassan@univ-rennes1.fr

M. Wühr
Department of Molecular Biology and the Lewis-Sigler Institute for Integrative Genomics, Princeton University, Princeton, NJ, USA

J.-P. Tassan, J.Z. Kubiak (eds.), *Asymmetric Cell Division in Development, Differentiation and Cancer*, Results and Problems in Cell Differentiation 61,
DOI 10.1007/978-3-319-53150-2_11

Xenopus laevis—a model used in developmental biology research for more than a century—allowed discovering basic principles of early development in vertebrates including asymmetric distribution of cell fate determinants, which leads to dorsalization of the embryo. Although the asymmetric cell-size division of the early Xenopus embryo was described long ago, it is still, in contrast to other systems such as *C. elegans* or *D. melanogaster,* poorly understood. This is mostly due to the large size of the embryo and its blastomeres—a fantastic advantage for micromanipulation studies but an obstacle for cell biology studies, especially those involving microscopy approaches. In the first part of this chapter, we present a speculative model on the origin of the asymmetry of the second embryonic cell division. We also discuss the apparently coincident asymmetric distribution of cell fate determinants and cell-size asymmetry of the 4-cell stage embryo. In the second part of this chapter, we describe the asymmetric furrowing in epithelial cell cytokinesis and its regulation during early embryo development.

11.2 Asymmetric Cell Division

11.2.1 Unfertilized Egg

Before discussing asymmetric cell division in the early Xenopus embryo, we will overview the organization of the fertilized egg because it profoundly influences subsequent zygotic divisions.

The unfertilized Xenopus egg is a very large (1.2 mm in diameter) spherical and highly polarized cell. In wild type (non-albino Xenopus), the egg polarity is readily visible thanks to the dark pigment, which is asymmetrically concentrated in the animal hemisphere. Thus, the animal hemisphere is black to light brown, whereas the vegetal hemisphere, which contains much less pigment, appears whitish. The egg displays symmetry around the animal–vegetal axis with all pole-to-pole meridians of the egg surface seemingly identical and indistinguishable before fertilization. This symmetry is often referred to as radial symmetry or cylindrical symmetry. The egg organization roughly overlaps with the embryo germ layers' organization: the region of the animal and vegetal hemispheres corresponds to the future ectoderm and endoderm, respectively, while the equator corresponds to the future mesoderm (reviewed in Gerhart and Keller 1986). The animal and vegetal hemispheres also differ in their organelles and molecular content. For example, the yolk platelets of the vegetal hemisphere are larger and have a much higher density (Danilchik and Gerhart 1987). Massive accumulation of yolk in the vegetal hemisphere displaces egg cytoplasm toward the animal hemisphere. This asymmetric distribution of yolk and cytoplasm has clear consequences on the positioning of the third zygotic division plane and potentially also on the two first cell divisions (see below). In fully grown oocytes the large cell nucleus (called germinal vesicle) is located at the vicinity of the animal pole. This position facilitates polar body extrusion during meiosis I and II. Xenopus oocyte meiotic divisions are among the most asymmetric cell divisions, not only in terms of the 200-fold difference in linear size between the egg and polar bodies but also in a functional sense—the oocyte survives beyond meiosis while the polar bodies do not. At the molecular level, the animal and vegetal hemispheres differ greatly. In the Xenopus egg, the cortex of the vegetal hemisphere is a repository for developmental information, much of which is in the form of localized RNAs (Kloc and Etkin 1994; King et al. 1999; Mowry and Cote 1999).

11.2.2 From Fertilization to the Third Division

The asymmetry between animal and vegetal hemisphere of the egg is not restricted to its constituents' distributions. There is also a functional asymmetry; the sperm enters the egg only at the animal hemisphere (Elinson 1980 cited in Gerhart et al. 1989). The position at which the sperm enters the egg is called "sperm entry point" (SEP). Shortly after fertilization, the egg vitelline envelope hardens to become the fertilization envelope, which protects egg against polyspermy. Because of the asymmetric distribution of yolk, the more dense vegetal hemisphere rotates downward until the animal–vegetal axis aligns with the gravity vector and becomes vertical.

During oogenesis, Xenopus oocytes lose their centrioles but maintain centrosomal proteins allowing acentriolar meiotic divisions. Once the sperm has entered the egg, its centrosome organizes a huge microtubular aster (sperm aster), which functions to draw the male pronucleus toward the center of the animal hemisphere (Stewart-Savage and Grey 1982). After completion of meiosis and extrusion of the second polar body, the female pronucleus is captured by the sperm aster and migrates toward the male pronucleus. To reach the common position, the two pronuclei generally move in the three-dimensional space of the egg: in the X–Y plane (plane parallel to the equator) and also along the animal–vegetal axis. They always end up within the animal hemisphere probably because of the much lower density of yolk platelets in this hemisphere. They thus become situated between the egg's geometric center and the animal pole. The mechanisms regulating the geometric centering in such a huge cell as one-cell embryo are not fully understood. In small-sized cells, e.g., in vitro cultured somatic cells, microtubule asters are believed to center via pulling forces from the cellular cortex (Glotzer 1997; Grill and Hyman 2005). However, these mechanisms can't explain centering of pronuclei in the one-cell Xenopus embryo. It was proposed that microtubule asters center via a dynein-dependent mechanism. Dynein, anchored to cytoplasm, exerts length-proportional pulling forces onto microtubules and thus tends to pull the microtubule aster in the direction of the longest microtubules (Wühr et al. 2009; Mitchison et al. 2015).

The sperm aster microtubules are also believed to bias the orientation of the subset of cortical microtubules on the opposite site of the sperm entry point (Gerhart et al. 1984). During the first cell cycle following fertilization, three concentric layers can be distinguished in the zygote. From the outside toward the inside, these layers are the cortex, the shear zone, and the deep core cytoplasm. Microtubules are involved in the cortical rotation, an event which corresponds to the loosening of the embryo cortex from the deep cytoplasm. The cortex rotates, about 30° relative to the core of cytoplasm and away from the SEP (Vincent et al. 1986). The resulting shear modifies the pigment distribution at the equator, producing a gray crescent, which corresponds to the embryo presumptive dorsal side. The gray crescent is also observed in some other amphibian species such as *Rana pipiens*. Although the position of the SEP, which can be simply visualized due to the pigment concentration, is a relatively good indicator of the prospective dorsal–ventral axis, it is not absolutely reliable (Black and Gerhart 1985). In fact, the direction of the cortical rotation more accurately predicts the dorsal side of the embryo (Vincent et al. 1986). In the absence of sperm, the egg can be activated parthenogenetically, for example, by pricking, and the cortical rotation still occurs, but in this case it happens randomly. This indicates that the sperm aster is dispensable for cortical rotation, but it is necessary to polarize and direct the rotation.

Cortical rotation is intimately linked to the establishment of the embryo dorsal–ventral asymmetry. Indeed, it allows asymmetric redistribution of developmental factors associated with the cortex from the vegetal pole toward the equator opposite to SEP and will determine the new dorsal–ventral axis of the embryo, which is generated orthogonally to the preexisting animal–vegetal axis of the oocyte (Vincent and Gerhart 1987). It is along these primary axes that the three germ layers form. Therefore, fertilization leads to a symmetry breakage, which transforms the cylindrical asymmetry of the unfertilized egg into a bilateral symmetry of the embryo (Gerhart et al. 1989). Similarly to the cortical rotation, the translocation of the dorsal determinants depends on a parallel array of microtubule bundles situated in the vegetal hemisphere. These microtubules appear in the shear zone of the egg about midway through the first cell cycle (Elinson and Rowning 1988). They are of multiple origins: some originate from the sperm centriole, others from unknown sources deep in the cytoplasm, and some appear to be assembled in the vegetal shear zone (Houliston and Elinson 1991; Schroeder and Gard 1992). The rotation involves dynein and kinesin microtubule motors (Marrari et al. 2004). Cortical rotation is essential for embryo dorsalization. Indeed, when it is experimentally blocked, for example, with microtubule inhibitors, then dorsal development is prevented (Vincent and Gerhart 1987). However, such embryos in which cortical rotation has been prevented can be rescued by tipping, which mimics the natural cortical rotation (Scharf and Gerhart 1980).

Cortical rotation coincides with the translocation of the maternal dorsalizing activity from the vegetal pole toward the prospective dorsal side of the embryo, in the same direction as cortical rotation (Kageura 1997; Kikkawa et al. 1996; Sakai 1996). It results in the asymmetrical activation of the Wnt/β-catenin signaling pathway and the enrichment of β-catenin in dorsal nuclei (Larabell et al. 1997; Schneider et al. 1996). In this chapter, we will not discuss the mechanisms of embryo dorsalization because pathways involved in the embryo dorsalization have been recently reviewed (Houston 2012; Carron and Shi 2016). Here, we will focus on a novel topic: the apparent relationship between the asymmetric distribution of cell fate determinants and the cell-size asymmetry in the 4-cell Xenopus embryo.

The asymmetric positioning of the mitotic apparatus is manifested externally by "Surface Contraction Waves" (SCWs) during embryo cell divisions. SCWs are waves of pigmentation changes at the embryo surface. They propagate from the animal toward the vegetal pole just prior to cytokinesis (Rankin and Kirschner 1997). These SWCs are associated with MPF (M-phase Promoting Factor, the complex of CDK1 with cyclin B) activation–deactivation waves within the cytoplasm (Pérez-Mongiovi et al. 1998; Beckhelling et al. 2000). The activation of MPF is initiated at the centrosomes (originating from the sperm centrosomes and duplicated during the first cell cycle within the zygote) and propagates centripetally through the cytoplasm toward the cell periphery (Chang and Ferrell 2013). As the centrosomes (together with the mitotic spindle) are not situated in the geometric center of the one-cell embryo, but are clearly displaced toward the animal pole, the radially propagating MPF activation wave reaches the cell cortex with a slight delay along the cell surface. This creates local reactions of the cytoskeleton and in consequence the appearance of wave of pigment responding to the MPF activation. Therefore, in the one-cell embryo, the SWCs are the manifestations of the cell asymmetry and especially of the displaced centering of the mitotic spindle along the A–V axis. Thus, SWCs represent an easily observable readout of displaced cell center at this stage of embryo development. However, the SWCs are not restricted to the first cell cycle (despite that they are very well visible during the first cell division of the embryo), which indicates that nuclei of blastomeres are not in their geometric centers also in the following cell cycles. This point is of special importance for our further analysis and hypotheses presented below.

The first three cell divisions of the Xenopus embryo are stereotyped as the majority of embryos display the same division pattern. The first division plane will tend to pass close to the sperm entry point. Whether the first division plane defines the embryonic bilateral symmetry of the embryo has been a matter of debate (Klein 1987; Danilchik and Black 1988). However, for most embryos, the first cleavage furrow identifies a meridian, which approximately coincides with the embryo midline (Masho 1990). It has been shown that the sperm aster has an elongated shape with its long axis parallel to a tangent (a straight line or plane that touches a curve or curved surface at a point) at the embryo surface at the SEP (Wühr et al. 2010). This particular shape results from the inability of microtubules to grow at the SEP because they are blocked by the cell cortex, whereas they are free to grow in other directions into the oblate volume of the animal hemisphere containing the majority of egg cytoplasm (Fig. 11.2). Thus, because of this peculiar geometry, the sperm aster senses the cell shape and defines the position and orientation of the first division plane by following the so-called long-axis rule (Hertwig 1893). This rule predicts that the division plane bisects the cell through its longer axis. Interestingly, the orientation of centrosomes during the preceding cell cycle defines the orientation of the division plane of the subsequent division. Together with the orientation of the sperm aster along the long axis, the two of them nicely explain the observation that the first cleavage furrow tends to bisect the one-cell embryo close to SEP.

The two first embryonic divisions, which are perpendicular to each other and parallel to the animal–vegetal axis, generate two pairs of blastomeres (4 blastomeres total), which usually differ in size and also frequently differ in pigmentation (Nieuwkoop and Faber 1967, Fig. 11.1). The pair of dorsal blastomeres is smaller

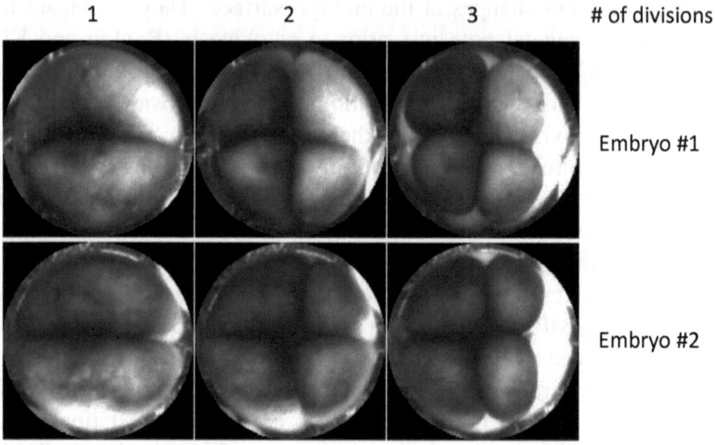

Fig. 11.1 The variability of cell divisions in the early Xenopus embryo in regard to the pigment distribution. Images are taken from a movie showing two embryos at the 2-, 4-, and 8-cell stage with the pigmented animal hemisphere in front of the viewer. The embryo shown in the *upper row* represents an asymmetric distribution of pigments at the 2-cell stage. The first cleavage furrow equally bisects the *lighter* and *darker* zones. The second division produces two darkly pigmented ventral blastomeres (left) and two lightly pigmented dorsal blastomeres (right). This embryo exemplifies a limited cell-size asymmetry of the first four blastomeres. This kind of embryos is usually selected for the cell fate studies (Grant et al. 2013). In the same clutch, some embryos, like the one shown in the *bottom row*, do not show a marked pigmentation asymmetry. Although the first division appears symmetric, or only slightly asymmetric, the second one results in clearly asymmetric embryo with two larger and two smaller blastomeres. This difference in cell size is maintained after the third division

and less pigmented and the ventral pair is larger and darker. At the 8-cell stage, the cell-size asymmetry can be visually even more pronounced. Following fertilization, the pigment concentrates toward the SEP indicating the future ventral side of the embryo. However, it is important to note that there can be considerable variations between clutches of embryos (Grant et al. 2013). The pigmentation asymmetry at the 2-, 4-, and 8-cell stage is easily perceptible by simple observation. The orientation of the first cleavage furrow relative to pigment distribution has been used as a selection criterion to produce fate maps (Moody and Kline 1990; Grant et al. 2013) or to study molecular mechanisms involved in the embryo asymmetry (Domenico et al. 2015). The cell-size asymmetry is also variable from clutch to clutch but can be frequently observed at the 4-cell stage (Fig. 11.1). However, the mechanisms leading to cell-size asymmetry of the two first cell divisions have been poorly studied. This is mainly due to the large size and opacity of the embryo at this stage, which allows analysis of fixed samples, but does not allow live imaging necessary to analyze and understand the highly dynamic process of cell division. Other biological models allowing live imaging, such as zebrafish, have been preferred (Wühr et al. 2010). One could assume that the cell-size asymmetry at the 4-cell stage could simply be randomly biased by a relatively neutral factor such as yolk distribution. However, this would be difficult to reconcile with the fact that this is observed with a very high frequency. Only a small proportion of embryos from the same clutch present a pronounced cell-size asymmetry at the 2-cell stage, which could be simply explained by an imperfect alignment of the sperm aster in relation to the SEP-center radius. The centering of the sperm aster in the X–Y axis is not perfect; it has been proposed that the sperm aster is placed closer to the site of sperm entry than to the opposite side (Wühr et al. 2009, 2010). Conversely, at the 4-cell stage, embryos often have blastomeres asymmetric in size (Fig. 11.1, Nieuwkoop and Faber 1967). The imperfect positioning of sperm aster and the resulting slightly eccentric position of the zygote nuclei after the first division could explain why the second cell division produces two smaller blastomeres closer to the SEP and two larger blastomeres on the opposite side. The two smaller blastomeres will transform into the dorsal and the larger into the ventral side of the embryo. Because the dorsal side is opposite to the SEP, the initial asymmetric positioning of the sperm aster farther from the SEP cannot explain the cell-size asymmetry of dorsal and ventral cells. Therefore, below, we put forward a speculative model that could reconcile these apparently conflicting observations. As discussed above, upon the cortical rotation the cortical cytoplasm rotates upward and away from the SEP, while the subcortical cytoplasm rotates in the opposite direction, upward and toward the SEP (Vincent et al. 1986; Hausen and Riebesell 1991) This cytoplasmic movement brings yolk-rich vegetal cytoplasm in the animal hemisphere close to the SEP. In contrast, on the future dorsal side the yolk moves away (Fig. 11.2a). This movement generates a ventral–dorsal yolk-density asymmetry. Staining experiments of yolk and cytoplasm at this stage (Fig. 11.2b, Danilchik and Denegre 1991) are roughly consistent with the proposed model shown in Fig. 11.2a. The microtubule aster's centering is biased against yolk granules, the size of which is apparently greater in the vegetal than it is in the animal hemisphere (see below). It seems plausible that the sperm aster and/or anaphase asters of mitotic spindles from the first and second cell division sense this asymmetry and center with a dorsal bias. In addition, the yolk asymmetry effectively reduces the amount of cytoplasm in the vegetal hemisphere. In conjunction with the cortex-induced asymmetry of the sperm aster discussed above, this cytoplasmic asymmetry could explain how the sperm aster directs the first cleavage plane, typically to cut through the sperm entry point. It finds the long axis of the animal hemisphere taking into consideration the yolk asymmetry.

How does the outlined hypothesis reconcile with the previous proposal that the sperm aster and anaphase asters' centering are biased toward the sperm entry point (Wühr et al. 2010)? These propositions were based on fixed images, and the location of the SEP relative to the asters was only inferred but not directly observed. Taking into consideration the arguments put forward here, it seems likely that this inference was mistaken.

The third division plane is orthogonal in relation to the two first division planes. In addition, because the nuclei resulting from the two first cell divisions are positioned eccentrically relatively to the geometric cell centers of blastomeres and displaced along the animal–vegetal axis (presumably due to differential yolk density), the third division plane becomes asymmetrically positioned and shifted toward the animal pole giving rise to four small animal blastomeres and four large vegetal blastomeres.

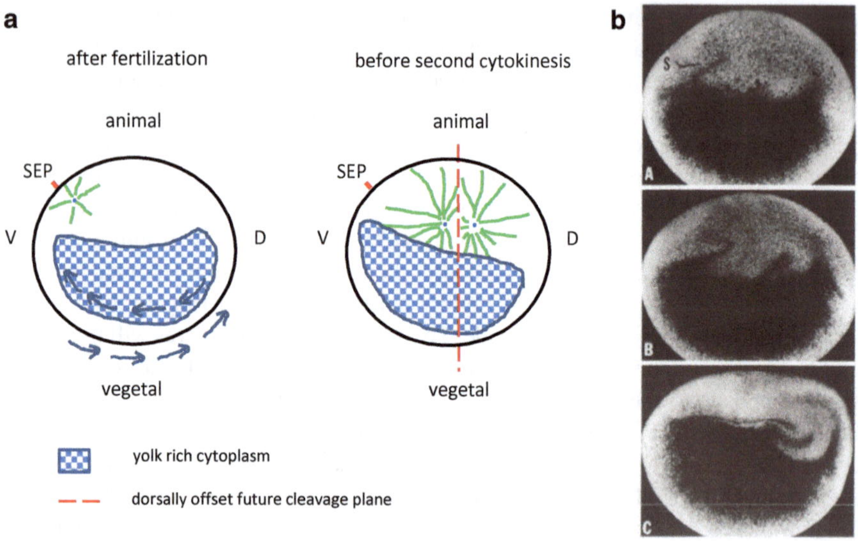

Fig. 11.2 (**a**) A model attempting to explain the typically observed dorsal shift of the second cleavage plane during *Xenopus laevis* first cell division. *Left*: While the egg is rotationally symmetric to the animal–vegetal axis, this symmetry is broken by the sperm entry and formation of the sperm aster. *Right*: The position of the sperm aster determines the orientation of cortical microtubules (not shown here for simplicity), which results in the rotation of cytoplasm relative to the cortex. We hypothesize that this rotation brings yolk-rich vegetal cytoplasm up and near the sperm entry point (*SEP*) and might also remove a significant part of yolk from the dorsal side. Such an asymmetry of a dorsal–vegetal yolk density might be sensed by the microtubule asters, and as a consequence, the second cleavage plane might shift dorsally. *V*: ventral, *D*: dorsal. (**b**) Observed yolk platelet movement in embryos seems consistent with this model. Shown are micrographs of yolk (*dark*) and cytoplasmic (*light*) rearrangements at 30 min (*top*), 45 min (*middle*), and 90 min (*bottom*) after fertilization. A major cytoplasmic swirl develops on the side opposite to the SEP. Orientation of the embryo as in (**a**). *S*: sperm trail which is a trace of pigment formed during the male pronucleus migration in the cytoplasm. Figure reprinted from Danilchik and Denegre (1991)

11.2.3 Asymmetric Divisions in Blastulae and Gastrulae

One of the earliest cell diversifications during Xenopus development is the production of polarized epithelial cells at the embryonic surface and nonpolar cells internally in the embryo. The outer, apical, membrane domains of these early polarized cells derive from the original egg membrane, while their basolateral membrane domains are newly formed (Fesenko et al. 2000). Asymmetric division along the apical–basal axis (divisions perpendicular to the embryo surface) occurs by rotation of the division plane (Chalmers et al. 2003). Again, this transition is stereotyped since it invariably occurs at the 32- to 64-cell stage (sixth zygotic division). The perpendicular divisions are detected until late blastula (stage 10, Nieuwkoop and Faber 1967) and are absent in gastrula (stage 11). Cell divisions plane can be perpendicular, tangential, or oblique to the embryo surface. Whereas divisions occurring perpendicular to the apical–basal axis symmetrically divide cells, those occurring tangentially give rise to an epithelial cell and an internal cell. In this case, only the daughter cell situated in the external cell layer (superficial cell) inherits the apical membrane from the mother cell. Because atypical protein kinase C (aPKC), an important phylogenetically conserved polarity protein (Martin-Belmonte and Perez-Moreno 2011), is localized at the egg membrane, it is preferentially inherited by the superficial daughter cells when the mother cells divide perpendicularly. Moreover, superficial cells express the bHLH gene ESR6e, whereas inner cells do not. Thus, perpendicular divisions, by allowing asymmetrical segregation of the polarity protein aPKC, and other factors such as ESR6e, generate diversification of the superficial and inner cell fates (Chalmers et al. 2003). The correlation between the orientation of cell divisions and the cell surface area (area of the apical membrane) suggests that the orientation of blastomere divisions (parallel or oblique versus orthogonal) is simply dictated by geometric rules. This indicates that the "long-axis rule" (Hertwig 1893), which controls division plane orientation in the early embryo, is also operant in blastulae. This also suggests that the "long-axis rule" could bring plasticity to the embryo and correct size asymmetry produced by the first two cell divisions, thus equilibrating cell size. This could also explain the robustness of the early developmental pattern in Xenopus, which would rely on geometric properties.

11.2.4 Does Cell-Size Asymmetry Produced by the Two First Cell Divisions Have a Role?

Cell-size asymmetry and the early dorsal–ventral polarization of the embryo appear intermingled: the small blastomeres correspond to the future dorsal structures, and, reversely, the large blastomeres correspond to the embryo ventral side. This raises the question whether the early embryonic cell-size asymmetry has a functional role or developmental implication. If dorsal–ventral polarization and cell-size asymmetry are indeed independent, then could it be possible to uncouple these two events? This question has not been directly raised by developmental biologists. However, studies on the establishment of the dorsal–ventral axis relative to the SEP might provide some clues to answer this question. Indeed, the dorsal–ventral axis appears to be topologically linked to the SEP since in normal conditions the dorsal side of the embryo will roughly correspond to the opposite of the SEP position. Although correlated, are the two asymmetries definitively linked? Based on results published by others, we again propose a completely speculative model, which, however, is consistent with results obtained in the insect model system.

Experiments of reorienting embryos along the gravity vector before the first cleavage may provide some important information. Indeed, as previously mentioned, shortly after fertilization the fertilized egg freely rotates in its fertilization envelope, thus aligning its animal–vegetal axis with gravity. Then, the 30° cortical rotation leads to redistribution of inner constituents relative to the zygote cortex. As already discussed, unfertilized eggs have a stereotyped distribution of yolk platelets, which is modified in the embryo (Danilchik and Denegre 1991; Ubbels et al. 1983; Gerhart et al. 1981). By reorienting the one-cell embryo, Gerhart and collaborators (Gerhart et al. 1981) prevented the naturally occurring redistribution of yolk and cytoplasm. Interestingly, this manipulation of the one-cell zygote apparently also displaced the dorsal structure of the embryo from the expected position, which is usually opposite to the SEP. When the SEP was oriented upward, the embryo dorsal side now coincided with the SEP side. Thus, by briefly reorienting the fertilized egg relatively to the gravity field it was possible to uncouple the natural topology of SEP and the embryo dorsal side. The effect of reorientation relative to gravity on embryos depended on when it was applied: the one-cell embryo became refractory to reorientation while progressing through the first cell cycle. This led the authors (Gerhart et al. 1981) to propose that during the early period of the first cell cycle, gravity might be able to interfere with the embryo dorsal–ventral polarization, which is naturally generated later during the first cell cycle. The low impact of gravity on Xenopus development was subsequently demonstrated by Souza et al. (1995). The embryos obtained in microgravity during a Space Shuttle mission showed minor developmental perturbations such as displacement of the blastocoel toward the vegetal hemisphere and thickening of the blastocoel roof. Both could be explained by the displacement of the third division plane toward the vegetal hemisphere. None of these perturbations led to apparent defect of embryo development. As previously suggested, in normal conditions, the embryo would be protected from adverse effects of gravity which could occur early during the first cell cycle by the free rotation of the embryo in the perivitelline space (Gerhart et al. 1981).

Although these authors did not examine cell-size asymmetry in the reoriented embryos, they described, in addition to yolk redistribution, the positions of aster and pronuclei in the fertilized egg. Interestingly, in the summarizing schemes they presented (Fig. 4 in Gerhart et al. 1981; Fig. 12 in Ubbels et al. 1983), they clearly showed that pronuclei were still displaced away from the SEP side (the side which should have formed the prospective dorsal side if the embryo would not have been reoriented). Although these embryos were fixed for microscopy analysis slightly before the first embryonic cleavage ($t = 0.7$ of normalized time in Gerhart et al. 1981) when the sperm aster depolymerizes and the first mitotic spindle forms (Wühr et al. 2009), it is reasonable to anticipate that the two pronuclei and centrosome deposited at this position away from SEP will determine the position of the subsequent cleavage plane. At the second cleavage, these embryos would be expected to cleave asymmetrically as normal embryos would do, thus producing two large and two small blastomeres. However, in this case cell-size asymmetry and fate would be uncoupled. Although clearly speculative, this interpretation raises the possibility that in the 4-cell Xenopus embryo, cell-size asymmetry would not be involved in the embryonic polarization but would be coincident with the asymmetric distribution of cell fate determinants. It would be possible that this early cell-size asymmetry is subsequently attenuated by divisions following the "long-axis rule" as discussed previously, thus reducing cell-size differences and homogenizing cell size in each hemisphere. Obviously, to be proven, the veracity of this speculative proposition needs experimental proof. Interestingly, such independence between asymmetric segregation of cell fate determinants and cell-size asymmetry, which we speculate for the Xenopus early embryo, was previously reported in Drosophila (Albertson and Doe 2003).

11.3 Asymmetric Furrowing During Epithelial Cytokinesis

In the second part of this chapter, we will discuss the asymmetric furrowing by which polarized epithelial cells divide in the Xenopus embryo.

11.3.1 Asymmetric Furrowing During Cytokinesis

In dividing one-cell embryos, cytokinesis starts with the formation of the division furrow at the zygote's animal pole. The oolemma ingresses progressively and the furrow expands circumferentially toward the vegetal pole. When the furrow has completed ingression at the cell surface, more deeply in the zygote the membrane is not yet fully closed because the furrow advances more rapidly through the animal than the vegetal hemisphere. Closure completes during the second cell division (Fesenko et al. 2000). By the end of the first embryonic division, the tight junction between blastomeres is implemented. This allows formation of a cavity located between blastomeres, which ultimately becomes the blastocoele. Consequently, all blastomeres until stage 16–32 cells, when internal cells are produced within the embryo (see below), are polarized. These cells inherit the "old" oolemma, which corresponds to the apical membrane. Concomitantly, their basal–lateral membranes become newly assembled (Bluemink and de Laat 1973; de Laat and Bluemink 1974), via the localized fusion of exocytotic vesicles near the furrow base (Danilchik et al. 1998, 2003). Thus, the external blastomeres form an epithelium at the surface of the embryo. From this time point forward, until the late blastula stage, during epithelial cells divisions, the cytokinetic furrow ingression proceeds from the apical to basal pole of these cells (Le Page et al. 2011). This pattern begins to change during pregastrular stages, and furrows begin to initiate on the basolateral membrane. In 16-cell stage embryos cytoplasmic bridges (midbodies) have been observed by electron microscopy at the proximity to the basal plasma membrane (see Fig. 2d in Danilchik et al. 2013). However, in 128-cell stage embryos, circumferential cytokinetic furrow could be observed in blastomeres of the animal hemisphere (Danilchik, personal communication). Cytoplasmic bridges with an approximately central position in the cell were also observed in 16-cell stage embryos (see Fig. 2d in Danilchik et al. 2013) and positioned more apically in 32-cell stage embryos, suggesting that in some cells the transition in the polarity of furrow ingression might occur early in development. In Le Page et al. (2011), only animal cells of blastula were studied by confocal microscopy. Thus, it is possible that

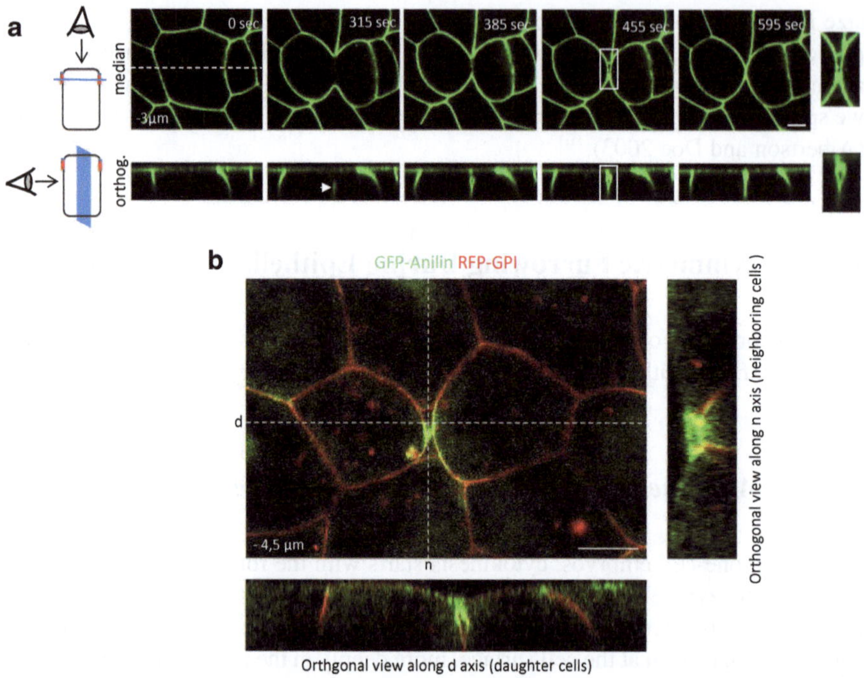

Fig. 11.3 The example of an asymmetric furrowing during epithelial cell cytokinesis. (**a**) Cell division of the epithelial cell expressing the fluorescent membrane marker GFP–GPI was followed by confocal microscopy. *Upper row* (*median*) shows a time-lapse *en face* view of the epithelium. The *dotted line* on the image at the time 0 s indicates the X–Y position of the plane used for the orthogonal projection shown in the *bottom row* (orthog.) (**a**). It shows that the cytokinetic furrow, indicated by a *white arrow* at the 315 s time point, ingresses asymmetrically from the basal side toward the apical side. White rectangles at the 455 s time point images indicate the regions, which are enlarged and shown on the *right*. This series of images shows that during cell division the gap forms between the two daughter cells. (**b**) GFP–anillin (*green*) localizes at the leading edge of the gap during epithelial cell cytokinesis in *Xenopus laevis* gastrula. The plasma membrane is visualized by the fluorescent membrane marker RFP–GPI (*red*). The orthogonal view along the neighboring cells' (*n*) axis shows the cytokinetic ring, while the orthogonal view along the daughter cells' (*d*) axis shows the gap between the two daughter cells. Scale bar: 10 µm

the change in the direction of furrowing takes place at different times in different region of blastula. Further analyses including both electron microscopy and live imaging will be necessary to determine the exact timing and location of the cytokinetic furrow polarity transition during the embryo development. In epithelial cells of the gastrula, the ingression of the cytokinesis furrow is still asymmetric. However, at this stage it progresses from the basal side toward the apical membrane (Fig. 11.3a). In blastulae, the tight junctions are not localized close to the apical membrane as typically found in epithelial tissue but localize laterally up to 200 µm away from the apical surface (Merzdorf et al. 1998). In contrast, in gastrulae, tight junctions occupy the typical position near the apical membrane. This correlation between the position of the tight junction along the apical–basal axis and the polarity of the furrowing allows speculating that the apical junctional complex is indeed involved in regulation of the cytokinetic furrow ingression polarity in epithelial cells.

The asymmetric furrowing occurring in Xenopus embryo is also encountered in other biological model animals including Drosophila (Founounou et al. 2013; Guillot and Lecuit 2013; Herszterg et al. 2013; Morais-de Sà and Sunkel 2013), ascidian embryo (Prodon et al. 2010), mouse intestine (Jinguji and Ishikawa 1992), and cultured MDCK cells (Reinsch and Karsenti 1994). During epithelial cells division, the asymmetric furrowing correlates with the asymmetric localization of apical junctional complexes. This suggests that the two processes are closely related. Indeed, in *Drosophila melanogaster* the knockdown of α-catenin, an important player in the adherent junction, interferes with the asymmetric furrow constriction (Guillot and Lecuit 2013). However, asymmetric furrowing also occurs independently of preexisting cell–cell junctions. It happens in one-cell Xenopus and *Caenorhabditis elegans* embryos (Audhya et al. 2005; Maddox et al. 2007). This indicates that the initiation of the asymmetric furrowing in the one-cell embryo does not involve cell–cell junctions. In the Xenopus embryo, apical junction appears soon after cytokinesis initiation but while the furrow still ingresses (Kalt 1971a, b). In the mouse oocyte, the polar body extrusion occurs through a unique combination of an asymmetric division and asymmetric furrowing (Ibanez et al. 2005; Wang et al. 2011). The question arises of whether this is correlated to the cell polarity and the proximity of the mitotic spindle creating a favorable cytoplasmic environment? The fact that asymmetric furrowing is observed in so many different systems suggests that this mode of cytokinesis may play an important role in embryo development and beyond (e.g., epithelial cells of adults).

11.3.2 Detailed Observations of Asymmetric Furrowing in Xenopus Embryo

Recently, we further characterized asymmetric furrowing during epithelial cell division in the ectoderm of *Xenopus laevis* gastrula (Hatte et al. 2014). We showed that the plasma membranes of the two daughter cells during cytokinesis are usually separated by a gap (Fig. 11.3a). In most cases, daughter cells form new contacts a few microns basally in relation to the furrow tip. This allows the formation of a volume tightly associated with the furrow, which we named “inverted teardrop” after the teardrop-shaped furrow canal during cellularization in *Drosophila melanogaster* (Lecuit and Wieschaus 2000). The complex organization of the cleavage furrow during the first Xenopus embryonic division and cleavages was described in detail by Kalt (1971a, b). These electron microscopy studies beautifully showed that formation of the presumptive blastocoele is not formed afterward but is intimately linked to the cytokinesis process as early as the first division. It also showed that the furrow continues to ingress below the presumptive blastocoele toward the embryonic vegetal pole. On the animal side, the plasma membranes are closely apposed, but between these contacts and the furrow tip, membranes diverge thus forming a dilatation. The strong similarity between cleaving embryos (Kalt 1971a, b) and gastrulae (Hatte et al. 2014) suggests that the "inverted teardrop" is not specific of gastrulae. It will be interesting to know if this organization is also found in epithelial cells at later developmental stages. Kalt (1971b) reported the presence of finely granular material in the dilated portion of the furrow tip. Whether a fluid fills the inverted teardrop of gastrulae and the composition of this putative fluid are currently unknown although during cytokinesis we could detect fibronectin below the furrow tip (JP Tassan, unpublished result). The cleavage furrow progresses with two different phases characterized by different speeds along the apical–basal axis: the fast one at the beginning of cytokinesis and the slow one when it reaches the neighborhood of the apical junctional complex. This difference in the dynamics of furrowing along the apical–basal axis could be linked to the fact that at some point the cytokinetic ring must cross the apical actin filaments. Also the size of the "inverted teardrop" evolves during its progression along the apical–basal axis. It is small when situated basally and enlarges progressively when it approaches the apical part of the dividing cell. It may be correlated with the presence of pulling forces, which spread plasma membranes and could be higher when the cytokinetic furrow reaches the apical adhesion belt. The inverted teardrop is resorbed at the end of the cell division. Therefore, it is a transient structure.

It is still unknown how the gap forms. However, several proteins actively involved in cytokinesis, including anillin (Fig. 11.3b), actin, and myosin II heavy chain, are localized both at the tip and lateral to the inverted teardrop, which suggests that membrane spreading is an active process. The presence of the teardrop gap during asymmetrically furrowing cell division raises the question of why the daughter cell membranes are transiently separated. The leading edge of the cytokinetic furrow could induce the plasma membrane-specific bending, thus creating a gap. It was shown that in Xenopus one-cell embryo, the leading edge of the cytokinetic furrow is mainly inherited from the surface plasma membrane, whereas the lateral side of the furrow is mainly formed by new membranes (Bluemink and de Laat 1973; de Laat and Bluemink 1974; Byers and Armstrong 1986; Danilchik et al. 1998). The membrane spreading at the leading edge of the cytokinetic furrow could create two independent membrane domains, where newly formed membranes are added. Thus, cytokinetic proteins localized at the leading edge of the inverted teardrop could be involved in the regulation of forces necessary to limit membrane spreading.

In gastrulae, during the phase of cytokinesis when epithelial cells are still dividing apically, the two daughter cells enter in contact basally, creating the "inverted teardrop." This mechanism coupled with asymmetric furrowing may contribute to the maintenance of epithelial integrity by ensuring cell–cell cohesion before completion of cytokinesis. Because the gap is transient, eventually the two daughter cells will be in close contact. However, this general scheme has an exception, which is the intercalation of neighboring cells. Although it may not be the unique mechanism, the intercalation of neighboring cells during cytokinesis participates in cell mixing during Xenopus gastrulation.

We also showed that intercalation of neighboring cells between daughter cells occasionally occurs during cytokinesis. The plasma membrane separation below the cytokinetic furrow forming the gap could be an active process and involves cytokinetic proteins. This suggests that these proteins could be involved in cell intercalation. It has been shown that remodeling of cell–cell contacts allows cell intercalation during convergent extension during elongation of the Xenopus embryo during later development (Keller 2002). Cytokinesis provides favorable conditions for cell intercalation and thus participates in cell mixing in the ectodermal epithelium of Xenopus gastrula. This raises the question of how cell–cell junctions are remodeled during intercalation. In Xenopus, cell mixing is restricted to clone borders (a group of cells stemming from the division of the same blastomere) (Bauer et al. 1994). This limited cell intermingling explains why the *Xenopus laevis* fate map is more consistent than in other model organisms including zebrafish and mouse, in which massive cell mixing happens during embryogenesis. Interestingly, it was shown in mouse small intestine that the neighboring cells can infiltrate the space in the cytokinetic furrow (Jinguji and Ishikawa 1992). This suggests that neighboring cell intercalation between dividing cells may be a general feature and occur also in mammalian tissues.

11.4 Conclusions

The Xenopus embryo displays multiple facets of asymmetric mechanisms during cell divisions. Some of these events, such as asymmetric localization of mRNAs and proteins, clearly participate in the development of the embryo. The understanding of mechanisms including asymmetric furrowing and its role during development and possibly in pathological processes still needs further investigation. The Xenopus early embryo offers a unique opportunity to study the role of these asymmetries in situ in a developing vertebrate embryo.

Acknowledgments We thank Michael V. Danilchik (Oregon Health and Sciences University, Portland, OR, USA) and Malgorzata Kloc (the Houston Methodist Hospital, Houston TX, USA) for helpful discussion and comments on the manuscript. We also thank the Microscopy Rennes Imaging Center (MRic, BIOSIT, IBiSA). Work in our lab was supported by le Centre National de la Recherche Scientifique (CNRS) and l'Agence Nationale de la Recherche (ANR, KinBioFRET). G.H. was supported by the MENESR and partly by a grant from the Ligue Nationale contre le Cancer. MW was supported by Princeton University start-up funding.

References

Albertson R, Doe CQ (2003) Dlg, Scrib and Lgl regulate neuroblast cell size and mitotic spindle asymmetry. Nat Cell Biol 5:166–170

Audhya A, Hyndman F, McLeod IX et al (2005) A complex containing the Sm protein CAR-1 and the RNA helicase CGH-1 is required for embryonic cytokinesis in *Caenorhabditis elegans*. J Cell Biol 171:267–279

Bauer DV, Huang S, Moody SA (1994) The cleavage stage origin of Spemann's organizer: analysis of the movements of blastomere clones before and during gastrulation in Xenopus. Development 120:1179–1189

Beckhelling C, Pérez-Mongiovi D, Houliston E (2000) Localised MPF regulation in eggs. Biol Cell 92:245–253

Black SD, Gerhart JC (1985) Experimental control of the site of embryonic axis formation in *Xenopus laevis* eggs centrifuged before first cleavage. Dev Biol 108:310–324

Bluemink JG, de Laat SW (1973) New membrane formation during cytokinesis in normal and cytochalasin B-treated eggs of *Xenopus laevis*. I. Electron microscope observations. J Cell Biol 59:89–108

Byers TJ, Armstrong PB (1986) Membrane protein redistribution during Xenopus first cleavage. J Cell Biol 102:2176–2184

Carron C, Shi DL (2016) Specification of anteroposterior axis by combinatorial signaling during Xenopus development. Wiley Interdiscip Rev Dev Biol 5:150–168

Chalmers AD, Strauss B, Papalopulu N (2003) Oriented cell divisions asymmetrically segregate aPKC and generate cell fate diversity in the early Xenopus embryo. Development 130:2657–2668

Chang JB, Ferrell JE Jr (2013) Mitotic trigger waves and the spatial coordination of the Xenopus cell cycle. Nature 500:603–607

Danilchik MV, Black SD (1988) The first cleavage plane and the embryonic axis are determined by separate mechanisms in *Xenopus laevis*. I. Independence in undisturbed embryos. Dev Biol 128:58–64

Danilchik MV, Denegre JM (1991) Deep cytoplasmic rearrangements during early development in *Xenopus laevis*. Development 111:845–856

Danilchik MV, Gerhart JC (1987) Differentiation of the animal-vegetal axis in *Xenopus laevis* oocytes. I. Polarized intracellular translocation of platelets establishes the yolk gradient. Dev Biol 122(1):101–112

Danilchik MV, Funk WC, Brown EE, Larkin K (1998) Requirement for microtubules in new membrane formation during cytokinesis of Xenopus embryos. Dev Biol 194:47–60. doi:10.1006/dbio.1997.8815

Danilchik MV, Bedrick SD, Brown EE, Ray K (2003) Furrow microtubules and localized exocytosis in cleaving *Xenopus laevis* embryos. J Cell Sci 116:273–283

Danilchik M, Williams M, Brown E (2013) Blastocoel-spanning filopodia in cleavage-stage *Xenopus laevis*: potential roles in morphogen distribution and detection. Dev Biol 382:70–81

De Domenico E, Owens ND, Grant IM, Gomes-Faria R, Gilchrist MJ (2015) Molecular asymmetry in the 8-cell stage Xenopus tropicalis embryo described by single blastomere transcript sequencing. Dev Biol 408:252–268

de Laat WS, Bluemink JG (1974) New membrane formation during cytokinesis in normal and cytochalasin B-treated eggs of *Xenopus laevis*. II. Electrophysiological observations. J Cell Biol 60:529–540

Elinson RP (1980) The amphibian egg cortex in fertilization and early development. In: The cell surface: mediator of developmental processes, pp 217–234

Elinson RP, Rowning B (1988) A transient array of parallel microtubules in frog eggs: potential tracks for a cytoplasmic rotation that specifies the dorso-ventral axis. Dev Biol 128:185–197

Fesenko I, Kurth T, Sheth B, Fleming TP, Citi S, Hausen P (2000) Tight junction biogenesis in the early Xenopus embryo. Mech Dev 96:51–65

Founounou N, Loyer N, Le Borgne R (2013) Septins regulate the contractility of the actomyosin ring to enable adherens junction remodeling during cytokinesis of epithelial cells. Dev Cell 24:242–255

Gerhart J, Keller R (1986) Region-specific cell activities in amphibian gastrulation. Annu Rev Cell Biol 2:201–229

Gerhart J, Ubbels G, Black S, Hara K, Kirschner M (1981) A reinvestigation of the role of the gray crescent in axis formation in *Xenopus laevis*. Nature 292:511–517

Gerhart JC et al (1984) Localization and induction in early development of Xenopus. Philos Trans R Soc Lond B Biol Sci 307:319–330

Gerhart J, Danilchik M, Doniach T, Roberts S, Rowning B, Stewart R (1989) Cortical rotation of the Xenopus egg: consequences for the anteroposterior pattern of embryonic dorsal development. Development 107(Suppl):37–51

Grant PA, Herold MB, Moody SA (2013) Blastomere explants to test for cell fate commitment during embryonic development. J Vis Exp. doi:10.3791/4458

Glotzer M (1997) The mechanism and control of cytokinesis. Curr Opin Cell Biol 9:815823

Grill SW, Hyman AA (2005) Spindle positioning by cortical pulling forces. Dev Cell 8:461465

Guillot C, Lecuit T (2013) Adhesion disengagement uncouples intrinsic and extrinsic forces to drive cytokinesis in epithelial tissues. Dev Cell 24:227–241

Hatte G, Tramier M, Prigent C, Tassan JP (2014) Epithelial cell division in the *Xenopus laevis* embryo during gastrulation. Int J Dev Biol 58:775–781

Hausen P, Riebesell M (1991) The early development of *Xenopus laevis*: an altlas of the histology. Springer, New York

Herszterg S, Leibfried A, Bosveld F, Martin C, Bellaiche Y (2013) Interplay between the dividing cell and its neighbors regulates adherens junction formation during cytokinesis in epithelial tissue. Dev Cell 24:256–270

Hertwig O (1893) Ueber den Werth der ersten Furchungszellen fuer die Organbildung des Embryo. Experimentelle Studien am Frosch- und Tritonei. Arch Mikr Anat xlii:662–807

Houliston E, Elinson RP (1991) Evidence for the involvement of microtubules, ER, and kinesin in the cortical rotation of fertilized frog eggs. J Cell Biol 114:1017–1028

Houston DW (2012) Cortical rotation and messenger RNA localization in Xenopus axis formation. Wiley Interdiscip Rev Dev Biol 1:371–388

Ibanez E, Albertini DF, Overstrom EW (2005) Effect of genetic background and activating stimulus on the timing of meiotic cell cycle progression in parthenogenetically activated mouse oocytes. Reproduction 129:27–38

Jinguji Y, Ishikawa H (1992) Electron microscopic observations on the maintenance of the tight junction during cell division in the epithelium of the mouse small intestine. Cell Struct Funct 17:27–37

Kageura H (1997) Activation of dorsal development by contact between the cortical dorsal determinant and the equatorial core cytoplasm in eggs of *Xenopus laevis*. Development 124:1543–1551

Kalt MR (1971a) The relationship between cleavage and blastocoel formation in *Xenopus laevis*. I. Light microscopic observations. J Embryol Exp Morphol 26:37–49

Kalt MR (1971b) The relationship between cleavage and blastocoel formation in *Xenopus laevis*. II. Electron microscopic observations. J Embryol Exp Morphol 26:51–66

Keller R (2002) Shaping the vertebrate body plan by polarized embryonic cell movements. Science 298:1950–1954

Kikkawa M, Takano K, Shinagawa A (1996) Location and behavior of dorsal determinants during first cell cycle in Xenopus eggs. Development 122:3687–3696

King ML, Zhou Y, Bubunenko M (1999) Polarizing genetic information in the egg: RNA localization in the frog oocyte. Bioessays 21:546–557

Klein SL (1987) The first cleavage furrow demarcates the dorsal-ventral axis in Xenopus embryos. Dev Biol 120:299–304

Kloc M, Etkin LD (1994) Delocalization of Vg1 mRNA from the vegetal cortex in Xenopus oocytes after destruction of Xlsirt RNA. Science 265:1101–1103

Larabell CA, Torres M, Rowning BA, Yost C, Miller JR, Wu M, Kimelman D, Moon RT (1997) Establishment of the dorso-ventral axis in Xenopus embryos is presaged by early asymmetries in beta-catenin that are modulated by the Wnt signaling pathway. J Cell Biol 136:1123–1136

Le Page Y, Chartrain I, Badouel C, Tassan JP (2011) A functional analysis of MELK in cell division reveals a transition in the mode of cytokinesis during Xenopus development. J Cell Sci 124:958–968

Lecuit T, Wieschaus E (2000) Polarized insertion of new membrane from a cytoplasmic reservoir during cleavage of the Drosophila embryo. J Cell Biol 150:849–860

Maddox AS, Lewellyn L, Desai A, Oegema K (2007) Anillin and the septins promote asymmetric ingression of the cytokinetic furrow. Dev Cell 12:827–835

Marrari Y, Rouvière C, Houliston E (2004) Complementary roles for dynein and kinesins in the Xenopus egg cortical rotation. Dev Biol 271:38–48

Martin-Belmonte F, Perez-Moreno M (2011) Epithelial cell polarity, stem cells and cancer. Nat Rev Cancer 12:23–38

Masho R (1990) Close correlation between the first cleavage plane and the body axis in early Xenopus embryos. Dev Growth Differ 32:57–64

Merzdorf CS, Chen YH, Goodenough DA (1998) Formation of functional tight junctions in Xenopus embryos. Dev Biol 195:187–203

Mitchison TJ, Ishihara K, Nguyen P, Wühr M (2015) Size scaling of microtubule assemblies in early Xenopus embryos. Cold Spring Harb Perspect Biol 7:a019182

Moody SA, Kline MJ (1990) Segregation of fate during cleavage of frog (*Xenopus laevis*) blastomeres. Anat Embryol (Berl) 182:347–362

Morais-de-Sá E, Sunkel C (2013) Adherens junctions determine the apical position of the midbody during follicular epithelial cell division. EMBO Rep 14:696–703

Mowry KL, Cote CA (1999) RNA sorting in Xenopus oocytes and embryos. Faseb J 13:435–445

Nieuwkoop PD, Faber J (1967) Normal table of *Xenopus laevis* (Daudin). North-Holland, Amsterdam

Pérez-Mongiovi D, Chang P, Houliston E (1998) A propagated wave of MPF activation accompanies surface contraction waves at first mitosis in Xenopus. J Cell Sci 111(Pt 3):385–393

Prodon F, Chenevert J, Hebras C et al (2010) Dual mechanism controls asymmetric spindle position in ascidian germ cell precursors. Development 137:2011–2021

Rankin S, Kirschner MW (1997) The surface contraction waves of Xenopus eggs reflect the metachronous cell-cycle state of the cytoplasm. Curr Biol 7:451–454

Reinsch S, Karsenti E (1994) Orientation of spindle axis and distribution of plasma membrane proteins during cell division in polarized MDCKII cells. J Cell Biol 126:1509–1526

Sakai M (1996) The vegetal determinants required for the Spemann organizer move equatorially during the first cell cycle. Development 122:2207–2214

Scharf SR, Gerhart JC (1980) Determination of the dorsal-ventral axis in eggs of *Xenopus laevis*: complete rescue of uv-impaired eggs by oblique orientation before first cleavage. Dev Biol 79:181–198

Schneider S, Steinbeisser H, Warga RM, Hausen P (1996) Beta-catenin translocation into nuclei demarcates the dorsalizing centers in frog and fish embryos. Mech Dev 57:191–198

Schroeder MM, Gard DL (1992) Organization and regulation of cortical microtubules during the first cell cycle of Xenopus eggs. Development 114:699–709

Souza KA, Black SD, Wassersug RJ (1995) Amphibian development in the virtual absence of gravity. Proc Natl Acad Sci U S A 92:1975–1978

Stewart-savage J, Grey RD (1982) The temporal and spatial relationships between cortical contraction, sperm trail formation and pronuclear migration in fertilized Xenopus eggs. Wilhelm Rouxs Arch Dev Biol 191:241–245

Ubbels GA, Hara K, Koster CH, Kirschner MW (1983) Evidence for a functional role of the cytoskeleton in determination of the dorsoventral axis in *Xenopus laevis* eggs. J Embryol Exp Morphol 77:15–37

Vincent JP, Gerhart JC (1987) Subcortical rotation in Xenopus eggs: an early step in embryonic axis specification. Dev Biol 123:526–539

Vincent JP, Oster GF, Gerhart JC (1986) Kinematics of gray crescent formation in Xenopus eggs: the displacement of subcortical cytoplasm relative to the egg surface. Dev Biol 113:484–500

Wang Q, Racowsky C, Deng M (2011) Mechanism of the chromosome-induced polar body extrusion in mouse eggs. Cell Div 6:17

Wühr M, Dumont S, Groen AC, Needleman DJ, Mitchison TJ (2009) How does a millimeter-sized cell find its center? Cell Cycle 8:1115–1121

Wühr M, Tan ES, Parker SK, Detrich HW 3rd, Mitchison TJ (2010) A model for cleavage plane determination in early amphibian and fish embryos. Curr Biol 20:2040–2045

Chapter 12
Asymmetric and Unequal Cell Divisions in Ascidian Embryos

Takefumi Negishi and Hiroki Nishida

Abstract Asymmetric cell division during embryogenesis contributes to cell diversity by generating daughter cells that adopt distinct developmental fates. In this chapter, we summarize current knowledge of three examples of asymmetric cell division occurring in ascidian early embryos: (1) Three successive cell divisions that are asymmetric in terms of cell fate and unequal in cell size in the germline lineage at the embryo posterior pole. A subcellular structure, the centrosome-attracting body (CAB), and maternal *PEM* mRNAs localized within it control both the positioning of the cell division planes and segregation of the germ cell fates. (2) Asymmetric cell divisions involving endoderm and mesoderm germ layer separation. Asymmetric partitioning of zygotically expressed mRNA for Not, a homeodomain transcription factor, promotes the mesoderm fate and suppresses the endoderm fate. This asymmetric partitioning is mediated by transient nuclear migration toward the mesodermal pole of the mother cell, where the mRNA is delivered. In this case, there is no special regulation of cleavage plane orientation. (3) Asymmetric cell divisions in the marginal region of the vegetal hemisphere. The directed extracellular FGF and ephrin signals polarize the mother cells, inducing distinct fates in a pair of daughter cells (nerve versus notochord and mesenchyme versus muscle). The directions of cell division are regulated and oriented but independently of FGF and ephrin signaling. In these examples, polarization of the mother cells is facilitated by localized maternal factors, by delivery of transcripts from the nucleus to one pole of each cell, and by directed extracellular signals. Two cellular processes—asymmetric fate allocation and orientation of the cell division plane—are coupled by a single factor in the first example, but these processes are regulated independently in the third example. Thus, various modes of asymmetric

T. Negishi
Division of Morphogenesis, National Institute for Basic Biology, Nishigonaka 38, Myodaiji, Okazaki, Aichi 444-8585, Japan
e-mail: negishi0525@gmail.com

H. Nishida (✉)
Department of Biological Sciences, Graduate School of Science, Osaka University, 1-1 Machikaneyama-Cho, Toyonaka, Osaka 560-0043, Japan
e-mail: hnishida@bio.sci.osaka-u.ac.jp

J.-P. Tassan, J.Z. Kubiak (eds.), *Asymmetric Cell Division in Development, Differentiation and Cancer*, Results and Problems in Cell Differentiation 61,
DOI 10.1007/978-3-319-53150-2_12

cell division operate even at the early developmental stages in this single type of organism.

12.1 Ascidians as Model Organisms for Studies of Asymmetric Cell Division

When one mother cell divides asymmetrically to generate two daughter cells with distinct developmental fates, the mother cell is polarized before the mother cell starts to divide in order to generate distinct daughters after the division. Then, the orientation of the division plane must be properly regulated, so that the division plane is perpendicular to the direction of the preestablished polarization of the mother cell. In many cases of asymmetric cell division, these two cellular processes (orienting cell polarization and spindle) are harmonized by common basic factors, although these processes may not necessarily be coupled, especially in early embryos. Thus, in this chapter, we use the term "asymmetric cell division" to refer to cell division that produces two different daughters from the time they are generated. Polarization of mother cells is directed by asymmetrically distributed cell-intrinsic factors or cell-extrinsic signaling molecules (Chen et al. 2016).

Ascidians are globally distributed marine invertebrates belonging to the subphylum Tunicata (Urochordata), constituting a sister group to vertebrates in the phylum Chordata (Delsuc et al. 2006). Ascidians spawn enormous numbers of eggs and exhibit an invariant embryonic cleavage pattern, which is also shared among diverse ascidian species, have long facilitated making them an ideal organism for investigations of embryogenesis up to the hatching tadpole larva stage (Fig. 12.1a) (Chabry 1887; Conklin 1905). Ascidian tadpole larvae possess the basic body plan of chordates, having an axial mesoderm including muscles and notochord, a dorsal central nerve system, and a brain. The embryonic cleavage pattern of the solitary ascidian, *Styela partita*, was documented for the first time in amazing detail by Conklin, and subsequent studies using modern techniques have reconfirmed Conklin's descriptions and traced the embryonic cell lineages (Nishida and Satoh 1983, 1985; Nishida 1986, 1987; Stach and Anselmi 2015).

The cleavage pattern of ascidian embryo is bilaterally symmetrical but not simple (Figs. 12.1 and 12.2a, b). The cell fates of most blastomeres are restricted to give rise to a single type of tissue by the 110-cell stage (after seven rounds of initial cell divisions) (Fig. 12.2a) (Nishida 1987; Kumano and Nishida 2007). The ascidian fate map (Fig. 12.2b) shows similarity to that of the frog in terms of the geographical topology of the tissue precursor cell territories (Lemaire et al. 2008). Both the cleavage pattern and fate map are highly conserved among ascidian species, which have become phylogenetically diversified (Hudson and Yasuo 2008; Lemaire 2009; Tsagkogeorga et al. 2009). We have confirmed that species spanning seven genera [*Styela* (Conklin 1905), *Halocynthia* (Nishida 1987), *Ciona* (Conklin 1905), *Phallusia* (Zalokar and Sardet 1984), *Ascidia* (H. Nishida, unpublished observation), and *Boltenia* (http://gvondassow.com/Research_Site/

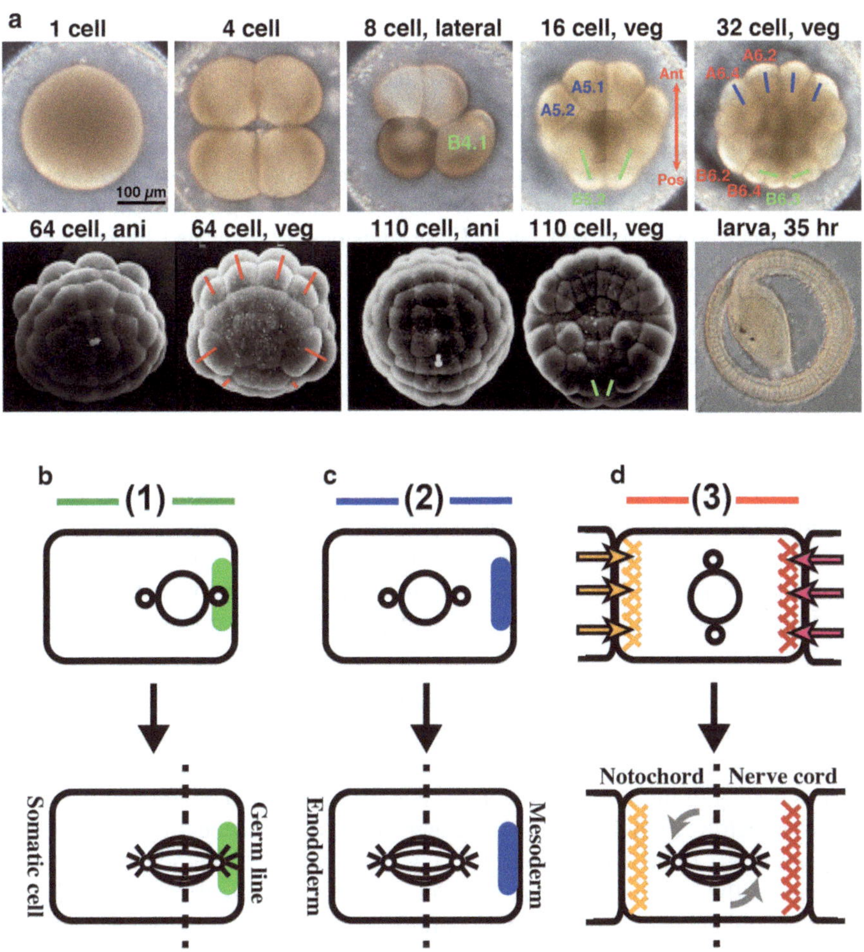

Fig. 12.1 Ascidian eggs develop into tadpole larvae through a relatively simple process of development involving a small number of constituent cells. (**a**) The photos show live and scanning electron microscopy images of *Halocynthia roretzi* embryos at various stages from the fertilized egg to the larva (35 h of development). The larvae exhibit the basic body plan of chordates, having a dorsal neural tube, a notochord flanked by bilateral muscle, and a brain with two sensory pigment cells. The notochord, a characteristic feature of chordates, is visible in the tail of the larva, consisting of exactly 40 disc-shaped cells arranged in a single line. The total number of cells in the hatched larva, 1.5 mm in length, is approximately 3000. The lines connecting the blastomeres show the directions of asymmetric cell division relevant to the three subheadings of this chapter. In 16- to 110-cell embryos, anterior is up. *Green bars* indicate three successive unequal and symmetric cell divisions resulting in generation of small primordial germ cells at the posterior pole. *Blue bars* indicate asymmetric divisions that segregate the endoderm and mesoderm cell fates at the 32-cell stage. *Red bars* correspond to asymmetric cell divisions that separate the notochord/nerve code fates in the anterior half and mesenchyme/muscle fates in the posterior half at the 64-cell stage. Names of some blastomeres are indicated. (**b–d**) Schematic representation of three types of asymmetric cell division that are described in this review. See details in text. (**b**) Asymmetric division in germ cell lineage. It corresponds to cell divisions designated by the *green*

Picture_of_the_week/Entries/2010/5/9_Embryogenesis_in_the_ascidian_Boltenia.html) and *Corella* (http://gvondassow.com/Research_Site/Video_-_Corella_early_cleavage.html)] share the common cleavage pattern. This implies that the cell division patterns in ascidian embryos are regulated through evolutionarily robust mechanisms.

In order to generate primordial tissue precursor cells during seven rounds of cell division after fertilization, many cell divisions take place to generate daughter cells with distinct prospective cell fates. These divisions are indicated by red bars in Fig. 12.2a. These red bars represent one out of a total of one at the 4-cell stage in the bilateral half, two out of two at the 8-cell stage, three out of four at the 16-cell stage, seven out of eight at the 32-cell stage, seven out of 16 at the 64-cell stage, and seven out of 23 at the 110-cell stage. However, these do not always represent asymmetric cell divisions because cell fates may be determined after the mother cell has divided. In asymmetric cell division, there should be evidence to indicate that the mother cell has already become polarized before division begins. In this chapter, we summarize current knowledge of three types of genuine asymmetric division occurring in ascidian early embryos as outlined below. Hereafter, we refer to the mitotic division responsible for production of daughter cells with different developmental potential as "asymmetric" cell division and that produce daughter cells differing in size through eccentric positioning of the division plane as "unequal" cell division.

1. Three successive asymmetric and unequal cell divisions at the posterior pole (green letters and bars in Fig. 12.1a, b). A subcellular structure, the centrosome-attracting body (CAB), and maternally localized mRNAs control both the positioning of the cell division planes and the segregation of cell fates.
2. Asymmetric cell divisions involving segregation of the endoderm and mesoderm germ layers (blue letters and bars in Fig. 12.1a, c). Asymmetric partitioning of zygotically expressed mRNA for Not, a homeodomain transcription factor, promotes the mesoderm fate and suppresses the endoderm fate. This asymmetric partitioning is mediated by nuclear migration toward the mesodermal pole within the mother cell. In this case, there is no special regulation of cleavage plane positioning, as the cell divides into daughters with approximately same size and the cleavage plane positioning simply follows Sach's rule that the new division plane is formed perpendicularly to the previous one.

Fig. 12.1 (continued) bars in (**a**). *Large circle* is nucleus, and *small circle* is centrosome. *Green oval* represents the centrosome-attracting body (*CAB*). *Broken line* indicates the next cell division plane. (**c**) Separation of endoderm and mesoderm. It corresponds to the *blue bars* in (**a**). *Blue oval* represents the asymmetrically localized mRNA of Not transcription factor in *Halocynthia.* (**d**) Asymmetric cell divisions in the marginal region of the vegetal hemisphere. It corresponds to the *red bars* in (**a**). *Yellow arrows* and *crisscrosses* indicate FGF signaling on one side. *Red arrows* and *crisscrosses* indicate Ephrin antagonizing signaling on the opposite side. Veg, vegetal view. Ani, animal view. Ant, anterior. Pos, posterior. SEM images have been reproduced from Nishida (1986) with permission

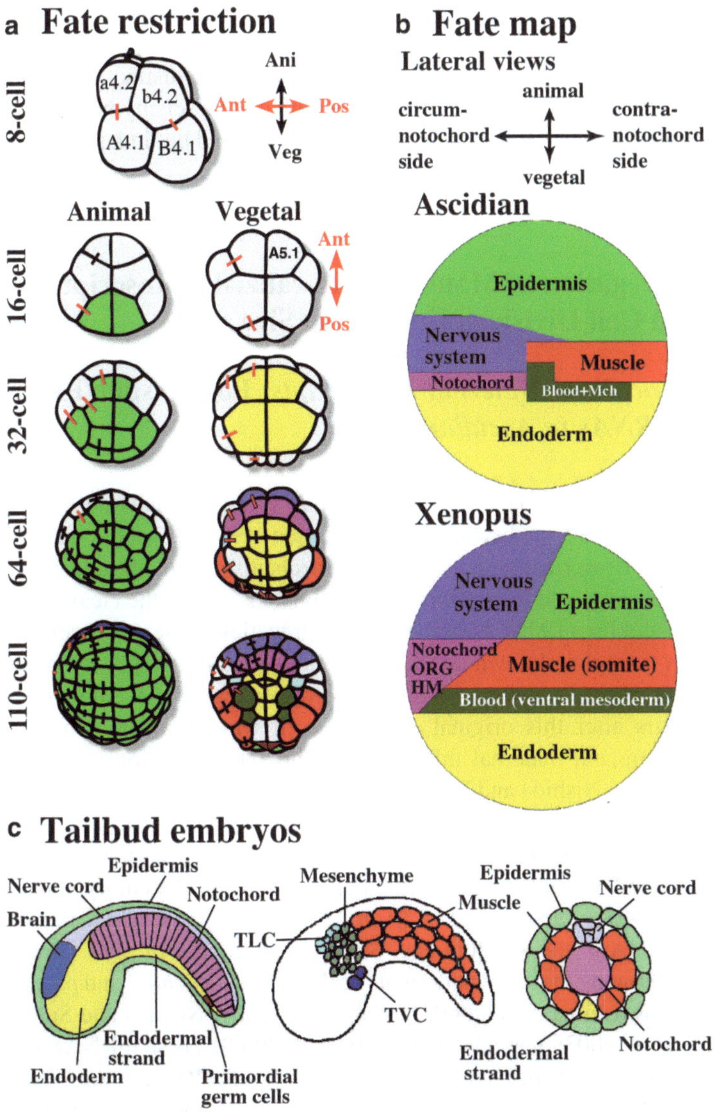

Fig. 12.2 Developmental fates of blastomeres of the ascidian embryo. Orientation of the each drawing is indicated. (**a**) Cell fate restriction during the cleavage stages. Blastomeres are colored when their fate is restricted to a single cell type. The colors correspond to those of the larval tissues indicated in (**c**). Fate restriction in ascidian early embryos proceeds quickly. Sister blastomeres are connected by bars. *Red bars* indicate that the prospective cell fates of the two sister blastomeres are distinct. The fate map is bilaterally symmetric. (**b**) Schematic fate maps of ascidians and *Xenopus* blastulae. Lateral views. Circum-notochord side is to the *left*. Note the topographic similarity of the presumptive tissue territories in the two fate maps. ORG, organizer; HM, head mesoderm; Mch, mesenchyme (precursors of adult tunic cells). (**c**) Organization of tailbud embryos. Mid-sagittal planes, sagittal planes, and transverse sections of the tail. TLC, trunk lateral cells (precursors of adult blood and body wall muscle); TVC, trunk ventral cells (precursors of adult heart and body wall muscle). Drawings have been reproduced with permission from Nishida (2005) and Lemaire et al. (2008)

3. Asymmetric cell divisions in the marginal region of the vegetal hemisphere (red letters and bars in Fig. 12.1a, d). The directed extracellular FGF and antagonizing ephrin signals polarize the mother cells, inducing distinct fates in a pair of daughter blastomeres (nerve versus notochord and mesenchyme versus muscle). The direction of cell division is regulated and oriented but independently of FGF and ephrin signaling.

12.2 Segregation of Maternally Localized Transcripts and Cell Divisions Unequal in Size

12.2.1 *Localized Maternal Transcripts:* Postplasmic/PEM *mRNAs in Ascidians*

Studies using many animals have revealed that localized maternal factors initiate establishment of the embryonic axes (Gurdon 1992; Johnston and Nüsslein-Volhard 1992; Bowerman et al. 1993). This mechanism would be a likely candidate for the process involved in asymmetric cell divisions during the cleavage stages. In ascidians, Conklin (1905) described the segregation pattern of yellow-colored ooplasm in a region known as the posterior–vegetal cytoplasm/cortex (PVC; myoplasm in Conklin's description) of fertilized eggs and embryos. About a hundred years after this original description, it was discovered that a genuine muscle determinant, maternal mRNA of macho-1 transcription factor, is present in the myoplasm (Nishida and Sawada 2001). *macho-1* is a member of the so-called *postplasmic/PEM* mRNAs.

In 1996, Yoshida et al. first reported that in *C. savignyi* a maternal mRNA was localized to the PVC of the egg and the posterior pole of the embryo, and the gene was named "*posterior end mark*" (*pem*). *pem* is the most abundant maternally localized mRNA in the ascidian egg. Since the discovery of *pem*, approximately 50 mRNAs showing the same pattern of localization as *macho-1* and *pem* have been identified (Kawashima et al. 2000; Makabe et al. 2001; Nishida and Sawada 2001; Nakamura et al. 2003; Yamada 2006; Paix et al. 2009). These mRNAs are referred to as *postplasmic/PEM* mRNAs (*macho-1*, *pem*, *Wnt5*, *POPK-1*, etc.) (reviewed in Sardet et al. 2005; Prodon et al. 2007; the most recently updated list of *postplasmic/PEM* RNAs is available in Makabe and Nishida 2012). They are first concentrated at the vegetal pole and then move to the posterior pole after fertilization (Fig. 12.3, 1st and 2nd phases), later becoming further concentrated into the centrosome-attracting body (CAB) at the posterior pole during the cleavage stages, as mentioned later (Fig. 12.3, the 2- to 110-cell embryos). The results of various types of screening, including large-scale in situ hybridization and microarray, suggest that the localization pattern shared by *postplasmic/PEM* RNAs is the sole pattern of localization of maternal mRNAs (Makabe et al. 2001; Yamada et al. 2005). There

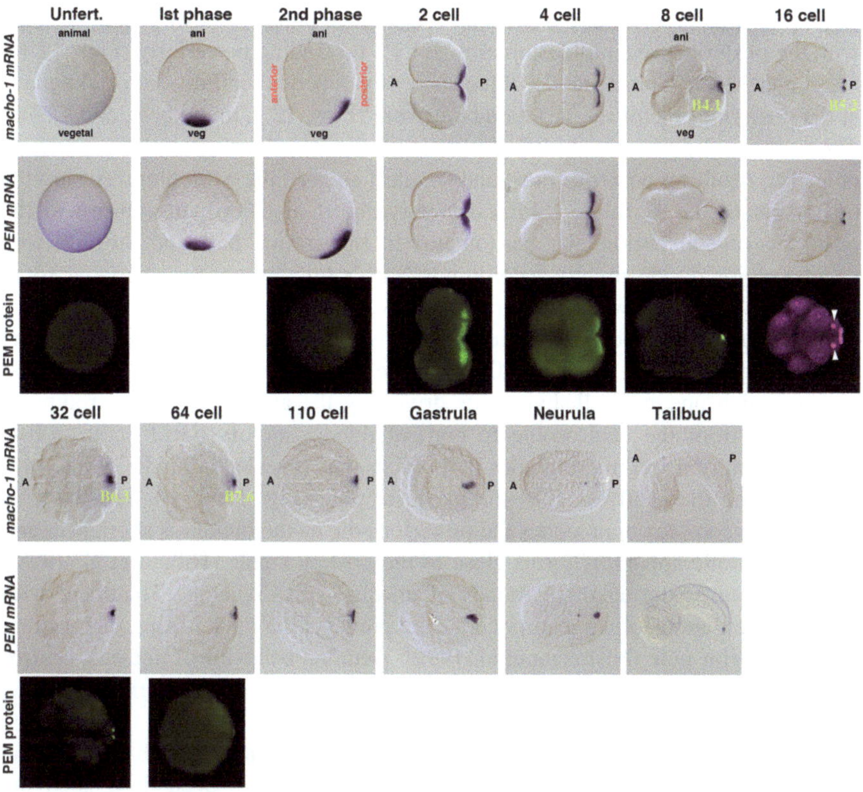

Fig. 12.3 Localization of two representative *postplasmic/PEM* mRNAs, *PEM*, and *macho-1*, during early embryogenesis. PEM protein localization is also shown. In the 16-cell-stage embryos, PEM protein is detected with higher sensitivity, revealing the presence of PEM protein in nuclei of the posterior blastomeres in addition to localization of the CAB (*arrowheads*). Egg and 8-cell embryos are lateral views. The 2-, 4-, 32- to neurula-stage embryos are vegetal views. Names of posterior-most blastomeres which inherit the *postplasmic/PEM* RNAs are indicated. Unfert, unfertilized egg. Ani, animal pole. Veg, vegetal pole. A, anterior. P, posterior. Images have been reproduced from Negishi et al. (2007) and Kumano et al. (2011) with permission

have been no reports of maternal RNA localized to the animal and anterior regions of the egg.

12.2.2 Successive Unequal Cell Divisions

Up to the 8-cell stage, the fertilized egg is cleaved equally in accordance with Sach's rule that the new division plane is formed perpendicularly to the previous one (Strome 1993). At the end of the four-cell stage, the third cleavage planes intersect the animal–vegetal axis, although the posterior cleavage planes are slightly tilted. As a result, the ascidian 8-cell-stage embryo shows a unique shape with the posterior–vegetal blastomeres (B4.1[1]) protruding posteriorly (Figs. 12.1a, 12.2a and 12.4a). Subsequently, three successive unequal cell divisions take place at the posterior pole of the vegetal hemisphere (Figs. 12.1, green bars, and 12.4a). These unequal divisions always generate smaller daughter cells at the posterior pole. The resulting posterior most and smallest cells of the 64-cell embryo are the precursors of the germline (Shirae-Kurabayashi et al. 2006). In contrast to the posterior–vegetal region, the blastomeres in the animal half and the anterior–vegetal region divide equally according to Sach's rule, except for a certain case at the sixth cleavage, as will be described later.

One strategy for achieving unequal cell division is eccentric placement of the mitotic spindle (Siller and Doe 2009). In the one-cell zygote of *Caenorhabditis elegans*, such unequal cell division is driven by the attachment of microtubules emanating from the centrosome to cortical dynein motors (Grill et al. 2001). Investigation of astral microtubules in ascidians has revealed a bundle of microtubules linking the centrosome to the posterior pole (Fig. 12.4b, c). The nucleus/centrosome complex approaches the posterior pole as the bundle is shortened, and then the mitotic apparatus forms close to the posterior pole (Hibino et al. 1998). This is repeated during three rounds of unequal cell divisions. Treatment with microtubule dissociation agent, nocodazole, prevents the nuclei from approaching to the posterior pole (Nishikata et al. 1999). Detailed behavior of nucleus, centrosome, and spindle during the B5.2 unequal cell division has been reported in Prodon et al. (2010).

In cytoplasm-extracted embryos, a brilliant structure has been observed at the posterior pole under light microscope. This structure has been designated the centrosome-attracting body (CAB) (Fig. 12.4b, d) (Hibino et al. 1998). The CAB at the posterior cortex is always inherited by the smaller blastomere at each division. The CAB is not present in the egg, and it is first recognized as a group of small dots with high refraction index at the late two-cell stage (Hibino et al. 1998). The location of this subcellular structure corresponds exactly to that of the *postplasmic*/*PEM* mRNAs (Yoshida et al. 1996; Hibino et al. 1998; Nakamura et al. 2003). Microsurgical removal and transplantation of the PVC of eggs have shown that the PVC is responsible for formation of the CAB and unequal cleavages (Nishikata et al. 1999).

Observations of the CAB using transmission electron microscopy (TEM) and isolation of the cell cortex have revealed that the CAB is characterized by an electron-dense matrix/granule (EDM) that resembles the germ plasm of other animals (Fig. 12.4d) and a network of cortical rough endoplasmic reticulum (cER) (Iseto and Nishida 1999; Sardet et al. 2003). Some *postplasmic*/*PEM*

[1]Nomenclature of ascidian blastomeres. Each blastomere is assigned its name such as B4.1, A5.2, and b5.6. Capital letters are for the cells of the vegetal hemisphere. Lower case letters are for the animal hemisphere cells. A and a for the anterior blastomere. B and b for the posterior blastomeres (see Fig. 12.2a, 8-cell embryo). First digit represents the generation. Second digit is a unique number based on its relative position after cell division.

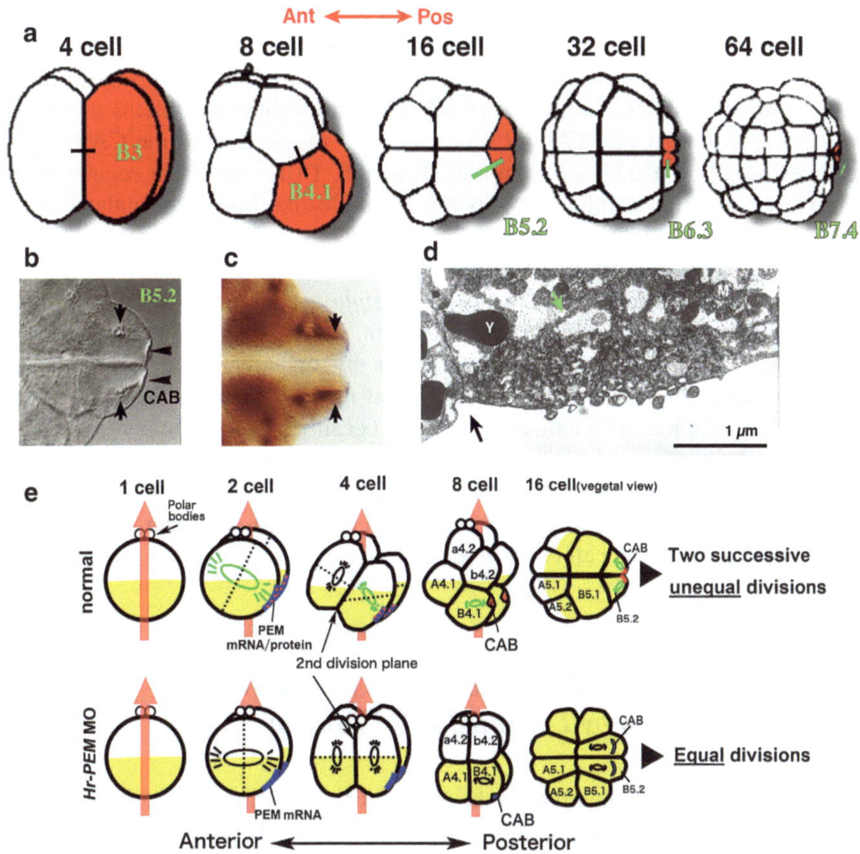

Fig. 12.4 Unequal cell divisions taking place at the posterior pole. (**a**) *Red* blastomeres divide unequally in size from the 8-cell to 64-cell stages, always generating smaller cells at the posterior pole. *Green bars* indicate three successive unequal cell divisions resulting in generation of small primordial germ cells at the posterior pole. The 4- and 8-cell embryos are lateral views. The 16- to 64-stage embryos are vegetal views. Name of relevant cells are indicated in *green letters*. Ant, anterior. Pos, posterior. (**b**) Posterior part of the extracted 16-cell embryo. These cells have divided unequally and are about to divide unequally again. The CAB is present at the posterior pole (*arrowheads*). Positions of the nuclei are indicated by *arrows*. (**c**) A bundle of microtubules (*arrows*) connects the posterior centrosome to the CAB. The bundle is shortening to pull the nucleus toward the CAB. (**d**) Ultrastructure of the CAB. The CAB is characterized by an electron-dense matrix, which is considered to be the putative germ plasm. *Black arrow* indicates the midline of the embryo. *Green arrow* indicates a microtubule. Y, yolk granule. M, mitochondrion. (**e**) Model for control of orientation and positioning of the cell division plane by the PEM. (*Upper*) In a normal embryo at the two-cell stage, the centrosome axis rotates as the posterior centrosome is attracted toward the posterior–vegetal cortex (*blue cortex*), where PEM protein is localized (*red dots* on *blue cortex*), and a tilted spindle (*green*) forms. The second division plane is oblique to the A–V axis (*large red vertical arrow*). *Broken lines* indicate the planes of the next division. At the 4-cell stage, similar events occur again, only in the posterior blastomeres. Therefore, the B4.1 blastomere protrudes posteriorly at the 8-cell stage. *Yellow color* shows the vegetal half of the cytoplasm. In the B4.1 blastomeres, the centrosome/nucleus complex approaches the CAB, and the blastomere divides unequally. At the 16-cell stage, a smaller B5.2 blastomere forms at the posterior pole. This posterior-most blastomere undergoes two further unequal cleavages. (*Lower*) Without PEM protein, the cleavage pattern is quite regular as there is no centrosome-attracting

mRNAs are known to be anchored to the cER and others to granules in the CAB (Paix et al. 2009). Their protein products, including *pem* and *vasa*, are mostly localized to the CAB (Fig. 12.3) (Shirae-Kurabayashi et al. 2006; Negishi et al. 2007; Paix et al. 2009; Kumano et al. 2010). Some additional proteins translated from "non"-*postplasmic/PEM* mRNAs also show localization to the CAB; these include the aPKC–Par3–Par6 complex, β-catenin, Dishevelled, the regulators of translation initiation (MnK-4EBP-S6 K), and Polo-like kinase1 (Patalano et al. 2006; Kawai et al. 2007; Negishi et al. 2011; Paix et al. 2011). Thus, the CAB is a multifunctional subcellular structure that attracts the centrosome, provides a structural scaffold for the localization of *postplasmic/PEM* mRNA and several proteins, and contains the putative germ plasm. The CAB plays a robust role in these asymmetric and unequal cell divisions by organizing the orientation of cleavage planes and asymmetric partitioning of the *postplasmic/PEM* mRNAs. In our knowledge, the CAB is unique to ascidians because in cell-autonomous unequal cell divisions in other animals such as *C. elegans*, *Drosophila*, and sea urchin, there has been no report on visible macroscopic structure like the CAB. In these examples, it seems that microtubules emanating from the centrosome interact with wide area of the cortex of the cell (Grill et al. 2001).

12.2.3 Control of Cell Division Planes and Germ Cell Formation by Localized PEM Protein

What are the molecules involved in the orientation of cell division planes and germ cell specification? *PEM* was the first *postplasmic/PEM* mRNA to be identified and is the most abundant, as mentioned previously. PEM protein is translated and localized in the CAB (Fig. 12.3). When the function of PEM is inhibited with the antisense morpholino oligo (MO), unequal cell divisions are converted to equal ones, and the posterior cleavage pattern becomes similar to the anterior pattern. In PEM-deficient embryos, the astral microtubules do not focus on the CAB, and consequently the cells divide equally (Fig. 12.4e) (Negishi et al. 2007). In addition, PEM-deficient eight-cell embryos show no protrusion of the posterior–vegetal blastomere (Fig. 12.4e). PEM localization also determines the orientation of mitotic planes at the second and third cleavages before the start of successive unequal cell divisions. Observations of the mitotic spindle by immunostaining and live imaging have supported the idea that from the 2- to the 32-cell stages the closest centrosomes are always attracted and pulled toward the cortex where PEM protein is localized (Negishi et al. 2007).

Fig. 12.4 (continued) activity. No specific regulation of the cleavage pattern occurs. Ant, anterior. Post, posterior. Drawings and images have been reproduced from Hibino et al. (1998), Iseto and Nishida (1999), and Negishi et al. (2007) with permission

Similar localization and function of the PEM have been observed in three ascidians, *Halocynthia*, *Phallusia*, and *Ciona*, and are thought to underlie the common cleavage pattern of ascidians (Prodon et al. 2010; Shirae-Kurabayashi et al. 2011). However, ascidian PEM has no known domain except the C-terminal WPRW sequence that is a binding site for groucho, a transcriptional corepressor, and there is no apparent homolog in other deuterostomes. It has shown that aa 258–341 of the PEM of *Halocynthia*, where some conserved amino acid stretches are shared among ascidian species, are required for this function, although the molecular mechanism responsible for pulling of the centrosome toward the CAB is still unclear (Kumano et al. 2011).

Interestingly, a certain amount of PEM protein is translocated into the nuclei of posterior-most cells, which are of germ cell lineage, and mediates transcriptional quiescence in the germline precursor (Kumano et al. 2011; Shirae-Kurabayashi et al. 2011). In the *Halocynthia* embryo, aa 342–426 in the C-terminal region of the PEM, where some conserved amino acid stretches are shared among ascidian species, are crucial for transcriptional quiescence. This region interacts with the P-TEFb complex, resulting in suppression of zygotic transcription (Kumano et al. 2011). Suppression of P-TEFb is a conserved mechanism for germline quiescence by localized germ plasm factors in ascidians, *Drosophila*, and *C. elegans* (Nakamura and Seydoux 2008). Having two distinct domains that mediate cell division and germ cell development, PEM protein ensures that proper asymmetric cell divisions occur by harmonizing both orientation of the cell division planes and germ cell fate specification.

12.3 Asymmetric Cell Divisions to Separate Endoderm and Mesoderm Cell Fates

In most bilaterians, one of the most important steps during early embryogenesis is the formation of three germ layers (Kimelman and Griffin 2000; Rodaway and Patient 2001; Davidson et al. 2008). Since fate restriction in ascidian embryos is accomplished with a small number of constituent cells, some studies have investigated the molecular mechanism responsible for separation of the mesoderm and endoderm lineages at the single-cell level (Takatori et al. 2010, 2015; Hudson et al. 2013). In the 16-cell ascidian embryo, three pairs of vegetal cells (A5.1, A5.2, and B5.1 of the 16-cell embryos in Fig. 12.1) divide asymmetrically into mesoderm daughter cells (A6.2, A6.4, and B6.2) that are located in the marginal zone and endoderm daughter cells (A6.1, A6.3, and B6.1) that are located in the vegetal pole region (Conklin 1905; Nishida 1987) (Fig. 12.1, blue bars in the 32-cell embryos).

In *Halocynthia*, asymmetrically distributed mRNA for *Not*, which encodes a conserved homeodomain transcription factor (Von Dassow et al. 1993; Utsumi et al. 2004), is preferentially partitioned into the daughter cell on the marginal side, where Not promotes the mesodermal fate and suppresses the endoderm fate

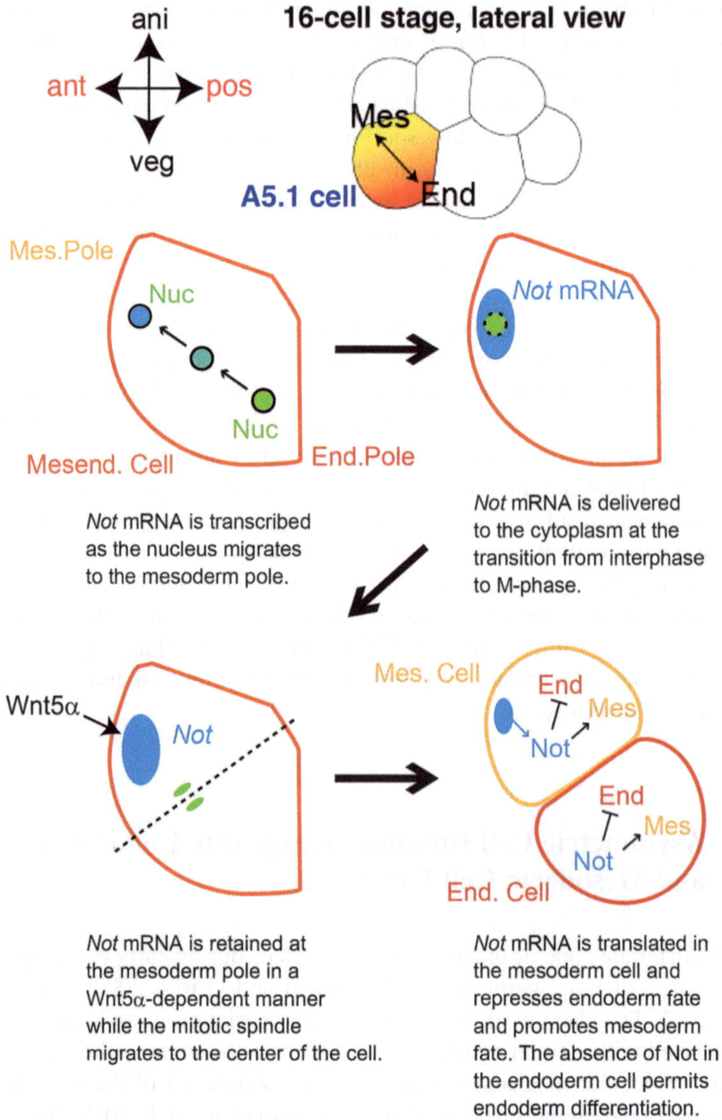

Fig. 12.5 Schematic diagram of *Not* mRNA asymmetric partitioning and mesendoderm fate segregation. Lateral view of the mesendoderm cell (A5.1 cell) from the 16- to 32-cell stage. Animal pole is *up*. *Not* mRNA (*blue*) is transcribed in the nucleus (*green*) as it migrates to the future mesoderm-forming side of the mesendoderm cell. *Not* mRNA is then delivered from the nucleus to the future mesoderm cell cytoplasm at the transition from interphase to M-phase. Wnt5α-dependent mechanism retains *Not* mRNA at the mesoderm pole as the mitotic spindle repositions to the center of the cell (*green*) and subsequent cell division partitions *Not* mRNA to the mesoderm precursor cell. *Dotted line* indicates the plane of the next cell division. After the cell division, Not protein is translated in the mesoderm cell (*yellow*), inhibiting the endoderm fate and promoting the mesoderm fate. Absence of Not in the endoderm (*red*) permits endoderm differentiation. Nuc, nucleus; ani, animal; veg, vegetal; ant, anterior; pos, posterior; Mesend, mesendoderm; Mes, mesoderm; End, endoderm. Drawings have been reproduced with permission from Takatori et al. (2010, 2015)

(Fig. 12.5) (Takatori et al. 2010). *Not* mRNA segregation is coupled to transient migration of the nucleus, which is zygotically transcribing *Not* mRNA, to the future mesodermal region (mesoderm pole) of the mother cell via a microtubule network (Fig. 12.5) (Takatori et al. 2010). After delivering *Not* mRNA to the cytoplasm of the mesoderm pole, the mitotic apparatus returns to the center, and consequently the mRNA is preferentially partitioned into a mesoderm daughter cell. It has been revealed that, subsequently, the direction of displacement of the nucleus within the mother cell is determined by the localized distribution of PtdIns(3,4,5)P_3 (PIP3), a plasma membrane-bound lipid. The region of PIP3 localization is determined by localization of the PIP3-producing enzyme, phosphatidylinositol 3-kinase (PI3K). It is also intriguing that the gradient of PI3K along the animal–vegetal axis is established by ooplasmic movement within 5 min after fertilization (Takatori et al. 2015). These results indicate that, in the ascidian embryo, some processes important for germ layer segregation have already been initiated in the early fertilized egg and that the mesoderm and endoderm fates are separated into two distinct cells following a series of early cleavages.

While the cleavage pattern and fate map are well conserved among the divergent ascidian species, the molecular pathways responsible for specifying cell fates occasionally vary (Hudson and Yasuo 2008; Lemaire 2009). A study using *Ciona* and *Phallusia* has shown that the mechanism responsible for mesoderm versus endoderm fate separation differs from that in *Halocynthia.* Differential nuclear β-catenin activity coupled with two rounds of mitotic divisions along the animal–vegetal axis separates the mesoderm and endoderm fates. The ON-to-OFF sequence of nuclear β-catenin accumulation during the 16- to 32-cell stage is sufficient to induce the mesoderm fate without the involvement of *Not* asymmetric segregation (Hudson et al. 2013), although it is still unclear how nuclear accumulation of β-catenin is lost only in the mesoderm precursor cell at the 32-cell stage. The difference in molecular mechanisms of mesoderm specification between *Halocynthia* and *Ciona*/*Phallusia* provides a good example of the developmental system drift (DSD) (True and Haag 2001; Burgess 2011). DSD is the theory that, despite the strong conservation of developmental processes across closely related species at the morphological level, the underlying molecular mechanisms may be diverse. In this case as well as secondary muscle and secondary notochord fate induction in ascidian embryos (Hudson and Yasuo 2008; Lemaire 2009), homologous cell fates involving homologous cells can nonetheless be specified by completely different molecular mechanisms in *Halocynthia* and *Ciona.* Thus, the constraints on embryo anatomy appear stronger than those on the choice of underlying regulatory molecules. In both cases of *Halocynthia* and *Ciona*/*Phallusia*, there is no special regulation of cleavage plane positioning, as the cleavage plane simply follows Sach's rule and forms perpendicularly to the previous one. Accordingly, depletion of key molecules for fate specification (Not, PI3K, or β-catenin) does not alter the pattern of cell division (Takatori et al. 2010, 2015; Hudson et al. 2013).

12.4 Asymmetric Cell Divisions in the Marginal Zone of the Vegetal Hemisphere

12.4.1 Localized FGF Signaling Polarizes Mother Cells that Divide into Daughter Cells with Distinct Fates

After separation of the mesoderm and endoderm lineages, subsequent asymmetric cell divisions along the animal–vegetal axis occur radially in the mesodermal blastomeres located in the marginal region (Fig. 12.1, red bars in the 64-cell embryo) (Nishida 2005; Kumano and Nishida 2007). In these cell divisions from the 32- to 64-cell stages, the posterior marginal cells (B6.2 and B6.4 cells of the 32-cell embryo in Fig. 12.1, namely, MM cells: mesenchyme/muscle mother blastomeres, shown as red stripes on a white background in Fig. 12.6a, c) divide into the marginal side muscle precursors (colored red in Fig. 12.6b, c) and the vegetal pole side mesenchyme precursors (colored green). Likewise, in the anterior region, marginal cells (A6.2 and B6.4 cells of the 32-cell embryo in Fig. 12.1, namely, NN cells: notochord/nerve cord mother blastomeres, white in Fig. 12.6a, c) divide into the marginal side nerve cord precursors (light purple) and the vegetal pole side notochord precursors (pink). Notochord and mesenchyme are the induced fates, while nerve cord and muscle are the default uninduced fates. In this system, the inducer blastomeres, stage of induction, signaling molecules and intracellular signal transduction, spatial and temporal regulation of cellular competence, and the identities of competence factors have been extensively revealed in the last two decades.

All of these asymmetric cell divisions in the marginal region are mediated by FGF signaling. The mother cells respond to the FGF signal, but the daughter cells are no longer able to respond (Nakatani et al. 1996; Hashimoto et al. 2011). Therefore, the mother cells are polarized by localized FGF signaling (Fig. 12.6, blue arrows) and then divide asymmetrically. The same FGF signal induces notochord in the anterior region and mesenchyme in the posterior region. The difference in response is regulated by the intracellular competence factors, zygotic FoxA and Zic within the anterior mother cells and maternal macho-1 within the posterior mother cells (Fig. 12.6c) (Nishida 2005). Extracellular FGF signaling is transduced through Ras, ERK/MAPK, and ETS transcription factor (Nakatani and Nishida 1997; Kim and Nishida 2001).

In the posterior region, when FGF is applied from both sides of the mother cell, both of the daughter cells assume a mesenchyme fate, whereas abrogation of the FGF signal results in formation of two muscle cells (Fig. 12.6d). Transplantation of ectodermal cells that artificially express FGF and simultaneous suppression of endogenous FGF production can reverse the direction of asymmetric cell division (Kim et al. 2007). This clearly demonstrates that polarity of the muscle/mesenchyme mother cell is determined solely by the direction from which the FGF9/16/20 signal is presented.

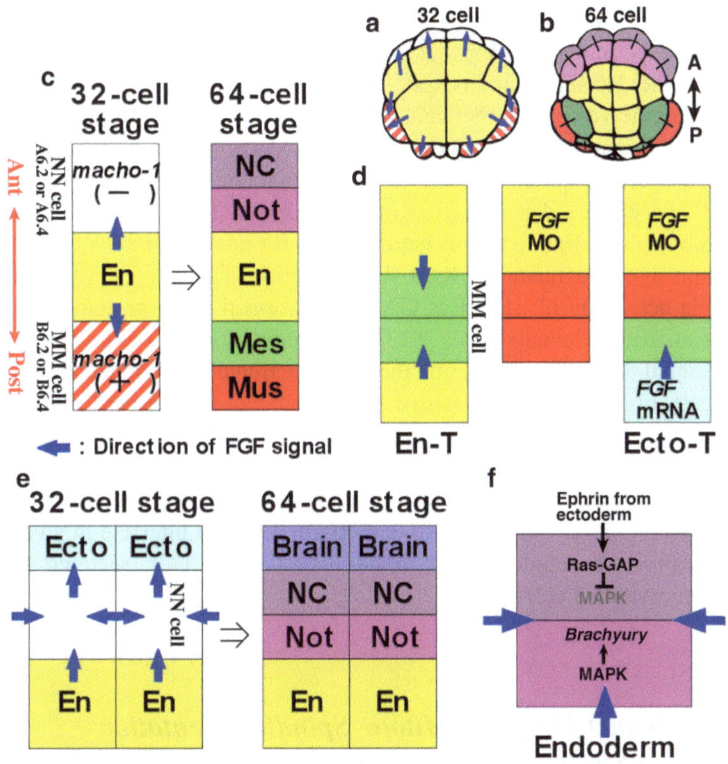

Fig. 12.6 Directed FGF signaling and asymmetric cell divisions during mesenchyme and notochord induction. (**a**) Position of each cell at the 32-cell stage. FGF signaling (*blue arrows*) takes place in the anterior and posterior margins of the vegetal hemisphere. (**b**) Then, asymmetric cell divisions occur to generate sister cells that assume induced and default cell fates. The bars connecting the cells show the directions of asymmetric cell divisions in the anterior and posterior regions. (**c**) Patterning in normal embryos shown as schematic diagrams of the arrangement of cells along the anterior–posterior axis. Cell types are highlighted by the same color code as in (**b**). NC, nerve cord. Not, notochord. En, endoderm. Mes, mesenchyme. Mus, muscle. At the division to the 64-cell stage, FGF-mediated asymmetric divisions take place in the anterior (NC vs. Not) and posterior (Mes vs. Mus) marginal zones. macho-1 is a maternal and intrinsic competence factor for mesenchyme induction. (**d**) Experimental results confirming that the direction of asymmetry for segregation of the muscle and mesenchyme fates is determined by the direction from which the FGF signal comes. FGF MO, FGF morpholino antisense oligo-injected endoderm cell. En-T, transplanted endoderm cell. Ecto-T transplanted ectoderm cell that has been injected with *FGF* mRNA. (**e**) In normal embryos, FGF is also expressed in the nerve cord/notochord precursor to induce brain in the ectoderm (Ect) marginal zone. (**f**) Polarity of asymmetric division that produces nerve cord and notochord precursors is determined by the direction from which the inhibitory ephrin signal comes. The ephrin signal from ectoderm inhibits activation of MAPK by FGF signaling via activation of Ras-GAP on the nerve cord side and consequently results in the inhibition of the notochord-specific *Brachyury* gene. Drawings have been reproduced with permission from Nishida (2002) and Kim et al. (2007)

Similarly in the anterior marginal region, the daughter cell in which ERK/MAPK is activated adopts a notochord fate, whereas the other adopts a nerve cord fate. However, in this system, the mechanism is more complex because the notochord/nerve cord mother cells themselves express FGF in order to induce a brain fate (dark purple in Fig. 12.6e) in the adjacent ectoderm region. Polarization of the mother cells requires not only the FGF signal but also an antagonizing signal from the adjacent ectodermal cell (Kim et al. 2007). This antagonizing action is mediated by ephrin–Eph, which is better known for its roles in axon guidance. The ephrin–Eph signal attenuates ERK/MAPK activation in the nerve cord-fated daughter cell via activation of p120 Ras GAP (GTPase-activating protein) (Fig. 12.6f) (Picco et al. 2007; Haupaix et al. 2013). In combination with a directional antagonizing signal from the ectoderm, the nondirectional FGF signal polarizes the mother cell. In this case, the polarity of the notochord/nerve cord mother cell is determined by the direction from which the ephrin signal is presented. The unique feature of this system is that an ectodermal GPI membrane-bound form of ephrin ligand (Arvanitis and Davy 2008) ensures one-way signaling from the ectoderm cell to the notochord/nerve cord cells, so that ephrin does not interfere in an autocrine manner within the ectoderm with brain induction by FGF secreted from the notochord/nerve cord cells.

12.4.2 Distinct Modes of Mitotic Spindle Orientation in a Sister Cell Pair for Proper Allocation of Blastomeres

The abovementioned studies of signaling for asymmetric cell fate allocation have suggested that the polarity of asymmetric cell division is determined by the direction from which the inducing or antagonizing signal is presented. Because the division plane of the notochord/nerve cord mother cell is parallel to the previous one, this cell division does not follow Sach's rule, suggesting the presence of some form of division plane control (Figs. 12.2a, 12.6a, b and 12.7a) (Conklin 1905; Kumano and Nishida 2007). In the asymmetric division of the EMS cell of the *C. elegans* 4-cell embryo, the Wnt signal as an extrinsic cue from the adjacent posterior cells controls both the specification of cell fate and orientation of the mitotic spindle (Goldstein 1995; Rocheleau et al. 1997; Thorpe et al. 1997). In ascidians, however, depletion of FGF and ephrin signaling does not alter the direction of cell division (Kim and Nishida 2001; Kumano et al. 2006; Kim et al. 2007; Picco et al. 2007). Thus, cell fates and the division plane are regulated independently in the notochord/nerve mother cell. The mechanism of spindle orientation has been investigated using *Phallusia*, which has a transparent egg and embryo, and is able to efficiently translate mRNA injected into unfertilized eggs, making it a suitable organism for live imaging (Prodon et al. 2010). This approach has revealed that the mitotic spindle in the notochord/nerve mother cell

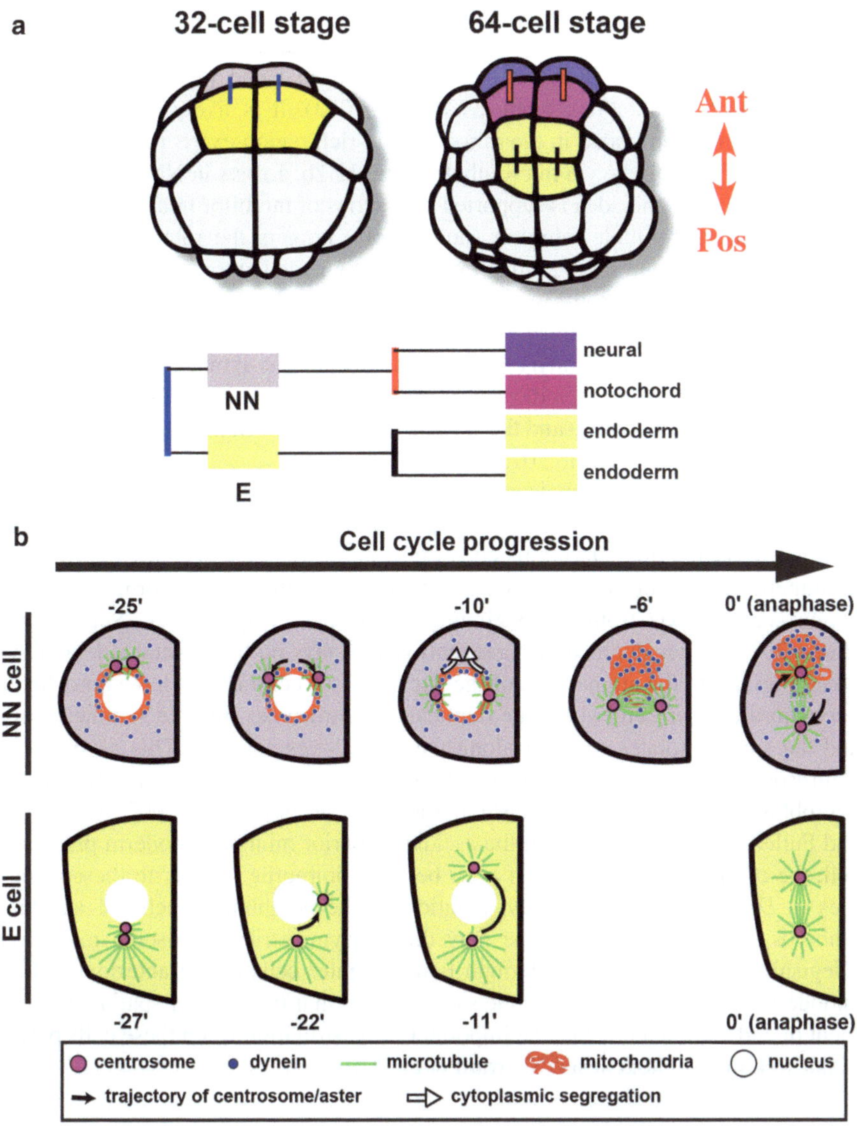

Fig. 12.7 Oriented cell division of the medial NN and E cells. (**a**) Drawings of the 32- and 64-cell-stage embryos highlighting the cell lineage of the sister cells. Embryos are in vegetal pole view with the anterior side up. Asymmetric cell division produces notochord/nerve cord mother cells (NN cell, in *gray*) and endodermal precursors (E cell in *yellow*) at the 32-cell stage. The first division axis is parallel to the anterior–posterior axis. These two daughter cells then divide again along the anterior–posterior axis to generate one nerve cord precursor (*purple*), one notochord precursor (*magenta*), and two endoderm precursors (*yellow*) at the 64-cell stage. *Green bars* indicate a pair of sister blastomeres. (**b**) Model of the mechanisms of spindle orientation in NN and E cells. Time is indicated with 0 min corresponding to the beginning of anaphase. In the NN cell, the centrosome is duplicated on the anterior side of the nucleus. The centrosomes migrate 90° in *opposite directions* and result in formation of the mitotic spindle orthogonal to the A–P axis. In

(NN cell in Fig. 12.7a) rotates 90° to align along the animal–vegetal axis (Fig. 12.7b) (Negishi and Yasuo 2015). This spindle rotation is coupled with segregation of the mitochondria toward the marginal end (Zalokar and Sardet 1984; Takatori et al. 2010; Kobayashi et al. 2013). It is likely that the dynein motor protein accumulated in the mitochondria-rich region exerts pulling force on both poles of the spindle via microtubules (Fig. 12.7b, arrows in NN cell) (Negishi and Yasuo 2015). This idea is supported by a series of inhibitor treatments: (1) low doses of the microtubule inhibitor (nocodazole) impairs the rotation of spindle, while allowing normal segregation of mitochondria; and (2) treatment with the two dynein inhibitors (erythro-9-[3-(2-hy-droxynonyl)]adenine and Ciliobrevin A) after mitochondrial segregation blocks the spindle rotation (Negishi and Yasuo 2015). The eventual selection of which of the two spindle poles is actually pulled is merely random. In addition, the spindle orientation is cell-autonomous process because segregation of mitochondria and the spindle rotation occur in the isolated cells from embryos (Negishi and Yasuo 2015).

The anterior midline endoderm precursor cell of the 32-cell embryo (E cell in Fig. 12.7a), which is a sister cell of the notochord/nerve mother cell, also divides contrary to Sach's rule. During this cell division, the duplicated centrosomes show asymmetric migration in interphase and then the mitotic spindle forms along the animal–vegetal axis from the beginning of M phase (Fig. 12.7b) (Negishi and Yasuo 2015). After duplication, one centrosome stays on the vegetal side and the other migrates to the animal side of the nucleus. This contrasts with the notochord/nerve cord mother cell, where the duplicated centrosomes migrate symmetrically and the spindle rotates to align along the animal–vegetal axis. This asymmetric centrosome behavior is reminiscent of that in male germline stem cells and in neuroblasts of *Drosophila* (Yamashita et al. 2003; Rebollo et al. 2007; Yamashita and Fuller 2008), and indeed, in the ascidian anterior midline endoderm precursor cell, the centrosome with a larger aster becomes immotile, similar to these examples in *Drosophila*. Interestingly, ablation of the neighboring cells of ascidian embryos suggests that cell shape may direct asymmetric centrosome migration (Negishi and Yasuo 2015). Although both daughter cells of the anterior midline endoderm precursor cell assume the same endodermal fate, proper spatial arrangement of endodermal cells may be important for gastrulation when these cells play a major role in morphogenesis (Sherrard et al. 2010).

Fig. 12.7 (continued) this cell, cytoplasmic dynein and mitochondria, which are initially distributed around the nucleus, segregate toward the anterior side in an actomyosin-dependent manner. This enrichment of cytoplasmic dynein exerts the pulling force for rotation of the spindle in the NN cell. In the E cell, the duplicated centrosomes are first positioned at the posterior side of the nucleus. After duplication, the centrosome having a larger aster shows less movement, and the other migrates to the opposite side of the nucleus. The spindle then forms along the A–P axis from the beginning of M phase. Images have been reproduced with permission from Negishi and Yasuo (2015)

12.5 Perspectives

Ascidian early embryogenesis involves several unequal and asymmetric cell divisions, with or without regulation of division plane control. In some cases, prelocalized maternal factors are involved. In endoderm and mesoderm segregation, nuclear migration and asymmetric partitioning of the transcript play a role. These are controlled by the intrinsic and cell-autonomous factors. In other cases, directed extracellular signals induce asymmetric allocation of cell fates. In unequal divisions of germline lineage cells, cell fate specification and division plane orientation are coupled by the CAB. In endoderm and mesoderm segregation, there is no specific regulation of the division plane. In other cases, cell fate specification and division plane orientation are regulated independently. Thus, various modes of asymmetric cell division may operate, even in the early developmental stages of any single organism. A specific series of the stereotypic asymmetric cell divisions allows ascidian embryos to establish reproducible cell fates from a small number of constituent cells.

In ascidian embryos, the functions of the two major protein complexes involved in asymmetric cell division and orientation of the division axis in a wide range of animals have not been elucidated. Although the aPKC–PAR3–PAR6 complex (Kemphues et al. 1988; Tabuse et al. 1998; Kemphues 2000; Ohno 2001; Macara 2004) is conserved in bilaterians and is localized to the CAB in ascidians (Patalano et al. 2006), its function is still unclear. It has only been confirmed recently that this complex has a role in the tubulogenesis of ascidian notochord cells (Denker et al. 2013). Another major complex, LGN–NuMA–Gαi, has also been shown to have a highly conserved role in asymmetric cell division and spindle orientation in many animals (Schaefer et al. 2000; Du and Macara 2004; Couwenbergs et al. 2007; Nguyen-Ngoc et al. 2007). In ascidians, however, its role has not yet been elucidated. Although LGN and Gαi are apparently present in ascidian genomes, a simple BLAST search fails to identify NuMA ortholog because of low degree of conservation in the sequences of NuMA proteins across taxa.

In addition to the three kinds of asymmetric cell divisions described in this chapter, there are two other examples that have been investigated in ascidian embryos. Both of them involved unequal cell divisions and the two daughter cells assume distinct cell fate. First, precursors of the secondary notochord cell are generated at the 110-cell stage. The precursor is smaller than its sister cell that assumes mesenchyme fate. FGF and BMP sequentially induce the secondary notochord in *Halocynthia*, while Nodal and Notch are players in the relay in *Ciona* (Darras and Nishida 2001; Hudson and Yasuo 2006). However, it is not known whether these extracellular signals polarize the mother cell prior to cell division and promote genuine asymmetric cell division. Unlike notochord and mesenchyme induction described in the fourth section in this chapter, the system is not a simple binary cell fate switch because the notochord progenitor does not adopt its sister fate (mesenchyme) in Nodal or Notch signaling-inhibited embryos in *Ciona* (Hudson and Yasuo 2006). It is currently not known whether the unequal cleavage is affected in those embryos. Second, asymmetric cell divisions that generate heart precursor cells take place at late neurula stage. Each of four cardiac founder cells divides into a smaller daughter (heart precursor) and a large daughter (tail muscle precursor). The mother cells are exposed uniformly to FGF, which is required for heart precursor fate, but matrix adhesion on ventral side of the cell polarizes the cell. In free plasma membrane, FGF receptors are internalized and degraded, while these processes are suppressed in adherent membrane, resulting in elevated levels of FGF signaling. In addition, membrane remodeling mediated by Caveolin enhances the receptor enrichment at the adherent membrane. The mechanism mediates both asymmetric fate specification and positioning of cell division plane (Davidson et al. 2006; Cooley et al. 2011; Cota and Davidson 2015). On the other hand, detailed quantification of the volume of every blastomere at each stage of *Ciona* embryo development has revealed other cases of unequal cell division, which do not involve the CAB (Tassy et al. 2006). Further investigation of the mechanisms and functions of these newly found unequal cell divisions would be intriguing.

References

Arvanitis D, Davy A (2008) Eph/ephrin signaling: networks. Genes Dev 22:416–429. doi:10.1101/gad.1630408

Bowerman B, Draper BW, Mello CC, Priess JR (1993) The maternal gene skn-1 encodes a protein that is distributed unequally in early *C. elegans* embryos. Cell 74:443–452. doi:10.1016/0092-8674(93)80046-H

Burgess DJ (2011) Hidden rewiring comes to light. Nat Rev Genet 12:586. doi:10.1038/nrg3060

Chabry L (1887) Contribution à l'embrologie normale tératologique des ascidies simples. J l'anatomie la Physiol Norm Pathol l'homme des animaux 23:167–321

Chen C, Fingerhut JM, Yamashita YM (2016) The ins(ide) and outs(ide) of asymmetric stem cell division. Curr Opin Cell Biol 43:1–6. doi:10.1016/j.ceb.2016.06.001

Conklin E (1905) The organization and cell lineage of the ascidian egg. J Acad Nat Sci 13:1–119. doi:10.1007/s13398-014-0173-7.2

Cooley J, Whitaker S, Sweeney S, Fraser S, Davidson B (2011) Cytoskeletal polarity mediates localized induction of the heart progenitor lineage. Nat Cell Biol 13:952–957. doi:10.1038/ncb2291

Cota CD, Davidson B (2015) Mitotic membrane turnover coordinates differential induction of the heart progenitor lineage. Dev Cell 34:505–519. doi:10.1016/j.devcel.2015.07.001

Couwenbergs C, Labbé JC, Goulding M et al (2007) Heterotrimeric G protein signaling functions with dynein to promote spindle positioning in *C. elegans*. J Cell Biol 179:15–22. doi:10.1083/jcb.200707085

Darras S, Nishida H (2001) The BMP signaling pathway is required together with the FGF pathway for notochord induction in the ascidian embryo. Development 128:2629–2638

Davidson B, Shi W, Beh J et al (2006) FGF signaling delineates the cardiac progenitor field in the simple chordate, *Ciona intestinalis*. Genes Dev 20:2728–2738. doi:10.1101/gad.1467706

Davidson EH, Davidson EH, Levine MS, Levine MS (2008) Properties of developmental gene regulatory networks. Proc Natl Acad Sci 105:20063–20066. doi:10.1073/pnas.0806007105

Delsuc F, Brinkmann H, Chourrout D, Philippe H (2006) Tunicates and not cephalochordates are the closest living relatives of vertebrates. Nature 439:965–968. doi:10.1038/nature04336

Denker E, Bocina I, Jiang D (2013) Tubulogenesis in a simple cell cord requires the formation of bi-apical cells through two discrete Par domains. Development 140:2985–2996. doi:10.1242/dev.092387

Du Q, Macara IG (2004) Mammalian Pins is a conformational switch that links NuMA to heterotrimeric G proteins. Cell 119:503–516. doi:10.1016/j.cell.2004.10.028

Goldstein B (1995) Cell contacts orient some cell division axes in the *Caenorhabditis elegans* embryo. J Cell Biol 129:1071–1080. doi:10.1111/j.1749-6632.2008.03624.x

Grill SW, Gönczy P, Stelzer EH, Hyman AA (2001) Polarity controls forces governing asymmetric spindle positioning in the *Caenorhabditis elegans* embryo. Nature 409:630–633. doi:10.1038/35054572

Gurdon JB (1992) The generation of diversity and pattern in animal development. Cell 68:185–199. doi:10.1016/0092-8674(92)90465-O

Hashimoto H, Enomoto T, Kumano G, Nishida H (2011) The transcription factor FoxB mediates temporal loss of cellular competence for notochord induction in ascidian embryos. Development 138:2591–2600. doi:10.1242/dev.053082

Haupaix N, Stolfi A, Sirour C et al (2013) p120RasGAP mediates ephrin/Eph-dependent attenuation of FGF/ERK signals during cell fate specification in ascidian embryos. Development 140:4347–4352. doi:10.1242/dev.098756

Hibino T, Nishikata T, Nishida H (1998) Centrosome-attracting body: a novel structure closely related to unequal cleavages in the ascidian embryo. Dev Growth Differ 40:85–95

Hudson C, Yasuo H (2006) A signalling relay involving Nodal and Delta ligands acts during secondary notochord induction in *Ciona* embryos. Development 133:2855–2864. doi:10.1242/dev.02466

Hudson C, Yasuo H (2008) Similarity and diversity in mechanisms of muscle fate induction between ascidian species. Biol Cell 100:265–277. doi:10.1042/BC20070144

Hudson C, Kawai N, Negishi T, Yasuo H (2013) β-Catenin-driven binary fate specification segregates germ layers in Ascidian embryos. Curr Biol 23:491–495. doi:10.1016/j.cub.2013.02.005

Iseto T, Nishida H (1999) Ultrastructural studies on the centrosome-attracting body: electron-dense matrix and its role in unequal cleavages in ascidian embryos. Dev Growth Differ 41:601–609. doi:10.1046/j.1440-169x.1999.00457.x

Johnston DS, Nüsslein-Volhard C (1992) The origin of pattern and polarity in the *Drosophila* embryo. Cell 68:201–219. doi:10.1016/0092-8674(92)90466-P

Kawai N, Iida Y, Kumano G, Nishida H (2007) Nuclear accumulation of β-catenin and transcription of downstream genes are regulated by zygotic Wnt5α and maternal Dsh in ascidian embryos. Dev Dyn 236:1570–1582. doi:10.1002/dvdy.21169

Kawashima T, Kawashima S, Kanehisa M et al (2000) MAGEST: MAboya gene expression patterns and sequence tags. Nucleic Acids Res 28:133–135. doi: gkd103 [pii]

Kemphues K (2000) PARsing embryonic polarity. Cell 101:345–348. doi:10.1016/S0092-8674(00)80844-2

Kemphues KJ, Priess JR, Morton DG, Cheng NS (1988) Identification of genes required for cytoplasmic localization in early *C. elegans* embryos. Cell 52:311–320. doi:10.1016/S0092-8674(88)80024-2

Kim GJ, Nishida H (2001) Role of the FGF and MEK signaling pathway in the ascidian embryo. Dev Growth Differ 43:521–533. doi:10.1046/j.1440-169X.2001.00594.x

Kim GJ, Kumano G, Nishida H (2007) Cell fate polarization in ascidian mesenchyme/muscle precursors by directed FGF signaling and role for an additional ectodermal FGF antagonizing signal in notochord/nerve cord precursors. Development 134:1509–1518. doi:10.1242/dev.02825

Kimelman D, Griffin KJ (2000) Vertebrate mesendoderm induction and patterning. Curr Opin Genet Dev 10:350–356. doi:10.1016/S0959-437X(00)00095-2

Kobayashi K, Yamada L, Satou Y, Satoh N (2013) Differential gene expression in notochord and nerve cord fate segregation in the *Ciona intestinalis* embryo. Genesis 51:647–659. doi:10.1002/dvg.22413

Kumano G, Nishida H (2007) Ascidian embryonic development: an emerging model system for the study of cell fate specification in chordates. Dev Dyn 236:1732–1747. doi:10.1002/dvdy.21108

Kumano G, Yamaguchi S, Nishida H (2006) Overlapping expression of FoxA and Zic confers responsiveness to FGF signaling to specify notochord in ascidian embryos. Dev Biol 300:770–784. doi:10.1016/j.ydbio.2006.07.033

Kumano G, Kawai N, Nishida H (2010) Macho-1 regulates unequal cell divisions independently of its function as a muscle determinant. Dev Biol 344:284–292. doi:10.1016/j.ydbio.2010.05.013

Kumano G, Takatori N, Negishi T et al (2011) A maternal factor unique to ascidians silences the germline via binding to P-TEFb and RNAP II regulation. Curr Biol 21:1308–1313. doi:10.1016/j.cub.2011.06.050

Lemaire P (2009) Unfolding a chordate developmental program, one cell at a time: invariant cell lineages, short-range inductions and evolutionary plasticity in ascidians. Dev Biol 332:48–60. doi:10.1016/j.ydbio.2009.05.540

Lemaire P, Smith WC, Nishida H (2008) Ascidians and the plasticity of the chordate developmental program. Curr Biol 18:R620–R631. doi:10.1016/j.cub.2008.05.039

Macara IG (2004) Parsing the polarity code. Nat Rev Mol Cell Biol 5:220–231. doi:10.1038/nrm1332

Makabe KW, Nishida H (2012) Cytoplasmic localization and reorganization in ascidian eggs: role of postplasmic/PEM RNAs in axis formation and fate determination. Wiley Interdiscip Rev Dev Biol 1:501–518. doi:10.1002/wdev.54

Makabe KW, Kawashima T, Kawashima S et al (2001) Large-scale cDNA analysis of the maternal genetic information in the egg of Halocynthia roretzi for a gene expression catalog of ascidian development. Development 128:2555–2567

Nakamura A, Seydoux G (2008) Less is more: specification of the germline by transcriptional repression. Development 135:3817–3827. doi:10.1242/dev.022434

Nakamura Y, Makabe KW, Nishida H (2003) Localization and expression pattern of type I postplasmic mRNAs in embryos of the ascidian *Halocynthia roretzi*. Gene Expr Patterns 3:71–75. doi:10.1016/S1567-133X(02)00069-8

Nakatani Y, Nishida H (1997) Ras is an essential component for notochord formation during ascidian embryogenesis. Mech Dev 68:81–89. doi:10.1016/S0925-4773(97)00131-7

Nakatani Y, Yasuo H, Satoh N, Nishida H (1996) Basic fibroblast growth factor induces notochord formation and the expression of As-T, a Brachyury homolog, during ascidian embryogenesis. Development 122:2023–2031

Negishi T, Yasuo H (2015) Distinct modes of mitotic spindle orientation align cells in the dorsal midline of ascidian embryos. Dev Biol 408:66–78. doi:10.1016/j.ydbio.2015.09.019

Negishi T, Takada T, Kawai N, Nishida H (2007) Localized PEM mRNA and protein are involved in cleavage-plane orientation and unequal cell divisions in ascidians. Curr Biol 17:1014–1025. doi:10.1016/j.cub.2007.05.047

Negishi T, Kumano G, Nishida H (2011) Polo-like kinase 1 is required for localization of posterior end mark protein to the centrosome-attracting body and unequal cleavages in ascidian embryos. Dev Growth Differ 53:76–87. doi:10.1111/j.1440-169X.2010.01231.x

Nguyen-Ngoc T, Afshar K, Gönczy P (2007) Coupling of cortical dynein and G alpha proteins mediates spindle positioning in *Caenorhabditis elegans*. Nat Cell Biol 9:1294–1302. doi:10.1038/ncb1649

Nishida H (1986) Cell division pattern during Gastrulation of the Ascidian, *Halocynthia roretzi*. Dev Growth Differ 28:191–201. doi:10.1111/j.1440-169X.1986.00191.x

Nishida H (1987) Cell lineage analysis in ascidian embryos by intracellular injection of a tracer enzyme. III. Up to the tissue restricted stage. Dev Biol 121:526–541. doi:10.1016/0012-1606(87)90188-6

Nishida H (2002) Patterning the marginal zone of early ascidian embryos: localized maternal mRNA and inductive interactions. Bioessays 24:613–624. doi:10.1002/bies.10099

Nishida H (2005) Specification of embryonic axis and mosaic development in ascidians. Dev Dyn 233:1177–1193

Nishida H, Satoh N (1983) Cell lineage analysis in ascidian embryos by intracellular injection of a tracer enzyme. I. Up to the eight-cell stage. Dev Biol 99:382–394. doi:10.1016/0012-1606(83)90288-9

Nishida H, Satoh N (1985) Cell lineage analysis in ascidian embryos by intracellular injection of a tracer enzyme. II. The 16- and 32-cell stages. Dev Biol 110:440–454. doi:10.1016/0012-1606(85)90102-2

Nishida H, Sawada K (2001) macho-1 encodes a localized mRNA in ascidian eggs that specifies muscle fate during embryogenesis. Nature 409:724–729. doi:10.1038/35055568

Nishikata T, Hibino T, Nishida H (1999) The centrosome-attracting body, microtubule system, and posterior egg cytoplasm are involved in positioning of cleavage planes in the ascidian embryo. Dev Biol 209:72–85. doi:10.1006/dbio.1999.9244

Ohno S (2001) Intercellular junctions and cellular polarity: the PAR-aPKC complex, a conserved core cassette playing fundamental roles in cell polarity. Curr Opin Cell Biol 13:641–648. doi:10.1016/S0955-0674(00)00264-7

Paix A, Yamada L, Dru P et al (2009) Cortical anchorages and cell type segregations of maternal postplasmic/PEM RNAs in ascidians. Dev Biol 336:96–111. doi:10.1016/j.ydbio.2009.09.001

Paix A, Le Nguyen PN, Sardet C (2011) Bi-polarized translation of ascidian maternal mRNA determinant pem-1 associated with regulators of the translation machinery on cortical Endoplasmic Reticulum (cER). Dev Biol 357:211–226. doi:10.1016/j.ydbio.2011.06.019

Patalano S, Prulière G, Prodon F et al (2006) The aPKC-PAR-6-PAR-3 cell polarity complex localizes to the centrosome attracting body, a macroscopic cortical structure responsible for asymmetric divisions in the early ascidian embryo. J Cell Sci 119:1592–1603. doi:10.1242/jcs.02873

Picco V, Hudson C, Yasuo H (2007) Ephrin-Eph signalling drives the asymmetric division of notochord/neural precursors in *Ciona* embryos. Development 134:1491–1497. doi:10.1242/dev.003939

Prodon F, Chenevert J, Hebras C et al (2010) Dual mechanism controls asymmetric spindle position in ascidian germ cell precursors. Development 137:2011–2021. doi:10.1242/dev.047845

Prodon F, Yamad L, Shirae-Kurabayashi M, Nakamura Y, Sasakura Y (2007) Postplasmic/PEM RNAs: a class of localized maternal mRNAs with multiple roles in cell polarity and development in ascidian embryos. Dev Dyn 236:1698–1715. doi:10.1002/dvdy.21109

Rebollo E, Sampaio P, Januschke J et al (2007) Functionally unequal centrosomes drive spindle orientation in asymmetrically dividing *Drosophila* neural stem cells. Dev Cell 12:467–474. doi:10.1016/j.devcel.2007.01.021

Rocheleau CE, Downs WD, Lin R et al (1997) Wnt signaling and an APC-related gene specify endoderm in early *C. elegans* embryos. Cell 90:707–716. doi:10.1016/S0092-8674(00)80531-0

Rodaway A, Patient R (2001) Mesendoderm: an ancient germ layer? Cell 105:169–172. doi:10.1016/S0092-8674(01)00307-5

Sardet C, Nishida H, Prodon F, Sawada K (2003) Maternal mRNAs of PEM and macho 1, the ascidian muscle determinant, associate and move with a rough endoplasmic reticulum network in the egg cortex. Development 130:5839–5849. doi:10.1242/dev.00805

Sardet C, Dru P, Prodon F (2005) Maternal determinants and mRNAs in the cortex of ascidian oocytes, zygotes and embryos. Biol Cell 97:35–49. doi:10.1042/BC20040126

Schaefer M, Shevchenko A, Shevchenko A, Knoblich JA (2000) A protein complex containing inscuteable and the Galpha-binding protein Pins orients asymmetric cell divisions in Drosophila. Curr Biol 10:353–362. doi:10.1016/S0960-9822(00)00401-2

Sherrard K, Robin F, Lemaire P, Munro E (2010) Sequential activation of apical and basolateral contractility drives ascidian endoderm invagination. Curr Biol 20:1499–1510. doi:10.1016/j.cub.2010.06.075

Shirae-Kurabayashi M, Nishikata T, Takamura K et al (2006) Dynamic redistribution of vasa homolog and exclusion of somatic cell determinants during germ cell specification in *Ciona intestinalis*. Development 133:2683–2193. doi:10.1242/dev.02446

Shirae-Kurabayashi M, Matsuda K, Nakamura A (2011) Ci-Pem-1 localizes to the nucleus and represses somatic gene transcription in the germline of *Ciona intestinalis* embryos. Development 138:2871–2881. doi:10.1242/dev.058131

Siller KH, Doe CQ (2009) Spindle orientation during asymmetric cell division. Nat Cell Biol 11:365–374. doi:10.1038/ncb0409-365

Stach T, Anselmi C (2015) High-precision morphology: bifocal 4D-microscopy enables the comparison of detailed cell lineages of two chordate species separated for more than 525 million years. BMC Biol 13:113. doi:10.1186/s12915-015-0218-1

Strome S (1993) Determination of cleavage planes. Cell 72:3–6. doi:10.1016/0092-8674(93)90041-N

Tabuse Y, Izumi Y, Piano F et al (1998) Atypical protein kinase C cooperates with PAR-3 to establish embryonic polarity in *Caenorhabditis elegans*. Development 125:3607–3614

Takatori N, Kumano G, Saiga H, Nishida H (2010) Segregation of germ layer fates by nuclear migration-dependent localization of not mRNA. Dev Cell 19:589–598. doi:10.1016/j.devcel.2010.09.003

Takatori N, Oonuma K, Nishida H, Saiga H (2015) Polarization of PI3K activity initiated by ooplasmic segregation guides nuclear migration in the Mesendoderm. Dev Cell 35:333–343. doi:10.1016/j.devcel.2015.10.012

Tassy O, Daian F, Hudson C et al (2006) A quantitative approach to the study of cell shapes and interactions during early chordate embryogenesis. Curr Biol 16:345–358. doi:10.1016/j.cub.2005.12.044

Thorpe CJ, Schlesinger A, Clayton Carter J, Bowerman B (1997) Wnt signaling polarizes an early *C. elegans* blastomere to distinguish endoderm from mesoderm. Cell 90:695–705. doi:10.1016/S0092-8674(00)80530-9

True JR, Haag ES (2001) Developmental system drift and flexibility in evolutionary trajectories. Evol Dev 3:109–119. doi:10.1046/j.1525-142x.2001.003002109.x

Tsagkogeorga G, Turon X, Hopcroft RR et al (2009) An updated 18S rRNA phylogeny of tunicates based on mixture and secondary structure models. BMC Evol Biol 9:187. doi:10.1186/1471-2148-9-187

Utsumi N, Shimojima Y, Saiga H (2004) Analysis of ascidian not genes highlights their evolutionarily conserved and derived features of structure and expression in development. Dev Genes Evol 214:460–465. doi:10.1007/s00427-004-0425-1

Von Dassow G, Schmidt JE, Kimelman D (1993) Induction of the *Xenopus* organizer: Expression and regulation of Xnot, a novel FGF and activin-regulated homeo box gene. Genes Dev 7:355–366. doi:10.1101/gad.7.3.355

Yamada L (2006) Embryonic expression profiles and conserved localization mechanisms of pem/postplasmic mRNAs of two species of ascidian, *Ciona intestinalis* and *Ciona savignyi*. Dev Biol 296:524–536. doi:10.1016/j.ydbio.2006.05.018

Yamada L, Kobayashi K, Satou Y, Satoh N (2005) Microarray analysis of localization of maternal transcripts in eggs and early embryos of the ascidian, *Ciona intestinalis*. Dev Biol 284:536–550. doi:10.1016/j.ydbio.2005.05.027

Yamashita YM, Fuller MT (2008) Asymmetric centrosome behavior and the mechanisms of stem cell division. J Cell Biol 180:261–266. doi:10.1083/jcb.200707083

Yamashita YM, Jones DL, Fuller MT (2003) Orientation of asymmetric stem cell division by the APC tumor suppressor and centrosome. Science 301:1547–1550. doi:10.1126/science.1087795

Yoshida S, Marikawa Y, Satoh N (1996) Posterior end mark, a novel maternal gene encoding a localized factor in the ascidian embryo. Development 122:2005–2012

Zalokar M, Sardet C (1984) Tracing of cell lineage in embryonic development of *Phallusia mammillata* (Ascidia) by vital staining of mitochondria. Dev Biol 102:195–205

Chapter 13
Asymmetries and Symmetries in the Mouse Oocyte and Zygote

Agathe Chaigne, Marie-Emilie Terret, and Marie-Hélène Verlhac

Abstract Mammalian oocytes grow periodically after puberty thanks to the dialogue with their niche in the follicle. This communication between somatic and germ cells promotes the accumulation, inside the oocyte, of maternal RNAs, proteins and other molecules that will sustain the two gamete divisions and early embryo development up to its implantation. In order to preserve their stock of maternal products, oocytes from all species divide twice minimizing the volume of their daughter cells to their own benefit. For this, they undergo asymmetric divisions in size where one main objective is to locate the division spindle with its chromosomes off-centred. In this chapter, we will review how this main objective is reached with an emphasis on the role of actin microfilaments in this process in mouse oocytes, the most studied example in mammals. This chapter is subdivided into three parts: I—General features of asymmetric divisions in mouse oocytes, II—Mechanism of chromosome positioning by actin in mouse oocytes and III—Switch from asymmetric to symmetric division at the oocyte-to-embryo transition.

13.1 General Features of Asymmetric Divisions in Mouse Oocytes

13.1.1 Importance of Asymmetric Divisions in the Female Gamete

As mentioned in the abstract, asymmetrically dividing in terms of size is a conserved feature of oocytes from all organisms. The products of the two asymmetric divisions of oocytes are so uneven that the daughter cells have been named polar

A. Chaigne
MRC Laboratory for Molecular Cell Biology, UCL, London, WC1E 6BT, UK
Institute for the Physics of Living Systems, UCL, London, WC1E 6BT, UK

M.-E. Terret • Marie-Hélène Verlhac (✉)
CIRB, Collège de France, CNRS-UMR7241, INSERM-U1050, Paris 75005, France
e-mail: marie-helene.verlhac@college-de-france.fr

J.-P. Tassan, J.Z. Kubiak (eds.), *Asymmetric Cell Division in Development, Differentiation and Cancer*, Results and Problems in Cell Differentiation 61,
DOI 10.1007/978-3-319-53150-2_13

bodies. Even though some species can use polar bodies as protective or nutritive tissues (Schmerler and Wessel 2011), in most organisms polar bodies contain so little cytoplasm that they degenerate. *Drosophila* oocytes do not even undergo cytokinesis during meiotic divisions and thus do not extrude polar bodies. Instead, they specifically degrade three products out of four coming from the two rounds of DNA segregation during female meiosis. These asymmetries in size are very important for these huge cells (ranging from 100 μm to 1 mm) to preserve important sources of energy, maternal RNAs and proteins to sustain early development. In mammals, oocytes are enclosed in a matrix of glycoproteins, called the zona pellucida (pale beige layer ensheathing the oocyte on all figures), which protects them until the embryo hatches and can implant in the uterus. When oocytes are artificially prevented to undergo asymmetric divisions, and instead cleave symmetrically, each of their halves can be fertilized by a different sperm, producing chimeric embryos (Otsuki et al. 2012). Chimeras result from the fusion of two fertlilized oocytes, and as such they carry two populations of genetically distinct cells originated from each zygote. A significant portion of the population might be affected by chimerism, with repercussions for transfusion and transplantation medicine as chimeras have a higher risk of graft rejection and transfusion incompatibility (Wolinsky 2007). This strongly argues that dividing asymmetrically is essential for oocytes.

After their growth in the follicle, oocytes are arrested in Prophase I of meiosis. In the mouse, meiosis resumption takes place in the ovary, in response to the LH pic and starts with nuclear envelope breakdown (NEBD; Fig. 13.1). The first meiotic spindle forms where the nucleus was more or less at the oocyte centre (Fig. 13.1). It will then migrate to the periphery of the cell to allow asymmetry in size of the division. Indeed, as cytokinesis is initiated by signals coming from the central spindle, the meiotic spindle has to be off-centred for cleavage to be unequal (Rappaport and Rappaport 1974; Green et al. 2012). After anaphase I, the second meiotic spindle forms parallel to the oocyte cortex (Fig. 13.1). Oocytes are arrested in metaphase II, waiting for the sperm. After fertilization, the second meiotic

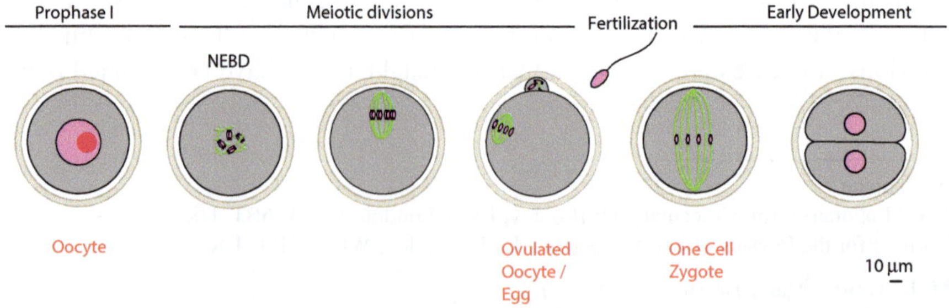

Fig. 13.1 Scheme of meiotic maturation and early steps of mouse development. Meiotic maturation starts from NEBD up to the arrest in metaphase II. The zona pellucida of oocytes is in *pale beige*, oocytes are *grey*, chromatin in *pink* (nucleolus *dark pink*) and microtubules in *green*

spindle rotates and the second polar body is extruded. Spindle migration to the cortex is thus a specificity of meiosis I. Interestingly, this is the case for many species, such as starfish, nematodes, *Drosophila* and Xenopus, where the positioning of the first meiotic spindle is also actively regulated by motion of either the oocyte nucleus or its spindle, whereas the second meiotic spindle is anchored to the cortex and is maintained there, ensuring the asymmetry of the second meiotic division.

13.1.2 Chromosome Positioning Depends on Actin

Importantly, in most somatic cells, spindle positioning is mediated by astral microtubules nucleated at the centrosomes (Gönczy 2002; Roubinet and Cabernard 2014). By interacting with the cell cortex, astral microtubules orient the mitotic spindle. Mouse oocytes, like many oocytes, lose their centrioles during oogenesis and therefore do not assemble their meiotic spindle in the presence of canonical centrosomes nor with astral microtubules (Szollosi et al. 1972). It is actually known for a long time that chromosome positioning during mouse meiosis I is microtubule independent but requires actin microfilaments (Longo and Chen 1985). Complete depolymerization of microtubules does not prevent the migration of the mass of chromosomes from the oocyte centre to its periphery (Verlhac et al. 2000). In this respect, mouse oocytes are very different from nematodes where the first meiotic spindle motion to the cortex depends on microtubules. Recently, it was also shown that nucleus centration in mouse oocytes does not require the presence of microtubules (Almonacid et al. 2015). Therefore, it seems that actin filaments are the main drivers of chromosome positioning in fully grown mouse oocytes.

In this chapter, we will detail the properties of actin filaments mediating chromosome positioning and which permit opposite types of movement, from centring to off-centring (Fig. 13.2).

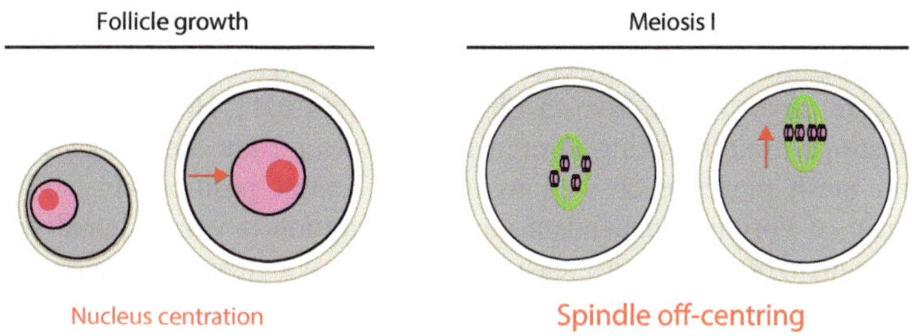

Fig. 13.2 Scheme of chromosome motion in Prophase I and Meiosis I. The *red arrow* indicates the directionality of chromosome movement

13.2 Mechanism for Chromosome Positioning by Actin in Mouse Oocytes

13.2.1 Nucleus Centring in Prophase I

Mouse oocytes start their growth in the follicle with off-centred nuclei. When they are fully grown, their nucleus is central (Luksza et al. 2013). Interestingly, in both mouse and human oocytes, a failure to put the nucleus at the oocyte centre correlates with poor developmental outcome, suggesting that nucleus centration is actually important prior to meiotic divisions (Brunet and Maro 2007; Levi et al. 2013). This situation is peculiar if compared to most oocytes where nucleus position is often asymmetric and is translated into embryo polarity. This is the case in *Drosophila* oocytes, where the nucleus migrates to the dorsal side of the oocyte and defines the future dorso–ventral axis of the embryo and then of the adult (Roth and Lynch 2009; Zhao et al. 2012). Mammals could be an exception since oocyte polarity is acquired during meiotic divisions, and there is no correlation between egg polarity and early embryo polarity (Hiiragi and Solter 2004; Motosugi et al. 2005; Halet and Carroll 2007).

Surprisingly, most mouse oocytes have central nuclei, except oocytes from the *formin 2* knockout mice (Leader et al. 2002; Dumont et al. 2007a; Azoury et al. 2011). Formin 2 is a straight F-actin nucleator and functions as a polymerase promoting actin monomer addition at the barbed end of filaments (Romero et al. 2004, 2007; Kovar 2006; Renault et al. 2008). It is an essential maternal gene, since females invalidated for Formin 2 are sterile producing aneuploid eggs (Leader et al. 2002). In its absence, oocyte does not display a cytoplasmic mesh of actin filaments (Azoury et al. 2008; Schuh and Ellenberg 2008). The cytoplasmic mesh present in oocytes is extremely dynamic, organized from Rab11a-positive vesicles and actually contains two nucleators: Formin 2 as well as Spire 1/2 (Azoury et al. 2008; Pfender et al. 2011; Schuh 2011). The nucleators are anchored into these vesicles thanks to the presence of N-myristoyl and poly-basic stretches in their N-terminal sequences. In *Drosophila* oocytes, there is also an actin cytoplasmic mesh, which has to disappear by late Prophase I for oocyte polarity to develop and which also depends on two types of nucleators, a Formin, Cappucino and Spire (Quinlan et al. 2005, 2007; Quinlan 2013; Yoo et al. 2015). Importantly, the vesicles that nucleate filaments are cargos for the Myosin Vb motor (Schuh 2011; Holubcová et al. 2013).

Interestingly, reintroducing Formin 2 into *formin 2* knockout oocytes restores the actin mesh presence as well as nucleus central position (Fig. 13.3). The mechanism of Formin 2-dependent nucleus centration is very original since it is due to a gradient of pressure decreasing towards the oocyte centre, exerted by actin-positive vesicles on the nucleus. Furthermore, the viscosity of the cytoplasm is decreased by global cytoplasmic vesicle motion due to Myosin Vb activity (Almonacid et al. 2015). Hence, a non-specific pressure gradient coupled to a global reduction in viscosity allows the moving of a big object (the mouse oocyte nucleus is 30 μm wide) in a directed manner towards the cell centre.

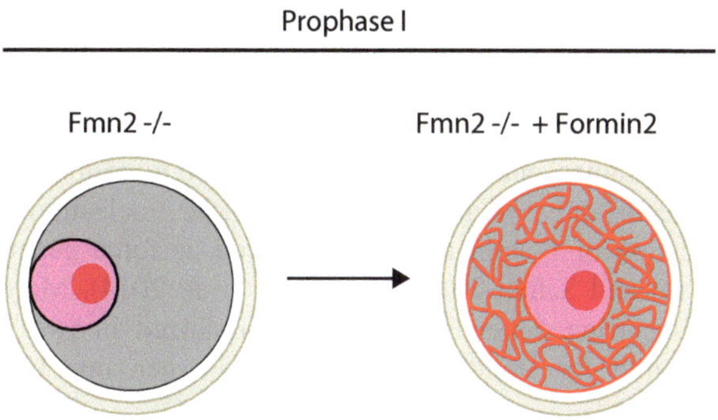

Fig. 13.3 Scheme of chromosome motion promoted by Formin 2 in Prophase I. Cytoplasmic actin filaments nucleated by Formin 2 (Fmn2) are in *red*

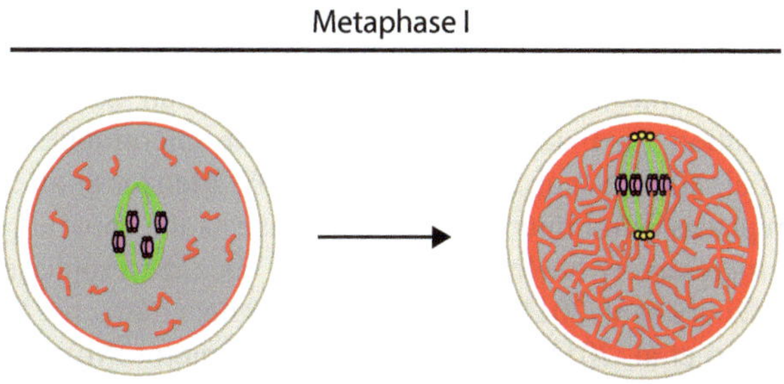

Fig. 13.4 Scheme of chromosome motion by two actin networks in Metaphase I. Cytoplasmic actin filaments are in *red*. Cortex thickening is highlighted in a *thick red circle* outlining the oocyte contour

Importantly, this cytoplasmic actin mesh will be dismantled at NEBD, allowing the formation of a new mesh now able to off-centre the chromosomes in meiosis I (Fig. 13.4; Azoury et al. 2011).

13.2.2 Meiosis I Spindle Off-Centring

Oocytes are expelled from the ovary inside their niche of follicular cells as an entity called the cumulus. Despite an important chemical dialogue between the oocyte and their niche which is essential for optimal gamete fitness (Chen et al. 2013; Cakmak et al. 2016), mouse oocytes undergoing meiotic divisions can be considered as physically isolated cells. This situtation is unique compared to most somatic cells embedded in a tissue and as such in close physical contact with other cells. Indeed, upon meiosis resumption, most physical contacts between the oocyte and its niche of follicular cells are abolished (Gilula et al. 1978; Sela-Abramovich et al. 2006; Norris et al. 2008). As a result, the symmetry-breaking event initiating spindle migration comes from within the oocyte.

In mitotic cells, spindle positioning relies on interactions between astral microtubules emanating from the centrosomes at spindle poles and the cortex (Théry et al. 2007; Fink et al. 2011; Kwon et al. 2015). In order for these interactions to be efficient in transmitting forces, the cortex has to be stiff, a consequence of the increase of cortical tension following mitosis entry (Stewart et al. 2011). The increase in cortical tension is important to ensure correct spindle positioning but also for correct spindle morphogenesis (Kunda et al. 2008; Lancaster et al. 2013; Lancaster and Baum 2014; Cadart et al. 2014). Since mouse oocytes lack true centrosomes and their associated astral microtubules, their spindle positioning depends only on microfilaments whose nucleation is regulated by the spatial location of F-actin nucleators as well as force generating elements (Myosin-II) (Clift and Schuh 2013; Almonacid et al. 2014). More precisely, meiotic spindle off-centring depends on F-actin, organized in two networks. Each network is required for spindle migration since when one is missing spindle motion is abolished.

The first one is a dynamic cytoplasmic actin meshwork comparable to the one present in Prophase I. It is nucleated by Formin 2 and Spire 1/2, both anchored into Rab11a-and Myosin Vb-positive vesicles. These vesicles organize the cytoplasmic F-actin meshwork as in Prophase I (Schuh 2011; Holubcová et al. 2013). In addition, an F-actin cage surrounds the microtubule spindle (Fig. 13.4; (Azoury et al. 2008; Schuh and Ellenberg 2008; Pfender et al. 2011). Thus, this meshwork replaces astral microtubules and connects the spindle poles to the cell cortex. It is progressively nucleated after NEBD once the Prophase I meshwork is dismantled (Azoury et al. 2011). This meshwork is not contractile by itself as in starfish oocytes (Mori et al. 2011). Instead, it requires a motor, Myosin-II, located at both poles of the actin cage surrounding the microtubule spindle to generate the forces required for spindle migration (Simerly et al. 1998; Schuh and Ellenberg 2008; Chaigne et al. 2013).

The second one is a cortical F-actin thickening nucleated by the Arp2/3 complex, a branched actin nucleator (Chaigne et al. 2013). Indeed, the Arp2/3 complex (Sun et al. 2011a) and many of its activators, the Nucleation Promoting Factors NPFs (such as Wave2, JMY, WHAMM), are implicated in spindle migration (Liu et al. 2012; Sun et al. 2011a, b; Huang et al. 2013). This network is progressively homogeneously nucleated after NEBD under the cortex. It is only at the end of spindle migration that it starts deforming in the region where the spindle is reaching it. The Mos/. . ./MAPK pathway is necessary and sufficient for the nucleation of this F-actin cortical thickening. Mos is a MAPKKK specifically expressed in the oocyte (Propst et al. 1987). It is synthesized progressively after NEBD and induces ERK1/2 activation around NEBD + 3 h (Verlhac et al. 1993, 1996), triggering the nucleation of the cortical F-actin thickening. Indeed, the Mos/. . ./MAPK pathway probably activates the Arp2/3 complex by regulation of its NPF Wave2 (Chaigne et al. 2013). Arp2/3, Wave2 and Mos inhibitions all lead to similar phenotypes, namely an absence of spindle migration (Verlhac et al. 2000; Sun et al. 2011a, b).

The nucleation of the cortical F-actin thickening excludes Myosin-II from the cortex, (Fig. 13.4), decreasing cortical tension (Larson et al. 2010; Chaigne et al. 2013, 2015). This change in cortex mechanics amplifies an initial imbalance of pulling forces exerted by Myosin-II at the poles of the actin cage, the forces being stronger at the pole closest to the cortex because of the initial slight asymmetry of nuclear position (Verlhac et al. 2000; Maro and Verlhac 2002; Brunet and Maro 2007; Brunet and Verlhac 2011). The motion is very slow but is amplified by the deformation of the cortex, which is rendered possible by the lowering of cortical tension, allowing recruitment of more filaments between the cortex and the spindle and therefore amplifying the initial forces. Although the drop in cortical tension is required for spindle migration in oocytes, as artificial stiffening of the cortex impairs spindle off-centring (Chaigne et al. 2013), spindle migration is also impaired by a too low tension (Chaigne et al. 2015). Thus, the geometry of the division of mouse oocytes depends on a narrow window of cortical tension, regulated by Myosin-II cortical localization, itself fine-tuned by actin nucleation.

Towards the end of spindle migration, when the chromosomes and their Ran-GTP associated gradient reach (Dumont et al. 2007b) and differentiate the cortex, actin flows mediated by the Arp2/3 complex generate a cytoplasmic streaming that accelerates spindle motion and finishes the process of spindle off-centring (Yi et al. 2011). This mechanism is reminiscent of the one responsible for spindle maintenance at the cortex in metaphase II and is described below.

13.2.3 Spindle Maintenance at the Cortex in Metaphase II

After anaphase I and first polar body extrusion, oocytes enter directly into their second meiotic division without going through a replicative phase. The spindle reforms parallel to the cortex and can be maintained there for days since oocytes arrest in metaphase II until fertilization. The chromosomes being near the cortex, their Ran-GTP-associated gradient differentiates the cortex above them, leading to actin accumulation and microvilli inhibition in a zone overlooking the spindle named actin cap, as at the end of the first meiotic division. One target of the Ran-GTP gradient is a Rho GTPase, Cdc42, that will activate N-WASP, an NPF of the Arp2/3 complex (Dehapiot et al. 2013). As a result, the Arp2/3 complex is activated and localized to the actin cap, promoting actin nucleation in this region (Yi et al. 2011). Another Rho GTPase, Rac, is the target of the Ran-GTP gradient (Dehapiot et al. 2013; Halet and Carroll 2007). It probably activates Wave2, another NPF of the Arp2/3 complex, resulting also in Arp2/3 complex activation and localization to the actin cap region (Yi et al. 2011). This pathway may serve as a parallel pathway to activate the Arp2/3 complex locally. Actin filaments nucleated by the Arp2/3 complex in this restricted region flow inwardly towards the cell interior. These actin flows drive cytoplasmic streaming in the oocyte, generating a pushing force responsible for spindle maintenance at the cortex in metaphase II (Yi et al. 2011). Interestingly, the same mechanism is at play at the end of the first meiotic division to push completely the spindle towards the cortex (Yi et al. 2013).

Hence, the chromosomes are responsible for establishing the cortical actin cap, but actin polymerization in this region, in turn, impacts chromosome positioning (Yi et al. 2011). A positive feedback therefore exists between chromosome positioning and cortical polarization.

13.3 Switch from Asymmetric to Symmetric Divisions at the Oocyte-to-Embryo Transition

While the two meiotic divisions are highly asymmetric, the first mitotic division is symmetric (Fig. 13.1). For this, it is essential that the first mitotic spindle be positioned at the perfect embryo centre so that the cleavage furrow starts forming in the middle of the zygote. However, contrary to human zygotes where the sperm brings a centrosome that contributes to the mitotic spindle, in mice the centrosome is gradually lost during spermatogenesis and is only nucleated de novo after the 32-cell stage (Calarco-Gillam et al. 1983; Gueth-Hallonet et al. 1993). Hence, spindle positioning cannot be ensured by centrosome and their associated astral microtubules.

13.3.1 Coarse Pronuclei Centring

After fertilization and the second meiotic division, the two pronuclei containing maternal and paternal sister chromatids form at the cortex of the embryo around 4–5 h after addition of the sperm in in vitro fertilization (Maro et al. 1984). A fertilization cone enriched in actin filaments transiently assembles at the site of sperm entry; subcortical actin, at the site of sperm entry, persists and gradually disappears while the two pronuclei slowly migrate towards the centre of the embryo (Maro et al. 1984; Chaigne et al. 2016). Importantly, the first cleavage plane coincides with the plane defined by the two apposing pronuclei once they have moved to the centre of the egg (Hiiragi and Solter 2004).

Even though microtubule nucleating foci are often apposed to the two pronuclei, these foci are never nucleating astral microtubules and do not seem involved in pronuclei migration (Schatten et al. 1985; Courtois et al. 2012). Differently, a dense actin meshwork that depends on Cofilin and the Subcortical Maternal Complex (a complex of at least four proteins expressed only in the mammalian oocyte and embryo) fills the cytoplasm (Yu et al. 2014; Chaigne et al. 2016), and depolymerization of actin filaments strongly impairs pronuclei migration (Maro et al. 1984; Chaigne et al. 2016). The dynamics of this mesh depends on the molecular motor Myosin Vb as in Prophase I and is crucial for the migration of the two pronuclei to the centre of the zygote (Fig. 13.5) (Almonacid et al. 2015; Chaigne et al. 2016). However, centring of the two pronuclei is rather coarse and needs to be followed by precise mitotic spindle centration to allow symmetric division of the zygote (Chaigne et al. 2016).

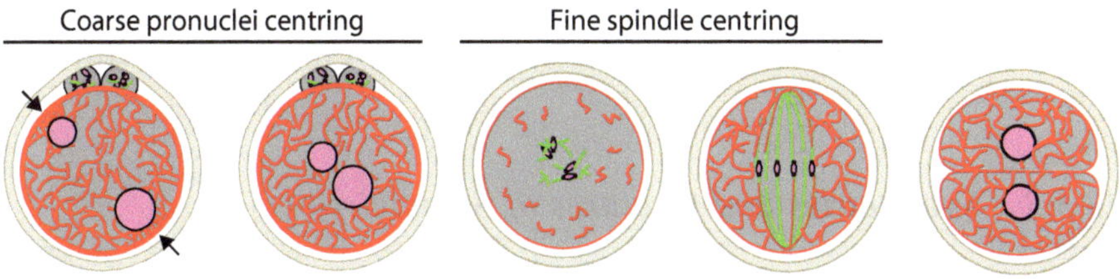

Fig. 13.5 Scheme of chromosome motion in two steps in the zygote. The *black arrows* indicate the directionality of pronuclei movement

13.3.2 Precise Spindle Positioning

The first mitotic division is peculiar because it involves not only building a metaphase spindle but also merging the two sets of parental chromosomes together. Similar to oocytes, a multipolar spindle progressively forms around the chromosomes facilitated by the presence of Ran-GTP (Schatten et al. 1985; Dumont et al. 2007b; Courtois et al. 2012). Kinesin-5 and Dynein are also crucial for subsequent bipolarization of the first mitotic spindle (Fitzharris 2009; Courtois et al. 2012). The formation of the spindle is required to bring the parental sets of DNA together (Chaigne et al. 2016).

The cytoplasmic actin meshwork gets extensively remodelled during NEBD and spindle formation, and an actin cage forms surrounding the microtubule spindle as in oocytes (Azoury et al. 2008, 2011; Chaigne et al. 2016; Schuh and Ellenberg 2008). This actin cage is crucial to centre the spindle because it relays the increase in tension experienced by the cell cortex (Fig. 13.5). Indeed, when cortical tension is artificially decreased or when actin is depolymerized at the onset of mitosis, the spindle is no longer properly centred (Chaigne et al. 2016). Similarly, when the spindle is artificially pushed off-centre in the absence of F-actin, the spindle does not reposition and the division is asymmetric (Chew et al. 2012).

Interestingly, during the duration of the following metaphase (around 1 h), the spindle does not experience extensive motion, but maintenance of its central position does not depend on actin, cortical tension or microtubules but might depend on a massive increase in cytoplasmic viscosity that remains to be investigated (Chaigne et al. 2016). At anaphase, the elongation of spindle or anaphase B is reduced and chromosomes are as a result positioned close to the centre of the future 2-cell embryo, which might also be a mechanism to ensure that the division is symmetric. Indeed, it has been suggested that a short spindle midzone would prevent mislocalization of the cleavage furrow (Yamagata and FitzHarris 2013).

Most zygotes, except rodents, do possess canonical centrioles, yet the mechanism previously described could be more general as for example in fish and frog early embryos, which are gigantic, and where the astral microtubules are too short to connect with the cortex (Wühr et al. 2010). Similarly, whereas centrioles are detectable at the poles of the spindle of human zygotes (Manandhar et al. 2006), parthenogenetic eggs can develop up to the blastocyst stage, suggesting that the centrioles might not be necessary for the first divisions (Clift and Schuh 2013).

13.3.3 Symmetries and Ruptures of Symmetry

How are the switches from symmetric to asymmetric and back to symmetric realized during mouse early development? Before spindle migration, the oocyte does not harbour any obvious polarity, none of the traditional proteins involved in cell polarity such as Par-3, Par-6, aPKC (Vinot et al. 2004), cytoskeleton determinants (Halet and Carroll 2007) or organelles (FitzHarris et al. 2007; Dalton and Carroll 2013) presenting an asymmetric distribution. The only asymmetry resides in the slight off-centring of the Prophase I nucleus (Brunet and Maro 2007; Brunet and Verlhac 2011), which is crucial for asymmetric spindle positioning (Chaigne et al. 2013, 2015). Interestingly, as discussed before, an F-actin/Myosin Vb meshwork is responsible for precise positioning of the nucleus at the geometrical centre of this huge cell (Almonacid et al. 2015). The fact that the spindles always end up slightly off-centred (either because the nucleus itself oscillates around an equilibrium position or because the spindle, being slightly smaller than the nucleus, will always form with one pole closer to the cortex) could thus represent an example of biological noise with strong biological relevance, since any slight asymmetry would be amplified and lead to a very asymmetric division.

Furthermore, another way to ensure the switch between asymmetric and symmetric division could be to control the duration of the division. Indeed, meiosis I is a very long division (8–10 h from meiosis resumption to polar body extrusion); any slight asymmetry due to the instable equilibrium mentioned before could be amplified and would lead to an asymmetric division. In contrast, the first mitosis is relatively short (2–3 h), which may not be sufficient to amplify any slight unbalance of forces that might exist.

In oocytes, the asymmetric division is controlled by the Mos/.../MAPK pathway, which ensures cortex thickening required for the drop in cortical tension and spindle migration to the cortex. However, the Mos/.../MAPK pathway is not present in the mouse zygote (Weber et al. 1991; Verlhac et al. 1994) which could explain the switch from asymmetric to symmetric division (Fig. 13.6). Indeed, this transition is fast and occurs in the absence of transcription; in oocytes and early embryos, spindle positioning is controlled by the same players (namely F-actin, Myosin Vb and Myosin-II). The Mos/.../MAPK signalling pathway could thus act as on/off switch allowing to shift quickly from an asymmetric to a symmetric mode of centrosome-independent spindle positioning.

Cortical actin mechanics, by regulating cortical tension and thus cell stiffness, are essential for spindle positioning in mouse oocytes and zygotes. Similar mechanisms may play important roles in humans. Indeed, it has been shown very recently that human oocytes' developmental potential is predicted (very accurately) by mechanical properties within hours after fertilization. Their stiffness, as for

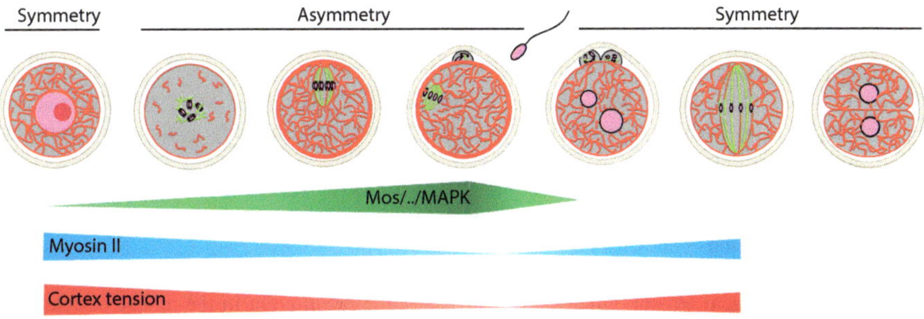

Fig. 13.6 Scheme of the evolution of symmetries and asymmetries during the oocyte to embryo transition. Intensity of Mos/. . ./MAPK signalling is in *green*, amount of cortical Myosin II is in *blue* and intensity of cortex tension is in *red*

mouse oocytes, has to be tightly gated in order for them to develop into blastocysts, if they are too stiff or too soft they will arrest (Yanez et al. 2016). These mechanical properties were measured using a micropipette aspiration technique, as performed before on mouse oocytes and embryos (Larson et al. 2010; Chaigne et al. 2013, 2015, 2016). This technique is not invasive and could very easily be used in IVF clinics where up to now embryo selection consists of highly subjective morphological assessments (Yanez et al. 2016).

Acknowledgements This work was supported by the Fondation pour la Recherche Médicale (Equipe FRM to MHV), the ANR (ANR-14-CE11-0002 to MHV) and the Fondation ARC (PJA20131200412 to MET). This work has received support from the Fondation Bettencourt Schueller, support under the programme "Investissements d'Avenir" launched by the French Government and implemented by the ANR, with the references: ANR-10-LABX-54 MEMO LIFE, ANR-11-IDEX-0001-02 PSL* Research University.

References

Almonacid M, Terret M-É, Verlhac M-H (2014) Actin-based spindle positioning: new insights from female gametes. J Cell Sci 127:477–483

Almonacid M, Ahmed WW, Bussonnier M, Mailly P, Betz T, Voituriez R, Gov NS, Verlhac M-H (2015) Active diffusion positions the nucleus in mouse oocytes. Nat Cell Biol 17:470–479

Azoury J, Lee K, Georget V, Rassinier P, Leader B, Verlhac M (2008) Spindle positioning in mouse oocytes relies on a dynamic meshwork of actin filaments. Curr Biol 18:1514–1519

Azoury J, Lee KW, Georget V, Hikal P, Verlhac M-H (2011) Symmetry breaking in mouse oocytes requires transient F-actin meshwork destabilization. Development 138:2903–2908

Brunet S, Maro B (2007) Germinal vesicle position and meiotic maturation in mouse oocyte. Reproduction 133:1069–1072

Brunet S, Verlhac MH (2011) Positioning to get out of meiosis: the asymmetry of division. Hum Reprod Update 17:68–75

Cadart C, Zlotek-Zlotkiewicz E, Le Berre M, Piel M, Matthews HK (2014) Exploring the function of cell shape and size during mitosis. Dev Cell 29:159–169

Cakmak H, Franciosi F, Zamah AM, Cedars MI, Conti M (2016) Dynamic secretion during meiotic reentry integrates the function of the oocyte and cumulus cells. Proc Natl Acad Sci USA 113:2424–2429

Calarco-Gillam PD, Siebert MC, Hubble R, Mitchison T, Kirschner M (1983) Centrosome development in early mouse embryos as defined by an autoantibody against pericentriolar material. Cell 35:621–629

Chaigne A, Campillo C, Gov NS, Voituriez R, Azoury J, Umaña-Diaz C, Almonacid M, Queguiner I, Nassoy P, Sykes C et al (2013) A soft cortex is essential for asymmetric spindle positioning in mouse oocytes. Nat Cell Biol 15:958–966

Chaigne A, Campillo C, Gov NS, Voituriez R, Sykes C, Verlhac MH, Terret ME (2015) A narrow window of cortical tension guides asymmetric spindle positioning in the mouse oocyte. Nat Commun 6:6027

Chaigne A, Campillo C, Voituriez R, Gov NS, Sykes C, Verlhac M-H, Terret M-E (2016) F-actin mechanics control spindle centring in the mouse zygote. Nat Commun 7:10253

Chen J, Torcia S, Xie F, Lin C-J, Cakmak H, Franciosi F, Horner K, Onodera C, Song JS, Cedars MI et al (2013) Somatic cells regulate maternal mRNA translation and developmental competence of mouse oocytes. Nat Cell Biol 15:1415–1423

Chew TG, Lorthongpanich C, Ang WX, Knowles BB, Solter D (2012) Symmetric cell division of the mouse zygote requires an actin network. Cytoskeleton 69:1040–1046

Clift D, Schuh M (2013) Restarting life: fertilization and the transition from meiosis to mitosis. Nat Rev Mol Cell Biol 14:549–562

Courtois A, Schuh M, Ellenberg J, Hiiragi T (2012) The transition from meiotic to mitotic spindle assembly is gradual during early mammalian development. J Cell Biol 198:357–370

Dalton CM, Carroll J (2013) Biased inheritance of mitochondria during asymmetric cell division in the mouse oocyte. J Cell Sci 126:2955–2964

Dehapiot B, Carrière V, Carroll J, Halet G (2013) Polarized Cdc42 activation promotes polar body protrusion and asymmetric division in mouse oocytes. Dev Biol 377:20212

Dumont J, Million K, Sunderland K, Rassinier P, Lim H, Leader B, Verlhac M-H (2007a) Formin-2 is required for spindle migration and for the late steps of cytokinesis in mouse oocytes. Dev Biol 301:254–265

Dumont J, Petri S, Pellegrin F, Terret M-E, Bohnsack MT, Rassinier P, Georget V, Kalab P, Gruss OJ, Verlhac M-H (2007b) A centriole- and RanGTP-independent spindle assembly pathway in meiosis I of vertebrate oocytes. J Cell Biol 176:295–305

Fink J, Carpi N, Betz T, Bétard A, Chebah M, Azioune A, Bornens M, Sykes C, Fetler L, Cuvelier D et al (2011) External forces control mitotic spindle positioning. Nat Cell Biol 13:771–778

Fitzharris G (2009) A shift from kinesin 5-dependent metaphase spindle function during preimplantation development in mouse. Development 136:2111–2119

FitzHarris G, Marangos P, Carroll J (2007) Changes in endoplasmic reticulum structure during mouse oocyte maturation are controlled by the cytoskeleton and cytoplasmic dynein. Dev Biol 305:133–144

Gilula NB, Epstein ML, Beers WH (1978) Cell-to-cell communication and ovulation. A study of the cumulus-oocyte complex. J Cell Biol 78:58–75

Gönczy P (2002) Mechanisms of spindle positioning: focus on flies and worms. Trends Cell Biol 12:332–339

Green RA, Paluch E, Oegema K (2012) Cytokinesis in animal cells. Annu Rev Cell Dev Biol 28:29–58

Gueth-Hallonet C, Antony C, Aghion J, Santa-Maria A, Lajoie-Mazenc I, Wright M, Maro B (1993) gamma-Tubulin is present in acentriolar MTOCs during early mouse development. J Cell Sci 105:157–166

Halet G, Carroll J (2007) Rac activity is polarized and regulates meiotic spindle stability and anchoring in mammalian oocytes. Dev Cell 12:309–317

Hiiragi T, Solter D (2004) First cleavage plane of the mouse egg is not predetermined but defined by the topology of the two apposing pronuclei. Nature 430:360–364

Holubcová Z, Howard G, Schuh M (2013) Vesicles modulate an actin network for asymmetric spindle positioning. Nat Cell Biol 15:937–947
Huang X, Ding L, Pan R, Ma P-F, Cheng P-P, Zhang C-H, Shen Y-T, Xu L, Liu Y, He X-Q et al (2013) WHAMM is required for meiotic spindle migration and asymmetric cytokinesis in mouse oocytes. Histochem Cell Biol 139:525–534
Kovar DR (2006) Molecular details of formin-mediated actin assembly. Curr Opin Cell Biol 18:11–17
Kunda P, Pelling AE, Liu T, Baum B (2008) Moesin controls cortical rigidity, cell rounding, and spindle morphogenesis during mitosis. Curr Biol 18:91–101
Kwon M, Bagonis M, Danuser G, Pellman D (2015) Direct microtubule-binding by myosin-10 orients centrosomes toward retraction fibers and subcortical actin clouds. Dev Cell 34:323–337
Lancaster OM, Baum B (2014) Shaping up to divide: coordinating actin and microtubule cytoskeletal remodelling during mitosis. Semin Cell Dev Biol 34:109–115
Lancaster OM, Le Berre M, Dimitracopoulos A, Bonazzi D, Zlotek-Zlotkiewicz E, Picone R, Duke T, Piel M, Baum B (2013) Mitotic rounding alters cell geometry to ensure efficient bipolar spindle formation. Dev Cell 25:270–283
Larson SM, Lee HJ, Hung P, Matthews LM, Robinson DN, Evans JP (2010) Cortical mechanics and meiosis II completion in mammalian oocytes are mediated by myosin-II and ezrin-radixin-moesin (ERM) proteins. Mol Biol Cell 21:3182–3192
Leader B, Lim H, Carabatsos MJ, Harrington A, Ecsedy J, Pellman D, Maas R, Leder P (2002) Formin-2, polyploidy, hypofertility and positioning of the meiotic spindle in mouse oocytes. Nat Cell Biol 4:921–928
Levi M, Ghetler Y, Shulman A, Shalgi R (2013) Morphological and molecular markers are correlated with maturation-competence of human oocytes. Hum Reprod 28:2482–2489
Liu J, Wang Q-C, Wang F, Duan X, Dai X-X, Wang T, Liu H-L, Cui X-S, Kim N-H, Sun S-C (2012) Nucleation promoting factors regulate the expression and localization of Arp2/3 complex during meiosis of mouse oocytes. PLoS One 7:e52277
Longo FJ, Chen DY (1985) Development of cortical polarity in mouse eggs: involvement of the meiotic apparatus. Dev Biol 107:382–394
Luksza M, Queguigner I, Verlhac M-H, Brunet S (2013) Rebuilding MTOCs upon centriole loss during mouse oogenesis. Dev Biol 382:48–56
Manandhar G, Feng D, Yi Y-J, Lai L, Letko J, Laurincik J, Sutovsky M, Salisbury JL, Prather RS, Schatten H et al (2006) Centrosomal protein centrin is not detectable during early pre-implantation development but reappears during late blastocyst stage in porcine embryos. Reproduction 132:423–434
Maro B, Verlhac M-H (2002) Polar body formation: new rules for asymmetric divisions. Nat Cell Biol 4:E281–E283
Maro B, Johnson MH, Pickering SJ, Flach G (1984) Changes in actin distribution during fertilization of the mouse egg. J Embryol Exp Morphol 81:211–237
Mori M, Monnier N, Daigle N, Bathe M, Ellenberg J, Lénárt P (2011) Intracellular transport by an anchored homogeneously contracting F-actin meshwork. Curr Biol 21:606–611
Motosugi N, Bauer T, Polanski Z, Solter D, Hiiragi T (2005) Polarity of the mouse embryo is established at blastocyst and is not prepatterned. Genes Dev 19:1081–1092
Norris RP, Freudzon M, Mehlmann LM, Cowan AE, Simon AM, Paul DL, Lampe PD, Jaffe LA (2008) Luteinizing hormone causes MAP kinase-dependent phosphorylation and closure of connexin 43 gap junctions in mouse ovarian follicles: one of two paths to meiotic resumption. Development 135:3229–3238
Otsuki J, Nagai Y, Lopata A, Chiba K, Yasmin L, Sankai T (2012) Symmetrical division of mouse oocytes during meiotic maturation can lead to the development of twin embryos that amalgamate to form a chimeric hermaphrodite. Hum Reprod 27:380–387
Pfender S, Kuznetsov V, Pleiser S, Kerkhoff E, Schuh M (2011) Spire-type actin nucleators cooperate with Formin-2 to drive asymmetric oocyte division. Curr Biol 21:955–960

Propst F, Rosenberg MP, Iyer A, Kaul K, Vande Woude GF (1987) c-mos proto-oncogene RNA transcripts in mouse tissues: structural features, developmental regulation, and localization in specific cell types. Mol Cell Biol 7:1629–1637

Quinlan ME (2013) Direct interaction between two actin nucleators is required in Drosophila oogenesis. Development 140:4417–4425

Quinlan ME, Heuser JE, Kerkhoff E, Mullins RD (2005) Drosophila spire is an actin nucleation factor. Nature 433:382–388

Quinlan ME, Hilgert S, Bedrossian A, Mullins RD, Kerkhoff E (2007) Regulatory interactions between two actin nucleators, Spire and Cappuccino. J Cell Biol 179:117–128

Rappaport R, Rappaport BN (1974) Establishment of cleavage furrows by the mitotic spindle. J Exp Zool 189:189–196

Renault L, Bugyi B, Carlier M-F (2008) Spire and cordon-bleu: multifunctional regulators of actin dynamics. Trends Cell Biol 18:494–504

Romero S, Le Clainche C, Didry D, Egile C, Pantaloni D, Carlier M-F (2004) Formin is a processive motor that requires profilin to accelerate actin assembly and associated ATP hydrolysis. Cell 119:419–429

Romero S, Didry D, Larquet E, Boisset N, Pantaloni D, Carlier M-F (2007) How ATP hydrolysis controls filament assembly from profilin-actin: implication for formin processivity. J Biol Chem 282:8435–8445

Roth S, Lynch JA (2009) Symmetry breaking during drosophila oogenesis. Cold Spring Harb Perspect Biol 1:a001891

Roubinet C, Cabernard C (2014) Control of asymmetric cell division. Curr Opin Cell Biol 31:84–91

Schatten G, Simerly C, Schatten H (1985) Microtubule configurations during fertilization, mitosis, and early development in the mouse and the requirement for egg microtubule-mediated motility during mammalian fertilization. Proc Natl Acad Sci U S A 82:4152–4156

Schmerler S, Wessel G (2011) Polar Bodies—more a lack of understanding than a lack of respect. Mol Reprod Dev 78:3–8

Schuh M (2011) An actin-dependent mechanism for long-range vesicle transport. Nat Cell Biol 13:1431–1436

Schuh M, Ellenberg J (2008) A new model for asymmetric spindle positioning in mouse oocytes. Curr Biol 18:1986–1992

Sela-Abramovich S, Edry I, Galiani D, Nevo N, Dekel N (2006) Disruption of gap junctional communication within the ovarian follicle induces oocyte maturation. Endocrinology 147:2280–2286

Simerly C, Nowak G, de Lanerolle P, Schatten G (1998) Differential expression and functions of cortical myosin IIA and IIB isotypes during meiotic maturation, fertilization, and mitosis in mouse oocytes and embryos. Mol Biol Cell 9:2509–2525

Stewart MP, Helenius J, Toyoda Y, Ramanathan SP, Muller DJ, Hyman AA (2011) Hydrostatic pressure and the actomyosin cortex drive mitotic cell rounding. Nature 469:226–230

Sun S-C, Wang Z-B, Xu Y-N, Lee S-E, Cui X-S, Kim N-H (2011a) Arp2/3 complex regulates asymmetric division and cytokinesis in mouse oocytes. PLoS One 6:e18392

Sun S-C, Xu Y-N, Li Y-H, Lee S-E, Jin Y-X, Cui X-S, Kim N-H (2011b) WAVE2 regulates meiotic spindle stability, peripheral positioning and polar body emission in mouse oocytes. Cell Cycle 10:1853–1860

Szollosi D, Calarco P, Donahue RP (1972) Absence of centrioles in the first and second meiotic spindles of mouse oocytes. J Cell Sci 11:521–541

Théry M, Jiménez-Dalmaroni A, Racine V, Bornens M, Jülicher F (2007) Experimental and theoretical study of mitotic spindle orientation. Nature 447:493–496

Verlhac MH, de Pennart H, Maro B, Cobb MH, Clarke HJ (1993) MAP kinase becomes stably activated at metaphase and is associated with microtubule-organizing centers during meiotic maturation of mouse oocytes. Dev Biol 158:330–340

Verlhac MH, Kubiak JZ, Clarke HJ, Maro B (1994) Microtubule and chromatin behavior follow MAP kinase activity but not MPF activity during meiosis in mouse oocytes. Development 120:1017–1025

Verlhac MH, Kubiak JZ, Weber M, Géraud G, Colledge WH, Evans MJ, Maro B (1996) Mos is required for MAP kinase activation and is involved in microtubule organization during meiotic maturation in the mouse. Development 122:815–822

Verlhac M-H, Lefebvre C, Guillaud P, Rassinier P, Maro B (2000) Asymmetric division in mouse oocytes: with or without Mos. Curr Biol 10:1303–1306

Vinot S, Le T, Maro B, Louvet-Vallée S (2004) Two PAR6 proteins become asymmetrically localized during establishment of polarity in mouse oocytes. Curr Biol 14:520–525

Weber M, Kubiak JZ, Arlinghaus RB, Pines J, Maro B (1991) c-mos proto-oncogene product is partly degraded after release from meiotic arrest and persists during interphase in mouse zygotes. Dev Biol 148:393–397

Wolinsky H (2007) A mythical beast. Increased attention highlights the hidden wonders of chimeras. EMBO Rep 8:212–214

Wühr M, Tan ES, Parker SK, Detrich HW, Mitchison TJ (2010) A model for cleavage plane determination in early amphibian and fish embryos. Curr Biol 20:2040–2045

Yamagata K, FitzHarris G (2013) 4D imaging reveals a shift in chromosome segregation dynamics during mouse pre-implantation development. Cell Cycle 12:157–165

Yanez LZ, Han J, Behr BB, Pera RAR, Camarillo DB (2016) Human oocyte developmental potential is predicted by mechanical properties within hours after fertilization. Nat Commun 7:10809

Yi K, Unruh JR, Deng M, Slaughter BD, Rubinstein B, Li R (2011) Dynamic maintenance of asymmetric meiotic spindle position through Arp2/3-complex-driven cytoplasmic streaming in mouse oocytes. Nat Cell Biol 13:1252–1258

Yi K, Rubinstein B, Unruh JR, Guo F, Slaughter BD, Li R (2013) Sequential actin-based pushing forces drive meiosis I chromosome migration and symmetry breaking in oocytes. J Cell Biol 200:567–576

Yoo H, Roth-Johnson EA, Bor B, Quinlan ME (2015) Drosophila Cappuccino alleles provide insight into formin mechanism and role in oogenesis. Mol Biol Cell 26:1875–1886

Yu X-J, Yi Z, Gao Z, Qin D, Zhai Y, Chen X, Ou-Yang Y, Wang Z-B, Zheng P, Zhu M-S et al (2014) The subcortical maternal complex controls symmetric division of mouse zygotes by regulating F-actin dynamics. Nat Commun 5:4887

Zhao T, Graham OS, Raposo A, St Johnston D (2012) Growing microtubules push the oocyte nucleus to polarize the Drosophila dorsal-ventral axis. Science 336:999–1003

Chapter 14
Symmetry Does not Come for Free: Cellular Mechanisms to Achieve a Symmetric Cell Division

Damian Dudka and Patrick Meraldi

Abstract During mitosis cells can divide symmetrically to proliferate or asymmetrically to generate tissue diversity. While the mechanisms that ensure asymmetric cell division have been extensively studied, it is often assumed that a symmetric cell division is the default outcome of mitosis. Recent studies, however, imply that the symmetric nature of cell division is actively controlled, as they reveal numerous mechanisms that ensure the formation of equal-sized daughter cells as cells progress through cell division. Here we review our current knowledge of these mechanisms and highlight possible key questions in the field.

14.1 Introduction

To build an organism, cells have to divide both asymmetrically, to increase diversity of the tissue, and symmetrically, to increase specific cell populations (Morin and Bellaiche 2011). The equilibrium between these two types of divisions shapes tissue architecture, and any imbalance may lead to developmental abnormalities or favour cancer formation in adult tissues (reviewed in Noatynska et al. 2012; Yamashita et al. 2010; Bajaj et al. 2015; Pease and Tirnauer 2011). A symmetric cell division is defined by the equal inheritance of cell-fate determinants and the equal size of the daughter cells (Knoblich 2008; Tzur et al. 2009; Sung et al. 2013), since cell size influences the behaviour and survival of daughter cells. Prescott observed already over 50 years ago that the products of an amoeba cell division will only divide again after reaching the size of their parental cell; an asymmetric cell division will thus lead to daughter cells that differ in their cell cycle timing (Prescott 1955, 1956). A more recent study on *Drosophila melanogaster* stem cells demonstrated that abnormally small daughter neuroblasts become

---

D. Dudka • P. Meraldi (✉)
Medical Faculty, Department of Physiology and Metabolism, University of Geneva, 1211 Geneva 4, Switzerland
e-mail: patrick.meraldi@unige.ch

J.-P. Tassan, J.Z. Kubiak (eds.), *Asymmetric Cell Division in Development, Differentiation and Cancer*, Results and Problems in Cell Differentiation 61,
DOI 10.1007/978-3-319-53150-2_14

dormant faster, resulting in developmental abnormalities due to insufficient number of neuronal progenitors (Kitajima et al. 2010). Similarly, in human tissue culture cells the bigger cell arising from an unequal cell division will enter the subsequent mitosis faster, while the smaller cell will undergo apoptosis more often (Kiyomitsu and Cheeseman 2013). Differences in daughter cell volumes may also lead to the unequal distribution of various organelles, since they are fragmented and uniformly partitioned within the cytoplasm as cells enter mitosis (reviewed in Jongsma et al. 2015). Even though it has been long assumed that symmetric cell divisions are a default state and that asymmetric divisions arise by breaking the symmetry of the spindle apparatus, recent studies uncovered cellular mechanisms that correct potential asymmetries in spindle or chromosome position and ensure symmetric cell division (Kiyomitsu and Cheeseman 2013; Tan et al. 2015). This implies that the symmetry of the cell division is not a default state, but rather that it needs to be actively established. Once symmetry is set, cells can either divide symmetrically or break the symmetry to obtain daughter cells of different cell size. Consistent with this hypothesis, the sizes of daughter cells in symmetric divisions do not differ by more than 15%, indicating a low variability that might be the result of mechanisms that favour symmetric cell division (Tzur et al. 2009; Sung et al. 2013). Here, we review these mechanisms, in order of their time of action, and explore key future questions.

14.2 The Bipolar Spindle Determines the Position of the Cytokinetic Furrow

The size of daughter cells is determined by the position of the cleavage furrow during cytokinesis. This position is in turn controlled by several elements of the mitotic spindle in the preceding mitosis. Rappaport first demonstrated in 1961 with his groundbreaking sand dollar egg experiments that the location of the two centrosomes at spindle poles plays a key role in positioning the cleavage furrow (Rappaport 1961). When he confined sand dollar eggs with a glass bead to obtain a toroid-shaped cell, he obtained a binucleate horseshoe-like cell after the first cell division. As this cell divided a second time, it yielded two spindles but four cleavage furrows—two between opposite spindle poles within a spindle and two between spindle poles of neighbouring spindles. Later experiments performed again in sand dollar eggs (Shuster and Burgess 2002), *C. elegans* (Baruni et al. 2008) and mammalian somatic cells (Rieder et al. 1997) confirmed that the cleavage furrow builds up in between adjacent centrosomes of two separate spindles. More recent studies demonstrated that the position of the cytokinetic furrow is, beyond centrosomes, also influenced by the position of the spindle mid-zone and of chromosomes (see below and Tan et al. 2015; Canman et al. 2003; Bringmann and Hyman 2005). The key role of centrosomes and of the bipolar spindle in positioning the cleavage furrow is reflected by the behaviour of cells possessing more than two centrosomes, which tend to form multipolar spindles. If cells with multipolar spindles proceed to anaphase, it will lead to the mispositioning of the furrow and to aberrant asymmetric cell divisions (Ganem et al. 2009) (Fig. 14.1). Such multipolar divisions are rare and most of the time lethal; instead, cells often cluster their centrosomes into two poles using the activity of microtubule motors such as dynein and HSET or more rarely inactivate some of their centrosomes (Ganem et al. 2009; Basto et al. 2008; Sabino et al. 2015; Quintyne et al. 2005; Kwon et al. 2008; Leber et al. 2010). While centrosome clustering allows the formation of a bipolar spindle, it nevertheless frequently causes chromosome segregation errors (Ganem et al. 2009). Furthermore in *Drosophila* neuroblasts, which normally divide in an asymmetric manner, supernumerary centrosome clustering leads to spindle orientation defects (see below) and the appearance of symmetric cell divisions that favour tumour formation (Basto et al. 2008). Overall this indicates that the formation and organization of the bipolar spindle is a key determinant for the (a)symmetric nature of cell division.

14.3 Cell Shape Influences the Symmetry of Cell Division

A first factor that influences the bipolar spindle and the outcome of cell division is cell shape. The rounding of mitotic cells is a key feature that is conserved in virtually all metazoans, pointing to an evolutionary important role (McConnell 1930; Harris 1973; Cramer and Mitchison 1997; Kunda et al. 2008; Luxenburg et al. 2011; Cadart et al. 2014). Cell rounding favours the formation of a bipolar spindle, as shown by experiments in which it was prevented genetically or by mechanical constraints (Fig. 14.2; Lancaster et al. 2013). A failure in cell rounding leads to spindle pole splitting, which may cause asymmetric cell divisions via the formation of multipolar spindles. The symmetry of cell division is also influenced by the cell shape in the preceding interphase, as shown by experimental manipulations and in silico modelling (Fink et al. 2011; Gibson et al. 2011; Minc et al. 2011). By placing urchin eggs in micro-chambers of different shapes, Minc and colleagues proved that the nucleus is positioned in the centre of the cell mass and is able to “perceive” cell shape. Similarly placing human culture cells in micro-channels resulted in frequent asymmetric cell division due to the displacement of nuclei. This displacement led to a spindle mispositioning that could not be corrected most likely because the cell cortex was beyond the reach of the astral microtubules (Cadart et al. 2014; Lancaster et al. 2013; Fig. 14.2). Computational modelling showed that nuclear positioning depends on microtubules that stretch and position the nucleus along the axis of the future mitotic spindle and thus determine the position of the cleavage furrow (Gibson et al. 2011). These studies indicate that cells normally sense their shape with the aim to position their nucleus in the centre of the cells, thus favouring a future symmetric position of the cytokinetic furrow.

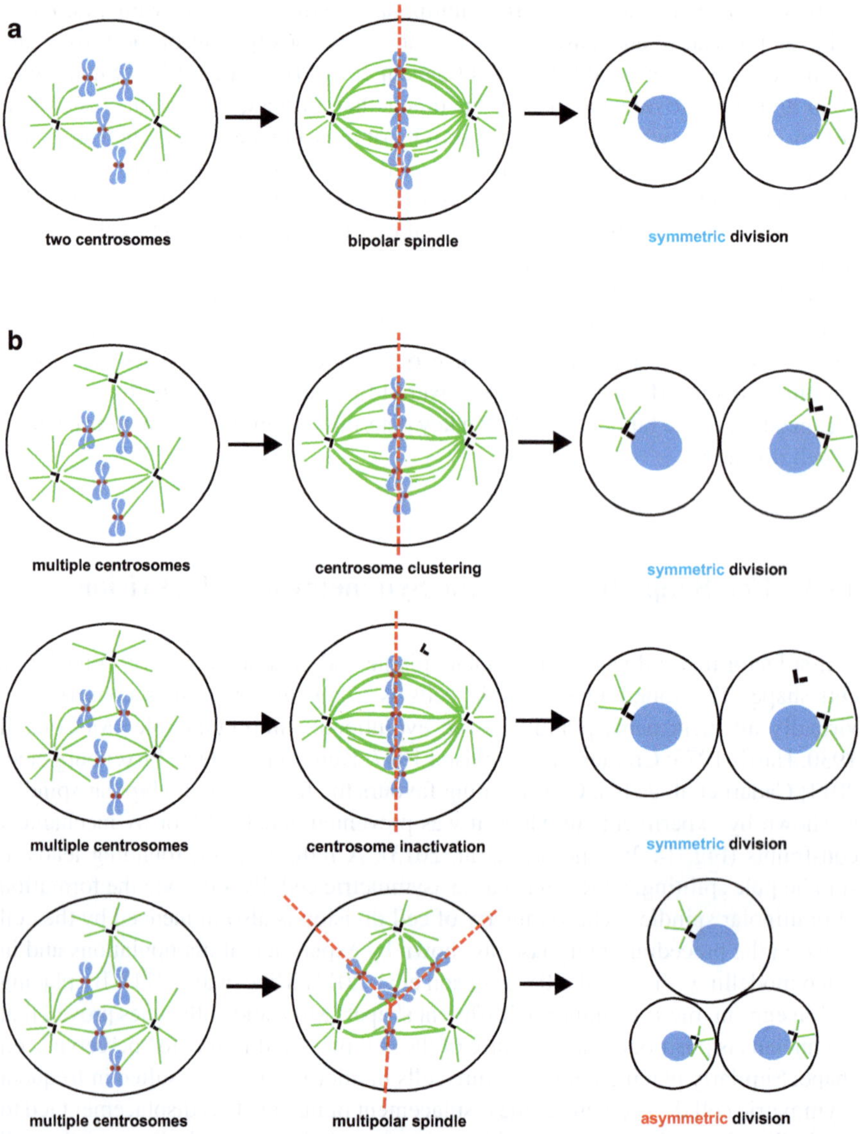

Fig. 14.1 Bipolar spindle assembly is a prerequisite of the symmetric cell division. (**a**) Cells with two centrosomes build a symmetric bipolar spindle and divide into two equal-sized daughter cells. Centrosomes are depicted in *black*, spindle microtubules in *green*, chromosomes in *blue* and kinetochores in *red*. (**b**) Cells with multiple centrosomes can also form a symmetric bipolar spindle by either centrosome clustering (*top panel*) or centrosome inactivation (*middle panel*) and divide symmetrically, or alternatively they can build an asymmetric, multipolar spindle (*lower panel*) and give rise to progeny of different size

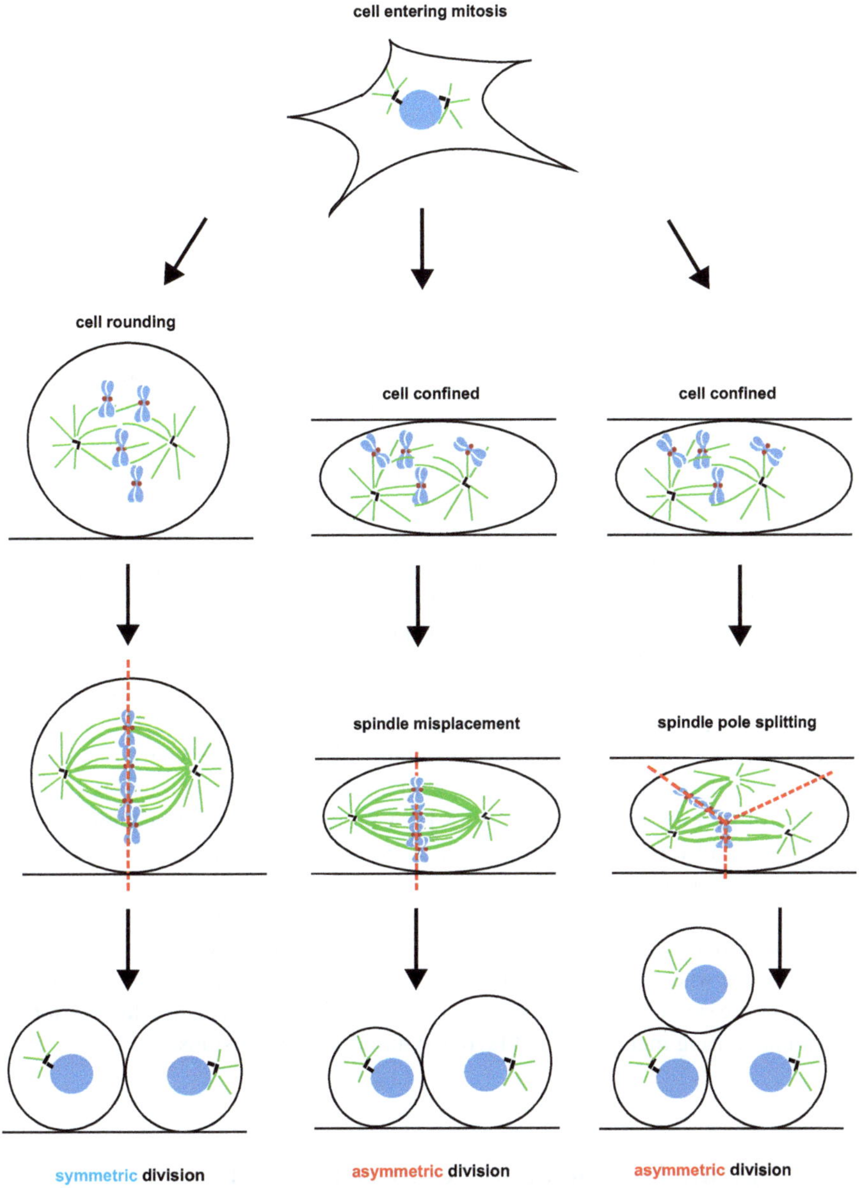

Fig. 14.2 Cell rounding combined with a central position of the nucleus at mitotic entry ensures a symmetric cell division. In a normal cell division (*left panel*), the nucleus has a central position, and mitotic rounding allows the formation of a bipolar spindle formation in the middle of the cell. In case mitotic rounding is prevented, cells might displace the nucleus (*middle panel*) or split the spindle poles leading to a multipolar spindles (*right panel*), two conditions that are each associated with asymmetric cell divisions

14.4 Controlling Spindle Positioning in Metaphase

Once a bipolar spindle is formed, it has to maintain a central positioning, as any displacement of the mitotic spindle might influence the plane of division, as shown almost 100 years ago by Conklin, who observed asymmetric division of centrifuged eggs of *Crepidula* in which the spindle had been displaced (Conklin 1917). Observation made 50 years later revealed that the mitotic spindle does not occupy a fix position within the cell, but instead is constantly moving around the centre of the cells; the mechanisms controlling this behaviour, however, remained unclear (Hughes 1952; Sato 1974). These movements, which consist of rapid back and forth displacements along the spindle axis, are called "spindle rocking". They are part of a mechanism that senses and actively corrects spindle positioning in metaphase. O'Connell and Wang first proposed this concept, after showing that cell deformations caused by micromanipulations lead to spindle movements and the repositioning of the spindle. This correction movement was blocked by injection of the anti-dynein antibodies, pointing to a central role of this microtubule motor protein in spindle positioning (O'Connell and Wang 2000). More recent work confirmed that spindle positioning and orientation is a readout of dynein-dependent pulling forces that emanate from the cell cortex via astral microtubules. Cytoplasmic dynein is recruited at the cell cortex by a tripartite complex comprised of Gα, LGN and NuMA (Kiyomitsu and Cheeseman 2013, 2012; Nguyen-Ngoc et al. 2007; Kotak et al. 2013). The cortical forces and spindle positioning are fine-tuned by a number of microtubule-associated proteins, such as EB1, LIS1, CLIP170 or NuDE, which interact with dynein complex on spindle microtubules (reviewed in Kardon and Vale 2009). Spindle rocking and dynein localization at the cell cortex is under the control of Polo-like kinase 1 (PLK1), a protein kinase that accumulates on centrosomes: PLK1 disrupts the interaction between dynein and the NuMA–LGN complex through phosphorylation and releases it from the cortex when the centrosome is in close proximity (Fig. 14.3a). This results in an asymmetric localization of dynein that creates a force imbalance, which pulls the mitotic spindle back towards the cell centre when it is positioned in an off-centred manner. As a result, the spindle undergoes the stereotypical rocking movements, which overall centre the position of the spindle during metaphase and favour symmetric cell divisions.

14.5 Spindle Positioning in Acentrosomal Cells

Spindle positioning also determines the (a)symmetry of acentrosomal cell divisions. Meiotic mouse oocytes are a classical example of how an asymmetric spindle position leads to an asymmetric cell division yielding a big oocyte and a small polar body (reviewed in Chaigne et al. 2012). The first mitotic division of the mouse embryo, however, is symmetric. In mice, the early embryos do not have centrioles;

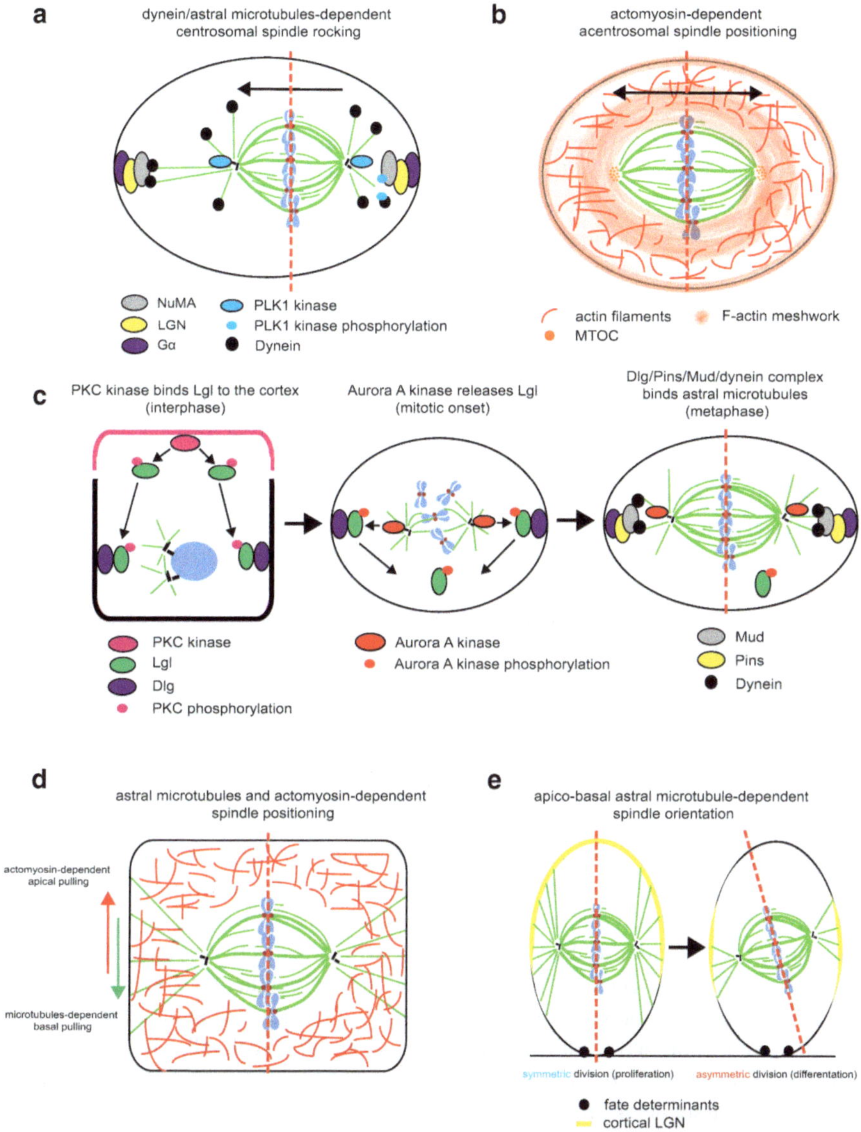

Fig. 14.3 Precise spindle positioning is essential for symmetric cell division in multiple contexts. (**a**) Spindle rocking in metaphase depends on the centrosomal kinase PLK1, which disrupts the interaction between dynein and the NuMA/LGN complex at the proximal cell cortex. This negative regulatory loop creates a differential dynein pulling gradient that centres the mitotic spindle. (**b**) Spindle positioning in acentrosomal mouse zygote depends on tension created by actomyosin cytoskeleton and F-actin cage surrounding the spindle, thus guiding it to the central position of the cell and ensuring symmetric cell division. In both meiosis and mitosis, myosin X mediates spindle orientation by directly linking cortical actin and astral microtubules. (**c**) To define the mitotic spindle axis, polarized epithelial cells break the PKC kinase-dependent polarity at the mitotic entry as Aurora A kinase phosphorylates Dlg protein to release it from Lgl. This allows the formation of the Dlg/Pins/Mud complex at the cell cortex, which will position the spindle in the centre of a cell via dynein-dependent microtubule pulling. (**d**) During gastrulation of frog embryos,

hence, the spindle poles lack centrosomes and astral microtubules. Instead, spindle poles form by clustering of microtubule-organizing centres (MTOCs; Schuh and Ellenberg 2007; Courtois et al. 2012). Symmetric division of the mouse zygote relays on two processes that both depend on the F-actin cytoskeleton: pronuclei meeting at the centre of the cell and the centralization of the mitotic spindle itself (Fig. 14.3b). Indeed, F-actin depolymerisation with Cytochalasin D results in asymmetric cell division and yields two blastomeres of unequal size (Chew et al. 2012). Moreover, Cytochalasin D-treated cells cannot restore the central position of the spindle, if it has been moved towards the cell cortex by a micromanipulation needle. At the molecular level, it was found that a F-actin/Myosin Vb-mediated cytoplasmic flow first drives a centred pronuclei meeting (Almonacid et al. 2015). In the second step, a spindle pole-localized pool of myosin II exerts tension on an F-actin cage surrounding the spindle and thus ensures the precise position of the spindle in the centre of a cell. The organization of the F-actin cage requires the presence of the Subcortical Maternal Complex (SCMC), which is composed of Mater, Filia, Floped and Tle6 (human NLRP5, KHDC3L, OOEP and TLE6) and is present only in oocytes and early embryos. Embryonic cells with a disrupted SCMC divide asymmetrically and die at the cleavage stage (Yu et al. 2014). Once centred, the symmetric position of the spindle does not depend on F-actin anymore as instead it relies on passive forces such as friction and the viscosity of the cytosol (Chaigne et al. 2016).

14.6 Controlling Spindle Orientation

In polarized tissues, symmetric cell divisions do not only rely on the central positioning of the spindle but also on its proper orientation in three dimensions. Polarized cells will divide in an asymmetric manner if the spindle axis and the polarity axis coincide, as fate determinants will be selectively inherited by one daughter cell; however, if these two axes are perpendicular to each other, a symmetric cell division will ensue (Haydar et al. 2003; Kosodo et al. 2004). Spindle orientation in polarized epithelial cells depends on cell-to-cell adhesion and cortical polarity proteins, which control astral microtubules and the spindle position via microtubule associated proteins (Reinsch and Karsenti 1994; Lu et al. 2001; den Elzen et al. 2009; Lu and Johnston 2013). Misorientation of the spindle changes the axis of division, which may cause developmental defects or tissue deformations, as shown in neuronal stem cells (Godin et al. 2010) and in skeletal muscle stem cells

Fig. 14.3 (continued) actomyosin cytoskeleton and astral microtubules pull the spindle in opposite sites, and this interplay leads to precise spindle positioning, to symmetric cell division and ultimately to the tissue thickening. (**e**) Neuronal progenitors relay on proper spindle orientation as a specific pool of astral microtubules stabilizes the spindle perpendicularly to the basal lamina, which secures the equal partitioning of the fate determinants and therefore symmetric cell division

(Dumont et al. 2015), which can all divide asymmetrically or symmetrically. Such cells must strictly regulate spindle orientation, in particular in the z-axis, to divide symmetrically, as any alteration in spindle orientation may result in asymmetric cell division.

Cells have developed various strategies to control the orientation of the spindle in the z-axis. A first mechanism relies on altering cell polarity at mitotic onset. Studies in flies have shown that the symmetric division in polarized epithelial cells is regulated via the localization of the Lgl protein (LLGL1 in human), which in polarized interphase cells is bound to the cortical Dlg protein (hDLG in human) (Guilgur et al. 2012). At mitotic onset, the aPKC kinase clears the apical cortex from Lgl protein and the Aurora A kinase relocates it to the cytoplasm (Fig. 14.3c). This alters the polarization of the cell and allows Pins protein (LGN in human) to bind to the cortical protein Dlg, which is no longer occupied by Lgl. In turn, Pins binds to spindle pole protein Mud (related to NuMA in human; Merdes et al. 1996), which itself recruits dynein. Cortical dynein then pulls onto the astral microtubules to stabilize the spindle in the planar position (Bell et al. 2015; Carvalho et al. 2015). Such a mechanism of phosphorylation-driven polarity breakage is not specific to flies, as it has been also seen in mammals (Jaffe et al. 2008; Hao et al. 2010; Durgan et al. 2011).

Another model of control over symmetric cell division in polarized cells was found in frog embryos. During *Xenopus laevis* early embryo gastrulation the polarized epithelial cells divide symmetrically to spread the epithelial tissue. An uncontrolled positioning of the spindle in these cells would lead to random changes in the division axis, impairing efficient tissue spreading and instead promoting tissue thickening. The mitotic spindle of these epithelial cells display dynamic movements along the x and y axis, allowing spreading of the epithelium; however the z-axis is fixed by pushing forces emanating from astral microtubules at the apical membrane and from the actomyosin cytoskeleton at the basal membrane to avoid tissue thickening (Fig. 14.3d; Woolner and Papalopulu 2012). Cooperation of astral microtubules and actin cytoskeleton is mediated by myosin X, which is a motor protein that links microtubules to actin and promote the bipolar spindle assembly in meiosis (Weber et al. 2004). Myosin X was also showed to help position the mitotic spindle in polarized and non-polarized cells by directly coupling centrosome-anchored astral microtubules to the subcortical actin clouds and retraction fibres (Kwon et al. 2015; Toyoshima and Nishida 2007; Liu et al. 2012).

Finally, during mammalian neurogenesis there is a pool of neuroepithelial cells, which proliferate and expand via symmetric division or differentiate into neurons upon asymmetric divisions (reviewed in Gotz and Huttner 2005; Martynoga et al. 2012). Several studies suggested that a subtle misorientation of the mitotic spindle may decide about the symmetry of the division during neurogenesis (Haydar et al. 2003; Kosodo et al. 2004), implying that cells had to develop mechanisms to keep the spindle orientation in check in order to preserve homeostasis of the tissue (Fig. 14.3e). These results should, however, be interpreted with caution, since other studies failed to find a correlation between spindle misorientation and tissue malformation (Konno et al. 2008; Noctor et al. 2008; Insolera et al. 2014). Spindle pole proteins such as ASPM (Fish et al. 2006) and Huntingtin (Godin et al. 2010), as well as cortical proteins LGN (Konno et al. 2008) and Inscutable (INSC in human; Postiglione et al. 2011) control spindle orientation during neurogenesis. Recently, using fixed and live cell imaging Mora-Bermudez et al. distinguished a specific subset of astral microtubules reaching the apical and basal part of the cell cortex in mouse neuronal progenitors. These microtubules anchor the spindle in a planar position, and their numbers correlate with abundance of LGN at the cell cortex. Consistently, neuronal progenitors of mice lacking LGN protein or overexpressing a dominant-negative form of LGN (which prevents microtubule binding to the cortex) had fewer apico-basal microtubules, while the number of central astral microtubules was not changed. Stabilization or destabilization of these microtubules by pico-molar concentrations of taxol or nocodazole resulted in asymmetric cell division, suggesting that the spindle orientation controls via cortical LGN and apico-basal microtubules an essential regulatory mechanism to ensure symmetric cell divisions (Mora-Bermudez et al. 2014).

14.7 Controlling Chromosome Positioning

Although centrosomes play a key role in determining the position of the cytokinetic furrow, other elements of the spindle, such as the chromosomes themselves, exert an influence (Fig. 14.4a). In Ptk2 cells that were forced into anaphase with monopolar spindles, chromosome-derived signals influenced cleavage furrow positioning by fortifying adjacent microtubules that reached the cell cortex (Fig. 14.4b; Canman et al. 2003). A more recent study in human cells confirmed that position of chromosomes in bipolar spindles could also influence the symmetry of the cell division (Tan et al. 2015). A partial impairment of centrosome duplication led to bipolar spindles with an unequal number of centrioles at each spindle pole, resulting in an asymmetric position of the chromosomes within the spindle. This asymmetry was normally corrected prior to anaphase via changes in microtubule dynamics. However, if cells with an asymmetric chromosome position were forced to enter anaphase, e.g. after inactivation of the spindle assembly checkpoint, an asymmetric cell division would ensue even if the mitotic spindle itself was positioned symmetrically. This indicated that the symmetry of the chromosome position provides an important cue for symmetric cell divisions (Fig. 14.4c). This work also postulated that the initial symmetry of the chromosome position and therefore the symmetry of the two half-spindles may provide a reference point, from which cells develop an asymmetry to develop an asymmetric cell division. While this hypothesis remains to be proven, there is anecdotal evidence supporting it. In *Drosophila melanogaster*, the asymmetric cell division of neuroblasts is driven by the asymmetry of the bipolar spindle. The symmetry is broken in anaphase as one of the half-spindles elongates faster than the other; this displaces the cleavage furrow from the equator of the cell and leads to an asymmetric cell division (Kaltschmidt et al. 2000). Furthermore, Ren and colleagues demonstrated that the first asymmetric

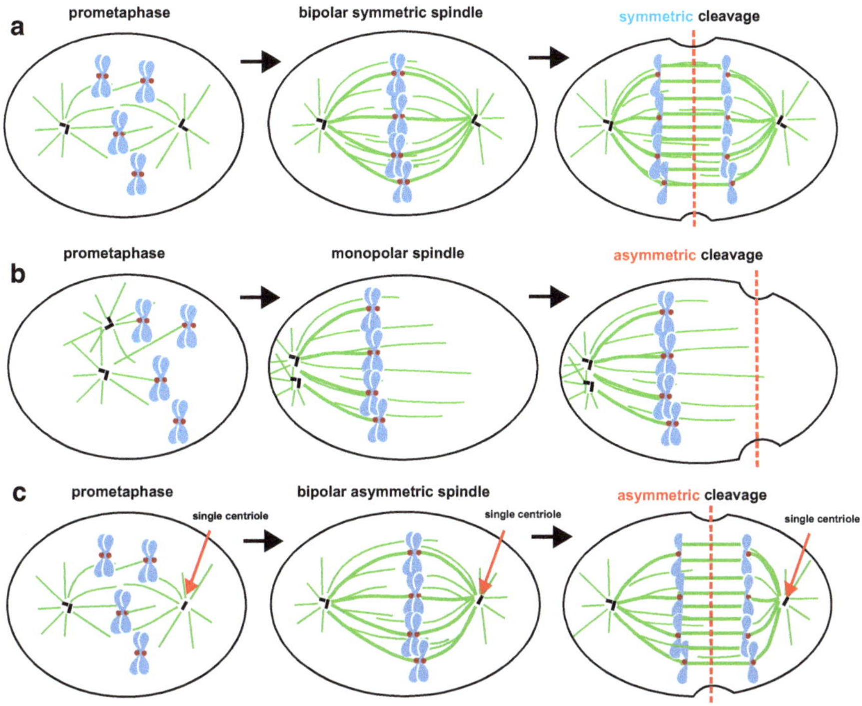

Fig. 14.4 Chromosome position within the mitotic spindle influences the symmetry of the cell division. (**a**) In normal conditions, mitotic cells put the metaphase plate in the middle of the mitotic spindle. (**b** and **c**) In cases cells are allowed to enter anaphase with (**b**) a monopolar spindle due to inhibition of Eg5 kinesin (Canman et al. 2003) or (**c**) with an asymmetrically positioned metaphase plate, which may arise if the centriole numbers in each spindle pole is not the same (Tan et al. 2015), chromosome position will modify the positioning of the cleavage furrow resulting in an asymmetric cell division

division in leech *Helobdella robusta* relies on breaking the symmetry of the mitotic spindle in metaphase, by downregulating the levels of γ-tubulin on one of the two spindle poles (Ren and Weisblat 2006). This downregulation results in unequal lengths of two half-spindles leading to the asymmetric cell division. Such asymmetry in half-spindle lengths can also occur in neural precursors of the mouse neural cortex, which divide asymmetrically yielding one progenitor and one differentiated cell (Delaunay et al. 2014). Neurons preferentially originated from cells formed from larger half-spindles, which raises the possibility that spindle asymmetry may direct the fate of daughter cells. This asymmetry depends on moesin, which is affected by the Wnt/planar cell polarity pathway, as moesin depletion enhances spindle asymmetry and altered the fate of daughter cells. Based on these observations, we conclude that the symmetry of the mitotic spindle provides an additional level of the control over the symmetry of the division.

14.8 Spindle Positioning in Anaphase

While cells have a dynein-dependent spindle-centring mechanism in metaphase, they do not monitor the position of the spindle before anaphase. Therefore excessive spindle motion in metaphase may lead to improper spindle positioning at anaphase onset (Kiyomitsu and Cheeseman 2013; Collins et al. 2012). The spindle asymmetry can also be corrected in anaphase in a dynein-dependent manner, as this correction was abolished when cells were microinjected with anti-dynein antibodies or treated with low doses of nocodazole (Collins et al. 2012). Several mechanisms have been proposed to control this dynein-dependent movements: first it has been proposed that dynein pulls harder on longer astral microtubules attached to the distal part of the cell, as those long astral microtubules would recruit more motor proteins (Collins et al. 2012; Tolic-Norrelykke 2010); second dynein is recruited to the cell cortex through LGN-independent pathways specifically in anaphase. This includes a pathway that requires the 4.1 cortical protein and a pathway that depends on the phospholipid PIP2, which recruits NuMA to the cell cortex once CDK1 kinase (the main mitotic-entry regulator) activity drops allowing the PP2A phosphatase to dephosphorylate NuMA (Fig. 14.5a; Kiyomitsu and Cheeseman 2013; Kotak et al. 2013). Therefore, even after sister chromatids started to segregate to the prospective future daughter cells, dynein-mediated mechanisms monitor and reinforce the symmetry of the dividing cell.

14.9 Balancing Cortical Forces in Late Mitosis

Assembly of the cytokinetic furrow depends on the position of the centrosomes, the spindle mid-zone and chromosomes. However, the cortical forces also contribute to the furrow positioning. Symmetrically dividing cells depleted of Anillin, an actomyosin cytoskeleton protein involved in blebs formation (Charras et al. 2006), fail to keep the furrow at the equatorial position of the cell due to improper polar contractions (Piekny and Glotzer 2008). More recent findings show that tension imbalance at the polar cortex in cytokinetic cells may lead to abnormal cell shape oscillations resulting in the furrow displacement and possibly to aneuploidy (Sedzinski et al. 2011). This study observed minor shape oscillations caused by cytoplasmic flow in early cytokinesis in mammalian cells that could be enhanced by local actin depolymerization or laser ablation of the cortex. These results implied that enhanced polar contractility is responsible for shape oscillations and that to prevent them, cells balance the tension on the poles during cytokinesis. Previous observations suggested that bleb formation at the poles of the dividing cell could serve as a release valve for polar tension (Charras et al. 2005; Tinevez et al. 2009). Consistent with this hypothesis, induction of blebs by laser ablation was sufficient to counterbalance polar contractions; moreover, inhibition of blebbing led to major shape oscillations (Sedzinski et al. 2011). Overall, this indicates that bleb formation

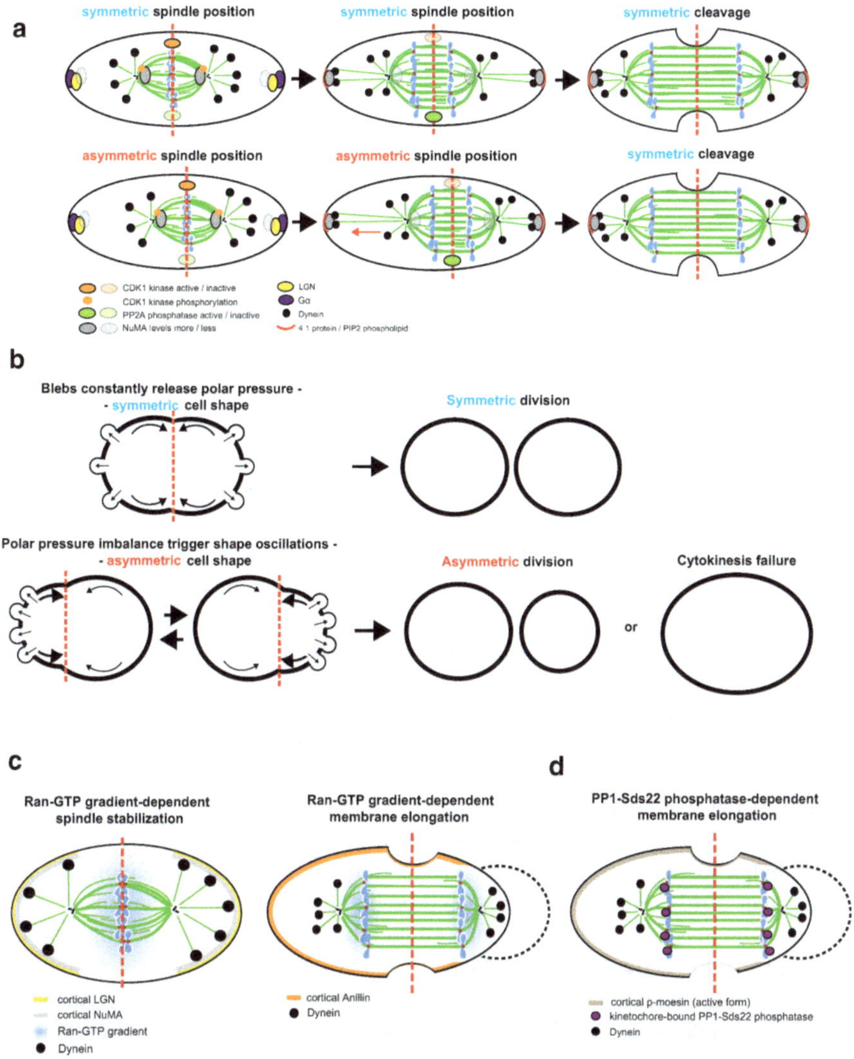

Fig. 14.5 Anaphase mechanisms correct asymmetries in spindle positioning to ensure symmetric cell division. (**a**) In anaphase, the dephosphorylation of CDK1 sites on NuMA and the binding to the 4.1 proteins allow an enhanced recruitment to the cell cortex generating dynein-dependent pulling forces that re-centre the mitotic spindle. (**b**) Cortical blebbing works as a polar pressure-releasing mechanism that stabilizes cell size oscillations and enables symmetric cell division. (**c**) Chromosome-associated Ran-GTP gradient stabilizes the spindle in metaphase by clearing LGN and NuMA from the equatorial cortex; later in anaphase, it causes membrane elongation by clearing Anilin from the lateral cortex in a proximity-dependent manner. (**d**) Finally, we hypothesize that similar to Ran-GTP gradient, the kinetochore-associated PP1-Sds22 phosphatase may also enable membrane elongation by dephosphorylating cortical moesin at the mitotic exit and thus provide a final mechanism for symmetric cell division

at the cell poles allows cell to compensate for shape oscillations during cytokinesis to ensure a stable position of the cytokinetic furrow and the symmetry of cell division (Fig. 14.5b).

14.10 Membrane Elongation in Anaphase

A final mechanism that ensures the symmetry of cell division is the chromatin-bound Ran gradient, which links the position of the mitotic spindle to the regulation of the cell cortex. Ran is a small GTPase, which forms a gradient around the chromatin in mitotic cells, due to the binding of its guanine exchange factor, RCC1, on chromosomes. The Ran-GTP gradient orchestrates bipolar spindle formation in acentrosomal mitoses, as originally found in *Xenopus* egg extracts (Carazo-Salas et al. 1999; Kalab et al. 1999; Ohba et al. 1999; Wilde and Zheng 1999; Zhang et al. 1999) and may contribute to spindle formation in somatic cells (Gruss et al. 2002; Moore et al. 2002). Ran-GTP releases spindle-supporting proteins from α- and β-importin to promote spindle formation (Gruss et al. 2001; Nachury et al. 2001; Wiese et al. 2001). The Ran-GTP gradient is also thought to polarize actin in mouse oocytes to ensure the asymmetric meiotic division by regulating localization of the cortical actomyosin proteins, such as myosin II (Deng et al. 2007; Yi et al. 2011; Dehapiot et al. 2013), Arp2/3 complex and the Cdc42 GTPase. It further controls spindle orientation by regulating the cortical localization of Pins and Mud in *Drosophila* neural stem cells (Wee et al. 2011; Bird et al. 2013) and their orthologs LGN and NuMA in human cells (Bird et al. 2013). In metaphase the Ran-GTP gradient specifically restricts the localization of LGN and NuMA to the lateral cortex by clearing them from the apical and basal membrane, which helps to maintain a correct axis of division (Fig. 14.5c; Kiyomitsu and Cheeseman 2012). Finally, the Ran-GTP gradient plays a role in anaphase if the chromosome masses are closer to one cell membrane, as it will reduce the cortical levels of Anilin (Silverman-Gavrila et al. 2008). This will cause a membrane relaxation and elongation that will compensate for the asymmetry of the anaphase spindle and yield equal-sized daughter cells (Kiyomitsu and Cheeseman 2013).

14.11 Conclusions and Outlook

The presence of numerous cellular mechanisms at the level of the mitotic spindle, chromosomes and the cell cortex ensuring the formation of equal-sized daughter cells indicates that cells actively impose symmetric cell divisions. Many of these mechanisms involve negative spatial regulatory systems, reminiscent of the Min system controlling the position of the FtsZ cytokinesis ring in prokaryotes, pointing to common strategies (Adams and Errington 2009; Kiekebusch and Thanbichler 2014; Haeusser and Margolin 2016). The principles of other mechanisms, such as the forces that ensure the symmetric position of the chromosomes within the spindle, remain to be precisely defined (Dumont and Mitchison 2009). Beyond these detailed mechanistic aspects, there are a number of general questions to address in the future. First, we postulate that the study of mechanisms ensuring symmetric cell division is also relevant to understand asymmetric cell division, as cells may use symmetricity as an initial reference point to break symmetry and thereafter impose a controlled asymmetry. Strikingly, many asymmetric cell divisions emerge from an initially symmetric configuration, be at the level of the spindle, such as *D. melanogaster* neuroblasts, or at the level of spindle positioning, such as *C. elegans* embryos. Such a hypothesis needs, however, to be proven; moreover, it will be essential to identify the mechanistic switches that break the symmetry during cell division.

A second key point will be to study how known mechanisms might work in other contexts. For example, it might be worthwhile to explore how F-actin cytoskeleton-mediated cortical tension, which is essential in acentrosomal cells, influences symmetric cell divisions in centrosomal cells. Since myosin X helps to position the spindle by linking actin and microtubules in centrosomal cells, it is possible that the actomyosin cytoskeleton provides redundancy to the astral microtubule/dynein-driven spindle positioning. Indeed, *Drosophila* embryos devoid of centrioles have only mild developmental defects probably due to the lack of cilia (Basto et al. 2008). One could speculate that the F-actin cytoskeleton provides centrosome-free flies sufficient compensatory mechanics to ensure symmetric cell divisions and develop a whole organism (N.B. these flies nevertheless fail at the asymmetric neuroblast divisions). Such an approach might be particularly relevant, since centrosomes have been found to also nucleate actin filaments (Farina et al. 2016). Another example could be membrane blebbing, which so far has been only studied in the context of non-polarized cells. It is tempting to hypothesize that this mechanism could also contribute to regulate the symmetric cell division in polarized cells. Therefore, it would be interesting to explore if the cortical tension release by blebs also contributes to cytokinetic furrow positioning in symmetrically dividing epithelial cells, where spindle position plays a major role in defining the size of daughter cells (Bell et al. 2015; Carvalho et al. 2015; Woolner and Papalopulu 2012).

Finally, it is likely that new mechanisms controlling the symmetricity of cell division might emerge. For example, it was recently found that segregating chromosomes induce polar relaxation during late anaphase (Rodrigues et al. 2015). This relaxation was known to be caused by the clearing of actin and phosphorylated ezrin/radixin/moesin (ERM) complex from the poles before cleavage furrow formation (Hickson et al. 2006). This recent study demonstrated that the proximity to chromatin in late anaphase is the key signal for relaxation and membrane blebbing at the poles. This pathway depends on the regulatory PP1 phosphatase subunit Sds22, which specifically binds to kinetochores until late anaphase. Depletion of Sds22 delays anaphase cell elongation and causes cytokinesis defects (Kunda et al. 2012). Even though PP1-Sds22-dependent polar relaxation has not been linked to symmetric cell division, it would be exciting to test whether it contributes to symmetric cell divisions in cases where the anaphase spindle is positioned asymmetrically. Such a mechanism would again link spindle positioning to the control of the cortical organization, providing an additional feedback loop to ensure symmetric cell division at every step of mitotic progression.

Acknowledgements We thank Patrick Viollier (University of Geneva) and the Meraldi lab members for critical discussions of the manuscript. P.M. is funded by an SNF-project grant (31003A_160006) and the University of Geneva.

References

Adams DW, Errington J (2009) Bacterial cell division: assembly, maintenance and disassembly of the Z ring. Nat Rev Microbiol 7:642–653

Almonacid M, Ahmed WW, Bussonnier M, Mailly P, Betz T, Voituriez R, Gov NS, Verlhac MH (2015) Active diffusion positions the nucleus in mouse oocytes. Nat Cell Biol 17:470–479

Bajaj J, Zimdahl B, Reya T (2015) Fearful symmetry: subversion of asymmetric division in cancer development and progression. Cancer Res 75:792–797

Baruni JK, Munro EM, von Dassow G (2008) Cytokinetic furrowing in toroidal, binucleate and anucleate cells in *C. elegans* embryos. J Cell Sci 121:306–316

Basto R, Brunk K, Vinadogrova T, Peel N, Franz A, Khodjakov A, Raff JW (2008) Centrosome amplification can initiate tumorigenesis in flies. Cell 133:1032–1042

Bell GP, Fletcher GC, Brain R, Thompson BJ (2015) Aurora kinases phosphorylate Lgl to induce mitotic spindle orientation in Drosophila epithelia. Curr Biol 25:61–68

Bird SL, Heald R, Weis K (2013) RanGTP and CLASP1 cooperate to position the mitotic spindle. Mol Biol Cell 24:2506–2514

Bringmann H, Hyman AA (2005) A cytokinesis furrow is positioned by two consecutive signals. Nature 436:731–734

Cadart C, Zlotek-Zlotkiewicz E, Le Berre M, Piel M, Matthews HK (2014) Exploring the function of cell shape and size during mitosis. Dev Cell 29:159–169

Canman JC, Cameron LA, Maddox PS, Straight A, Tirnauer JS, Mitchison TJ, Fang G, Kapoor TM, Salmon ED (2003) Determining the position of the cell division plane. Nature 424:1074–1078

Carazo-Salas RE, Guarguaglini G, Gruss OJ, Segref A, Karsenti E, Mattaj IW (1999) Generation of GTP-bound Ran by RCC1 is required for chromatin-induced mitotic spindle formation. Nature 400:178–181

Carvalho CA, Moreira S, Ventura G, Sunkel CE, Morais-de-Sa E (2015) Aurora A triggers Lgl cortical release during symmetric division to control planar spindle orientation. Curr Biol 25:53–60

Chaigne A, Verlhac MH, Terret ME (2012) Spindle positioning in mammalian oocytes. Exp Cell Res 318:1442–1447

Chaigne A, Campillo C, Voituriez R, Gov NS, Sykes C, Verlhac MH, Terret ME (2016) F-actin mechanics control spindle centring in the mouse zygote. Nat Commun 7:10253

Charras GT, Yarrow JC, Horton MA, Mahadevan L, Mitchison TJ (2005) Non-equilibration of hydrostatic pressure in blebbing cells. Nature 435:365–369

Charras GT, Hu CK, Coughlin M, Mitchison TJ (2006) Reassembly of contractile actin cortex in cell blebs. J Cell Biol 175:477–490

Chew TG, Lorthongpanich C, Ang WX, Knowles BB, Solter D (2012) Symmetric cell division of the mouse zygote requires an actin network. Cytoskeleton 69:1040–1046

Collins ES, Balchand SK, Faraci JL, Wadsworth P, Lee WL (2012) Cell cycle-regulated cortical dynein/dynactin promotes symmetric cell division by differential pole motion in anaphase. Mol Biol Cell 23:3380–3390

Conklin EG (1917) Effects of centrifugal force on the structure and development of the eggs of crepidula. J Exp Zool 22:311–419

Courtois A, Schuh M, Ellenberg J, Hiiragi T (2012) The transition from meiotic to mitotic spindle assembly is gradual during early mammalian development. J Cell Biol 198:357–370

Cramer LP, Mitchison TJ (1997) Investigation of the mechanism of retraction of the cell margin and rearward flow of nodules during mitotic cell rounding. Mol Biol Cell 8:109–119

Dehapiot B, Carriere V, Carroll J, Halet G (2013) Polarized Cdc42 activation promotes polar body protrusion and asymmetric division in mouse oocytes. Dev Biol 377:202–212

Delaunay D, Cortay V, Patti D, Knoblauch K, Dehay C (2014) Mitotic spindle asymmetry: a Wnt/PCP-regulated mechanism generating asymmetrical division in cortical precursors. Cell Rep 6:400–414

den Elzen N, Buttery CV, Maddugoda MP, Ren G, Yap AS (2009) Cadherin adhesion receptors orient the mitotic spindle during symmetric cell division in mammalian epithelia. Mol Biol Cell 20:3740–3750

Deng M, Suraneni P, Schultz RM, Li R (2007) The Ran GTPase mediates chromatin signaling to control cortical polarity during polar body extrusion in mouse oocytes. Dev Cell 12:301–308

Dumont S, Mitchison TJ (2009) Force and length in the mitotic spindle. Curr Biol 19:R749–R761

Dumont NA, Wang YX, von Maltzahn J, Pasut A, Bentzinger CF, Brun CE, Rudnicki MA (2015) Dystrophin expression in muscle stem cells regulates their polarity and asymmetric division. Nat Med 21:1455–1463

Durgan J, Kaji N, Jin D, Hall A (2011) Par6B and atypical PKC regulate mitotic spindle orientation during epithelial morphogenesis. J Biol Chem 286:12461–12474

Farina F, Gaillard J, Guerin C, Coute Y, Sillibourne J, Blanchoin L, Thery M (2016) The centrosome is an actin-organizing centre. Nat Cell Biol 18:65–75

Fink J, Carpi N, Betz T, Betard A, Chebah M, Azioune A, Bornens M, Sykes C, Fetler L, Cuvelier D et al (2011) External forces control mitotic spindle positioning. Nat Cell Biol 13:771–778

Fish JL, Kosodo Y, Enard W, Paabo S, Huttner WB (2006) Aspm specifically maintains symmetric proliferative divisions of neuroepithelial cells. Proc Natl Acad Sci USA 103:10438–10443

Ganem NJ, Godinho SA, Pellman D (2009) A mechanism linking extra centrosomes to chromosomal instability. Nature 460:278–282

Gibson WT, Veldhuis JH, Rubinstein B, Cartwright HN, Perrimon N, Brodland GW, Nagpal R, Gibson MC (2011) Control of the mitotic cleavage plane by local epithelial topology. Cell 144:427–438

Godin JD, Colombo K, Molina-Calavita M, Keryer G, Zala D, Charrin BC, Dietrich P, Volvert ML, Guillemot F, Dragatsis I et al (2010) Huntingtin is required for mitotic spindle orientation and mammalian neurogenesis. Neuron 67:392–406

Gotz M, Huttner WB (2005) The cell biology of neurogenesis. Nat Rev Mol Cell Biol 6:777–788

Gruss OJ, Carazo-Salas RE, Schatz CA, Guarguaglini G, Kast J, Wilm M, Le Bot N, Vernos I, Karsenti E, Mattaj IW (2001) Ran induces spindle assembly by reversing the inhibitory effect of importin alpha on TPX2 activity. Cell 104:83–93

Gruss OJ, Wittmann M, Yokoyama H, Pepperkok R, Kufer T, Sillje H, Karsenti E, Mattaj IW, Vernos I (2002) Chromosome-induced microtubule assembly mediated by TPX2 is required for spindle formation in HeLa cells. Nat Cell Biol 4:871–879

Guilgur LG, Prudencio P, Ferreira T, Pimenta-Marques AR, Martinho RG (2012) Drosophila aPKC is required for mitotic spindle orientation during symmetric division of epithelial cells. Development 139:503–513

Haeusser DP, Margolin W (2016) Splitsville: structural and functional insights into the dynamic bacterial Z ring. Nat Rev Microbiol 14:305–319

Hao Y, Du Q, Chen X, Zheng Z, Balsbaugh JL, Maitra S, Shabanowitz J, Hunt DF, Macara IG (2010) Par3 controls epithelial spindle orientation by aPKC-mediated phosphorylation of apical Pins. Curr Biol 20:1809–1818

Harris A (1973) Location of cellular adhesions to solid substrata. Dev Biol 35:97–114

Haydar TF, Ang E Jr, Rakic P (2003) Mitotic spindle rotation and mode of cell division in the developing telencephalon. Proc Natl Acad Sci USA 100:2890–2895

Hickson GR, Echard A, O'Farrell PH (2006) Rho-kinase controls cell shape changes during cytokinesis. Curr Biol 16:359–370

Hughes AFW (1952) The mitotic cycle; the cytoplasm and nucleus during interphase and mitosis, 1st edn. Butterworths Scientific Publications, London

Insolera R, Bazzi H, Shao W, Anderson KV, Shi SH (2014) Cortical neurogenesis in the absence of centrioles. Nat Neurosci 17:1528–1535

Jaffe AB, Kaji N, Durgan J, Hall A (2008) Cdc42 controls spindle orientation to position the apical surface during epithelial morphogenesis. J Cell Biol 183:625–633

Jongsma ML, Berlin I, Neefjes J (2015) On the move: organelle dynamics during mitosis. Trends Cell Biol 25:112–124

Kalab P, Pu RT, Dasso M (1999) The ran GTPase regulates mitotic spindle assembly. Curr Biol 9:481–484

Kaltschmidt JA, Davidson CM, Brown NH, Brand AH (2000) Rotation and asymmetry of the mitotic spindle direct asymmetric cell division in the developing central nervous system. Nat Cell Biol 2:7–12

Kardon JR, Vale RD (2009) Regulators of the cytoplasmic dynein motor. Nat Rev Mol Cell Biol 10:854–865

Kiekebusch D, Thanbichler M (2014) Spatiotemporal organization of microbial cells by protein concentration gradients. Trends Microbiol 22:65–73

Kitajima A, Fuse N, Isshiki T, Matsuzaki F (2010) Progenitor properties of symmetrically dividing Drosophila neuroblasts during embryonic and larval development. Dev Biol 347:9–23

Kiyomitsu T, Cheeseman IM (2012) Chromosome- and spindle-pole-derived signals generate an intrinsic code for spindle position and orientation. Nat Cell Biol 14:311–317

Kiyomitsu T, Cheeseman IM (2013) Cortical dynein and asymmetric membrane elongation coordinately position the spindle in anaphase. Cell 154:391–402

Knoblich JA (2008) Mechanisms of asymmetric stem cell division. Cell 132:583–597

Konno D, Shioi G, Shitamukai A, Mori A, Kiyonari H, Miyata T, Matsuzaki F (2008) Neuroepithelial progenitors undergo LGN-dependent planar divisions to maintain self-renewability during mammalian neurogenesis. Nat Cell Biol 10:93–101

Kosodo Y, Roper K, Haubensak W, Marzesco AM, Corbeil D, Huttner WB (2004) Asymmetric distribution of the apical plasma membrane during neurogenic divisions of mammalian neuroepithelial cells. EMBO J 23:2314–2324

Kotak S, Busso C, Gonczy P (2013) NuMA phosphorylation by CDK1 couples mitotic progression with cortical dynein function. EMBO J 32:2517–2529

Kunda P, Pelling AE, Liu T, Baum B (2008) Moesin controls cortical rigidity, cell rounding, and spindle morphogenesis during mitosis. Curr Biol 18:91–101

Kunda P, Rodrigues NT, Moeendarbary E, Liu T, Ivetic A, Charras G, Baum B (2012) PP1-mediated moesin dephosphorylation couples polar relaxation to mitotic exit. Curr Biol 22:231–236

Kwon M, Godinho SA, Chandhok NS, Ganem NJ, Azioune A, Thery M, Pellman D (2008) Mechanisms to suppress multipolar divisions in cancer cells with extra centrosomes. Genes Dev 22:2189–2203

Kwon M, Bagonis M, Danuser G, Pellman D (2015) Direct microtubule-binding by myosin-10 Orients Centrosomes toward Retraction Fibers and Subcortical Actin Clouds. Dev Cell 34:323–337

Lancaster OM, Le Berre M, Dimitracopoulos A, Bonazzi D, Zlotek-Zlotkiewicz E, Picone R, Duke T, Piel M, Baum B (2013) Mitotic rounding alters cell geometry to ensure efficient bipolar spindle formation. Dev Cell 25:270–283

Leber B, Maier B, Fuchs F, Chi J, Riffel P, Anderhub S, Wagner L, Ho AD, Salisbury JL, Boutros M, et al. (2010) Proteins required for centrosome clustering in cancer cells. Sci Transl Med 2:33ra38

Liu KC, Jacobs DT, Dunn BD, Fanning AS, Cheney RE (2012) Myosin-X functions in polarized epithelial cells. Mol Biol Cell 23:1675–1687

Lu MS, Johnston CA (2013) Molecular pathways regulating mitotic spindle orientation in animal cells. Development 140:1843–1856

Lu B, Roegiers F, Jan LY, Jan YN (2001) Adherens junctions inhibit asymmetric division in the Drosophila epithelium. Nature 409:522–525

Luxenburg C, Pasolli HA, Williams SE, Fuchs E (2011) Developmental roles for Srf, cortical cytoskeleton and cell shape in epidermal spindle orientation. Nat Cell Biol 13:203–214

Martynoga B, Drechsel D, Guillemot F (2012) Molecular control of neurogenesis: a view from the mammalian cerebral cortex. Cold Spring Harb Perspect Biol 4:a008359

McConnell CH (1930) The mitosis found in Hydra Science. Science 72:170

Merdes A, Ramyar K, Vechio JD, Cleveland DW (1996) A complex of NuMA and cytoplasmic dynein is essential for mitotic spindle assembly. Cell 87:447–458

Minc N, Burgess D, Chang F (2011) Influence of cell geometry on division-plane positioning. Cell 144:414–426

Moore W, Zhang C, Clarke PR (2002) Targeting of RCC1 to chromosomes is required for proper mitotic spindle assembly in human cells. Curr Biol 12:1442–1447

Mora-Bermudez F, Matsuzaki F, Huttner WB (2014) Specific polar subpopulations of astral microtubules control spindle orientation and symmetric neural stem cell division. eLife 3:e02875

Morin X, Bellaiche Y (2011) Mitotic spindle orientation in asymmetric and symmetric cell divisions during animal development. Dev Cell 21:102–119

Nachury MV, Maresca TJ, Salmon WC, Waterman-Storer CM, Heald R, Weis K (2001) Importin beta is a mitotic target of the small GTPase Ran in spindle assembly. Cell 104:95–106

Nguyen-Ngoc T, Afshar K, Gonczy P (2007) Coupling of cortical dynein and G alpha proteins mediates spindle positioning in *Caenorhabditis elegans*. Nat Cell Biol 9:1294–1302

Noatynska A, Gotta M, Meraldi P (2012) Mitotic spindle (DIS)orientation and DISease: cause or consequence? J Cell Biol 199:1025–1035

Noctor SC, Martinez-Cerdeno V, Kriegstein AR (2008) Distinct behaviors of neural stem and progenitor cells underlie cortical neurogenesis. J Comp Neurol 508:28–44

O'Connell CB, Wang YL (2000) Mammalian spindle orientation and position respond to changes in cell shape in a dynein-dependent fashion. Mol Biol Cell 11:1765–1774

Ohba T, Nakamura M, Nishitani H, Nishimoto T (1999) Self-organization of microtubule asters induced in Xenopus egg extracts by GTP-bound Ran. Science 284:1356–1358

Pease JC, Tirnauer JS (2011) Mitotic spindle misorientation in cancer--out of alignment and into the fire. J Cell Sci 124:1007–1016

Piekny AJ, Glotzer M (2008) Anillin is a scaffold protein that links RhoA, actin, and myosin during cytokinesis. Curr Biol 18:30–36

Postiglione MP, Juschke C, Xie Y, Haas GA, Charalambous C, Knoblich JA (2011) Mouse inscuteable induces apical-basal spindle orientation to facilitate intermediate progenitor generation in the developing neocortex. Neuron 72:269–284

Prescott DM (1955) Relations between cell growth and cell division. I. Reduced weight, cell volume, protein content, and nuclear volume of *Amoeba proteus* from division to division. Exp Cell Res 9:328–337

Prescott DM (1956) Relation between cell growth and cell division. II. The effect of cell size on cell growth rate and generation time in *Amoeba proteus*. Exp Cell Res 11:86–94

Quintyne NJ, Reing JE, Hoffelder DR, Gollin SM, Saunders WS (2005) Spindle multipolarity is prevented by centrosomal clustering. Science 307:127–129

Rappaport R (1961) Experiments concerning the cleavage stimulus in sand dollar eggs. J Exp Zool 148:81–89

Reinsch S, Karsenti E (1994) Orientation of spindle axis and distribution of plasma membrane proteins during cell division in polarized MDCKII cells. J Cell Biol 126:1509–1526

Ren X, Weisblat DA (2006) Asymmetrization of first cleavage by transient disassembly of one spindle pole aster in the leech *Helobdella robusta*. Dev Biol 292:103–115

Rieder CL, Khodjakov A, Paliulis LV, Fortier TM, Cole RW, Sluder G (1997) Mitosis in vertebrate somatic cells with two spindles: implications for the metaphase/anaphase transition checkpoint and cleavage. Proc Natl Acad Sci USA 94:5107–5112

Rodrigues NT, Lekomtsev S, Jananji S, Kriston-Vizi J, Hickson GR, Baum B (2015) Kinetochore-localized PP1-Sds22 couples chromosome segregation to polar relaxation. Nature 524:489–492

Sabino D, Gogendeau D, Gambarotto D, Nano M, Pennetier C, Dingli F, Arras G, Loew D, Basto R (2015) Moesin is a major regulator of centrosome behavior in epithelial cells with extra centrosomes. Curr Biol 25:879–889

Sato H, Izutsu K (1974) Birefringence in mitosis of the spermatocyte of grasshopper chrysochraon japonicus. Time-lapse motion picture. George W. Colburn Laboratory, Chicago, IL

Schuh M, Ellenberg J (2007) Self-organization of MTOCs replaces centrosome function during acentrosomal spindle assembly in live mouse oocytes. Cell 130:484–498

Sedzinski J, Biro M, Oswald A, Tinevez JY, Salbreux G, Paluch E (2011) Polar actomyosin contractility destabilizes the position of the cytokinetic furrow. Nature 476:462–466

Shuster CB, Burgess DR (2002) Transitions regulating the timing of cytokinesis in embryonic cells. Curr Biol 12:854–858

Silverman-Gavrila RV, Hales KG, Wilde A (2008) Anillin-mediated targeting of peanut to pseudocleavage furrows is regulated by the GTPase Ran. Mol Biol Cell 19:3735–3744

Sung Y, Tzur A, Oh S, Choi W, Li V, Dasari RR, Yaqoob Z, Kirschner MW (2013) Size homeostasis in adherent cells studied by synthetic phase microscopy. Proc Natl Acad Sci USA 110:16687–16692

Tan CH, Gasic I, Huber-Reggi SP, Dudka D, Barisic M, Maiato H, Meraldi P (2015) The equatorial position of the metaphase plate ensures symmetric cell divisions. eLife 4:e05124

Tinevez JY, Schulze U, Salbreux G, Roensch J, Joanny JF, Paluch E (2009) Role of cortical tension in bleb growth. Proc Natl Acad Sci USA 106:18581–18586

Tolic-Norrelykke IM (2010) Force and length regulation in the microtubule cytoskeleton: lessons from fission yeast. Curr Opin Cell Biol 22:21–28

Toyoshima F, Nishida E (2007) Integrin-mediated adhesion orients the spindle parallel to the substratum in an EB1- and myosin X-dependent manner. EMBO J 26:1487–1498

Tzur A, Kafri R, LeBleu VS, Lahav G, Kirschner MW (2009) Cell growth and size homeostasis in proliferating animal cells. Science 325:167–171

Weber KL, Sokac AM, Berg JS, Cheney RE, Bement WM (2004) A microtubule-binding myosin required for nuclear anchoring and spindle assembly. Nature 431:325–329

Wee B, Johnston CA, Prehoda KE, Doe CQ (2011) Canoe binds RanGTP to promote Pins(TPR)/Mud-mediated spindle orientation. J Cell Biol 195:369–376

Wiese C, Wilde A, Moore MS, Adam SA, Merdes A, Zheng Y (2001) Role of importin-beta in coupling Ran to downstream targets in microtubule assembly. Science 291:653–656

Wilde A, Zheng Y (1999) Stimulation of microtubule aster formation and spindle assembly by the small GTPase Ran. Science 284:1359–1362

Woolner S, Papalopulu N (2012) Spindle position in symmetric cell divisions during epiboly is controlled by opposing and dynamic apicobasal forces. Dev Cell 22:775–787

Yamashita YM, Yuan H, Cheng J, Hunt AJ (2010) Polarity in stem cell division: asymmetric stem cell division in tissue homeostasis. Cold Spring Harb Perspect Biol 2:a001313

Yi K, Unruh JR, Deng M, Slaughter BD, Rubinstein B, Li R (2011) Dynamic maintenance of asymmetric meiotic spindle position through Arp2/3-complex-driven cytoplasmic streaming in mouse oocytes. Nat Cell Biol 13:1252–1258

Yu XJ, Yi Z, Gao Z, Qin D, Zhai Y, Chen X, Ou-Yang Y, Wang ZB, Zheng P, Zhu MS et al (2014) The subcortical maternal complex controls symmetric division of mouse zygotes by regulating F-actin dynamics. Nat Commun 5:4887

Zhang C, Hughes M, Clarke PR (1999) Ran-GTP stabilises microtubule asters and inhibits nuclear assembly in Xenopus egg extracts. J Cell Sci 112(Pt 14):2453–2461

Chapter 15
A Comparative Perspective on Wnt/β-Catenin Signalling in Cell Fate Determination

Clare L. Garcin and Shukry J. Habib

Abstract The Wnt/β-catenin pathway is an ancient and highly conserved signalling pathway that plays fundamental roles in the regulation of embryonic development and adult homeostasis. This pathway has been implicated in numerous cellular processes, including cell proliferation, differentiation, migration, morphological changes and apoptosis. In this chapter, we aim to illustrate with specific examples the involvement of Wnt/β-catenin signalling in cell fate determination. We discuss the roles of the Wnt/β-catenin pathway in specifying cell fate throughout evolution, how its function in patterning during development is often reactivated during regeneration and how perturbation of this pathway has negative consequences for the control of cell fate.

The origin of all life was a single cell that had the capacity to respond to cues from the environment. With evolution, multicellular organisms emerged, and as a result, subsets of cells arose to form tissues able to respond to specific instructive signals and perform specialised functions. This complexity and specialisation required two types of messages to direct cell fate: intra- and intercellular. A fundamental question in developmental biology is to understand the underlying mechanisms of cell fate choice. Amongst the numerous external cues involved in the generation of cellular diversity, a prominent pathway is the Wnt signalling pathway in all its forms.

15.1 The Wnt Pathway

Nineteen known Wnt ligands have been identified in mammalian cells to date (Widelitz 2005). Wnt ligands are secreted signalling molecules of approximately 350 amino acids (Janda et al. 2012) that have several key shared features including multiple cysteine residues, a peptide directing the protein for secretion and

C.L. Garcin • S.J. Habib (✉)
Centre for Stem Cells and Regenerative Medicine, King's College, London, UK
e-mail: shukry.habib@kcl.ac.uk

J.-P. Tassan, J.Z. Kubiak (eds.), *Asymmetric Cell Division in Development, Differentiation and Cancer*, Results and Problems in Cell Differentiation 61,
DOI 10.1007/978-3-319-53150-2_15

numerous glycosylation sites (Janda et al. 2012; Barker 2008). Wnt ligands are subject to post-translational modifications, including the addition of a palmitate molecule to a conserved serine residue (e.g. Ser209 in Wnt3a) (Takada et al. 2006; Willert et al. 2003), which confers hydrophobicity and is essential for activity.

Wnt proteins are often secreted locally and presented to responsive cells in a spatially restricted manner. Specific instances where this has been clearly demonstrated include during homeostasis of mammalian intestinal epithelium (Farin et al. 2016; Sato et al. 2011), embryonic and post-embryonic cell polarity in *C. elegans* (Goldstein et al. 2006) and the developing *Drosophila* embryo (van den Heuvel et al. 1989; Strand and Micchelli 2011; Alexandre et al. 2014), reviewed in Clevers et al. (2014). Wnt ligand availability is mediated by a number of proteins (Table 15.1), followed by binding to the cell surface receptor Frizzled (Fz) in association with the low-density-lipoprotein-receptor-related proteins 5 and 6 (Lrp5/6) (Niehrs 2012). Frizzled proteins are seven-pass transmembrane receptors that possess a cysteine-rich domain at the N-terminus, to which Wnt binds (Janda et al. 2012). Recently, the crystallisation of Xenopus Wnt8 (XWnt8) in complex with the mouse Frizzled 8 receptor revealed that the ligand-associated palmitate and Frizzled receptor cysteine-rich domain are largely responsible for the interaction of Wnt ligands with their receptors (Janda et al. 2012). Additionally, Chu et al. (2013) demonstrated that the N-terminal domain linker region in WntD (*Drosophila*), which contains six α-helices and two antiparallel β strands that form a hairpin, is an Lrp6 binding site. Lrp6 binds via its β-propeller region, where Dickkopf 1 (Dkk1) can also bind (Chu et al. 2013).

The (canonical) Wnt/β-catenin pathway (Fig. 15.1) centres around the accumulation or degradation of β-catenin. The translocation of β-catenin to the nucleus appears to be a critical step for the transduction of Wnt signals. In the absence of a Wnt ligand, β-catenin is phosphorylated by two kinases, which are part of the 'destruction complex': casein kinase-1 (CK-1) and glycogen synthase kinase-3 (GSK-3) (Rubinfeld et al. 1996; Behrens et al. 1998; Aberle et al. 1997). Also in the 'destruction complex' are Axin and adenomatous polyposis coli (APC), which are thought to act as scaffold proteins to facilitate the interaction between CK-1, GSK-3 and β-catenin (Ikeda et al. 1998; Ha et al. 2004). Whilst traditionally it was thought that the destruction complex sequestered β-catenin, more recently it has been suggested that it cycles β-catenin. It has been suggested that GSK-3 and CK1 first phosphorylate β-catenin and subsequently phosphorylate APC. This allows APC to displace Axin from β-catenin, freeing Axin to interact with another β-catenin molecule (Xing et al. 2003). However, the precise mechanism of β-catenin destruction remains unclear, as there is no direct evidence that β-catenin must dissociate from the complex in order to be degraded (Li et al. 2012). Future research will help to elucidate the mechanisms underlying the precise control of cellular β-catenin levels.

Once phosphorylated β-catenin is a target for ubiquitination by the E3 ubiquitin ligase βTrCP, which may act within the 'destruction complex' and facilitate the direct removal of β-catenin via proteasomal degradation (Li et al. 2012). Recently, it has been shown that Hippo transducers YAP/TAZ bind directly to Axin and are

Table 15.1 Wnt signalling inhibitors/enhancers

Inhibitor	Mode of action	Selected references
Secreted Frizzled-Related Proteins (sFRPs)	Interact with ligands via the cysteine-rich domain to prevent receptor binding to Wnt ligands	Lin et al. (1997)
Dickkopfs (DKKs)	High-affinity ligands for Lrp6. Bind to Lrp6 and disrupt Wnt-induced Frizzled–Lrp complexes. Does not bind Wnt or Frizzled	Moon et al. (2004) and Semënov et al. (2001)
Notum	Removes palmitate moiety from Wnt ligands, preventing Wnt from interacting with Frizzled receptors	Kakugawa et al. (2015)
Wnt inhibitory factor (WIF)	Disrupts interaction between Wnt and Frizzled by binding to Wnt ligands	Hsieh et al. (1999)
Wise/SOST (Sclerostin)	Wise: In vitro has been shown to block Wnt3a and Wnt1 by competing with Wnt ligands for Lrp6 binding. Can also reduce cell surface Lrp6. Binds Lrp6 receptor to modulate Lrp5/6-mediated Wnt signalling SOST: Binds to the first YWTD-EGF repeat domains of Lrp5/6. Can also disrupt the Fz8–Lrp6 complex	Yanagita et al. (2004), Blish et al. (2008), Guidato and Itasaki (2007), Semënov et al. (2005) and Li et al. (2005)
Enhancer	**Mode of action**	**Selected references**
R-spondins	Inhibition of E3 ubiquitin ligases, reduction of β-catenin degradation. Lgr5–Rspondin complexes also neutralise Znrf3 and Rnf43, which remove receptors from the cell surface. Therefore, this complex increases receptor availability	de Lau et al. (2014) and Snippert et al. (2010a)
Norrin	Binds to Frizzled4/Lrp5 and initiates gene transcription	Xu et al. (2004)

essential for βTrCP recruitment to the 'destruction complex' and subsequent inactivation (Azzolin et al. 2012, 2014).

Disruption of the 'destruction complex' occurs when a Wnt/Fz/LRP complex forms. The intracellular tail of Lrp6 becomes phosphorylated along with Dishevelled (Dsh), a protein that can interact with APC, and is important for signal transduction (MacDonald and He 2012; Gao and Chen 2010; Metcalfe et al. 2010). Axin is also recruited to this complex at the cell membrane via Dsh (He 2004; Tolwinski and Wieschaus 2004; Cliffe et al. 2003). Although the recruitment of APC and Axin to the cell membrane is presumably sufficient to prevent GSK3-mediated phosphorylation of β-catenin (Li et al. 2012), it has been suggested that this in turn leads to inhibition of β-catenin ubiquitination (Hernández et al. 2012; Li et al. 2012). Therefore, β-catenin accumulates in the cytoplasm,

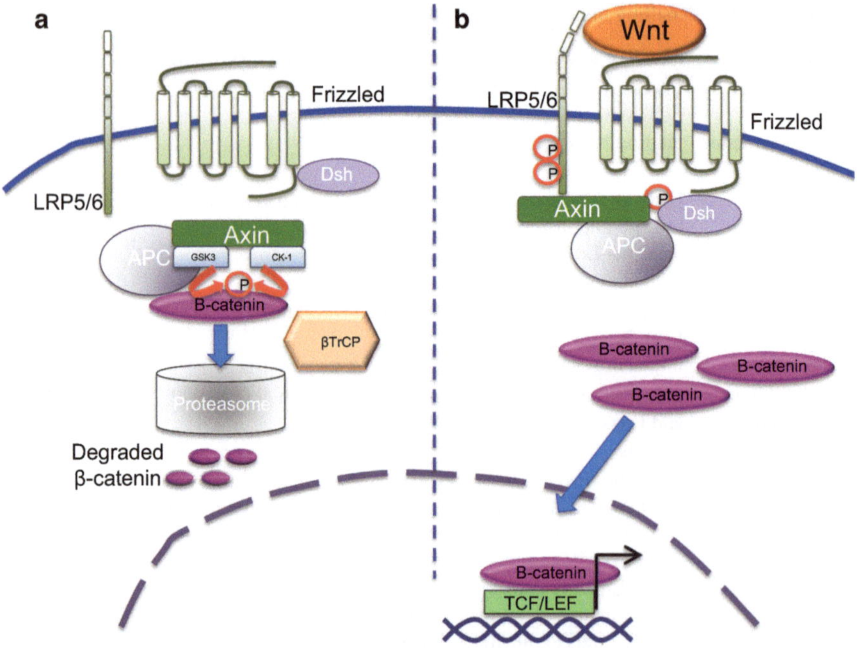

Fig. 15.1 The basics of Wnt/β catenin signalling. (**a**) In the absence of Wnt binding to the receptors, β-catenin is phosphorylated by GSK-3 and CK1, kinases present in the 'destruction complex', which also contains the scaffolding proteins APC and Axin. Phosphorylation targets β-catenin for ubiquitination by βTrCP. βTrCP may act directly within the destruction complex to facilitate the removal of β-catenin and its subsequent destruction by the proteasome. (**b**) Wnt binding causes the formation of a Wnt–Frizzled–LRP5/6 complex. The intracellular tail of LRP5/6 and Dsh become phosphorylated and Axin and APC are recruited to the cell membrane, causing disruption of the 'destruction complex'. β-catenin can now accumulate in the cytoplasm and translocate to the nucleus, where it associates with Tcf/Lef and can initiate target gene transcription

allowing it to translocate to the nucleus, where it can interact with the DNA-binding proteins lymphoid enhancer factor/T cell factor (Lef/Tcf) and initiate the transcription of target genes. This infers that a fold change in the level of β-catenin (pre- vs. post-Wnt) can control a β-catenin-dependent binary 'on/off' switch for gene transcription. Since β-catenin is very sensitive to small fluctuations in the levels of other pathway components, this allows for β-catenin levels to act as a signalling buffer (Goentoro and Kirschner 2009).

Apart from signalling through β-catenin, Wnt ligands are also thought to signal via a number of alternative pathways, termed 'non-canonical'. With time and increasing research into these pathways, it has become less easy to separate them so linearly: Wnt5a, classically considered a 'non-canonical' ligand, can in fact activate β-catenin (van Amerongen et al. 2008; Mikels and Nusse 2006). An example of 'non-canonical' Wnt signalling is Wnt/Ca^{2+} signalling, during which Wnt binding activates phospholipase C (PLC) and initiates calcium efflux from the endoplasmic reticulum (ER). This initiates the activity of protein kinase C effectors (Lien and Fuchs 2014), such as calcium–calmodulin-dependent kinase II (CamKII) or calcium-dependent phosphatase calcineurin (CaN) (Clevers 2006; Ma and Wang 2007). Downstream of this, Lef/Tcf become phosphorylated, preventing β-catenin binding and, ultimately, gene transcription (Liu et al. 2011). More recently, it has been suggested that β-catenin and Ca^{2+} may coordinate signalling downstream of Wnt ligand binding, interdependently (Thrasivoulou et al. 2013).

Wnt/Planar cell polarity (PCP) signalling is also considered to be 'non-canonical'. Here, for example, a Wnt/Fz/Ror2 complex forms, which results in Dishevelled being targeted to the membrane. This pathway largely activates proteins involved in the organisation of actin filaments, such as the GTPases Rho and Rac, Jun and Cdc42 (Camilli and Weeraratna 2010). This interaction was first described in *Drosophila*, where there is a possibility that no Wnt ligand is involved (van Amerongen et al. 2008; Chen et al. 2008; Lawrence et al. 2002). During zebrafish gastrulation, Wnt/PCP signalling is important for the orientation of cell divisions in specification of the body axis (Gong et al. 2004).

15.2 Wnt/β-Catenin Pathway: From Body Axes Specification to Cell Fate Determination, an Evolutionary Perspective

From the simplest, single-celled origins, evolution has generated complex, multicellular organisms with varying degrees of asymmetry. Single-celled eukaryotic organisms, commonly termed protozoa, are largely symmetric around a central axis. Despite this, it has been demonstrated that some Wnt/β-catenin pathway components are asymmetrically distributed across the axis, including β-catenin, indicating some degree of polarity (Petersen and Reddien 2009a). Although individual pathway components (β-catenin, GSK and Frizzleds) are present in protozoans (Harwood 2008a, b), Wnt signalling centres appear to have evolved at the transition from unicellular to multicellular eukaryotes. Indeed, in early metazoan organisms such as Cnidarians and Poriferans, the beginnings of a Wnt signalling pathway can be found: Wnt ligands, Frizzled, sFRPs, Lrp5/6, Dishevelled, Axin, GSK-3, β-catenin, TCF and CK1 (Adamska et al. 2007, 2010; Pang et al. 2010). APC protein, on the other hand, appeared later in evolution and can be found in organisms with secondary body axes (Adamska et al. 2010; Petersen and Reddien 2009a). Body axes are used to describe the position and orientation of body structures. During development, embryos polarise along two axes, defining head and tail (anterio-posterior axis) and front and back (dorsoventral) (Petersen and Reddien 2009a). Throughout evolution, Wnt signals have provided positional cues required for the orientation and proper establishment of body axes and tissue patterning (Fig. 15.2a–c). This is evident during embryogenesis, development, adult homeostasis and tissue regeneration. In this section, we aim to illustrate

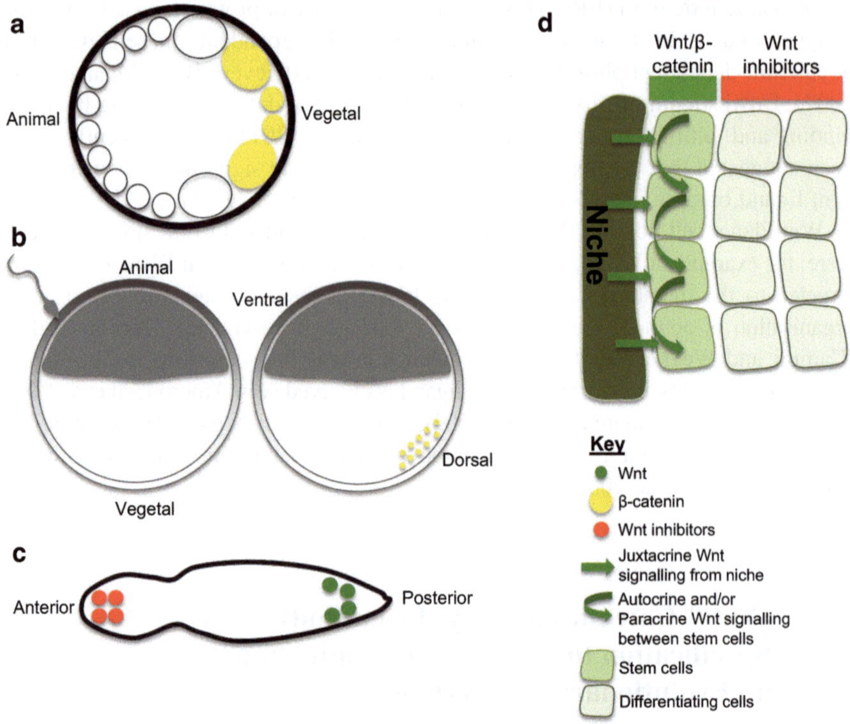

Fig. 15.2 Examples of components of Wnt/β catenin signalling in establishing body axes, cell polarisation and cell fate during development. (**a**) At the 16-cell stage in the sea urchin embryo, β-catenin localises to the vegetal pole, specifying endoderm cell fate. (**b**) Sperm entry in the Xenopus egg establishes the dorsal–ventral axis, with Dishevelled and β-catenin localising to the dorsal 'organising centre'. (**c**) In many examples in evolution, Wnt also specifies the anterior–posterior axis, shown here in Planaria. Wnt is present in the posterior, whereas Wnt inhibitors are found at the anterior. (**d**) Potential mechanism of action of Wnt signalling for stem cell maintenance in adult niches

some examples of this role of the pathway from various developmental stages across different organisms.

15.2.1 Wnt Signalling in Cell Fate During Embryogenesis and Axis Formation

The cell has a remarkable ability to confer polarity cues, such as members of the Wnt pathway, into instructive signals for cell fate specification. Throughout evolution, the Wnt pathway has shown restricted spatial distribution that appears essential for the specification of cell fate along an axis. In early metazoans such as Cnidarians and Poriferans, Wnt ligands and/or β-catenin have a polarised distribution along the oral–aboral axis during development (Wikramanayake et al. 2003; Leclère et al. 2016; Momose et al. 2008; Kusserow et al. 2005; Adamska et al. 2010). The importance of this localised distribution is evident in experiments in which the spatial orientation of Wnt components is disrupted. For example, disruption of Wnt signalling in Cnidarians causes alterations to oral–aboral tissue specification (Momose et al. 2008; Wikramanayake et al. 2003). Specifically, forced localised expression of Wnt3 in the cnidarian *Clytia hemisphaerica* resulted in the appearance of ectopic oral poles, indicating its importance in directing cell fate (Momose et al. 2008).

In bilaterian organisms, β-catenin is also important in establishing embryonic polarity, including axis specification: β-catenin is localised asymmetrically along the animal–vegetal axis of the developing *Platynereis dumerilii* (Schneider and Bowerman 2007). From the 8-cell stage to 16-cell stage, β-catenin expression is low in animal-pole tissues, and high in vegetal-pole tissues, which will form cells of the endoderm lineage (Schneider and Bowerman 2007). In the widely studied bilaterian nematode *Caenorhabditis elegans,* both Wnt and β-catenin show polarised expression to influence cell fate along the anterior–posterior axis. Specifically, the β-catenin homologues WRM-1 and SYS1 localise to the posterior side of the cell and, following division, are specific to the posterior daughter cell (Petersen and Reddien 2009b; Nakamura et al. 2005; Takeshita and Sawa 2005; Huang et al. 2007b). This is particularly important in the division of the EMS cell, which divides asymmetrically in response to a polarised exposure to Wnt (MOM-2). The anterior daughter is Wnt^{lo} and will adopt a mesodermal fate, whereas the posterior Wnt^{hi} cell forms the endoderm. Here, spatially localised MOM-2 orients the mitotic spindle in the EMS cell and induces asymmetric cell division to produce different cell fates along an axis (Goldstein et al. 2006).

In the developing Sea Urchin *Strongylocentrotus purpuratus*, β-catenin becomes localised to the nucleus of cells in the vegetal region of the embryo at the 16-cell stage (Wikramanayake et al. 2004). Dishevelled was found to be critical for β-catenin stabilisation, and its localisation to the vegetal pole may occur as early as fertilisation (Weitzel et al. 2004). Nuclear localisation of β-catenin in cells at the vegetal pole has been shown to be required for the establishment of all vegetal cell fates (Logan et al. 1999). Wnt8 is thought to control the vegetally localised β-catenin during gastrulation and endoderm specification (Wikramanayake et al. 2004).

In Drosophila, the Wnt homologue *Wingless (Wg)* has multiple roles during development, including patterning the imaginal disc, the future wing. Expression of *Wg* is initially ubiquitous across the imaginal disc, until it becomes localised to a narrow stripe along the dorsoventral boundary (Zecca et al. 1996). *Wg* proteins continue to control gene expression and growth by spreading in a gradient from the producing cells up to 15–20 cell diameters away (Morata and Struhl 2014). *Wg* gradients are thought to be dependent on *Wg* ligand-associated vesicles. For example, *Wg* ligands that associate with lipoprotein particles or secreted Wingless-interacting molecule (SWIM) have long-range signal capacities (Mulligan et al. 2012; Holstein 2012). It was originally thought that a *Wg* gradient allowed for differential gene expression (Neumann and Cohen 1997): high-level signalling initiates wing margin cell fate at the boundary (Couso et al. 1994; Nolo et al. 2000), whereas low-level signalling further from the source initiates expression of other patterning genes such as *Vestigial, Distal-less* and *Frizzled* (Zecca et al. 1996; Neumann and Cohen 1997; Sivasankaran et al. 2000; Alexandre et al. 2014). Importantly, it has also been shown that *Wg* signals are presented locally to the target cells (van den Heuvel et al. 1989). More recently, it has been demonstrated through a series of elegant studies involving the restriction of *Wg* to the cell membrane that *Wg* gradients were not essential for wing patterning (Alexandre et al. 2014). The authors suggest that an initial ubiquitous expression *of Wg* throughout the wing provides a sufficient signal to initiate further patterning of the wing.

Another important example of Wnt/β-catenin pathway activity during early development is in the specification of the Xenopus dorsoventral body axis, which is specified following fertilisation. Here, sperm entry causes accumulation of β-catenin and Dishevelled at the dorsal side of the early embryo, creating the Spemann organiser (Funayama et al. 1995; Sokol et al. 1995; Larabell et al. 1997; Schneider et al. 1996). Localised activity of Wnt/β-catenin mediates microtubule-dependent relocalisation of some cortical components to the future dorsal side of the egg, thus specifying distinct fates along the dorsoventral body axis (Heasman et al. 1994; Smith and Harland 1991; Sokol et al. 1991). Wnt/β-catenin is also involved at the late blastula stage of *Xenopus* development. In the ectoderm, Lrp6 localises to the basolateral domain of ectodermal cells, causing this pathway component to be preferentially inherited by 'deep' ectodermal cells. These cells display higher β-catenin signalling than superficial 'outer' cells and acquire a distinct cell fate: outer cells differentiate to goblet cells, which are responsible for mucus secretion, whereas deep cells differentiate to ciliated cells. Importantly, Dishevelled appears to be required for Lrp6 localisation, as in its absence, Lrp6 does not display polarisation (Huang and Niehrs 2014). It has also been shown that polarisation of atypical protein kinase C (aPKC) (mediated by external Wnt cues) and an oriented mitotic spindle is responsible for the generation of deep and superficial cells (Chalmers et al. 2003; Ohno 2007). This is an elegant example of polarised Wnt pathway components leading to inheritance of different cell fates in daughter cells.

Wnt signalling is implicated during mouse gastrulation. Here, Wnt becomes localised to the posterior region of the embryo and is crucial for patterning along the anterior–posterior axis (Fossat et al. 2011). Although contentious, Wnt signalling potentially plays a role in mammalian pre-implantation embryogenesis (Lloyd et al. 2003; Kemp et al. 2005; Mohamed et al. 2004; Wang et al. 2004; De Vries et al. 2004; Biechele et al. 2013; Haegel et al. 1995; Huelsken and Birchmeier 2001; Xie et al. 2008). It is probable that Wnt signalling is responsible for specifying the primitive streak, as Wnt3a and β-catenin knockout mice fail to form this structure, which is critical for the process of gastrulation (Haegel et al. 1995; Liu et al. 1999; Huelsken et al. 2000).

Overall, a large body of evidence across a variety of organisms implicates the Wnt/β-catenin pathway in the specification of cell fate throughout development, from setting up polarisation of the embryo in the earliest stages of development, to establishing the body axes, to specifying the identity of specific subsets of cells later during development. In this sense, the Wnt/β-catenin pathway can be seen to facilitate the variety of cell types that have evolved within metazoan organisms. This role appears to be evolutionarily conserved in adult organisms during tissue homeostasis and tissue regeneration. For example, head regeneration in *Hydra*, which are simple Cnidarians, involves Wnt/β-catenin from very early in the process (Hobmayer et al. 2000; Lengfeld et al. 2009). Removal of the head causes apoptosis in the remaining tissue, which results in caspase-mediated Wnt3a release from these cells. This serves to promote proliferation and regeneration (Chera et al. 2009). Mutants that are incapable of head regeneration do not display Wnt or β-catenin within the same time frame as wild-type *Hydra*, suggesting that upregulation of this pathway within a specific window is critical for complete regeneration (Hobmayer et al. 2000). Spatio-temporal localisation of Wnt ligands is also observed during regeneration following injury in planarians (Herman et al. 1995; Petersen and Reddien 2009a). Aberrant expression of Wnt/β-catenin pathway components can prevent regeneration; ectopic expression of Wnt in the anterior of several planarians prevents regeneration, but inhibition of the pathway can rescue this effect (Liu et al. 2013; Sikes and Newmark 2013). Posteriorly expressed Wnts are balanced with anteriorly expressed extracellular signal-related kinase (ERK) during regeneration to ensure formation of a complete anterior–posterior axis (Lander and Petersen 2016; Umesono et al. 2013; Sureda-Gómez et al. 2015).

Tissue maintenance and regeneration in vertebrates often involves the activation of the stem cell compartment to produce differentiated cells required to replenish the tissue in homeostasis or replace the damaged tissue upon injury. However, mechanisms to maintain regenerative capacity of tissues require the maintenance of a stem cell population within the tissue. Recently, a clear role for Wnt/β-catenin signalling has emerged in controlling stem cell activity and function.

15.2.2 Regulation of the Stem Cell Identity by Wnt/β-Catenin Pathway

Stem cells have the unique capacity to both self-renew and produce differentiated daughter cells to contribute to tissue maintenance. Stem cells reside in specialised environments referred to as ‘niches’, which are thought to provide the signals necessary to support their function. Signals arising from the niche must be spatially restricted in order to strictly control the number and distribution of stem cells (Losick et al. 2011). A tethered ligand or a molecule with short-range signalling fits this purpose. Upon cell division one daughter cell will move out of range of the signal that maintains ‘stemness’ and begin to differentiate. A large body of evidence suggest that localised Wnt ligands fulfil these criteria and Wnt/β-catenin signalling plays a crucial part in stem cell maintenance and differentiation. Importantly, stem cells can produce Wnt ligands to regulate their function in an autocrine manner or/and receive it from neighbouring cells (paracrine), reviewed in Clevers et al. (2014). A potential mechanism for Wnt/β-catenin signalling in the adult stem cell niche is depicted in Fig. 15.2d.

The microenvironment is particularly important in adult stem cell systems, such as the skin and the intestine, where rapid turnover takes place. The intestinal stem cell niche has been most extensively investigated with regard to Wnt signalling. The intestine comprises crypts, which house the least differentiated cells at the base, and display a degree of differentiation upwards. Within the crypts are Paneth cells, believed to act as niche cells and provide essential signals to intestinal stem cells (Tan and Barker 2014). A subpopulation of intestinal stem cells is marked by Lgr5 (Barker et al. 2007; Haegebarth and Clevers 2009), a Wnt target gene that also interacts with the Wnt agonist R-spondin (de Lau et al. 2011). It has recently been demonstrated that Lgr5+ve cells respond to Wnt signals through the Frizzled7 receptor (Flanagan et al. 2015). Expression of Lgr5 has also been correlated with cell cycle stage and level of 'stemness', with the continuously cycling Lgr5 + ve cells being considered higher up in stem cell hierarchy (Basak et al. 2014). Wnt signals appear to be presented only to specific cells within the intestinal stem cell niche, suggesting a role in cell identity control. This has been contested by other suggestions: that Lgr5 + ve cells undergo symmetric cell division and compete with niche cells to maintain crypt homeostasis (Snippert et al. 2010b). More recently, it has been demonstrated that a local Wnt3 signal (from Paneth cells) coordinates 'stemness' by distributing membrane-tethered Wnt ligands through cell division, up to a 2-cell radius from the source. Using an N-terminal haemagglutinin-tagged Wnt3, Farin and colleagues show that Paneth cells at the base of the crypt can transfer Wnt3 to adjacent Lgr5 + ve stem cells, much like the juxtacrine signalling described in the *Drosophila* imaginal disc (Alexandre et al. 2014). The authors proposed a model in which Frizzled receptors tether Wnt3 to the membranes of stem cells. Wnt3 is then distributed through stem cell divisions, generating a decreased Wnt response as the amount received by subsequent daughters decreases (Farin et al. 2016). This implies a 'dose-dependent' effect of Wnt on the daughter cells and mirrors the distribution of Wnt signals during development, with localisation at a specific point along a body axis. The technique of successfully 'tagging' Wnt proteins has also been performed by MacDonald and colleagues, who found that the addition of a V5 tag to the carboxyl terminus of Wnt3 did not reduce its activity (MacDonald et al. 2014). Given the difficulty of detecting Wnt proteins in mammalian systems, it is encouraging that tagging at either terminal does not disrupt protein activity. Employing these techniques as a means of visualisation will be invaluable.

Whilst Lgr5 marks a population of Wnt-responsive stem cells within the intestine, Axin2 (Behrens et al. 1998) has become a widely used reporter of Wnt activity. It was first used as an inducible lineage tracing reporter of Wnt activity to identify Wnt-responsive stem cells in the mammary gland (van Amerongen et al. 2012) and has since been adapted to a number of systems, including the intestine (Li et al.

Table 15.2 Summary of Wnt-responsive stem cells in murine tissues

Tissue	Stem cell	Reporter mouse for linage tracing	Selected references
Intestine	Intestinal crypt base cell	Lgr5-lacZ mice, Wnt3HA allele	Farin et al. (2016), Jaks et al. (2008) and Barker et al. (2007)
Skin—interfollicular epidermis	Basal cells	Axin2-CreERT2	Lim et al. (2013), Haegebarth and Clevers (2009)
Hair follicle	Outer bulge cell	Lgr5-EGFP-Ires-CreERT2 Axin2-CreERT2	Lim et al. (2016) and Jaks et al. (2008)
Brain	Sub-ventricular zone neural stem cells	Axin2-LacZ reporter Axin2CreERT2/+; R26RmTmG/+	Kalani et al. (2008) and Bowman et al. (2013)
Prostate	Luminal epithelial cells	Axin2-CreERT2;Rosa26-mTmGflox	Lee et al. (2015)
Testis	Spermatogonial stem cell	Axin2-CreERT2;Rosa26-mTmGflox	Takase and Nusse (2016)
Mammary gland	Basal cell	Axin2CreERT2/+;R26RlacZ/+ or Axin2CreERT2/+; R26RmTmG/+ ProcrCreERT2-IRES-tdTomato	van Amerongen et al. (2012) and Wang et al. (2015b)
Kidney	Renal precursor	Axin2CreERT2/+; R26RmTmG/+	Rinkevich et al. (2014a)
Liver	Peri-central hepatocyte	Axin2-CreERT2;Rosa26-mTmGflox	Wang et al. (2015a) and Planas-Paz et al. (2016)
Ear	Tympanic border cell (cochlea)	Axin2lacZ Wnt reporter mouse Lgr5-GFP-IRES-Cre-ER	Jan et al. (2013) and Shi et al. (2013)
Eye	Ciliary marginal zone (retina)	TCF/Lef-LacZ	Liu et al. (2003, 2006, 2007)
Bone	Osteoblast	Axin2CreERT2;Rosa26-STOP $^{fl/fl}$ZsGreen	Tan et al. (2014)
Stomach	Basal pyloric cell	Lgr5-EGFP-ires-CreERT2	Barker et al. (2010)
Tongue	Taste progenitors	Lgr5+/−, Rosa26-LacZ+/− and Lgr5+/−, Rosa26-tdTomato+/−	Yee et al. (2013)

2016), liver (Wang et al. 2015a), brain (Kalani et al. 2008), prostate (Lee et al. 2015), testis (Takase and Nusse 2016), kidney (Rinkevich et al. 2014a) and skin (Lim et al. 2013) (see Table 15.2).

Skin renewal occurs in the interfollicular epidermis, which is comprised of numerous keratinocyte layers, the outermost of which is a cornified layer called the stratum corneum. Cells from the stratum corneum are continually sloughed off and are periodically replaced by epidermal stem cells, which reside in the basal layer of the epidermis. For a comprehensive review on skin structure, see Blanpain and Fuchs (2009). In the skin, undifferentiated cells in the basal layer of the epidermis express the Wnt-reporter Axin2 (Behrens et al. 1998) and are also a source of Wnt in the skin. Lim and colleagues showed that Wnt activity is responsible for maintaining the stem cell population within the basal layer of the skin: labelled clones were observed throughout the interfollicular epidermis and persisted for up to 1 year (Lim et al. 2013; van Amerongen et al. 2012). Further support for Wnt as a niche signal in the skin is provided by experiments in which β-catenin is inactivated, leading to an increase in differentiation, and the detection of Dkk3 (Wnt inhibitor) in the supra-basal layers by immunofluorescence, which may restrict Wnt responsiveness to the basal niche. This bears similarity to patterning events during development, where Wnt proteins are expressed posteriorly, and inhibitors are expressed anteriorly, restricting activity of the pathway (Kusserow et al. 2005). Wnt as a niche signal has also recently been discussed in the context of hair follicle niche formation; within the Wnt-expressing placode, initial divisions are perpendicular to the basement membrane, positioning one daughter outside of the Wnt–signal range and generating two separate cell fates (Ouspenskaia et al. 2016). Axin2 has also been shown to mark a population of Wnt-responsive stem cells in the hair follicle bulge (Lim et al. 2016). Importantly, activation of β-catenin in interfollicular epidermal stem cells can alter the fate of these cells from epidermal cells to hair follicle and sebaceous lineages (Lo Celso et al. 2004).

Recently, Axin2+ Wnt-responsive liver cells were shown to reside adjacent to the central vein, of which the endothelial cells appear to be a Wnt source (Wang et al. 2015a). Under homeostasis, the Axin2+ cells possess the ability to self-renew and produce more differentiated progeny. Wang and colleagues showed that peri-central Axin2+ cells produced populations of hepatocytes that could persist within the liver for up to a year. Interestingly, whilst peri-central Axin2+ cells maintained expression of Wnt target genes, their descendants acquired a new expression pattern as they moved away from the Wnt source, indicating a degree of differentiation (Wang et al. 2015a). R-spondins and Lgr4 (see Table 15.1) are also important regulators of Wnt-responsive liver cells. Lgr4 expression mirrors that of Axin2, and Lgr4 knockout mice showed reduced Axin2 expression in peri-central hepatocytes, leading to a reduction in liver weight that suggests reduced cell turnover (Planas-Paz et al. 2016).

The identification of Wnt-responsive stem cells in vivo allowed researchers to isolate these stem cells from various tissues and expand them in vitro in the presence of purified Wnt ligands, which act as self-renewing signals. These include haematopoietic stem cells (Reya et al. 2003; Huang et al. 2012), mammary gland (Zeng and Nusse 2010; Shackleton et al. 2006), neuroprogenitors (Bowman et al. 2013; Kalani et al. 2008) and embryonic stem cells (ten Berge et al. 2011). Culturing primary stem cells in vitro provided unique opportunities to genetically profile them (e.g. Wang et al. 2015b) and study their differentiation potential. Additionally it paved the way to develop protocols to generate organoids. Organoids represent mini-organs that can form from a population of stem cells in vitro and display a similar level of intricacy and complexity to in vivo niches of organs. Organoids have a promising potential to be used for human disease modelling, drug testing, gene editing and regenerative medicine (Clevers 2016). This is important because the majority of organs in mammals have limited regenerative capacity upon severe injuries. Notably, the liver has a remarkable ability to regenerate even after losing ca. 75% of its original size (Fausto 2000). Although the involvement of a stem cell compartment in liver regeneration is not entirely clear (Fausto 2000; Huch et al. 2013; also see review Kopp et al. 2016), murine studies have revealed a key role for Wnt/β-catenin signalling in liver regeneration. For example, following partial hepatectomy, liver-specific β-catenin knock-out as well as Lrp5/6 (Wnt co-receptors) knock-out mice showed reduced regenerative capacity (Yang et al. 2014).

The liver appears to be unique in its ability to regenerate extensively amongst all vertebrates (Stoick-Cooper et al. 2007a). Limb regeneration is another conserved example, but it is more limited in mammals to digits (Takeo et al. 2013). For example, in the mouse digit, Wnt signals produced by the nail epithelium in response to amputation stabilise β-catenin and are essential for attracting nerves (Takeo et al. 2013) to the injured area that are involved in tissue patterning and subsequent regeneration (Rinkevich et al. 2014b). Wnt signalling may also be required for the specification of the various tissues that comprise the regenerating digit inducing bone, nail and muscle (Lehoczky and Tabin 2015).

Lower vertebrates such as Salamander and Xenopus are capable of full limb regeneration (Yokoyama et al. 2007; Kawakami et al. 2006), which is mediated by committed tissue-specific stem cells. In Xenopus, Wnt3a expression has been observed specifically in the apical epithelium of the regenerating limb bud, and its inhibition by inducible Dkk (Yokoyama et al. 2007) or the expression of Axin (Kawakami et al. 2006) during the early stages following amputation resulted in reduced limb regeneration. Accordingly, it is postulated that Wnt signalling provides positional information as well as promoting growth. Interestingly, this was also observed in Zebrafish, where several Wnt pathway components are upregulated during fin regeneration (Poss et al. 2000; Kawakami et al. 2006; Stoick-Cooper et al. 2007b). Comparative analysis of the spatio-temporal activation of Wnt/β-catenin signalling in tissue stem cells from salamanders, fish, mice and human during appendage regeneration might reveal some molecular aspects that underlie the observed differences in regenerative capacities between various taxon groups.

The Wnt pathway has a clear role in regulating stem cell fate, and in some cases, it can be linked to the evolutionarily conserved role for Wnt signalling in control of cell division. A number of studies have linked Wnt signalling to several crucial processes during mitosis (Niehrs and Acebron 2012). In particular, Wnt signalling has been suggested to promote entry to G1 phase, by upregulating Cyclin D1 and c-myc, a β-catenin target gene (He et al. 1998; Shtutman et al. 1999; Tetsu and McCormick 1999). As discussed previously, Wnt-responsive Lgr5 + ve intestine stem cells have been shown to be continuously cycling, further illustrating a link between this pathway and cell division (Basak et al. 2014). In addition, several Wnt pathway components directly bind to the mitotic machinery: Axin2, GSK3 and β-catenin accumulate at the centrosome and are involved in the regulation of microtubule growth (Huang et al. 2007a). APC and Dishevelled have been suggested to facilitate mitotic spindle binding to the kinetochores, which is critical for controlling the appropriate segregation of chromosomes (Fodde et al. 2001a, b). Axin2 has also been implicated in this process, and disruptions to these proteins can be particularly detrimental for aberrant chromosomal inheritance (Hadjihannas et al. 2006, 2010). Orientation of the spindle is also critical for determining the plane of cell division, and often this can determine whether a cell remains within a particular environment, such as the niche, or exits it. Dishevelled has been implicated in controlling spindle orientation, providing further strong evidence that the Wnt pathway may play a critical role in controlling the mechanics of cell division and thus cell fate determination (Kikuchi et al. 2010).

To date, there is no in vivo study in mammals clearly linking Wnt signalling, the maintenance of stem cell fate and control of the plane of cell division. However, Wnt3a immobilised to a bead has demonstrated that a localised Wnt signal can orient the mitotic spindle and direct cell fate in vitro (Habib et al. 2013). Here, mouse embryonic stem cells in contact with Wnt3a beads showed polarisation of Wnt pathway components towards the bead: β-catenin, LRP6, Frizzled and APC. Following an oriented cell division based on bead location, the Wnt-proximal cell retained an embryonic stem cell fate (with higher levels of Wnt components and pluripotency markers), whilst the Wnt-distal cell exhibited a more 'primed' signature, similar to epiblast-like stem cells.

Habib et al. (2013) also showed the asymmetric inheritance of centrosomes in relation to the Wnt bead: the Wnt-proximal stem cell inherited the 'mother' centrosome, whereas the Wnt-distal cell inherited the daughter centrosome. Centrosomes have previously been linked to cell fate, with asymmetric centrosome inheritance having been shown in the mouse neocortex. Older, 'mother' centrosomes are inherited by neural progenitor cells, and daughters that inherit the newer centrosome exit the niche and begin to differentiate (Wang et al. 2009). In this system, it appears that asymmetric centrosome inheritance is essential for cell fate choice. Interestingly, Wnt pathway components have been linked to centrosome segregation in *Drosophila* male germline stem cells. Accordingly, it has been reported that centrosomes and APC orient the spindle perpendicular to the stem cell niche (Yamashita et al. 2003).

Whilst it is clear from current literature that Wnt/β-catenin signalling plays an important role in the maintenance of 'stemness', future studies will help to elucidate the precise roles of the various Wnt/β-catenin pathway components in the specific mechanics of cell polarisation, spindle orientation and ultimately cell fate. Translating these mechanistic studies to an in vitro system that mirrors proper localised Wnt signals, as during development and adult tissue homeostasis, is invaluable for tissue engineering and an accurate representation of the stem cell niche. By immobilising Wnt proteins onto a functionalised platform, Lowndes et al. (2016) produced a stable Wnt surface that can be adapted accordingly to mimic various tissue microenvironments. The authors demonstrate that the Wnt platform can enrich and maintain embryonic and adult stem cells in 2D culture. Additionally, in 3D culture the Wnt platform can direct the formation of a human osteogenic-like tissue from human mesenchymal stem cells within 1 week, whilst maintaining a stem cell population (Lowndes et al. 2016).

Collectively, Wnt/β-catenin signalling plays a crucial role in instructing cell fate in a wide range of organisms. Accordingly, Wnt signals provide critical instructive cues during development, tissue homeostasis and regeneration. Future technologies that aim to deliver purified Wnt proteins to specific cell populations within tissues in a controlled manner will have benefits for tissue regeneration. Additionally, Wnt ligands can be used in vitro as positional cues to engineer tissues for wide applications including pharmaceutical testing and regenerative medicine. Therefore, the spatio-temporal presentation of Wnt ligands and the controlled activation of Wnt/β-catenin are crucial so as not to result in incorrect acquisition of cell fate and/or aberrant growth

15.3 Dysregulation of the Wnt/β-Catenin Pathway: Acquisition of a Tumorigenic State

Wnt/β-catenin signalling plays critical roles in controlling cell division and cell fate, and therefore, it is essential for tissue homeostasis. Mutations that cause either constitutive or dysregulated activation of the Wnt pathway lead to a loss of this control and potentially tumour development. In fact, a starting point of the Wnt field was the discovery that mice that developed mammary tumours following infection with the murine mammary tumour virus (MMTV) also often displayed activation of the murine int-1 gene, later called Wnt1 (Nusse and Varmus 1982). There are a number of examples of mutated Wnt pathway components playing a major role in tumour development (Polakis 2012).

The involvement of the Wnt pathway has been best defined in colon cancer, where epithelial cells often acquire early mutations in a number of Wnt pathway components, including β-catenin and APC (Fearon and Vogelstein 1990), resulting in increased tumorigenesis. Mutations in β-catenin have been shown in other cancer types, including cutaneous and blood cancers (Vermeulen et al. 2010; Malanchi et al. 2008; Reya et al. 2003). Increased β-catenin levels have been reported to increase cell 'stemness' and their likelihood to undergo epithelial–mesenchymal transition (EMT), i.e. a change in cell fate (Brabletz et al. 2001; Jung et al. 2001). However, this transition may be largely due to a loss of E-cadherin, to which β-catenin binds at adherens junctions and which is also frequently associated with cancers (Huels et al. 2015). In the skin, β-catenin-stabilising mutations lead to the development of pilomatricomas, which are characterised by high expression of Lef1. In this context, the constitutive activation of β-catenin results in keratinocytes acquiring a highly proliferative nature not normally associated with their cell type (Chan et al. 1999). Aberrant β-catenin in epidermal stem cells was shown not only to change the properties of this cell type but could also induce niche remodelling through paracrine signalling in the stroma (Lichtenberger et al. 2016). Indeed, in some cancers, such as breast cancer and colon cancer, β-catenin is being considered a prognostic marker for tumours, with high levels of β-catenin correlating with poor patient prognosis or increased tumour severity (Lin et al. 2000; Martinez et al. 2010).

Similar to β-catenin, APC has many cellular functions, including interactions with microtubules and the microtubule-binding protein EB-1, which regulates the cytoskeleton (Green et al. 2005). Importantly, specific mutations to APC can result in a dominant negative form of this protein that prevents wild-type APC from making these interactions. This may represent a mechanism by which the spindle checkpoint is overridden, and tumour cells do not respond appropriately to errors in chromosome segregation (Green et al. 2005). Mutations in APC also cause loss of stem cell divisional control within the niche, specifically altered spindle alignment. In normal intestinal stem cells, spindle alignment is mostly perpendicular to the niche, whereas spindle alignment appears random in precancerous and cancerous tissues. This appeared to correlate with a loss of asymmetric inheritance of un-replicated DNA (Quyn et al. 2010).

Wnt is also proposed to play a role in generating the 'hallmark' metabolic features of cancer cells: high dependency on glycolysis, little involvement of oxidative phosphorylation and the Warburg effect (Warburg 1956). This may be achieved through the differential segregation of Wnt pathway components during division, or indeed, the metabolic organelles themselves. One kinase found to be a direct Wnt-target gene, pyruvate dehydrogenase kinase (PDK) is elevated in a number of different cancers (Pate et al. 2014; Koukourakis et al. 2006; Wigfield et al. 2008; Baumunk et al. 2013) and may promote respiration through glycolysis. Pate and colleagues suggest that aberrant Wnt signalling may promote respiration through the glycolytic pathway by upregulating key components in this pathway including cytochrome c oxidase (Lee et al. 2012) and pyruvate dehydrogenase kinase 1 (PDK1) (Esen et al. 2013). Thus, cancer cells would be more able to survive in different environments (Pate et al. 2014). More recently, it was also shown that mammary cancer cells overexpressing Wnt1 in vitro display elevated levels of mitochondrial proteins and glycolytic enzymes and showed elevated mitochondrial mass and activity. This suggests that Wnt can regulate the metabolic status of a cell, and this is often coupled with the cause of pre-malignancy (Lamb et al. 2015). In further support of this, APC has been shown to bind to components of the mitochondrial motor complex, promoting their traffic to the cell membrane that may provide energy for processes like cell migration, a key feature of malignancy (Mills et al. 2016).

Interestingly, but perhaps not surprisingly, embryonic stem cells display a number of the same features as cancerous cells with regard to metabolism. For example, whilst ESCs have low respiratory capacity, they display a higher mitochondrial membrane potential (Jang et al. 2015; Shyh-Chang et al. 2013). The metabolism of ESCs, like cancer cells, must also be highly adaptable: these cells must be able to survive in low oxygen conditions upon implantation (Ryall et al. 2015; Houghton et al. 1996; Leese and Barton 1984; Leese 2012). The metabolic status of a cell has also been shown to influence its epigenetic status, through post-translational modifications of both DNA and transcription factors (Ryall et al. 2015; Moussaieff et al. 2015); thus, metabolism may be directly related to cell fate. In addition, the kind of metabolism that occurs in stem cells may reduce the accumulation of reactive oxygen species, which can damage the genome, and stem cell metabolism may therefore serve as a protective mechanism (Jang et al. 2015). It remains to be seen whether Wnt signalling plays a direct role in the control of stem cell metabolism, but much evidence suggests that it may.

The vast involvement of the Wnt/β-catenin pathway in cellular processes means that pathway dysregulation can be disastrous. Recent research into the metabolic status of tumorigenic cells in relation to Wnt/β-catenin signalling may reveal novel targets for future therapies and elucidate the parallels that can be drawn between stem cells and cancer cells.

As described, Wnt/β-catenin signalling can influence numerous biological processes that regulate cell fate. The emergence of new techniques enabling more targeted administration of Wnt signals and better visualisation of pathway components in vivo will greatly assist in dissecting the influence of Wnt/β-catenin signalling on intra- and intercellular processes in health and disease. Understanding the specific mechanisms will also facilitate advances in tissue bioengineering with the goal of using the in vitro studies to reveal promising targets for improving in vivo tissue regeneration. A deeper understanding of how Wnt/β-catenin signalling can influence cell fate in tumorigenesis may also present means of restricting the growth of cancerous cells, potentially by promoting their differentiation.

Acknowledgements We would like to thank Dr. Molly Lowndes, Dr. Ignacio Sancho-Martienz, Dr. Kifayathullah Liakath-ali, Dr. Ajay Mishra and Dr. Beate Lichtenberger for their critical reading of the manuscript.

References

Aberle H, Bauer A, Stappert J, Kispert A, Kemler R (1997) beta-catenin is a target for the ubiquitin-proteasome pathway. EMBO J 16(13):3797–3804. doi:10.1093/emboj/16.13.3797

Adamska M, Degnan SM, Green KM, Adamski M, Craigie A, Larroux C, Degnan BM (2007) Wnt and TGF-beta expression in the sponge Amphimedon queenslandica and the origin of metazoan embryonic patterning. PLoS One 2(10):e1031. doi:10.1371/journal.pone.0001031

Adamska M, Larroux C, Adamski M, Green K, Lovas E, Koop D, Richards GS, Zwafink C, Degnan BM (2010) Structure and expression of conserved Wnt pathway components in the demosponge Amphimedon queenslandica. Evol Dev 12(5):494–518. doi:10.1111/j.1525-142X.2010.00435.x

Alexandre C, Baena-Lopez A, Vincent JP (2014) Patterning and growth control by membrane-tethered Wingless. Nature 505(7482):180–185. doi:10.1038/nature12879

Azzolin L, Zanconato F, Bresolin S, Forcato M, Basso G, Bicciato S, Cordenonsi M, Piccolo S (2012) Role of TAZ as mediator of Wnt signaling. Cell 151(7):1443–1456. doi:10.1016/j.cell.2012.11.027

Azzolin L, Panciera T, Soligo S, Enzo E, Bicciato S, Dupont S, Bresolin S, Frasson C, Basso G, Guzzardo V, Fassina A, Cordenonsi M, Piccolo S (2014) YAP/TAZ incorporation in the β-catenin destruction complex orchestrates the Wnt response. Cell 158(1):157–170. doi:10.1016/j.cell.2014.06.013

Barker N (2008) The canonical Wnt/beta-catenin signalling pathway. Methods Mol Biol 468:5–15. doi:10.1007/978-1-59745-249-6_1

Barker N, van Es JH, Kuipers J, Kujala P, van den Born M, Cozijnsen M, Haegebarth A, Korving J, Begthel H, Peters PJ, Clevers H (2007) Identification of stem cells in small intestine and colon by marker gene Lgr5. Nature 449(7165):1003–1007. doi:10.1038/nature06196

Barker N, Huch M, Kujala P, van de Wetering M, Snippert HJ, van Es JH, Sato T, Stange DE, Begthel H, van den Born M, Danenberg E, van den Brink S, Korving J, Abo A, Peters PJ, Wright N, Poulsom R, Clevers H (2010) Lgr5(+ve) stem cells drive self-renewal in the stomach and build long-lived gastric units in vitro. Cell Stem Cell 6(1):25–36. doi:10.1016/j.stem.2009.11.013

Basak O, van de Born M, Korving J, Beumer J, van der Elst S, van Es JH, Clevers H (2014) Mapping early fate determination in Lgr5+ crypt stem cells using a novel Ki67-RFP allele. EMBO J 33(18):2057–2068. doi:10.15252/embj.201488017

Baumunk D, Reichelt U, Hildebrandt J, Krause H, Ebbing J, Cash H, Miller K, Schostak M, Weikert S (2013) Expression parameters of the metabolic pathway genes pyruvate dehydrogenase kinase-1 (PDK-1) and DJ-1/PARK7 in renal cell carcinoma (RCC). World J Urol 31(5):1191–1196. doi:10.1007/s00345-012-0874-5

Behrens J, Jerchow BA, Würtele M, Grimm J, Asbrand C, Wirtz R, Kühl M, Wedlich D, Birchmeier W (1998) Functional interaction of an axin homolog, conductin, with beta-catenin, APC, and GSK3beta. Science 280(5363):596–599

Biechele S, Cockburn K, Lanner F, Cox BJ, Rossant J (2013) Porcn-dependent Wnt signaling is not required prior to mouse gastrulation. Development 140(14):2961–2971. doi:10.1242/dev.094458

Blanpain C, Fuchs E (2009) Epidermal homeostasis: a balancing act of stem cells in the skin. Nat Rev Mol Cell Biol 10(3):207–217. doi:10.1038/nrm2636

Blish KR, Wang W, Willingham MC, Du W, Birse CE, Krishnan SR, Brown JC, Hawkins GA, Garvin AJ, D'Agostino RB, Torti FM, Torti SV (2008) A human bone morphogenetic protein antagonist is down-regulated in renal cancer. Mol Biol Cell 19(2):457–464. doi:10.1091/mbc.E07-05-0433

Bowman AN, van Amerongen R, Palmer TD, Nusse R (2013) Lineage tracing with Axin2 reveals distinct developmental and adult populations of Wnt/β-catenin-responsive neural stem cells. Proc Natl Acad Sci USA 110(18):7324–7329. doi:10.1073/pnas.1305411110

Brabletz T, Jung A, Reu S, Porzner M, Hlubek F, Kunz-Schughart LA, Knuechel R, Kirchner T (2001) Variable beta-catenin expression in colorectal cancers indicates tumor progression driven by the tumor environment. Proc Natl Acad Sci USA 98(18):10356–10361. doi:10.1073/pnas.171610498

Camilli TC, Weeraratna AT (2010) Striking the target in Wnt-y conditions: intervening in Wnt signaling during cancer progression. Biochem Pharmacol 80(5):702–711. doi:10.1016/j.bcp.2010.03.002

Chalmers AD, Strauss B, Papalopulu N (2003) Oriented cell divisions asymmetrically segregate aPKC and generate cell fate diversity in the early Xenopus embryo. Development 130(12):2657–2668

Chan EF, Gat U, McNiff JM, Fuchs E (1999) A common human skin tumour is caused by activating mutations in beta-catenin. Nat Genet 21(4):410–413. doi:10.1038/7747

Chen WS, Antic D, Matis M, Logan CY, Povelones M, Anderson GA, Nusse R, Axelrod JD (2008) Asymmetric homotypic interactions of the atypical cadherin flamingo mediate intercellular polarity signaling. Cell 133(6):1093–1105. doi:10.1016/j.cell.2008.04.048

Chera S, Ghila L, Dobretz K, Wenger Y, Bauer C, Buzgariu W, Martinou JC, Galliot B (2009) Apoptotic cells provide an unexpected source of Wnt3 signaling to drive hydra head regeneration. Dev Cell 17(2):279–289. doi:10.1016/j.devcel.2009.07.014

Chu ML, Ahn VE, Choi HJ, Daniels DL, Nusse R, Weis WI (2013) structural studies of Wnts and identification of an LRP6 binding site. Structure 21(7):1235–1242. doi:10.1016/j.str.2013.05.006

Clevers H (2006) Wnt/beta-catenin signaling in development and disease. Cell 127(3):469–480. doi:10.1016/j.cell.2006.10.018

Clevers H (2016) Modeling development and disease with organoids. Cell 165(7):1586–1597. doi:10.1016/j.cell.2016.05.082
Clevers H, Loh KM, Nusse R (2014) Stem cell signaling. An integral program for tissue renewal and regeneration: Wnt signaling and stem cell control. Science 346(6205):1248012. doi:10.1126/science.1248012
Cliffe A, Hamada F, Bienz M (2003) A role of Dishevelled in relocating Axin to the plasma membrane during wingless signaling. Curr Biol 13(11):960–966
Couso JP, Bishop SA, Martinez Arias A (1994) The wingless signalling pathway and the patterning of the wing margin in Drosophila. Development 120(3):621–636
de Lau W, Barker N, Low TY, Koo BK, Li VS, Teunissen H, Kujala P, Haegebarth A, Peters PJ, van de Wetering M, Stange DE, van Es JE, Guardavaccaro D, Schasfoort RB, Mohri Y, Nishimori K, Mohammed S, Heck AJ, Clevers H (2011) Lgr5 homologues associate with Wnt receptors and mediate R-spondin signalling. Nature 476(7360):293–297. doi:10.1038/nature10337
de Lau W, Peng WC, Gros P, Clevers H (2014) The R-spondin/Lgr5/Rnf43 module: regulator of Wnt signal strength. Genes Dev 28(4):305–316. doi:10.1101/gad.235473.113
De Vries WN, Evsikov AV, Haac BE, Fancher KS, Holbrook AE, Kemler R, Solter D, Knowles BB (2004) Maternal beta-catenin and E-cadherin in mouse development. Development 131 (18):4435–4445. doi:10.1242/dev.01316
Esen E, Chen J, Karner CM, Okunade AL, Patterson BW, Long F (2013) WNT-LRP5 signaling induces Warburg effect through mTORC2 activation during osteoblast differentiation. Cell Metab 17(5):745–755. doi:10.1016/j.cmet.2013.03.017
Farin HF, Jordens I, Mosa MH, Basak O, Korving J, Tauriello DV, de Punder K, Angers S, Peters PJ, Maurice MM, Clevers H (2016) Visualization of a short-range Wnt gradient in the intestinal stem-cell niche. Nature. doi:10.1038/nature16937
Fausto N (2000) Liver regeneration. J Hepatol 32(1 Suppl):19–31
Fearon ER, Vogelstein B (1990) A genetic model for colorectal tumorigenesis. Cell 61 (5):759–767
Flanagan DJ, Phesse TJ, Barker N, Schwab RH, Amin N, Malaterre J, Stange DE, Nowell CJ, Currie SA, Saw JT, Beuchert E, Ramsay RG, Sansom OJ, Ernst M, Clevers H, Vincan E (2015) Frizzled7 functions as a Wnt receptor in intestinal epithelial Lgr5(+) stem cells. Stem Cell Reports 4(5):759–767. doi:10.1016/j.stemcr.2015.03.003
Fodde R, Kuipers J, Rosenberg C, Smits R, Kielman M, Gaspar C, van Es JH, Breukel C, Wiegant J, Giles RH, Clevers H (2001a) Mutations in the APC tumour suppressor gene cause chromosomal instability. Nat Cell Biol 3(4):433–438. doi:10.1038/35070129
Fodde R, Smits R, Clevers H (2001b) APC, signal transduction and genetic instability in colorectal cancer. Nat Rev Cancer 1(1):55–67. doi:10.1038/35094067
Fossat N, Jones V, Khoo PL, Bogani D, Hardy A, Steiner K, Mukhopadhyay M, Westphal H, Nolan PM, Arkell R, Tam PP (2011) Stringent requirement of a proper level of canonical WNT signalling activity for head formation in mouse embryo. Development 138(4):667–676. doi:10.1242/dev.052803
Funayama N, Fagotto F, McCrea P, Gumbiner BM (1995) Embryonic axis induction by the armadillo repeat domain of beta-catenin: evidence for intracellular signaling. J Cell Biol 128 (5):959–968
Gao C, Chen YG (2010) Dishevelled: the hub of Wnt signaling. Cell Signal 22(5):717–727. doi:10.1016/j.cellsig.2009.11.021
Goentoro L, Kirschner MW (2009) Evidence that fold-change, and not absolute level, of beta-catenin dictates Wnt signaling. Mol Cell 36(5):872–884. doi:10.1016/j.molcel.2009.11.017
Goldstein B, Takeshita H, Mizumoto K, Sawa H (2006) Wnt signals can function as positional cues in establishing cell polarity. Dev Cell 10(3):391–396. doi:10.1016/j.devcel.2005.12.016
Gong Y, Mo C, Fraser SE (2004) Planar cell polarity signalling controls cell division orientation during zebrafish gastrulation. Nature 430(7000):689–693. doi:10.1038/nature02796
Green RA, Wollman R, Kaplan KB (2005) APC and EB1 function together in mitosis to regulate spindle dynamics and chromosome alignment. Mol Biol Cell 16(10):4609–4622. doi:10.1091/mbc.E05-03-0259

Guidato S, Itasaki N (2007) Wise retained in the endoplasmic reticulum inhibits Wnt signaling by reducing cell surface LRP6. Dev Biol 310(2):250–263. doi:10.1016/j.ydbio.2007.07.033
Ha NC, Tonozuka T, Stamos JL, Choi HJ, Weis WI (2004) Mechanism of phosphorylation-dependent binding of APC to beta-catenin and its role in beta-catenin degradation. Mol Cell 15(4):511–521. doi:10.1016/j.molcel.2004.08.010
Habib SJ, Chen BC, Tsai FC, Anastassiadis K, Meyer T, Betzig E, Nusse R (2013) A localized Wnt signal orients asymmetric stem cell division in vitro. Science 339(6126):1445–1448. doi:10.1126/science.1231077
Hadjihannas MV, Brückner M, Jerchow B, Birchmeier W, Dietmaier W, Behrens J (2006) Aberrant Wnt/beta-catenin signaling can induce chromosomal instability in colon cancer. Proc Natl Acad Sci USA 103(28):10747–10752. doi:10.1073/pnas.0604206103
Hadjihannas MV, Brückner M, Behrens J (2010) Conductin/axin2 and Wnt signalling regulates centrosome cohesion. EMBO Rep 11(4):317–324. doi:10.1038/embor.2010.23
Haegebarth A, Clevers H (2009) Wnt signaling, lgr5, and stem cells in the intestine and skin. Am J Pathol 174(3):715–721. doi:10.2353/ajpath.2009.080758
Haegel H, Larue L, Ohsugi M, Fedorov L, Herrenknecht K, Kemler R (1995) Lack of beta-catenin affects mouse development at gastrulation. Development 121(11):3529–3537
Harwood AJ (2008a) Dictyostelium development: a prototypic Wnt pathway? Methods Mol Biol 469:21–32. doi:10.1007/978-1-60327-469-2_2
Harwood AJ (2008b) Use of the Dictyostelium stalk cell assay to monitor GSK-3 regulation. Methods Mol Biol 469:39–43. doi:10.1007/978-1-60327-469-2_4
He X (2004) Wnt signaling went derailed again: a new track via the LIN-18 receptor? Cell 118 (6):668–670. doi:10.1016/j.cell.2004.09.009
He TC, Sparks AB, Rago C, Hermeking H, Zawel L, da Costa LT, Morin PJ, Vogelstein B, Kinzler KW (1998) Identification of c-MYC as a target of the APC pathway. Science 281 (5382):1509–1512
Heasman J, Crawford A, Goldstone K, Garner-Hamrick P, Gumbiner B, McCrea P, Kintner C, Noro CY, Wylie C (1994) Overexpression of cadherins and underexpression of beta-catenin inhibit dorsal mesoderm induction in early Xenopus embryos. Cell 79(5):791–803
Herman MA, Vassilieva LL, Horvitz HR, Shaw JE, Herman RK (1995) The *C. elegans* gene lin-44, which controls the polarity of certain asymmetric cell divisions, encodes a Wnt protein and acts cell nonautonomously. Cell 83(1):101–110
Hernández AR, Klein AM, Kirschner MW (2012) Kinetic responses of β-catenin specify the sites of Wnt control. Science 338(6112):1337–1340. doi:10.1126/science.1228734
Hobmayer B, Rentzsch F, Kuhn K, Happel CM, von Laue CC, Snyder P, Rothbächer U, Holstein TW (2000) WNT signalling molecules act in axis formation in the diploblastic metazoan Hydra. Nature 407(6801):186–189. doi:10.1038/35025063
Holstein TW (2012) The evolution of the Wnt pathway. Cold Spring Harb Perspect Biol 4(7): a007922. doi:10.1101/cshperspect.a007922
Houghton FD, Thompson JG, Kennedy CJ, Leese HJ (1996) Oxygen consumption and energy metabolism of the early mouse embryo. Mol Reprod Dev 44(4):476–485. doi:10.1002/(SICI)1098-2795(199608)44:4<476::AID-MRD7>3.0.CO;2-I
Hsieh JC, Kodjabachian L, Rebbert ML, Rattner A, Smallwood PM, Samos CH, Nusse R, Dawid IB, Nathans J (1999) A new secreted protein that binds to Wnt proteins and inhibits their activities. Nature 398(6726):431–436. doi:10.1038/18899
Huang YL, Niehrs C (2014) Polarized Wnt signaling regulates ectodermal cell fate in Xenopus. Dev Cell 29(2):250–257. doi:10.1016/j.devcel.2014.03.015
Huang P, Senga T, Hamaguchi M (2007a) A novel role of phospho-beta-catenin in microtubule regrowth at centrosome. Oncogene 26(30):4357–4371. doi:10.1038/sj.onc.1210217
Huang S, Shetty P, Robertson SM, Lin R (2007b) Binary cell fate specification during *C. elegans* embryogenesis driven by reiterated reciprocal asymmetry of TCF POP-1 and its coactivator beta-catenin SYS-1. Development 134(14):2685–2695. doi:10.1242/dev.008268
Huang J, Nguyen-McCarty M, Hexner EO, Danet-Desnoyers G, Klein PS (2012) Maintenance of hematopoietic stem cells through regulation of Wnt and mTOR pathways. Nat Med 18 (12):1778–1785. doi:10.1038/nm.2984

Huch M, Dorrell C, Boj SF, van Es JH, Li VS, van de Wetering M, Sato T, Hamer K, Sasaki N, Finegold MJ, Haft A, Vries RG, Grompe M, Clevers H (2013) In vitro expansion of single Lgr5+ liver stem cells induced by Wnt-driven regeneration. Nature 494(7436):247–250. doi:10.1038/nature11826

Huels DJ, Ridgway RA, Radulescu S, Leushacke M, Campbell AD, Biswas S, Leedham S, Serra S, Chetty R, Moreaux G, Parry L, Matthews J, Song F, Hedley A, Kalna G, Ceteci F, Reed KR, Meniel VS, Maguire A, Doyle B, Söderberg O, Barker N, Watson A, Larue L, Clarke AR, Sansom OJ (2015) E-cadherin can limit the transforming properties of activating β-catenin mutations. EMBO J 34(18):2321–2333. doi:10.15252/embj.201591739

Huelsken J, Birchmeier W (2001) New aspects of Wnt signaling pathways in higher vertebrates. Curr Opin Genet Dev 11(5):547–553

Huelsken J, Vogel R, Brinkmann V, Erdmann B, Birchmeier C, Birchmeier W (2000) Requirement for beta-catenin in anterior-posterior axis formation in mice. J Cell Biol 148(3):567–578

Ikeda S, Kishida S, Yamamoto H, Murai H, Koyama S, Kikuchi A (1998) Axin, a negative regulator of the Wnt signaling pathway, forms a complex with GSK-3beta and beta-catenin and promotes GSK-3beta-dependent phosphorylation of beta-catenin. EMBO J 17(5):1371–1384. doi:10.1093/emboj/17.5.1371

Jaks V, Barker N, Kasper M, van Es JH, Snippert HJ, Clevers H, Toftgård R (2008) Lgr5 marks cycling, yet long-lived, hair follicle stem cells. Nat Genet 40(11):1291–1299. doi:10.1038/ng.239

Jan TA, Chai R, Sayyid ZN, van Amerongen R, Xia A, Wang T, Sinkkonen ST, Zeng YA, Levin JR, Heller S, Nusse R, Cheng AG (2013) Tympanic border cells are Wnt-responsive and can act as progenitors for postnatal mouse cochlear cells. Development 140(6):1196–1206. doi:10.1242/dev.087528

Janda CY, Waghray D, Levin AM, Thomas C, Garcia KC (2012) Structural basis of Wnt recognition by Frizzled. Science 337(6090):59–64. doi:10.1126/science.1222879

Jang H, Yang J, Lee E, Cheong JH (2015) Metabolism in embryonic and cancer stemness. Arch Pharm Res 38(3):381–388. doi:10.1007/s12272-015-0558-y

Jung A, Schrauder M, Oswald U, Knoll C, Sellberg P, Palmqvist R, Niedobitek G, Brabletz T, Kirchner T (2001) The invasion front of human colorectal adenocarcinomas shows co-localization of nuclear beta-catenin, cyclin D1, and p16INK4A and is a region of low proliferation. Am J Pathol 159(5):1613–1617

Kakugawa S, Langton PF, Zebisch M, Howell SA, Chang TH, Liu Y, Feizi T, Bineva G, O'Reilly N, Snijders AP, Jones EY, Vincent JP (2015) Notum deacylates Wnt proteins to suppress signalling activity. Nature 519(7542):187–192. doi:10.1038/nature14259

Kalani MY, Cheshier SH, Cord BJ, Bababeygy SR, Vogel H, Weissman IL, Palmer TD, Nusse R (2008) Wnt-mediated self-renewal of neural stem/progenitor cells. Proc Natl Acad Sci USA 105(44):16970–16975. doi:10.1073/pnas.0808616105

Kawakami Y, Rodriguez Esteban C, Raya M, Kawakami H, Martí M, Dubova I, Izpisúa Belmonte JC (2006) Wnt/beta-catenin signaling regulates vertebrate limb regeneration. Genes Dev 20(23):3232–3237. doi:10.1101/gad.1475106

Kemp C, Willems E, Abdo S, Lambiv L, Leyns L (2005) Expression of all Wnt genes and their secreted antagonists during mouse blastocyst and postimplantation development. Dev Dyn 233(3):1064–1075. doi:10.1002/dvdy.20408

Kikuchi K, Niikura Y, Kitagawa K, Kikuchi A (2010) Dishevelled, a Wnt signalling component, is involved in mitotic progression in cooperation with Plk1. EMBO J 29(20):3470–3483. doi:10.1038/emboj.2010.221

Kopp JL, Grompe M, Sander M (2016) Stem cells versus plasticity in liver and pancreas regeneration. Nat Cell Biol 18(3):238–245. doi:10.1038/ncb3309

Koukourakis MI, Giatromanolaki A, Harris AL, Sivridis E (2006) Comparison of metabolic pathways between cancer cells and stromal cells in colorectal carcinomas: a metabolic survival role for tumor-associated stroma. Cancer Res 66(2):632–637. doi:10.1158/0008-5472.CAN-05-3260

Kusserow A, Pang K, Sturm C, Hrouda M, Lentfer J, Schmidt HA, Technau U, von Haeseler A, Hobmayer B, Martindale MQ, Holstein TW (2005) Unexpected complexity of the Wnt gene family in a sea anemone. Nature 433(7022):156–160. doi:10.1038/nature03158

Lamb R, Bonuccelli G, Ozsvári B, Peiris-Pagès M, Fiorillo M, Smith DL, Bevilacqua G, Mazzanti CM, McDonnell LA, Naccarato AG, Chiu M, Wynne L, Martinez-Outschoorn UE, Sotgia F, Lisanti MP (2015) Mitochondrial mass, a new metabolic biomarker for stem-like cancer cells: understanding WNT/FGF-driven anabolic signaling. Oncotarget 6 (31):30453–30471. doi:10.18632/oncotarget.5852

Lander R, Petersen CP (2016) Wnt, Ptk7, and FGFRL expression gradients control trunk positional identity in planarian regeneration. Elife 5. doi:10.7554/eLife.12850

Larabell CA, Torres M, Rowning BA, Yost C, Miller JR, Wu M, Kimelman D, Moon RT (1997) Establishment of the dorso-ventral axis in Xenopus embryos is presaged by early asymmetries in beta-catenin that are modulated by the Wnt signaling pathway. J Cell Biol 136 (5):1123–1136

Lawrence PA, Casal J, Struhl G (2002) Towards a model of the organisation of planar polarity and pattern in the Drosophila abdomen. Development 129(11):2749–2760

Leclère L, Bause M, Sinigaglia C, Steger J, Rentzsch F (2016) Development of the aboral domain in Nematostella requires β-catenin and the opposing activities of Six3/6 and Frizzled5/8. Development 143(10):1766–1777. doi:10.1242/dev.120931

Lee SY, Jeon HM, Ju MK, Kim CH, Yoon G, Han SI, Park HG, Kang HS (2012) Wnt/Snail signaling regulates cytochrome C oxidase and glucose metabolism. Cancer Res 72 (14):3607–3617. doi:10.1158/0008-5472.CAN-12-0006

Lee SH, Johnson DT, Luong R, Yu EJ, Cunha GR, Nusse R, Sun Z (2015) Wnt/β-catenin-responsive cells in prostatic development and regeneration. Stem Cells 33(11):3356–3367. doi:10.1002/stem.2096

Leese HJ (2012) Metabolism of the preimplantation embryo: 40 years on. Reproduction 143 (4):417–427. doi:10.1530/REP-11-0484

Leese HJ, Barton AM (1984) Pyruvate and glucose uptake by mouse ova and preimplantation embryos. J Reprod Fertil 72(1):9–13

Lehoczky JA, Tabin CJ (2015) Lgr6 marks nail stem cells and is required for digit tip regeneration. Proc Natl Acad Sci USA 112(43):13249–13254. doi:10.1073/pnas.1518874112

Lengfeld T, Watanabe H, Simakov O, Lindgens D, Gee L, Law L, Schmidt HA, Ozbek S, Bode H, Holstein TW (2009) Multiple Wnts are involved in Hydra organizer formation and regeneration. Dev Biol 330(1):186–199. doi:10.1016/j.ydbio.2009.02.004

Li X, Zhang Y, Kang H, Liu W, Liu P, Zhang J, Harris SE, Wu D (2005) Sclerostin binds to LRP5/6 and antagonizes canonical Wnt signaling. J Biol Chem 280(20):19883–19887. doi:10.1074/jbc.M413274200

Li VS, Ng SS, Boersema PJ, Low TY, Karthaus WR, Gerlach JP, Mohammed S, Heck AJ, Maurice MM, Mahmoudi T, Clevers H (2012) Wnt signaling through inhibition of β-catenin degradation in an intact Axin1 complex. Cell 149(6):1245–1256. doi:10.1016/j.cell.2012.05.002

Li N, Yousefi M, Nakauka-Ddamba A, Tobias JW, Jensen ST, Morrisey EE, Lengner CJ (2016) Heterogeneity in readouts of canonical wnt pathway activity within intestinal crypts. Dev Dyn 245(8):822–33. doi:10.1002/dvdy.24423

Lichtenberger BM, Mastrogiannaki M, Watt FM (2016) Epidermal β-catenin activation remodels the dermis via paracrine signalling to distinct fibroblast lineages. Nat Commun 7:10537. doi:10.1038/ncomms10537

Lien WH, Fuchs E (2014) Wnt some lose some: transcriptional governance of stem cells by Wnt/β-catenin signaling. Genes Dev 28(14):1517–1532. doi:10.1101/gad.244772.114

Lim X, Tan SH, Koh WL, Chau RM, Yan KS, Kuo CJ, van Amerongen R, Klein AM, Nusse R (2013) Interfollicular epidermal stem cells self-renew via autocrine Wnt signaling. Science 342 (6163):1226–1230. doi:10.1126/science.1239730

Lim X, Tan SH, Yu KL, Lim SB, Nusse R (2016) Axin2 marks quiescent hair follicle bulge stem cells that are maintained by autocrine Wnt/β-catenin signaling. Proc Natl Acad Sci USA 113 (11):E1498–E1505. doi:10.1073/pnas.1601599113

Lin K, Wang S, Julius MA, Kitajewski J, Moos M, Luyten FP (1997) The cysteine-rich frizzled domain of Frzb-1 is required and sufficient for modulation of Wnt signaling. Proc Natl Acad Sci USA 94(21):11196–11200
Lin SY, Xia W, Wang JC, Kwong KY, Spohn B, Wen Y, Pestell RG, Hung MC (2000) Beta-catenin, a novel prognostic marker for breast cancer: its roles in cyclin D1 expression and cancer progression. Proc Natl Acad Sci USA 97(8):4262–4266. doi:10.1073/pnas.060025397
Liu P, Wakamiya M, Shea MJ, Albrecht U, Behringer RR, Bradley A (1999) Requirement for Wnt3 in vertebrate axis formation. Nat Genet 22(4):361–365. doi:10.1038/11932
Liu H, Mohamed O, Dufort D, Wallace VA (2003) Characterization of Wnt signaling components and activation of the Wnt canonical pathway in the murine retina. Dev Dyn 227(3):323–334. doi:10.1002/dvdy.10315
Liu H, Thurig S, Mohamed O, Dufort D, Wallace VA (2006) Mapping canonical Wnt signaling in the developing and adult retina. Invest Ophthalmol Vis Sci 47(11):5088–5097. doi:10.1167/iovs.06-0403
Liu H, Xu S, Wang Y, Mazerolle C, Thurig S, Coles BL, Ren JC, Taketo MM, van der Kooy D, Wallace VA (2007) Ciliary margin transdifferentiation from neural retina is controlled by canonical Wnt signaling. Dev Biol 308(1):54–67. doi:10.1016/j.ydbio.2007.04.052
Liu N, Shi S, Deng M, Tang L, Zhang G, Ding B, Liu W, Liu Y, Shi H, Liu L, Jin Y (2011) High levels of β-catenin signaling reduce osteogenic differentiation of stem cells in inflammatory microenvironments through inhibition of the noncanonical Wnt pathway. J Bone Miner Res 26 (9):2082–2095. doi:10.1002/jbmr.440
Liu SY, Selck C, Friedrich B, Lutz R, Vila-Farré M, Dahl A, Brandl H, Lakshmanaperumal N, Henry I, Rink JC (2013) Reactivating head regrowth in a regeneration-deficient planarian species. Nature 500(7460):81–84. doi:10.1038/nature12414
Lloyd S, Fleming TP, Collins JE (2003) Expression of Wnt genes during mouse preimplantation development. Gene Expr Patterns 3(3):309–312
Lo Celso C, Prowse DM, Watt FM (2004) Transient activation of beta-catenin signalling in adult mouse epidermis is sufficient to induce new hair follicles but continuous activation is required to maintain hair follicle tumours. Development 131(8):1787–1799. doi:10.1242/dev.01052
Logan CY, Miller JR, Ferkowicz MJ, McClay DR (1999) Nuclear beta-catenin is required to specify vegetal cell fates in the sea urchin embryo. Development 126(2):345–357
Losick VP, Morris LX, Fox DT, Spradling A (2011) Drosophila stem cell niches: a decade of discovery suggests a unified view of stem cell regulation. Dev Cell 21(1):159–171. doi:10.1016/j.devcel.2011.06.018
Lowndes M, Rotherham M, Price JC, El Haj AJ, Habib SJ (2016) Immobilized WNT proteins act as a stem cell niche for tissue engineering. Stem Cell Reports 7(1):126–137. doi:10.1016/j.stemcr.2016.06.004
Ma L, Wang HY (2007) Mitogen-activated protein kinase p38 regulates the Wnt/cyclic GMP/Ca2 + non-canonical pathway. J Biol Chem 282(39):28980–28990. doi:10.1074/jbc.M702840200
MacDonald BT, He X (2012) A finger on the pulse of Wnt receptor signaling. Cell Res 22 (10):1410–1412. doi:10.1038/cr.2012.91
MacDonald BT, Hien A, Zhang X, Iranloye O, Virshup DM, Waterman ML, He X (2014) Disulfide bond requirements for active Wnt ligands. J Biol Chem 289(26):18122–18136. doi:10.1074/jbc.M114.575027
Malanchi I, Peinado H, Kassen D, Hussenet T, Metzger D, Chambon P, Huber M, Hohl D, Cano A, Birchmeier W, Huelsken J (2008) Cutaneous cancer stem cell maintenance is dependent on beta-catenin signalling. Nature 452(7187):650–653. doi:10.1038/nature06835
Martinez NP, Kanno DT, Pereira JA, Cardinalli IA, Priolli DG (2010) Beta-catenin and E-cadherin tissue "content" as prognostic markers in left-side colorectal cancer. Cancer Biomark 8 (3):129–135. doi:10.3233/DMA-2011-0843
Metcalfe C, Mendoza-Topaz C, Mieszczanek J, Bienz M (2010) Stability elements in the LRP6 cytoplasmic tail confer efficient signalling upon DIX-dependent polymerization. J Cell Sci 123 (Pt 9):1588–1599. doi:10.1242/jcs.067546
Mikels AJ, Nusse R (2006) Purified Wnt5a protein activates or inhibits beta-catenin-TCF signaling depending on receptor context. PLoS Biol 4(4):e115. doi:10.1371/journal.pbio.0040115

Mills KM, Brocardo MG, Henderson BR (2016) APC binds the Miro/Milton motor complex to stimulate transport of mitochondria to the plasma membrane. Mol Biol Cell 27(3):466–482. doi:10.1091/mbc.E15-09-0632

Mohamed OA, Dufort D, Clarke HJ (2004) Expression and estradiol regulation of Wnt genes in the mouse blastocyst identify a candidate pathway for embryo-maternal signaling at implantation. Biol Reprod 71(2):417–424. doi:10.1095/biolreprod.103.025692

Momose T, Derelle R, Houliston E (2008) A maternally localised Wnt ligand required for axial patterning in the cnidarian *Clytia hemisphaerica*. Development 135(12):2105–2113. doi:10.1242/dev.021543

Moon RT, Kohn AD, De Ferrari GV, Kaykas A (2004) WNT and beta-catenin signalling: diseases and therapies. Nat Rev Genet 5(9):691–701. doi:10.1038/nrg1427

Morata G, Struhl G (2014) Developmental biology: tethered wings. Nature 505(7482):162–163. doi:10.1038/nature12848

Moussaieff A, Kogan NM, Aberdam D (2015) Concise review: energy metabolites: key mediators of the epigenetic state of pluripotency. Stem Cells 33(8):2374–2380. doi:10.1002/stem.2041

Mulligan KA, Fuerer C, Ching W, Fish M, Willert K, Nusse R (2012) Secreted Wingless-interacting molecule (Swim) promotes long-range signaling by maintaining Wingless solubility. Proc Natl Acad Sci USA 109(2):370–377. doi:10.1073/pnas.1119197109

Nakamura K, Kim S, Ishidate T, Bei Y, Pang K, Shirayama M, Trzepacz C, Brownell DR, Mello CC (2005) Wnt signaling drives WRM-1/beta-catenin asymmetries in early *C. elegans* embryos. Genes Dev 19(15):1749–1754. doi:10.1101/gad.1323705

Neumann CJ, Cohen SM (1997) Long-range action of Wingless organizes the dorsal-ventral axis of the Drosophila wing. Development 124(4):871–880

Niehrs C (2012) The complex world of WNT receptor signalling. Nat Rev Mol Cell Biol 13 (12):767–779. doi:10.1038/nrm3470

Niehrs C, Acebron SP (2012) Mitotic and mitogenic Wnt signalling. EMBO J 31(12):2705–2713. doi:10.1038/emboj.2012.124

Nolo R, Abbott LA, Bellen HJ (2000) Senseless, a Zn finger transcription factor, is necessary and sufficient for sensory organ development in Drosophila. Cell 102(3):349–362

Nusse R, Varmus HE (1982) Many tumors induced by the mouse mammary tumor virus contain a provirus integrated in the same region of the host genome. Cell 31(1):99–109

Ohno S (2007) Extrinsic Wnt signalling controls the polarity component aPKC. Nat Cell Biol 9 (7):738–740. doi:10.1038/ncb0707-738

Ouspenskaia T, Matos I, Mertz AF, Fiore VF, Fuchs E (2016) WNT-SHH antagonism specifies and expands stem cells prior to niche formation. Cell 164(1–2):156–169. doi:10.1016/j.cell.2015.11.058

Pang K, Ryan JF, Mullikin JC, Baxevanis AD, Martindale MQ, Program NCS (2010) Genomic insights into Wnt signaling in an early diverging metazoan, the ctenophore Mnemiopsis leidyi. Evodevo 1(1):10. doi:10.1186/2041- 9139-1-10

Pate KT, Stringari C, Sprowl-Tanio S, Wang K, TeSlaa T, Hoverter NP, McQuade MM, Garner C, Digman MA, Teitell MA, Edwards RA, Gratton E, Waterman ML (2014) Wnt signaling directs a metabolic program of glycolysis and angiogenesis in colon cancer. EMBO J 33 (13):1454–1473. doi:10.15252/embj.201488598

Petersen CP, Reddien PW (2009a) Wnt signaling and the polarity of the primary body axis. Cell 139(6):1056–1068. doi:10.1016/j.cell.2009.11.035

Petersen CP, Reddien PW (2009b) A wound-induced Wnt expression program controls planarian regeneration polarity. Proc Natl Acad Sci USA 106(40):17061–17066. doi:10.1073/pnas.0906823106

Planas-Paz L, Orsini V, Boulter L, Calabrese D, Pikiolek M, Nigsch F, Xie Y, Roma G, Donovan A, Marti P, Beckmann N, Dill MT, Carbone W, Bergling S, Isken A, Mueller M, Kinzel B, Yang Y, Mao X, Nicholson TB, Zamponi R, Capodieci P, Valdez R, Rivera D, Loew A, Ukomadu C, Terracciano LM, Bouwmeester T, Cong F, Heim MH, Forbes SJ, Ruffner H, Tchorz JS (2016) The RSPO-LGR4/5-ZNRF3/RNF43 module controls liver zonation and size. Nat Cell Biol 18(5):467–479. doi:10.1038/ncb3337

Polakis P (2012) Wnt signaling in cancer. Cold Spring Harb Perspect Biol 4(5). doi:10.1101/cshperspect.a008052

Poss KD, Shen J, Keating MT (2000) Induction of lef1 during zebrafish fin regeneration. Dev Dyn 219(2):282–286. doi:10.1002/1097-0177(2000)9999:9999<::AID-DVDY1045>3.0.CO;2-C

Quyn AJ, Appleton PL, Carey FA, Steele RJ, Barker N, Clevers H, Ridgway RA, Sansom OJ, Näthke IS (2010) Spindle orientation bias in gut epithelial stem cell compartments is lost in precancerous tissue. Cell Stem Cell 6(2):175–181. doi:10.1016/j.stem.2009.12.007

Reya T, Duncan AW, Ailles L, Domen J, Scherer DC, Willert K, Hintz L, Nusse R, Weissman IL (2003) A role for Wnt signalling in self-renewal of haematopoietic stem cells. Nature 423 (6938):409–414. doi:10.1038/nature01593

Rinkevich Y, Montoro DT, Contreras-Trujillo H, Harari-Steinberg O, Newman AM, Tsai JM, Lim X, Van-Amerongen R, Bowman A, Januszyk M, Pleniceanu O, Nusse R, Longaker MT, Weissman IL, Dekel B (2014a) In vivo clonal analysis reveals lineage-restricted progenitor characteristics in mammalian kidney development, maintenance, and regeneration. Cell Rep 7 (4):1270–1283. doi:10.1016/j.celrep.2014.04.018

Rinkevich Y, Montoro DT, Muhonen E, Walmsley GG, Lo D, Hasegawa M, Januszyk M, Connolly AJ, Weissman IL, Longaker MT (2014b) Clonal analysis reveals nerve-dependent and independent roles on mammalian hind limb tissue maintenance and regeneration. Proc Natl Acad Sci USA 111(27):9846–9851. doi:10.1073/pnas.1410097111

Rubinfeld B, Albert I, Porfiri E, Fiol C, Munemitsu S, Polakis P (1996) Binding of GSK3beta to the APC-beta-catenin complex and regulation of complex assembly. Science 272 (5264):1023–1026

Ryall JG, Cliff T, Dalton S, Sartorelli V (2015) Metabolic Reprogramming of Stem Cell Epigenetics. Cell Stem Cell 17(6):651–662. doi:10.1016/j.stem.2015.11.012

Sato T, van Es JH, Snippert HJ, Stange DE, Vries RG, van den Born M, Barker N, Shroyer NF, van de Wetering M, Clevers H (2011) Paneth cells constitute the niche for Lgr5 stem cells in intestinal crypts. Nature 469(7330):415–418. doi:10.1038/nature09637

Schneider SQ, Bowerman B (2007) beta-Catenin asymmetries after all animal/vegetal- oriented cell divisions in *Platynereis dumerilii* embryos mediate binary cell-fate specification. Dev Cell 13(1):73–86. doi:10.1016/j.devcel.2007.05.002

Schneider S, Steinbeisser H, Warga RM, Hausen P (1996) Beta-catenin translocation into nuclei demarcates the dorsalizing centers in frog and fish embryos. Mech Dev 57(2):191–198

Semënov MV, Tamai K, Brott BK, Kühl M, Sokol S, He X (2001) Head inducer Dickkopf-1 is a ligand for Wnt coreceptor LRP6. Curr Biol 11(12):951–961

Semënov M, Tamai K, He X (2005) SOST is a ligand for LRP5/LRP6 and a Wnt signaling inhibitor. J Biol Chem 280(29):26770–26775. doi:10.1074/jbc.M504308200

Shackleton M, Vaillant F, Simpson KJ, Stingl J, Smyth GK, Asselin-Labat ML, Wu L, Lindeman GJ, Visvader JE (2006) Generation of a functional mammary gland from a single stem cell. Nature 439(7072):84–88. doi:10.1038/nature04372

Shi F, Hu L, Edge AS (2013) Generation of hair cells in neonatal mice by β-catenin overexpression in Lgr5-positive cochlear progenitors. Proc Natl Acad Sci USA 110 (34):13851–13856. doi:10.1073/pnas.1219952110

Shtutman M, Zhurinsky J, Simcha I, Albanese C, D'Amico M, Pestell R, Ben-Ze'ev A (1999) The cyclin D1 gene is a target of the beta-catenin/LEF-1 pathway. Proc Natl Acad Sci USA 96 (10):5522–5527

Shyh-Chang N, Daley GQ, Cantley LC (2013) Stem cell metabolism in tissue development and aging. Development 140(12):2535–2547. doi:10.1242/dev.091777

Sikes JM, Newmark PA (2013) Restoration of anterior regeneration in a planarian with limited regenerative ability. Nature 500(7460):77–80. doi:10.1038/nature12403

Sivasankaran R, Calleja M, Morata G, Basler K (2000) The Wingless target gene Dfz3 encodes a new member of the Drosophila Frizzled family. Mech Dev 91(1–2):427–431

Smith WC, Harland RM (1991) Injected Xwnt-8 RNA acts early in Xenopus embryos to promote formation of a vegetal dorsalizing center. Cell 67(4):753–765

Snippert HJ, Haegebarth A, Kasper M, Jaks V, van Es JH, Barker N, van de Wetering M, van den Born M, Begthel H, Vries RG, Stange DE, Toftgård R, Clevers H (2010a) Lgr6 marks stem cells in the hair follicle that generate all cell lineages of the skin. Science 327 (5971):1385–1389. doi:10.1126/science.1184733

Snippert HJ, van der Flier LG, Sato T, van Es JH, van den Born M, Kroon-Veenboer C, Barker N, Klein AM, van Rheenen J, Simons BD, Clevers H (2010b) Intestinal crypt homeostasis results from neutral competition between symmetrically dividing Lgr5 stem cells. Cell 143 (1):134–144. doi:10.1016/j.cell.2010.09.016

Sokol S, Christian JL, Moon RT, Melton DA (1991) Injected Wnt RNA induces a complete body axis in Xenopus embryos. Cell 67(4):741–752

Sokol SY, Klingensmith J, Perrimon N, Itoh K (1995) Dorsalizing and neuralizing properties of Xdsh, a maternally expressed Xenopus homolog of dishevelled. Development 121(10):3487

Stoick-Cooper CL, Moon RT, Weidinger G (2007a) Advances in signaling in vertebrate regeneration as a prelude to regenerative medicine. Genes Dev 21(11):1292–1315. doi:10.1101/gad.1540507

Stoick-Cooper CL, Weidinger G, Riehle KJ, Hubbert C, Major MB, Fausto N, Moon RT (2007b) Distinct Wnt signaling pathways have opposing roles in appendage regeneration. Development 134(3):479–489. doi:10.1242/dev.001123

Strand M, Micchelli CA (2011) Quiescent gastric stem cells maintain the adult Drosophila stomach. Proc Natl Acad Sci USA 108(43):17696–17701. doi:10.1073/pnas.1109794108

Sureda-Gómez M, Pascual-Carreras E, Adell T (2015) Posterior Wnts Have Distinct Roles in Specification and Patterning of the Planarian Posterior Region. Int J Mol Sci 16 (11):26543–26554. doi:10.3390/ijms161125970

Takada R, Satomi Y, Kurata T, Ueno N, Norioka S, Kondoh H, Takao T, Takada S (2006) Monounsaturated fatty acid modification of Wnt protein: its role in Wnt secretion. Dev Cell 11(6):791–801. doi:10.1016/j.devcel.2006.10.003

Takase HM, Nusse R (2016) Paracrine Wnt/β-catenin signaling mediates proliferation of undifferentiated spermatogonia in the adult mouse testis. Proc Natl Acad Sci USA 113(11): E1489–E1497. doi:10.1073/pnas.1601461113

Takeo M, Chou WC, Sun Q, Lee W, Rabbani P, Loomis C, Taketo MM, Ito M (2013) Wnt activation in nail epithelium couples nail growth to digit regeneration. Nature 499 (7457):228–232. doi:10.1038/nature12214

Takeshita H, Sawa H (2005) Asymmetric cortical and nuclear localizations of WRM-1/beta-catenin during asymmetric cell division in *C. elegans*. Genes Dev 19(15):1743–1748. doi:10.1101/gad.1322805

Tan DW, Barker N (2014) Intestinal stem cells and their defining niche. Curr Top Dev Biol 107:77–107. doi:10.1016/B978-0-12-416022-4.00003-2

Tan SH, Senarath-Yapa K, Chung MT, Longaker MT, Wu JY, Nusse R (2014) Wnts produced by Osterix-expressing osteolineage cells regulate their proliferation and differentiation. Proc Natl Acad Sci USA 111(49):E5262–E5271. doi:10.1073/pnas.1420463111

ten Berge D, Kurek D, Blauwkamp T, Koole W, Maas A, Eroglu E, Siu RK, Nusse R (2011) Embryonic stem cells require Wnt proteins to prevent differentiation to epiblast stem cells. Nat Cell Biol 13(9):1070–1075. doi:10.1038/ncb2314

Tetsu O, McCormick F (1999) Beta-catenin regulates expression of cyclin D1 in colon carcinoma cells. Nature 398(6726):422–426. doi:10.1038/18884

Thrasivoulou C, Millar M, Ahmed A (2013) Activation of intracellular calcium by multiple Wnt ligands and translocation of β-catenin into the nucleus: a convergent model of Wnt/Ca2+ and Wnt/β-catenin pathways. J Biol Chem 288(50):35651–35659. doi:10.1074/jbc.M112.437913

Tolwinski NS, Wieschaus E (2004) Rethinking WNT signaling. Trends Genet 20(4):177–181. doi:10.1016/j.tig.2004.02.003

Umesono Y, Tasaki J, Nishimura Y, Hrouda M, Kawaguchi E, Yazawa S, Nishimura O, Hosoda K, Inoue T, Agata K (2013) The molecular logic for planarian regeneration along the anterior-posterior axis. Nature 500(7460):73–76. doi:10.1038/nature12359

van Amerongen R, Mikels A, Nusse R (2008) Alternative wnt signaling is initiated by distinct receptors. Sci Signal 1(35):re9. doi:10.1126/scisignal.135re9 van Amerongen R, Bowman AN, Nusse R (2012) Developmental stage and time dictate the fate of Wnt/β-catenin-responsive stem cells in the mammary gland. Cell Stem Cell 11(3):387–400. doi:10.1016/j.stem.2012.05.023

van den Heuvel M, Nusse R, Johnston P, Lawrence PA (1989) Distribution of the wingless gene product in Drosophila embryos: a protein involved in cell-cell communication. Cell 59 (4):739–749

Vermeulen L, De Sousa E Melo F, van der Heijden M, Cameron K, de Jong JH, Borovski T, Tuynman JB, Todaro M, Merz C, Rodermond H, Sprick MR, Kemper K, Richel DJ, Stassi G, Medema JP (2010) Wnt activity defines colon cancer stem cells and is regulated by the microenvironment. Nat Cell Biol 12(5):468–476. doi:10.1038/ncb2048

Wang QT, Piotrowska K, Ciemerych MA, Milenkovic L, Scott MP, Davis RW, Zernicka-Goetz M (2004) A genomewide study of gene activity reveals developmental signaling pathways in the preimplantation mouse embryo. Dev Cell 6(1):133–144

Wang X, Tsai JW, Imai JH, Lian WN, Vallee RB, Shi SH (2009) Asymmetric centrosome inheritance maintains neural progenitors in the neocortex. Nature 461(7266):947–955. doi:10.1038/nature08435

Wang B, Zhao L, Fish M, Logan CY, Nusse R (2015a) Self-renewing diploid Axin2(+) cells fuel homeostatic renewal of the liver. Nature 524(7564):180–185. doi:10.1038/nature14863

Wang D, Cai C, Dong X, Yu QC, Zhang XO, Yang L, Zeng YA (2015b) Identification of multipotent mammary stem cells by protein C receptor expression. Nature 517(7532):81–84. doi:10.1038/nature13851

Warburg O (1956) On respiratory impairment in cancer cells. Science 124(3215):269–270

Weitzel HE, Illies MR, Byrum CA, Xu R, Wikramanayake AH, Ettensohn CA (2004) Differential stability of beta-catenin along the animal-vegetal axis of the sea urchin embryo mediated by dishevelled. Development 131(12):2947–2956. doi:10.1242/dev.01152

Widelitz R (2005) Wnt signaling through canonical and non-canonical pathways: recent progress. Growth Factors 23(2):111–116. doi:10.1080/08977190500125746

Wigfield SM, Winter SC, Giatromanolaki A, Taylor J, Koukourakis ML, Harris AL (2008) PDK-1 regulates lactate production in hypoxia and is associated with poor prognosis in head and neck squamous cancer. Br J Cancer 98(12):1975–1984. doi:10.1038/sj.bjc.6604356

Wikramanayake AH, Hong M, Lee PN, Pang K, Byrum CA, Bince JM, Xu R, Martindale MQ (2003) An ancient role for nuclear beta-catenin in the evolution of axial polarity and germ layer segregation. Nature 426(6965):446–450. doi:10.1038/nature02113

Wikramanayake AH, Peterson R, Chen J, Huang L, Bince JM, McClay DR, Klein WH (2004) Nuclear beta-catenin-dependent Wnt8 signaling in vegetal cells of the early sea urchin embryo regulates gastrulation and differentiation of endoderm and mesodermal cell lineages. Genesis 39(3):194–205. doi:10.1002/gene.20045

Willert K, Brown JD, Danenberg E, Duncan AW, Weissman IL, Reya T, Yates JR, Nusse R (2003) Wnt proteins are lipid-modified and can act as stem cell growth factors. Nature 423 (6938):448–452. doi:10.1038/nature01611

Xie H, Tranguch S, Jia X, Zhang H, Das SK, Dey SK, Kuo CJ, Wang H (2008) Inactivation of nuclear Wnt-betacatenin signaling limits blastocyst competency for implantation. Development 135(4):717–727. doi:10.1242/dev.015339

Xing Y, Clements WK, Kimelman D, Xu W (2003) Crystal structure of a beta-catenin/axin complex suggests a mechanism for the beta-catenin destruction complex. Genes Dev 17 (22):2753–2764. doi:10.1101/gad.1142603

Xu Q, Wang Y, Dabdoub A, Smallwood PM, Williams J, Woods C, Kelley MW, Jiang L, Tasman W, Zhang K, Nathans J (2004) Vascular development in the retina and inner ear: control by Norrin and Frizzled-4, a high-affinity ligand-receptor pair. Cell 116(6):883–895

Yamashita YM, Jones DL, Fuller MT (2003) Orientation of asymmetric stem cell division by the APC tumor suppressor and centrosome. Science 301(5639):1547–1550. doi:10.1126/science.1087795

Yanagita M, Oka M, Watabe T, Iguchi H, Niida A, Takahashi S, Akiyama T, Miyazono K, Yanagisawa M, Sakurai T (2004) USAG-1: a bone morphogenetic protein antagonist abundantly expressed in the kidney. Biochem Biophys Res Commun 316(2):490–500. doi:10.1016/j.bbrc.2004.02.075

Yang J, Mowry LE, Nejak-Bowen KN, Okabe H, Diegel CR, Lang RA, Williams BO, Monga SP (2014) β-catenin signaling in murine liver zonation and regeneration: a. Wnt-Wnt situation! Hepatology 60(3):964–976. doi:10.1002/hep.27082

Yee KK, Li Y, Redding KM, Iwatsuki K, Margolskee RF, Jiang P (2013) Lgr5-EGFP marks taste bud stem/progenitor cells in posterior tongue. Stem Cells 31(5):992–1000. doi:10.1002/stem.1338

Yokoyama H, Ogino H, Stoick-Cooper CL, Grainger RM, Moon RT (2007) Wnt/beta-catenin signaling has an essential role in the initiation of limb regeneration. Dev Biol 306(1):170–178. doi:10.1016/j.ydbio.2007.03.014

Zecca M, Basler K, Struhl G (1996) Direct and long-range action of a wingless morphogen gradient. Cell 87(5):833–844

Zeng YA, Nusse R (2010) Wnt proteins are self-renewal factors for mammary stem cells and promote their long-term expansion in culture. Cell Stem Cell 6(6):568–577. doi:10.1016/j.stem.2010.03.020

Chapter 16
Extracellular Regulation of the Mitotic Spindle and Fate Determinants Driving Asymmetric Cell Division

Prestina Smith, Mark Azzam, and Lindsay Hinck

Abstract Stem cells use mode of cell division, symmetric (SCD) versus asymmetric (ACD), to balance expansion with self-renewal and the generation of daughter cells with different cell fates. Studies in model organisms have identified intrinsic mechanisms that govern this process, which involves partitioning molecular components between daughter cells, frequently through the regulation of the mitotic spindle. Research performed in vertebrate tissues is revealing both conservation of these intrinsic mechanisms and crucial roles for extrinsic cues in regulating the frequency of these divisions. Morphogens and positional cues, including planar cell polarity proteins and guidance molecules, regulate key signaling pathways required to organize cell/ECM contacts and spindle pole dynamics. Noncanonical WNT7A/VANGL2 signaling governs asymmetric cell division and the acquisition of cell fates through spindle pole orientation in satellite stem cells of regenerating muscle fibers. During cortical neurogenesis, the same pathway regulates glial cell fate determination by regulating spindle size, independent of its orientation. Sonic hedgehog (SHH) stimulates the symmetric expansion of cortical stem and cerebellar progenitor cells and contributes to cell fate acquisition in collaboration with Notch and Wnt signaling pathways. SLIT2 also contributes to stem cell homeostasis by restricting ACD frequency through the regulation of spindle orientation. The capacity to influence stem cells makes these secreted factors excellent targets for therapeutic strategies designed to enhance cell populations in degenerative disease or restrict cell proliferation in different types of cancers.

P. Smith • M. Azzam • L. Hinck (✉)
Department of Molecular, Cell and Developmental Biology, University of California, Santa Cruz, CA 95064, USA
e-mail: lhinck@ucsc.edu

J.-P. Tassan, J.Z. Kubiak (eds.), *Asymmetric Cell Division in Development, Differentiation and Cancer*, Results and Problems in Cell Differentiation 61,
DOI 10.1007/978-3-319-53150-2_16

16.1 Introduction

Key characteristics of stem/progenitor cells are their long-term capacity to expand, self-renew, and differentiate, attributes that serve as the foundation for the contribution of stem cells to tissue growth during morphogenesis, the maintenance of tissue homeostasis over time, and response to injury under critical circumstances. In accomplishing these tasks, stem cells undergo different modes of cell division. With asymmetric cell divisions (ACDs), stem cells self-renew, reproducing the stem cell while also generating a daughter progenitor that will adopt a different cell fate. In contrast, symmetric cell divisions (SCDs) result in identical daughters, either two stem cells or two differentiating daughter cells.

The largest range of stem cell responses occurs when population dynamics control stem cell behavior and stem cells divide via both SCD and ACD to maintain homeostasis or respond to injury depending on intrinsic and extrinsic cues. The intrinsic ability of stem cells to drive ACD was initially defined in model organisms (Morin and Bellaiche 2011). In these systems, the intrinsic nature of cell fate specification varies among stem/progenitor cell types based on their location and history of cell contacts. For example, in studies in which individual *Drosophila* central nervous system progenitors were isolated and cultured, differences were observed in the capacity of cells to self-renew and generate appropriate progeny based on their origin in the embryo (Ceron et al. 2006; Luer and Technau 2009). This suggests that cells are primed for cell fate acquisition during development based on environment cues. Thus, even in model organisms, intrinsic fate determination is influenced by extrinsic factors.

Mechanisms underlying intrinsic ACD depend on the acquisition of cellular asymmetry during interphase, which is subsequently used in mitosis to polarize the distribution of proteins that determine cell fate. The mitotic spindle is reoriented in reference to this polarity axis to produce an asymmetric division. The molecular requirements for these intrinsic ACDs have been determined and include the Par3/Par6/atypical protein kinase C complex that establishes and maintains apico-basal polarity and the microtubule-associated nuclear mitotic apparatus protein (NuMA)/LGN/Gαi complex that reorients the mitotic spindle along this apico-basal axis.

The basic process of spindle reorientation is conserved in mammalian tissues; however, there is mounting evidence that the increased complexity of higher organisms generates additional regulatory requirements. This includes extrinsic mechanisms in the form of secreted cues to regulate the mode of stem and progenitor cell division. Here, we describe recent studies identifying such cues and how they govern the crucial balance between SCD and ACD. These extracellular cues govern the choice between cell expansion and differentiation during development and in response to injury.

16.2 WNTs Function in Planar Cell Polarity

WNTs are secreted proteins that regulate various aspects of development and signal through canonical and noncanonical pathways. The canonical pathway, which is responsible for the regulation and subcellular localization of the transcription factor β-catenin (CTNNB1), is being covered elsewhere in this volume. Here, we describe WNT signaling in governing ACD through one of the two noncanonical pathways. The second of these noncanonical pathways results in the release of intracellular calcium, which is, in turn, associated with the activation of various enzymes such as Ca^{2+}/calmodulin-dependent (CamKII) protein kinase and Protein Kinase C (PKC) (De 2011). This pathway has not been implicated in ACD. Instead, the noncanonical WNT pathway regulating ACD is the planar cell polarity (PCP) pathway that is activated by noncanonical WNTs (WNT7 or WNT5A). Traditionally, this pathway functions to uniformly orient cells in a sheet of epithelium by establishing proximal–distal polarity in each cell (Devenport 2014). This is achieved by regulation of both cytoplasmic and membrane proteins. Originally characterized in *Drosophila* and later found in mammals, the core PCP genes include Van Gogh (*Vang*), *Flamingo* (*Fmi*) (*Celsr* in mammals), *Frizzled* (*Fzd*), *Prickled* (*Pk*), and *Disheveled* (*Dsh*). *Van Gogh* and *Frizzled* encode multi-pass transmembrane proteins, while *Disheveled* and *Prickled* encode cytoplasmic proteins. PCP protein localization follows a stereotypic pattern in postmitotic cells and is characterized by the asymmetric localization of these core PCP proteins. Within each individual epithelial cell, VANG and PK partition to the proximal side, while FZD and DSH localize to the cell's distal region (Fig. 16.1a). At the membrane of this cell, a proximal complex forms between VANG and FMI, whereas FZD and FMI interact on the opposite membrane. Across adjacent cells, junctions form when FMI of each complex binds to the other, bringing VANG and FZD into close proximity (Devenport 2014). The result of this repetitive patterning of PCP components in sheets of epithelia is the uniform orientation of global directional cues that can be employed to produce locally polarized cell behaviors. One of these polarized behaviors occurs in mitotic cells when noncanonical extracellular WNTs regulate the orientation of cell division and consequently the fate of daughter cells through the PCP pathway. WNTs take advantage of the asymmetric localization of these core PCP proteins to drive ACD by two major interrelated mechanisms: (1) cooperation with cell fate determinants and (2) regulation of the spindle pole.

16.2.1 Noncanonical WNT/PCP Governs the Balance Between SCD and ACD in Muscle Satellite Stem Cells

Noncanonical WNTs have the capacity to drive ACD through the PCP pathway by regulating spindle orientation. Specifically, studies on satellite cells in adult muscle have identified a mechanism by which WNT7A, VANGL2, and FZD control ACD

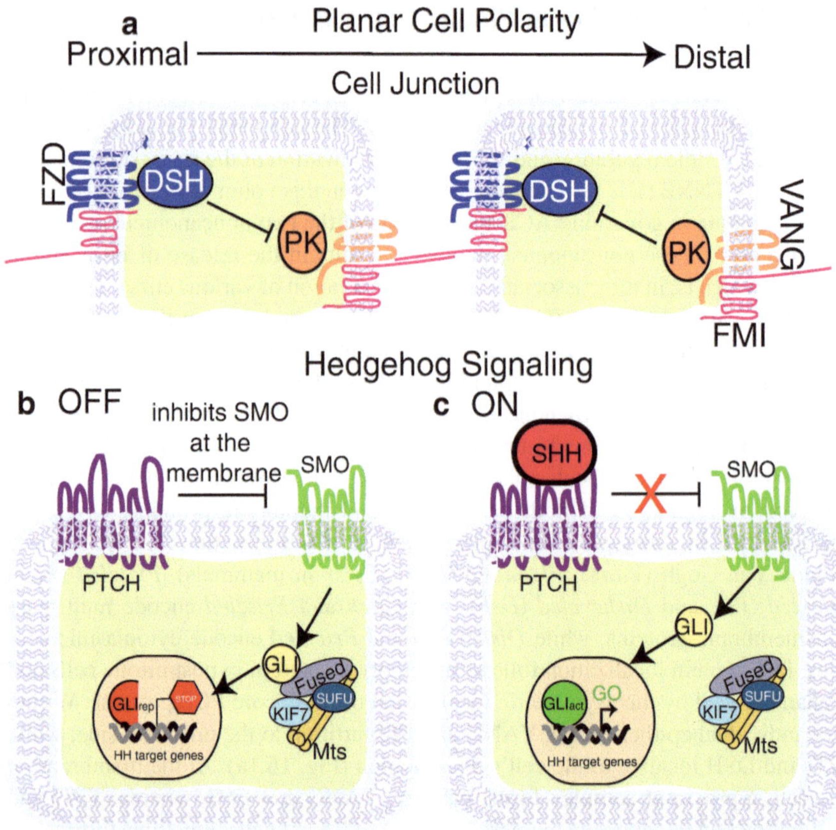

Fig. 16.1 PCP and SHH Signaling pathways. (**a**) Core planar cell polarity proteins asymmetrically distributed in two epithelial cells. Epithelial cells in contact with each other partition one group of proteins (FZD, DSH shown in *blue*) to the proximal side of the cell and another group (VANG, PK shown in *orange*) to the distal. FMI (shown in *pink*) forms a complex with FZD at the proximal surface of the membrane and VANG on the distal. Across cell junctions, FMI from one cell binds to FMI on the adjacent cell, which connects each cell within a sheet of epithelium. DSH and PK maintain asymmetry in a cell by functioning as inhibitors of each other. (**b**) In the absence of SHH, the pathway is in an "off" state in which PTCH (*purple*) inhibits SMO (*green*) at the cell membrane. SMO is then held captive on the membrane of vesicles (not depicted). This activates a complex of FUSED, SUFU, KIF7, and microtubules to process GLI into its transcriptional repressive form. The result is the repression of Hedgehog target genes. (**c**) In the "on" state, Sonic Hedgehog (*pink*) binds to PTCH, which relieves the repression on SMO. After SMO activation, GLI is left unprocessed and is able to enter the nucleus to stimulate Hedgehog target gene expression

versus SCD in satellite stem cells by orienting the plane of cell division. Satellite cells are a well-characterized, mixed population of stem and progenitor cells that are present in the adult tissue and are responsible for muscle repair. They are located between the basal lamina and sarcolemmal membranes of a muscle fiber and contained within small membrane depressions (Dumont et al. 2015) (Fig. 16.2a, b). In order to maintain their number, the cells remain quiescent until activated. Once activated, a process that is initiated by either traumatic injury or daily wear and tear, muscle satellite stem cells proliferate via ACD or SCD. ACDs allow self-renewal and population maintenance, whereas SCDs generate precursor cells that undergo multiple rounds of division before terminally differentiating and fusing onto host fibers. Repeat injury models reveal a remarkable ability of these cells to balance these processes to ensure lifelong upkeep of muscle (Dumont et al. 2015). Recent studies have focused on understanding how these satellite stem cells divide via ACD to self-renew, while producing progenitors that maintain the heterogeneity of the satellite cell population. During an ACD, satellite stem cells divide in the apical–basal plane, with one daughter cell self-renewing in contact with the basal lamina and the other daughter becoming a progenitor cell in contact with the sarcolemma. In contrast during SCD, which occurs in response to injury, the dividing satellite stem cell maintains a connection to the basal lamina; both daughters maintain their stem cell identity, resulting in expansion of the stem cell population (Dumont et al. 2015). Le Grand and colleagues have recently demonstrated that this symmetric expansion occurs via PCP signaling (Le Grand et al. 2009). PCP proteins regulate the plane of stem cell division so that both daughters maintain their contact with the basal lamina, thereby preserving their niche and supporting SCD.

The authors investigated SCD in the satellite cell population by isolating genes specifically expressed in satellite stem cells (Le Grand et al. 2009). One identified gene was the WNT receptor *Fzd7*, and subsequent immunohistochemistry experiments on fixed muscle tissue showed FZD7 specifically expressed in a subpopulation of quiescent satellite cells. To determine whether FZD7 is regulated when satellite stem cells are stimulated, the authors injured myofibers in culture using cardiotoxin. In response to this damage, *Fzd7* was upregulated along with *Wnt7a*, a known ligand, suggesting a role for this signaling pathway in regulating the regeneration of muscle fibers in response to injury. This result supported previous research demonstrating the expression and activity of WNTs and their FZD receptors in the satellite stem cell population during muscle regeneration (Polesskaya et al. 2003). However, these previously published studies did not address the underlying mechanism of WNT signaling in this context.

Bringing new insight into the role of WNTs in regulating satellite stem cell in response to injury, Le Grand and colleagues show that FZD7 forms a complex with co-expressed PCP pathway component, VANGL2 (Le Grand et al. 2009). To further address the intersection between noncanonical WNT and PCP signaling, the authors stimulated the proliferation of quiescent stem cells in culture using WNT7A and examined the outcome of the first division by immunostaining. They assayed both cell fate by quantifying *Myf5* expression and the plane of division by investigating whether cell doublets were oriented parallel (SCD) or perpendicular (ACD) to each other with respect to the surrounding myofiber. Upon WNT7A stimulation, satellite cells divided primarily via SCD, with spindle poles oriented parallel to the basal lamina. These dividing satellite cells also contained increased VANGL2 that was localized at opposite poles of each daughter cell (Fig. 16.2c).

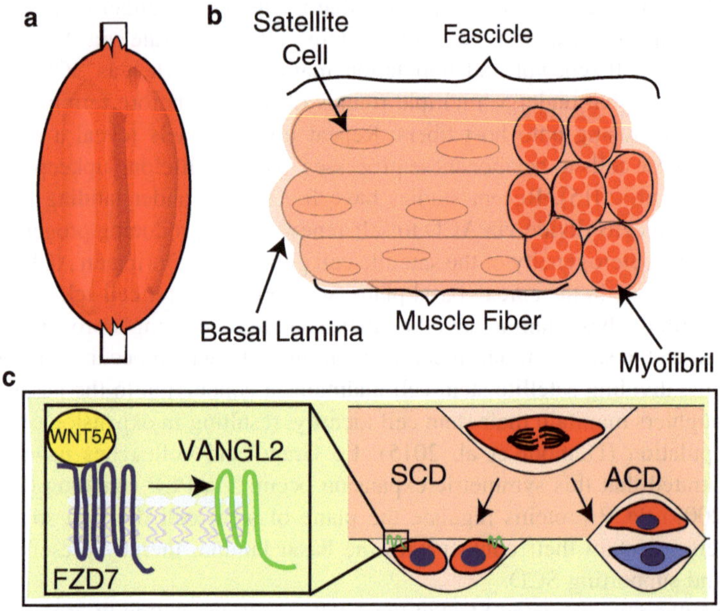

Fig. 16.2 WNT/PCP signals during SCD in satellite stem cells. (**a**) Cartoon of a limb comprising skeletal muscle and bone. (**b**) Skeletal muscle composed of seven fascicles, each of which contains muscle fibers (seven depicted here). Within each muscle fiber are multiple myofibrils. The satellite stem cells (*red*) reside in between the encapsulating basal lamina and the sarcolemma (not shown) of each muscle fiber. (**c**) Satellite stem cells undergo both SCD and ACD. The *left arrow* indicates the outcome of an SCD where both daughter cells are in contact with the sarcolemma anchored by VANGL2 (*green*). The *right arrow* indicates an ACD in which only one daughter cell is in contact with the sarcolemma, remaining a satellite stem cell (*red*), and the other daughter contacts the basal lamina, becoming a progenitor cell (*blue*). The enlarged inset depicts WNT5A (*yellow*) bound to FZD7 (*blue*) during an SCD, triggering the accumulation of VANGL2 on opposite poles of the daughter cells

Loss of *Fzd7* or knockdown of *Vangl2* impaired the ability of WNT7A to stimulate SCDs.

Taken together, this study supports a model in which satellite cells employ a classic PCP pathway, WNT7A signaling via FZD7 to VANGL2 that controls the orientation of satellite cell division and, as a result, their cell fate within the niche. This is achieved by polarizing VANGL2 to the opposite ends of daughter cells as they undergo SCD, an organization of VANGL2 that is distinct from the planar asymmetry of VANGL2 in sheets of cells. This finding suggests that VANGL2 localizes differently, depending on whether a cell is contacted on both sides or, in the case of a lone dividing satellite stem cell, along only one side. VANGL2 is distributed to the same cellular side in cells aligned in sheet (Fig. 16.1a) but to the side of no cell contact in cells that share a single border (Fig. 16.2c). Opposite polarization of VANGL2 in a couplet of satellite stem cells ensures a continuous border with the basal lamina and preserves their localization relative to the niche (Fig. 16.2c). Thus, differential orientation of VANGL2 to distinct cell membranes generates cellular contacts required to stabilize cells in orientations that regulate the fundamental process of stem cell division.

16.2.2 Noncanonical WNT/PCP Regulates ACD and Cell Fate Acquisition During Cortical Neurogenesis Through Spindle Size

The ability of WNT7A to regulate ACD frequency and cell fate through the PCP pathway is not limited to the cells of the regenerating muscle. In a process called spindle size asymmetry, the same combination of factors also functions during corticogenesis to regulate the relative size and shape of the mitotic spindle (Delaunay et al. 2014). The cerebral cortex is stratified into layers, with each layer containing a characteristic distribution of neuronal cell types (Fig. 16.3a). During embryogenesis, many precursor cells in the cortex are responsible for generating this diverse array of differentiated cells. One of these precursors, called the apical precursor, is a major type of radial glial cell that resides along the ventricular surface (Fig. 16.3b). During cortical neurogenesis, apical precursors undergo ACD to self-renew while generating a cell fate restricted neuron. While spindle pole orientation during apical precursor ACD may provide one level of regulation (Gauthier-Fisher et al. 2009), PCP does not appear to determine the axis of cell division in this context. Instead, PCP governs cell fate determination in these ACDs by signaling through WNT7A/VANGL2 to regulate the position of the metaphase plate and consequently the relative size of each spindle (Fig. 16.3c). The result of this spindle size asymmetry is that the daughter cell issued from the larger spindle becomes a neuron, whereas the daughter generated from the smaller spindle maintains its status as an apical progenitor cell.

In the murine cortex, apical precursors undergo the bulk of their ACDs from embryonic day (E) E11.5–E16.5 (Delaunay et al. 2014). To track spindles in apical precursors at metaphase, Delaunay and colleagues used confocal imaging to acquire high-resolution images of spindle poles. Examining different stages of cortical neurogenesis, they found that spindle size asymmetry was initially minimal, peaked at mid-corticogenesis, and decreased at the close of this developmental period, a time course in which the peak of ACD is correlated with the highest level of spindle size asymmetry. To investigate the mechanism underlying this asymmetry, the authors examined apical precursors, isolated from E14.5 embryos and grown in conditioned media containing either WNT7A or control WNT3A. Apical precursors that were treated with WNT7A, but not those treated with WNT3A, showed a reduced frequency of spindle size asymmetry. To understand how WNT7A regulates this behavior, the authors knocked down *Vangl2* using RNAi in cultured apical precursors and observed an increase in spindle size asymmetry frequency that was rescued by *Vangl2* overexpression. These results suggest that

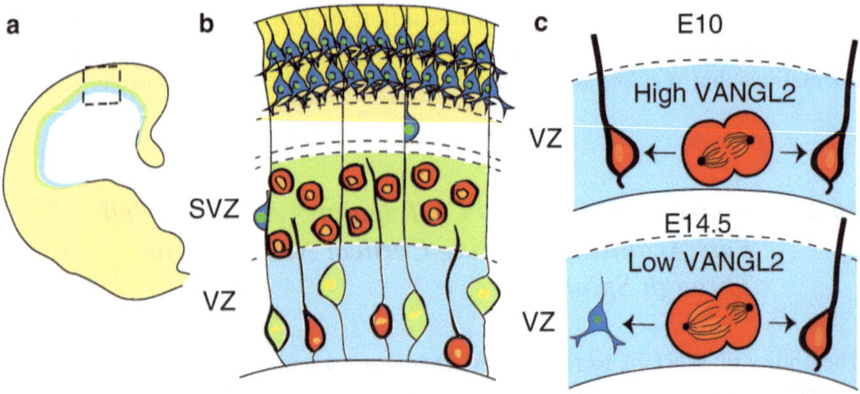

Fig. 16.3 Spindle size asymmetry and PCP in the cortex. (**a**) Cartoon of a section through a neonatal cortex. The ventricular zone (VZ) is in *blue*, the subventricular zone (SVZ) is in *green*, and the mature cortical layer is in *yellow*. (**b**) Enlarged view of the boxed area in (**a**) showing that each layer is composed of different cell types. Residing in the VZ are the apical precursors (*red*) and glial cells (*green*). The apical precursors remain quiescent or undergo cell division while the glial cells serve as a migratory scaffold for neurons moving up to the mature cortical layer. The SVZ contains precursor cells (*red*). The cortical plate (CP) in *yellow* houses mature neurons. (**c**) In the VZ, the level of VANGL2 expression determines ACD versus SCD. At embryonic day 10.5, high VANGL2 in dividing apical precursors leads to spindles with equal size resulting in an SCD. This generates two apical precursor cells. By embryonic day 14.5, VANGL2 expression decreases, leading to spindle size asymmetry and an increase in ACDs. This produces a precursor and a neuron

spindle size asymmetry is associated with ACD and that noncanonical WNT signaling through PCP proteins maintains symmetrical spindles in dividing apical precursors.

In order to understand how WNT7A/VANGL2 signaling maintains spindle pole symmetry, the authors investigated the role of ezrin, radixin, and moesin (ERM) proteins that are associated with actin microtubules (Delaunay et al. 2014). ERM proteins act as scaffolding proteins during cell division and when phosphorylated can tether actin filaments to the cell membrane (Clucas and Valderrama 2014). Using an antibody specific to the phosphorylated versions of ERM proteins (P-ERM), a high level of P-ERM was observed during metaphase in a ringlike structure on the inner face of the plasma membrane of apical precursors. Focusing on moesin, the authors knocked down this ERM using RNAi, dissolving P-ERM immunostaining, while significantly increasing the frequency of spindle size asymmetry. Loss of *Vangl2* also decreased P-ERM immunostaining, whereas treatment with WNT7A increased P-ERM immunostaining. Taken together, the data demonstrate a role for WNT7/VANGL2/P-MOESIN in preserving spindle pole symmetry, possibly by anchoring astral microtubules to the cell cortex.

Next, Delaunay and colleagues addressed the consequences of Vangl2 downregulation on division mode using an ex vivo clonal analysis on brain slices to track both spindle size and cell fate over time. Embryonic (E14) cortices were electroporated with plasmids encoding *GFP* along with either *Vangl2* or *Vangl2*$^{Lp/Lp}$. The latter construct, *Vangl2*$^{Lp/Lp}$, encodes a mutant protein lacking VANGL2 function due to a point mutation that causes mistargeting of the protein (Kibar et al. 2001). Individual daughter cells were tracked and their fate determined on the basis of their behavior and location. Expression of *Vangl2* reduced the frequency of ACDs, resulting in a higher proportion of SCDs yielding two neurons. In contrast, expression of *Vangl2*$^{Lp/Lp}$ increased the proportion of cells undergoing ACD, favoring divisions that yielded an apical precursor cell and a neuron. To further explore the events occurring during these ACDs, the daughter closest to the ventricle was denoted the lower cell, whereas the basally located cell was denoted the upper cell. Under the condition of *Vangl2*$^{Lp/Lp}$ overexpression, the majority of lower cells became APs, whereas upper cells became neurons. This is reverse of the results under wild-type (*WT*) conditions. Significant to the analysis of spindle size asymmetry, this change in identity under the mutant condition was associated with a concomitant change in the size and shape of the spindle; the smaller spindle consistently correlated with the generation of an apical precursor cell. In contrast, neurons arose from the larger spindle, suggesting that cell fate outcomes are regulated by spindle pole size.

There is a well-studied link between the mode of division and the plane of division. Indeed, the previously described paper by Le Grande and colleagues demonstrated such a role for WNT7A/FZD7/VANGL2 in specifying spindle pole orientation during the division of muscle satellite stem cells, a process that directed cell fate determination by tethering either both or only one daughter cell to the basal lamina. In this vein, Delauney and colleagues also examined the consequences of *Vangl2* downregulation on the plane of division of individual apical precursors during cortical neurogenesis (Delaunay et al. 2014). Early studies on this topic indicated that during corticogenesis, divisions of cortical progenitors perpendicular to the ventricular surface usually result in SCDs, whereas horizontally shifted division planes lead to asymmetric outcomes and neurogenic differentiation (Ang et al. 2003; Chenn and McConnell 1995). Since then, however, it has become clear that a key determinant of spindle orientation is whether it results in a cleavage plane that bisects the dividing cell in ways that influence inheritance of the apical membrane that attaches the daughter to the ventricular surface (Konno et al. 2008; Kosodo et al. 2008; Noctor et al. 2008). Thus, symmetric divisions that result in both daughters inheriting apical attachments to the ventricular surface are self-renewing, whereas asymmetric divisions, even just slightly asymmetric, in which only a single daughter inherits the small apical attachment, result in differentiative divisions with one daughter adopting neuronal cell fate.

With this in mind, Delauney examined division plane orientation of apical precursors at E11.5 and E14.5 in *WT Vangl2* knockdown and *Vangl2*$^{Lp/Lp}$ mice (Delaunay et al. 2014). In all genotypes, the majority of cell divisions occurred with the spindle aligned parallel to the ventricular surface and the division plane within 15 degrees of vertical, although loss of *Vangl2* slightly randomized the divisions. This result indicated that spindle size asymmetry, which is regulated by WNT7A/VANGL2 signaling, regulates cell fate acquisition independent of division plane, although it is important to note that the apical attachment of cells was not directly monitored. Nevertheless, this was a surprising result and was further investigated by examining the relationship between spindle pole orientation/division plane and spindle size asymmetry using a dominant negative form of LGN. LGN regulates spindle positioning during asymmetric cell divisions. Loss of *LGN* randomizes the plane of division (Konno et al. 2008), a result that was also observed by Delaunay and colleagues, who further showed that this randomization did not alter spindle size asymmetry (Delaunay et al. 2014). Taken together, these results show that spindle size asymmetry is independent of division plane because loss of *Vangl2* does not substantially affect the latter while dramatically impacting the former, whereas the opposite is true for LGN.

Taken together, current studies on noncanonical WNT signaling demonstrate distinct roles in influencing cell fate determination by regulating ACD. In the muscle, WNT7A/FZD7/VANGL2 signals to orient the plane of satellite stem cell division, fostering SCD at the expense of ACD. In contrast, during cortical neurogenesis, WNT7/VANGL2/MOESIN regulates spindle size asymmetry and governs cell fate determination independent of cell division plane or spindle orientation. These examples reveal tissue-specific deployment of signals affecting differential outcomes. From current knowledge, it seems likely that upstream activators such as different ECM components, as well as downstream effectors, such as various cytoskeletal proteins, may be the key to generating specific signaling outcomes. Additional examples of how this core noncanonical signaling pathway functions during stem cell division will be required to fully understand how WNT7/VANGL2 governs tissue morphogenesis and homeostasis.

16.3 Hedgehog Signals Regulate Cell Fate via Oriented Cell Division

WNTs are not the only family of extracellular cues presiding over stem cell division and cell fate acquisition during development. The Hedgehog signaling pathway also functions to control developmental programs in this manner (Alman 2015). The pathway is composed of secreted ligands, receptors, and transcription factors and exists in an "on or off" state (Fig. 16.1b). In mammals, when the pathway is off, the inhibitory receptor, Patched (PTCH), prevents the signaling receptor, Smoothened (SMO), from reaching the plasma membrane; instead, inactive SMO is held inside the cytoplasm within a vesicle. This results in the transcription factor GLI being processed into its repressive form by the FUSED/SUFU/KIF complex. GLI then travels to the nucleus to act as a transcriptional repressor of hedgehog target genes. When the pathway is in the "on," state, one of the Hedgehog ligands binds the PTCH receptor, which then gets degraded, and releases SMO from vesicles, allowing it to carry out signaling at the plasma membrane. SMO signaling halts the processing of GLI, leaving it in its active form. Uncleaved GLI then translocates to the nucleus where it promotes the expression of Hedgehog target genes (Fig. 16.1c). Hedgehog, well known for establishing embryonic patterning, has recently been shown to enhance proliferation in the developing cortex and cerebellum by stimulating nonterminal SCD of stem and progenitor cells. In performing this function, mounting evidence shows that SHH influences the subcellular localization of cell fate determinants by acting together with collaborating pathways such as those mediated by NOTCH (NOTC) and WNT. The ability to sequester intracellular proteins, for example, cell fate determinants to a single daughter cell, may require the precise deployment of these ligands via targeted, rather than global, delivery mechanisms that restrict the interaction between the extracellular cue and their cell surface receptor. This, in turn, has the potential to localize the influence of these signals to distinct regions of the receiving cell's cytoplasm.

16.3.1 Sonic Hedgehog Signaling Regulates Cell Fate by Promoting Neuronal SCD in Radial Glial Cells During Cortical Neurogenesis

Hedgehog signaling, specifically sonic hedgehog (SHH), promotes cortical neurogenesis by stimulating SCD in radial glial cells (Dave et al. 2011). It has been long understood that SHH is important for cortical development (Palma and Ruiz i Altaba 2004). As described above, radial glial cells serve as neural precursors, in addition to their role as migratory scaffolds, and can either undergo one of two different modes of SCD or ACD. To expand their pool, radial glial cells divide via SCD to generate two radial glial cells. Radial glial cells, however, can also undergo a differentiative, neurogenic SCD that results in two neurons. In contrast during ACD, radial glial cells produce one neuron and one radial glial cell (Fig. 16.4a). Therefore, by regulating both the type of SCD (proliferative versus neurogenic) and the mode of division (SCD versus ACD), these cells have the capacity to regulate the number of multiple cell types in the ventricular zone. In recent studies, Dave and colleagues investigated the role of SHH signaling in this process by examining the consequences of *Ptch* deletion (*Ptchl*$^{lox/lox}$ mice), and thus activation of SHH signaling, using a clonal pair cell assay. This assay monitors the division of single radial glial cells using immunocytochemistry to track the production of either radial glial cells (GLAST-positive) or postmitotic neurons (TuJ-1-positive). An SCD yields two GLAST-positive cells (proliferative SCD) or two TuJ-1-positive cells (neurogenic SCD), whereas an ACD results in one GLAST and one TuJ-1 positive cell. The authors observed that cortical cells from *Ptc1*$^{lox/lox}$ animals, in which SHH signaling is derepressed due to the absence of this negative regulator, divided primarily via SCD, generating either two radial glial

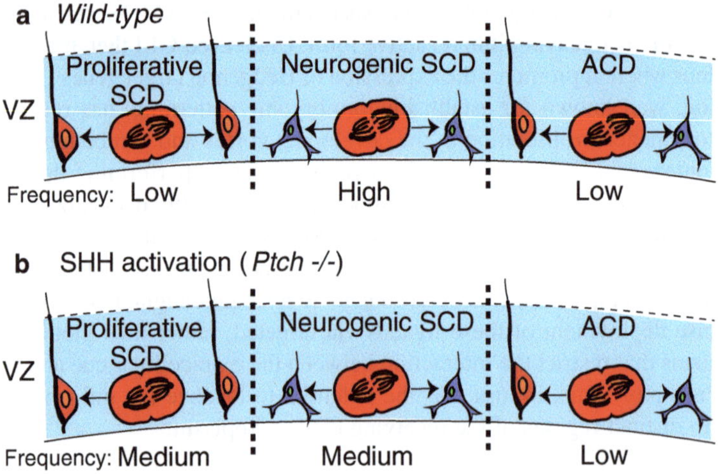

Fig. 16.4 SHH and cell division in the VZ of a neocortex. (**a**) In a wild-type brain, radial glial cells (*red*) divide via proliferative SCD (*left*) to generate two radial glial cells, neurogenic SCD (*middle*) to create two neurons (*blue*) or ACD (*right*) to produce a radial glial cell and a neuron. (**b**) Upon SHH activation in a $Ptch^{-/-}$ mutant, radial glial cells increase the frequency of proliferative SCD. The frequency of each division type is indicated

cells or two neurons (Fig. 16.4b). In contrast, *WT* control cells divided mainly by neurogenic SCD that promoted differentiation. In both cases, the frequency of ACD was low and remained unchanged in *WT* control and $Ptc1^{lox/lox}$ cells. Thus, SHH signaling stimulates proliferative SCD of radial glial neurons at the expense of neurogenic SCD.

To further understand the link between extracellular HH signaling in regulating division mode, the authors examined Notch signaling. This pathway has been previously implicated in regulating the balance between SCD/ACD in neural stem and progenitor cells, with activation of NOTC1 and NOTC3 found to promote radial glia cell identity (Gaiano et al. 2000). Indeed, Dave and colleagues showed that activation of SHH signaling, via *Ptch1* deletion, resulted in an upregulation of Notch effector proteins HES1 and BLBP, a result that connects extracellular Hedgehog with the cytoplasmic components of Notch signaling during SCD. The authors further explored the interaction between Notch and Hedgehog signaling during cortical neogenesis by deleting *Rbpj*, which encodes a DNA binding protein that acts with cleaved Notch intracellular domain to regulate the transcription of Notch target genes. Performing the clonal pair assay, the authors found that loss of *Rbpj* in an SHH-signaling environment significantly increased the number of cells undergoing neurogenic SCD, compared to the proliferative SCD that occurs in response to SHH when Notch signaling is intact. These results suggest that SHH and NOTCH signaling pathways cooperate to enhance symmetric proliferative divisions of neocortical stem cells.

16.3.2 Global Sonic Hedgehog Promotes SCD in the Developing Cerebellum: Targeted Sonic Hedgehog Promotes ACD

In addition to promoting SCD in the cortex, recent studies demonstrate that SHH also stimulates this mode of division to expand the population of granule neuron progenitors (GNPs) in the cerebellum. There are multiple proliferative zones in the embryonic and early postnatal brain; one is the ventricular zone (discussed in the previous section) and another is the external granule layer (EGL) of the cerebellum. Granule neurons are a large class of neurons with a unique pattern of differentiation, occurring in two migratory phases that are divided by a proliferative phase. During the first phase, GNPs arise from the rhombic lip located next to the ventricular zone and migrate to the surface of the cerebellum, forming the EGL (Miale and Sidman 1961). In this layer, GNPs are highly proliferative. The second phase of migration begins when the GNPs become postmitotic and migrate from the EGL inwards on Bergmann glial fibers through the Purkinje cell layer to form mature granule neurons of the internal granule layer (IGL) (Komuro et al. 2001). Studies have shown that a variety of factors affect the thickness of the EGL, including SHH, which increases EGL thickness by fostering GNP proliferation (Wechsler-Reya and Scott 1999). GNPs, like muscle satellite and other types of stem cells, divide into two orientations with respect to the pial or outer surface of the cerebellum: parallel or perpendicular. Yet, it is unclear whether factors like SHH, which stimulate GNP proliferation, also regulate spindle pole orientation and cell fate determination.

Increasingly clear, however, is the fact that networks of local signals, both soluble and membrane-bound niche factors, regulate these stem cell divisions. It is no surprise, then, that SHH and WNT/β-catenin signaling pathways have emerged as two such intertwined pathways governing spindle pole orientation and cell fate determination (Haldipur et al. 2015). In their recent paper, Haldipur and colleagues used immunohistochemistry with antibodies against phosphohistone-3 to examine the division plane of mitotic GNPs, from which spindle orientation can be determined. The authors found that between postnatal days 0 and 4, half of GNPs divided with spindle poles parallel and half with spindle poles perpendicular to the pial surface. The number of cell division with spindles perpendicular to the pial surface rose gradually between postnatal days 5 and 14 (Fig. 16.5a). Next, the authors investigated the effect of SHH on division orientation by treating pups with cyclopamine, a drug that inhibits SHH signaling, or an agonist of smoothened (SAG), which stimulates the pathway. The EGL of cyclopamine treated animals was thinner compared to control and a significantly higher percentage of cells divided with spindles perpendicular to the pial surface, corresponding to an increase in cells expressing the differentiation marker NeuroD1. In contrast, SAG-treated animals displayed a concordant increase in the percentage of parallel spindles and a reduced number of NeuroD1-positive cells. Thus, SHH appears to favor symmetric GNP divisions in which both cells are aligned along the outer pial surface and retain their progenitor status. This is the

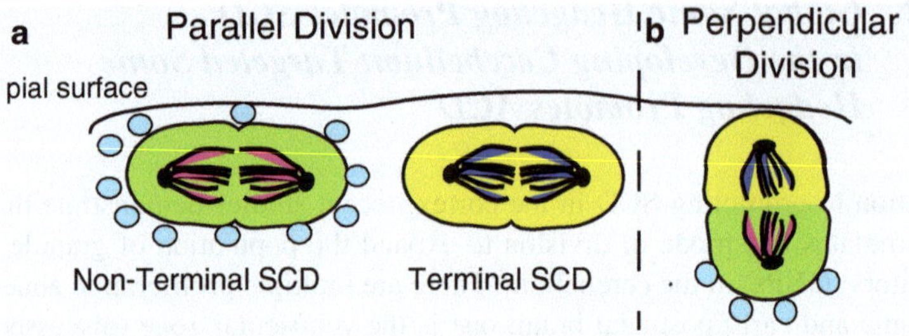

Fig. 16.5 Model for regulation of cell division and fate specification by SHH in the cerebellum. In the cerebellum, GNPs can divide so that their spindles are parallel to the pial surface (*left of dashed line*) or perpendicular to the pial surface (*right*). (**a**) During a parallel division, both daughter cells are exposed to the same level of SHH (*blue circles*) and retain their GNP fate, resulting in a nonterminal SCD (*green cells*). In the absence of SHH, both daughter cells undergo a fate change, causing terminal SCD (*yellow cells*). (**b**) During a perpendicular division, SHH is exposed locally to one daughter cell and not the other, resulting in an ACD that produces daughter cells with different cell fates (*yellow* versus *green*)

same SHH effect observed in the cortex (Dave et al. 2011) and similar to the effect of noncanonical WNT7A signaling on muscle satellite stem cells in which aligned cell divisions maintain stem cell contact to the outer tissue surface, resulting in population expansion via SCD.

In order to elucidate a possible mechanism for the influence of SHH on GNP oriented division, the authors first probed for β-catenin in anaphase cells of the EGL (Haldipur et al. 2015). They observed an asymmetric cellular distribution of β-catenin but no correlation between this asymmetric distribution and the plane of cell division. A recent study in murine embryonic stem cells, however, has showed that an asymmetric distribution of β-catenin can be produced by contact of the cell with a point source of WNT3A (Habib et al. 2013). To investigate whether GNPs are similarly influenced by focal contact with WNT3A or SHH, the authors seeded GNPs at low density on coverslips that were printed with stripes of either SHH or WNT3A and immunostained for β-catenin. Paired cells, in which one cell was in contact with the stripe and the other was outside the striped region, were examined for the intracellular distribution of β-catenin. Similar to the previous results (Habib et al. 2013), the authors found that β-catenin preferentially and asymmetrically localized to the cell that was in contact with WNT3A stripe. In contrast, when both cell nuclei were in contact with the stripe, β-catenin was symmetrically distributed in both daughters. SHH stripes produced similar, but not as dramatic, results. Taken together, these experiments indicate that SHH, locally and asymmetrically presented, may intersect with WNT signaling and have the capacity to regulate the subcellular localization of β-catenin and generate ACDs. However, as observed in the developing cortex, global stimulation by SHH enhances expansion of GNPs via SCD.

16.3.3 Sonic Hedgehog Signaling Determines Granule Neuron Progenitor Fate by Maintaining the Balance Between SCD and ACD in the Cerebellum

The studies by Haldipur and colleagues demonstrated the importance of SHH in favoring GNP expansion via SCD and showed how reduced or focal delivery of SHH shifted the division balance toward differentiation. However, these experiments did not track the acquisition of cell fate by individual GNPs as they proliferate in response to SHH. This investigation was recently tackled by Yang and colleagues using mice that express dual reporters: progenitor specific (*Math1-GFP*) and neuronal specific (*Dcx-DsRed*) (Yang et al. 2015). In this model, daughter cells that fluoresce the same color were the product of an SCD as evidenced by the production of two cells expressing the same fate determinant, whereas daughter cells of different colors were generated via ACD and have different cell fates (Fig. 16.5). In order to explore the role of SHH signaling on GNP fate acquisition, the authors first established a baseline by quantifying the *WT* frequency of SCD/ACD. The authors visualized GNPs, both in dissociated culture and in freshly dissected whole-mounted cerebella using time-lapsed imaging. This analysis defined three different cell fate outcomes produced by two division modes. There were two types of SCD; one produced two progenitor cells (both cells positive for *Math1-GFP*) and was therefore considered nonterminal. The second type of SCD was considered terminal because it produced two intermediate cells (both positive for both *Math1-GFP* and *Dcx-DsRed*) that subsequently differentiate into granule cells. In addition, ACD was observed, producing one terminal intermediate cell (*Math1-GFP/Dcx-DsRed* positive) and one nonterminal progenitor (only *Math1-GFP* positive) (Fig. 16.5). At P4, corresponding to an early stage of neurogenesis, the vast majority of divisions were SCD and expansive, producing two progenitor cells. However, at a later stage (P10), the balance shifted to terminal SCD, producing two intermediate cells that differentiated into granule cells. A minority (<5%) of cells underwent ACD, renewing the progenitor cell while also forming an intermediate cell that will differentiate into a granule cell. Taken together, these data show that GNPs divide primarily by SCD and differentiation to granule cells occurs through an intermediate cell that expresses both progenitor and neuronal markers.

The authors used two methods to stimulate SHH signaling in the dual reporter mice (*Math1-GFP;Dcx-DsRed*) (Yang et al. 2015). They either activated SHH signaling in culture using recombinant SHH (C25II) or they crossed the reporter mice to a *Ptch1*$^{-/-}$ line in which SHH signaling is enhanced. In response to SHH activation, the number of *Math-GFP* progenitors increased in proportion to the other cell types (intermediate and differentiated cells) over the time course of neurogenesis, suggesting an increase in progenitor SCD that delays cerebellar neurogenesis. Blockade of the SHH pathway, by either adding cyclopamine to GNP cultures or injecting it into *Math1-GFP;Dcx-DsRed;Ptch*$^{+/-}$ mice, reversed the expansive division mode, increasing both ACD and the terminal type of SCD that produces granule cells, at the expense of expansive, nonterminal SCDs that produce progenitor cells. Analysis of division planes by phosphohistone-3 immunostaining in *WT* versus *Ptch1*$^{+/-}$ mice revealed a decrease in parallel divisions (i.e., with spindle poles perpendicular to the pial surface). This result is consistent with the increase in SCD observed by these authors and by Halipur and colleagues who stimulated SHH signaling using SAG treatment. Together, the data demonstrate that SHH enhances proliferation by regulating both division mode (ACD/SCD) and type (nonterminal/terminal).

These examples of SHH activity in the cortex and cerebellum show how widespread SHH stimulation expands stem and progenitor populations by favoring SCD. However, the mechanism by which SHH controls the ratio of cells undergoing SCD versus ACD and how it contributes to cell fate acquisition is still mysterious. Conceivably, division mode and type may be regulated by the concentration and localization of SHH in the proliferative zone of the neocortex and in the EGL of the cerebellum (Komada et al. 2008; Martinez et al. 2013; Wallace 1999). During secretion, SHH is modified by palmitoylation as well as by a cholesterol modification, which occurs during an autoprocessing cleavage event. These modifications ensure that the extracellular movement of SHH is highly regulated as it is trafficked to the plasma membrane, stabilized on the cell surface, transferred to carrier lipoproteins, and released from the cell in large soluble lipoprotein complexes or spread by filopodia-like extensions (Briscoe and Therond 2013). This raises the possibility that Hedgehog, rather than globally bathing GNPs, is delivered in a highly regulated manner that governs division mode and type. In this model, regulated release would be developmentally controlled to provide for ample SHH distribution during GNP expansion via SCD at early time periods. In contrast, highly localized delivery at later stages of development would specify differentiation. These events may occur in collaboration with WNT and NOTCH signaling to further refine the ways stem and progenitor cells divide and acquire cell fates by regulating the subcellular localization of fate determinants.

16.4 Axon Guidance Molecules Driving ACD Through the Spindle

Neural development is complex and numerous secreted factors are likely to regulate the proliferation and cell fate acquisition of neurons as well as their migration to new environments. One type of signaling involves cues that were originally identified as axon guidance molecules, directing neurons and their growth cones to proper targets in the embryo. These guidance cues govern migration by regulating cytoskeletal dynamics, but more recently they have a newly respected role in regulating cell proliferation, including ACD, not only during neural development (Borrell et al. 2012) but also in epithelial organs such as the breast (Ballard et al. 2015).

SLITs are secreted ligands that bind to ROBO receptors to regulate multiple developmental programs. There are three SLIT ligands (*Slits* 1–3) and four ROBO receptors (*Robo* 1–4). Activation of the SLIT/ROBO pathway leads to a cascade of signaling events that range from axon guidance to cell proliferation (Ballard and Hinck 2012). Recently, SLIT/ROBO signaling has been implicated in the maintenance of mammary stem cells. The mammary gland is a dynamic organ that undergoes many rounds of growth and expansion with each estrus cycle and pregnancy (Macias and Hinck 2012). At the onset of puberty, the mammary gland grows from a rudimentary ductal structure to a fully arborized tree over a span of 6 weeks. Fueling this growth are the highly proliferative cells in the terminal end buds that generate ductal cells as they push through the fat pad, directing growth toward the unfilled space (Fig. 16.6a). Once the terminal end buds reach the outer edge of the fat pad, they dissolve leaving behind a bilayered ductal structure containing a heterogeneous luminal population and a basal layer, which contains stem cells. To fuel proliferative cycles, the mammary gland maintains a reserve of stem cells into adulthood. The inside of a terminal end bud serves as the hub for stem and progenitor cells during development and is made of inner body cells and an outer layer of cap cells (Fig. 16.6b). It is in the terminal end buds where the vast majority of ACD occurs (Ballard et al. 2015). Mammary stem cells utilize ACD to both self-renew and differentiate into the various cell types that compose the mammary duct. Stem cell self-renewal is fundamental to the function

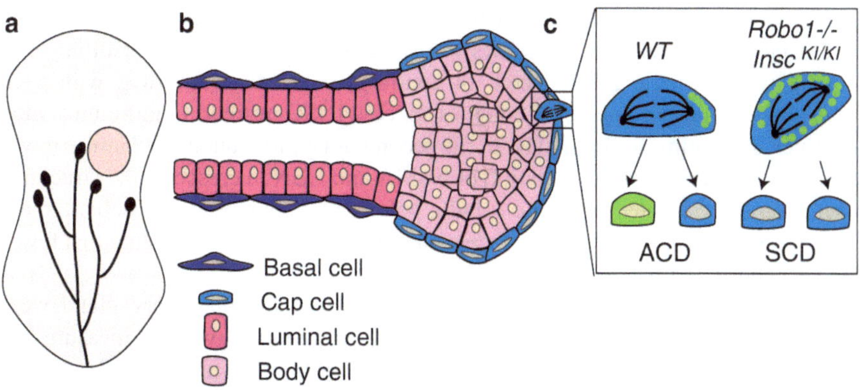

Fig. 16.6 SLIT2/ROBO1 signaling in the mammary gland. (**a**) Cartoon of developing mammary gland. Mammary ducts (*black*) grow postnatally from the nipple into the fat pad (*beige*) and at 5.5 weeks reach the lymph node (*pink circle*). Capping each duct is a terminal end bud, which serves as a site of cell proliferation. (**b**) Longitudinal section through a terminal end bud and subtending duct. Each duct is bilayered with an outer layer of basal cells and an inner layer of luminal cells. The terminal end bud is a spherical structure with an outer layer of cap cells and inner layers of luminal body cells. (**c**) SLIT2/ROBO1 signaling in the cap cells of the terminal end bud regulates cell division. In a *WT* terminal end bud, cap cells undergo ACD and are renewed (*blue*) while generating a progenitor cell (*green*). This is due to the expression of INSC (*green circles*) that accumulates on one side of the dividing cell. Loss of *Robo1* ($Robo1^{-/-}$) or overexpression of Insc ($Insc^{KI/KI}$), both of which increase INSC levels in the cell, results in a switch in division mode from ACD to SCD

of the mammary gland, and recent evidence shows that SLIT/ROBO signaling plays a role in regulating this process.

In a study by Ballard and colleagues, SLIT2/ROBO1 signaling was shown to regulate the expression of Inscuteable (INSC), which in turn governs the balance between ACD and SCD (Ballard et al. 2015). INSC is a central component of the spindle orientation complex and is initially recruited by the PAR complex and engages with LGN. Acting as a molecular baton, INSC hands off LGN to NuMA, resulting in the asymmetric co-localization of LGN and NuMA at the apical pole (Mapelli and Gonzalez 2012). This LGN/NuMA complex facilitates spindle pole tethering, thereby contributing to the unequal distribution of cell fate determinants. The authors first showed that loss of *Robo1* increases *Insc* levels, with no change in the expression of *Lgn* or *NuMA*. SLIT/ROBO signaling regulates *Insc* expression by governing the subcellular localization of the transcription factor Snail (SNAI1) through the PI3kinase/AKT/GSK-3β pathway. Loss of *Robo1* sends SNAI1 to the nucleus where it directly enhances *Insc* transcription.

Ballard and colleagues used two assays to demonstrate how excess INSC, generated by either loss of *Robo1* ($Robo1^{-/-}$) or transgenic overexpression of *Insc* ($Insc^{KI/KI}$), decreases the frequency of ACDs in mammary end buds (Ballard et al. 2015). In the first assay, the stem cell containing basal population of mammary cells was isolated by fluorescence activated cell sorting (FACS), labeled with a membrane permeable dye, PKH26, and plated at single cell density in Matrigel. PKH26 binds to cell membranes and is distributed to daughter cells upon division. ACDs generate a quiescent daughter stem cell, which maintains its fluorescence, and a progenitor cell that continues to divide and dilute the dye, resulting in a colony with a single labeled cell. In contrast, SCDs dilute the dye with each division, resulting in unlabeled colonies. Excess INSC resulted in more unlabeled colonies, indicating fewer ACDs. In the second assay, immunohistochemistry was used to visualize the co-localization of NuMA and LGN in crescent-like structures asymmetrically located over one spindle pole in mammary end bud cells. Excess INSC resulted in fewer of these crescent-like structures, again suggesting that fewer ACDs occurred in $Robo1^{-/-}$ and $Insc^{KI/KI}$ tissue.

The consequences of excess INSC expression on the mammary gland were investigated by serial passaging and limiting dilution assays that measure the robustness and number of mammary stem cells (Ballard et al. 2015). Mammary stem cells were enriched using FACS to isolate basal cells. For serial passaging, cells were plated at single cell density in Matrigel and passaged every 7 days. Cells expressing elevated levels of *Insc* generated larger colonies that passage longer compared to *WT* cells. For limiting dilution assays, a serially reduced number of enriched basal cells were transplanted into precleared mammary fat pads. After 8 weeks, the frequency and size of mammary outgrowths were measured, allowing an estimation of mammary stem cell number. Cells containing excess *Insc*, harvested from either $Robo1^{-/-}$ or $Insc^{KI/KI}$ mammary tissue, contained threefold more mammary stem cells compared to *WT* tissue. Together, these experiments demonstrate that SLIT/ROBO/SNAI1 signaling through INSC promotes ACD, with the loss of *Robo1* shifting the balance toward SCD due to upregulated *Insc* expression, which interferes with correct spindle pole positioning (Fig. 16.6c).

With this example, a number of themes come into focus in vertebrate tissue. Increasingly, evidence suggests that extracellular cues send signals to the mitotic spindle to regulate its orientation and size. These changes in the mitotic spindle govern the asymmetric distribution of cell fate determinants. Focal, rather than global, extrinsic signaling may be required to regulate spindle orientation and generate disproportionate signaling in daughter cells. For example, in the satellite stem cell niche, rotation of the spindle places one daughter in contact with the sarcolemma and the other in contact with the basal lamina of the muscle fiber. One explanation is that each location provides a distinct niche characterized by a different set of extracellular factors. Not all stem cell niches, however, are as geometrically constrained and allow for such discrete daughter cell interactions. This means that elaborate, and still largely unknown, mechanisms are likely required to regulate the extracellular distributions of extrinsic factors. Like WNT and Hedgehog proteins, the extracellular availability of SLIT is regulated by a number of extracellular matrix components, for example, heparin sulfate proteoglycans (Ballard and Hinck 2012). Such control has the capacity to create discrete niches and govern the delivery and differential activation of these ligands in temporally and spatially restricted manners. Future studies to improve our understanding of the extracellular environments governing the availability of signaling molecules that function in controlling stem cell division will further our knowledge of stem cell population dynamics in vertebrate tissues.

16.5 Conclusion: Implications of Extracellular Cues that Govern the Mode of Stem Cell Division in Medicine

In this chapter, we reviewed current research that describes the extracellular cues—WNT, SHH, and SLIT—and how they regulate cell division mode in the muscle, brain, and breast. Due to their critical role in regulating stem cell renewal and homeostasis, these extracellular cues and morphogens may be specific therapeutic targets for the development of treatments for diseases characterized by aberrant cell division mode. For example, WNT7A promotes expansion of the stem cell pool and therefore should be explored for developing novel therapies to combat degenerative diseases. In contrast, the loss of *Slit2* drives tumor proliferation, making SLIT2 pathway components potential therapeutic targets for translational efforts to fight cancer.

16.5.1 WNT7A as a Useful Agent in Duchenne Muscular Dystrophy

The studies performed by Le Grand, Delauney, and colleagues demonstrate the role of WNT7A signaling in regulating ACD through the PCP pathway. Specifically, Le Grand showed that WNT7A drives the expansion of satellite stem cells in skeletal muscle, thereby demonstrating its property as a growth factor. There is now emerging evidence for WNT7A in treating Duchenne Muscular Dystrophy (DMD), a genetic childhood disorder that results in progressive muscle weakness leading to death by age 30. In a recent study by Maltzahn and colleagues, WNT7A was used as a treatment agent in a mouse model of DMD (von Maltzahn et al. 2012). Animals treated with WNT7A increased their satellite stem cell number, which subsequently resulted in increased muscle strength and reduced damage in response to injury. These positive results make WNT7A a prime candidate for DMD therapeutics.

16.5.2 SLIT2 as a Therapeutic Breast Cancer Agent

Another molecule with promising therapeutic uses is SLIT2 in the treatment of breast cancer. In the study done by Ballard and colleagues, SLIT2 signaling through ROBO1 maintained the balance between ACD and SCD. Loss of SLIT2 signaling resulted in both an increase in SCD (Ballard et al. 2015) and the formation of hyperplastic lesions in normal breast tissue (Marlow et al. 2008). Taken together, these data suggest that loss of *Slit2*, which occurs in 40% percent of basal breast tumors (Cancer Genome Atlas 2012), may lead to increased proliferation due to the symmetric expansion of cancer stem/progenitor cells. To this end, a recent review by Gu and colleagues outlines multiple studies that use SLIT2 in cultured breast cancer cells or mouse tumor models to prevent the proliferation and metastasis of breast tumors (Gu et al. 2015).

16.5.3 SHH as a Target in Medulloblastoma

SHH, like SLIT, has signaling properties that can be exploited in cancer treatment. Overactivation of SHH signaling in the cerebellum leads to expansion of progenitor cells and cancers such as medulloblastoma (Kieran 2014; Yang et al. 2015); therefore, drugs inhibiting SHH signaling have garnered a great deal of attention and may represent promising therapeutic agents. Medulloblastoma is a pediatric cancer that affects the cerebellum and presents as four distinct subtypes. SHH overexpression is a marker for one subtype characterized by an intermediate prognosis. Recent efforts in drug design have been aimed at developing inhibitors that block SHH signaling in this subtype. Robinson and colleagues show that treatment of patients with the SHH inhibitor, Vismodegib, prevented cancer recurrence in a phase II clinical trial (Robinson et al. 2015). These results demonstrate that silencing the SHH pathway may halt cancer progression.

In conclusion, SHH, WNT, and SLIT are extracellular cues that have important developmental roles in multiple tissues and represent promising targets in drug discovery. Furthermore, these molecules are examples of how understanding the function of proteins during development leads to progress in combating disease.

Acknowledgments This work was supported by the NIH: GMI-098897 R01 to LH and HGRI-R25HG006836 predoctoral support to PS.

References

Alman BA (2015) The role of hedgehog signalling in skeletal health and disease. Nat Rev Rheumatol 11(9):552–560

Ang ES Jr, Haydar TF, Gluncic V, Rakic P (2003) Four-dimensional migratory coordinates of GABAergic interneurons in the developing mouse cortex. J Neurosci 23(13):5805–5815

Ballard MS, Hinck L (2012) A roundabout way to cancer. Adv Cancer Res 114:187–235

Ballard MS, Zhu A, Iwai N, Stensrud M, Mapps A, Postiglione MP, Knoblich JA, Hinck L (2015) Mammary stem cell self-renewal is regulated by Slit2/Robo1 signaling through SNAI1 and mINSC. Cell Rep 13(2):290–301

Borrell V, Cardenas A, Ciceri G, Galceran J, Flames N, Pla R, Nóbrega-Pereira S, García-Frigola C, Peregrín S, Zhao Z, Ma L, Tessier-Lavigne M, Marin O (2012) Slit/Robo signaling modulates the proliferation of central nervous system progenitors. Neuron 76(2):338–352

Briscoe J, Therond PP (2013) The mechanisms of Hedgehog signalling and its roles in development and disease. Nat Rev Mol Cell Biol 14(7):416–429

Cancer Genome Atlas Network (2012) Comprehensive molecular portraits of human breast tumours. Nature 490(7418):61–70

Ceron J, Tejedor FJ, Moya F (2006) A primary cell culture of Drosophila postembryonic larval neuroblasts to study cell cycle and asymmetric division. Eur J Cell Biol 85(6):567–575

Chenn A, McConnell SK (1995) Cleavage orientation and the asymmetric inheritance of Notch1 immunoreactivity in mammalian neurogenesis. Cell 82(4):631–641

Clucas J, Valderrama F (2014) ERM proteins in cancer progression. J Cell Sci 127(Pt 2):267–275

Dave RK, Ellis T, Toumpas MC, Robson JP, Julian E, Adolphe C, Bartlett PF, Cooper HM, Reynolds BA, Wainwright BJ (2011) Sonic hedgehog and notch signaling can cooperate to regulate neurogenic divisions of neocortical progenitors. PLoS One 6(2):e14680

De A (2011) Wnt/Ca^{2+} signaling pathway: a brief overview. Acta Biochim Biophys Sin (Shanghai) 43(10):745–756

Delaunay D, Cortay V, Patti D, Knoblauch K, Dehay C (2014) Mitotic spindle asymmetry: a Wnt/PCP-regulated mechanism generating asymmetrical division in cortical precursors. Cell Rep 6 (2):400–414

Devenport D (2014) The cell biology of planar cell polarity. J Cell Biol 207(2):171–179

Dumont NA, Wang YX, Rudnicki MA (2015) Intrinsic and extrinsic mechanisms regulating satellite cell function. Development 142(9):1572–1581

Gaiano N, Nye JS, Fishell G (2000) Radial glial identity is promoted by Notch1 signaling in the murine forebrain. Neuron 26(2):395–404

Gauthier-Fisher A, Lin DC, Greeve M, Kaplan DR, Rottapel R, Miller FD (2009) Lfc and Tctex-1 regulate the genesis of neurons from cortical precursor cells. Nat Neurosci 12(6):735–744

Gu F, Ma Y, Zhang J, Qin F, Fu L (2015) Function of Slit/Robo signaling in breast cancer. Front Med 9(4):431–436

Habib SJ, Chen BC, Tsai FC, Anastassiadis K, Meyer T, Betzig E, Nusse R (2013) A localized Wnt signal orients asymmetric stem cell division in vitro. Science 339(6126):1445–1448

Haldipur P, Sivaprakasam I, Periasamy V, Govindan S, Mani S (2015) Asymmetric cell division of granule neuron progenitors in the external granule layer of the mouse cerebellum. Biol Open 4 (7):865–872

Kibar Z, Vogan KJ, Groulx N, Justice MJ, Underhill DA, Gros P (2001) Ltap, a mammalian homolog of Drosophila Strabismus/Van Gogh, is altered in the mouse neural tube mutant Loop-tail. Nat Genet 28(3):251–255

Kieran MW (2014) Targeted treatment for sonic hedgehog-dependent medulloblastoma. Neuro Oncol 16(8):1037–1047. doi:10.1093/neuonc/nou109

Komada M, Saitsu H, Kinboshi M, Miura T, Shiota K, Ishibashi M (2008) Hedgehog signaling is involved in development of the neocortex. Development 135(16):2717–2727

Komuro H, Yacubova E, Yacubova E, Rakic P (2001) Mode and tempo of tangential cell migration in the cerebellar external granular layer. J Neurosci 21(2):527–540

Konno D, Shioi G, Shitamukai A, Mori A, Kiyonari H, Miyata T, Matsuzaki F (2008) Neuroepithelial progenitors undergo LGN-dependent planar divisions to maintain self-renewability during mammalian neurogenesis. Nat Cell Biol 10(1):93–101

Kosodo Y, Toida K, Dubreuil V, Alexandre P, Schenk J, Kiyokage E et al (2008) Cytokinesis of neuroepithelial cells can divide their basal process before anaphase. EMBO J 27 (23):3151–3163

Le Grand F, Jones AE, Seale V, Scime A, Rudnicki MA (2009) Wnt7a activates the planar cell polarity pathway to drive the symmetric expansion of satellite stem cells. Cell Stem Cell 4 (6):535–547

Luer K, Technau GM (2009) Single cell cultures of Drosophila neuroectodermal and mesectodermal central nervous system progenitors reveal different degrees of developmental autonomy. Neural Dev 4:30

Macias H, Hinck L (2012) Mammary gland development. Wiley Interdiscip Rev Dev Biol 1 (4):533–557

Mapelli M, Gonzalez C (2012) On the inscrutable role of Inscuteable: structural basis and functional implications for the competitive binding of NuMA and Inscuteable to LGN. Open Biol 2(8):120102

Marlow R, Strickland P, Lee JS, Wu X, Pebenito M, Binnewies M et al (2008) SLITs suppress tumor growth in vivo by silencing Sdf1/Cxcr4 within breast epithelium. Cancer Res 68 (19):7819–7827

Martinez C, Cornejo VH, Lois P, Ellis T, Solis NP, Wainwright BJ, Palma V (2013) Proliferation of murine midbrain neural stem cells depends upon an endogenous sonic hedgehog (Shh) source. PLoS One 8(6):e65818

Miale IL, Sidman RL (1961) An autoradiographic analysis of histogenesis in the mouse cerebellum. Exp Neurol 4:277–296

Morin X, Bellaiche Y (2011) Mitotic spindle orientation in asymmetric and symmetric cell divisions during animal development. Dev Cell 21(1):102–119

Noctor SC, Martinez-Cerdeno V, Kriegstein AR (2008) Distinct behaviors of neural stem and progenitor cells underlie cortical neurogenesis. J Comp Neurol 508(1):28–44

Palma V, Ruiz i Altaba A (2004) Hedgehog-GLI signaling regulates the behavior of cells with stem cell properties in the developing neocortex. Development 131(2):337–345

Polesskaya A, Seale P, Rudnicki MA (2003) Wnt signaling induces the myogenic specification of resident CD45+ adult stem cells during muscle regeneration. Cell 113(7):841–852

Robinson GW, Orr BA, Wu G, Gururangan S, Lin T, Qaddoumi I, Packer RJ, Goldman S, Prados MD, Desjardins A, Chintagumpala M, Takebe N, Kaste SC, Rusch M, Allen SJ, Onar-Thomas A, Stewart CF, Fouladi M, Boyett JM, Gilbertson RJ, Curran T, Ellison DW, Gajjar A (2015) Vismodegib exerts targeted efficacy against recurrent sonic hedgehog-subgroup medulloblastoma: results from phase II pediatric brain tumor consortium studies PBTC-025B and PBTC-032. J Clin Oncol 33(24):2646–2654

von Maltzahn J, Renaud JM, Parise G, Rudnicki MA (2012) Wnt7a treatment ameliorates muscular dystrophy. Proc Natl Acad Sci USA 109(50):20614–20619

Wallace VA (1999) Purkinje-cell-derived Sonic hedgehog regulates granule neuron precursor cell proliferation in the developing mouse cerebellum. Curr Biol 9(8):445–448

Wechsler-Reya RJ, Scott MP (1999) Control of neuronal precursor proliferation in the cerebellum by Sonic Hedgehog. Neuron 22(1):103–114

Yang R, Wang M, Wang J, Huang X, Yang R, Gao WQ (2015) Cell division mode change mediates the regulation of cerebellar granule neurogenesis controlled by the sonic hedgehog signaling. Stem Cell Rep 5(5):816–828

Chapter 17
Regulation of Asymmetric Cell Division in Mammalian Neural Stem and Cancer Precursor Cells

Mathieu Daynac and Claudia K. Petritsch

Abstract Stem and progenitor cells are characterized by their abilities to self-renew and produce differentiated progeny. The balance between self-renewal and differentiation is achieved through control of cell division mode, which can be either asymmetric or symmetric. Failure to properly control cell division mode may result in premature depletion of the stem/progenitor cell pool or abnormal growth and impaired differentiation. In many tissues, including the brain, stem cells and progenitor cells undergo asymmetric cell division through the establishment of cell polarity. Cell polarity proteins are therefore potentially critical regulators of asymmetric cell division. Decrease or loss of asymmetric cell division can be associated with reduced differentiation common during aging or impaired remyelination as seen in demyelinating diseases. Progenitor-like glioma precursor cells show decreased asymmetric cell division rates and increased symmetric divisions, which suggests that asymmetric cell division suppresses brain tumor formation. Cancer stem cells, on the other hand, still undergo low rates of asymmetric cell division, which may provide them with a survival advantage during therapy. These findings led to the hypotheses that asymmetric cell divisions are not always tumor suppressive but can also be utilized to maintain a cancer stem cell population. Proper control of cell division mode is therefore not only deemed necessary to generate cellular diversity during development and to maintain adult tissue homeostasis but may also prevent disease and determine disease progression. Since brain

M. Daynac
Department of Neurological Surgery, Brain Tumor Research Center, University of California, San Francisco, CA, USA

C.K. Petritsch (✉)
Department of Neurological Surgery, Brain Tumor Research Center, University of California, San Francisco, CA, USA
Eli and Edythe Broad Center of Regeneration Medicine and Stem Cell Research, University of California, San Francisco, CA, USA
Helen Diller Comprehensive Cancer Research Center, University of California, San Francisco, CA, USA
e-mail: Claudia.Petritsch@ucsf.edu

J.-P. Tassan, J.Z. Kubiak (eds.), *Asymmetric Cell Division in Development, Differentiation and Cancer*, Results and Problems in Cell Differentiation 61,
DOI 10.1007/978-3-319-53150-2_17

cancer is most common in the adult and aging population, we review here the current knowledge on molecular mechanisms that regulate asymmetric cell divisions in the neural and oligodendroglial lineage during development and in the adult brain.

17.1 Developmental and Temporal Dynamics of Cell Division Mode in Neural Stem and Progenitor Cells

From embryonic stages to adult, the development of the mammalian cortex relies on a temporal and tight control of cell division mode. Stem and progenitor cells across species undergo symmetric self-renewing divisions (SSD) to expand their pool, symmetric differentiating divisions (SDD) to generate two differentiated progeny, or asymmetric cell divisions (ACD) to generate one self-renewing and one differentiating progeny in a single division (Florio and Huttner 2014) (Fig. 17.1). The control of cell division mode affects development and size of the brain and sustains the formation of new neurons and oligodendrocytes in the adult brain. The regulation of ACD has been intensively studied in *Drosophila* neuroblasts (NBs), showing that deregulation of ACD leads to aberrant proliferation and genomic instability and thus might be causal to tumor formation (Gomez-Lopez et al. 2013). Recent evidence shows that adult neural stem cells (NSCs) and oligodendrocyte precursor cells (OPCs), which are cells of origin of glioma in mouse models, undergo ACD (Noctor et al. 2004). As decreased ACD has been evidenced in stem and progenitor cells during aging, in brain tumorigenesis (Sugiarto et al. 2011), and in other types of cancers (Cicalese et al. 2009; Wu et al. 2007), the studies of ACD will undoubtedly improve our understanding of cancerogenesis (for review see Gomez-Lopez et al. 2013). The study of the regulation of ACD in mammalian cells is at very early stages and in order to understand how cell division mode changes affect disease, it is critical to elucidate the function, regulation, and significance of ACD during normal development and in tissue homeostasis. Cell division modes of stem and progenitor cells during neurodevelopment have been well characterized, whereas less is known about the role of ACD in the adult and aging brain. In the following paragraphs, we will first describe cell division mode of neural stem cells during embryonic neurogenesis, followed by a discussion of adult neurogenesis in the hippocampus and subventricular zone of the lateral ventricle. We will then describe cell division mode of OPCs. In the following paragraphs, we will discuss the changes that neural stem and progenitor cells undergo during aging and how these may predispose them to contribute to diseases such as brain tumors and multiple sclerosis. Lastly, we will discuss how cell polarity is regulated and how it affects ACD and whether cell polarity proteins can act as cell fate determinants.

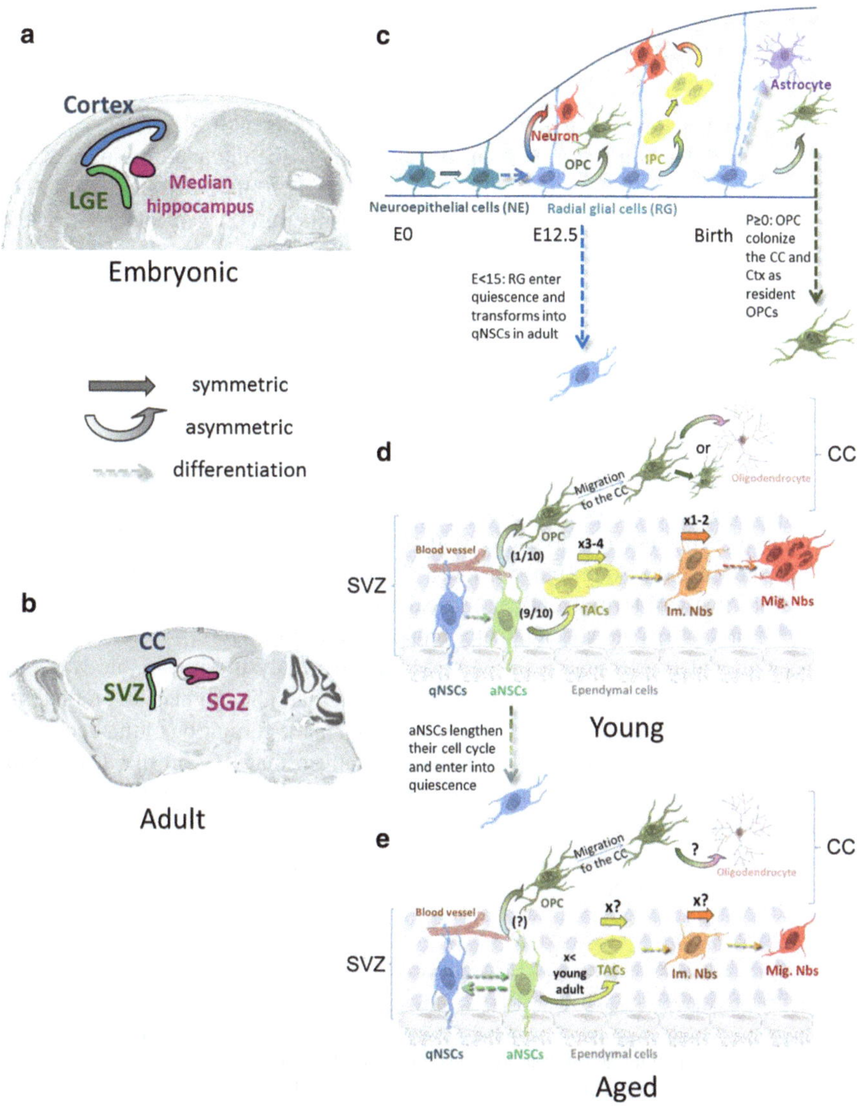

Fig. 17.1 Evolution of cell division mode of neuronal and oligodendrocyte precursors throughout life in the rodent brain. (**a**, **b**) RG cells from the ventral hippocampus migrate from the temporal to septal poles and remain as NSCs at adult ages (*purple*) (Li et al. 2013), while NSCs in the adult SVZ emerge from the embryonic lateral ganglionic eminences (LGE; *green*) (Fuentealba et al. 2015). Adult OPCs from the CC are first formed in the developing cortex (*blue*) (Kessaris et al. 2006). (**c**) Neuroepithelial (NE) cells, before the onset of neurogenesis (E11–E12), divide symmetrically to amplify the pool of NE cells. As the developing brain gets thicker, NE processes elongate and transform into radial glia cells (RGs). Around E12.5, RGs divide asymmetrically to generate oligodendrocytes through oligodendrocyte precursor cells (OPCs) or neurons through intermediate progenitor cells (IPCs). Around birth, most RGs convert into astrocytes while OPC production continues. (**d**) Between E13.5 and E15.5, RGs become the origin of quiescent NSCs in the adult SVZ. These quiescent NSCs have the ability to re-enter proliferation and transform into

17.2 Division Mode of NSPCs During Embryonic Brain Development

In the mammalian developing neocortex, neural stem cells (NSCs) are the source for the three functional cell types that will populate and shape the adult brain: neurons, astrocytes, and oligodendrocytes (Alvarez-Buylla et al. 2001; Doetsch et al. 1999). Before the onset of neurogenesis ($E \leq 11$–12 in mice), NSCs are located in the monolayer epithelium that constitutes the neural plate and are therefore called neuroepithelial cells (NECs). To expand their pool, NECs will first undergo symmetric, self-renewing divisions, resulting in a fast thickening of the neuroepithelium (Rakic 1995). At E11–12, NECs switch their division mode from symmetric to asymmetric and they divide to generate one NEC and one radial glia cell (RG) (Gao et al. 2014; Gotz and Huttner 2005). At peak neurogenesis (E13-E18), RGs undergo symmetric self-renewing cell divisions at first but progressively switch to ACDs, producing a self-renewing RG and either a postmitotic neuron or an intermediate progenitor (IP) cell (Noctor et al. 2008). Some IPCs can undergo one or more rounds of symmetric self-renewing divisions to increase their pool and enlarge the SVZ and form the intermediate zone (IZ), but most of them will divide and differentiate into two neurons that will form the upper cortical layers (Hansen et al. 2010; Noctor et al. 2004). All these steps are schematically represented in Fig. 17.1c.

Recent clonal analyses using the MADM (*M*osaic *A*nalysis with *D*ouble *M*arkers) technique give an unprecedented look into the remarkably tight and predictable control of the cell division mode of RG progenitors in the formation of the neocortex (Gao et al. 2014). These more recent studies confirm that RGs in the developing mouse cortex transit from symmetric, self-renewing (SSD) division to asymmetric neurogenic division around E11–E12. These data further show that RGs produce a defined number of neurons at the onset of neurogenesis, under a tight control by the transcription factor OTX1 (Gao et al. 2014; Greig et al. 2013).

Fig. 17.1 (continued) activated NSCs, which give rise to transit-amplifying neuronal progenitors (TACs). After an amplification of neuronal progenitors by 3–4 symmetrical TAC divisions and 1–2 symmetrical immature neuroblast divisions, the latter differentiate into migrating neuroblasts that will reach the olfactory bulbs (OB) to produce neurons. Ten percent of the NSCs have an oligodendrocyte fate (Ortega et al. 2013) and produce OPCs that will migrate to the corpus callosum (CC). In the corpus callosum, OPCs divide symmetrically or asymmetrically depending on intrinsic and extrinsic factors (Boda et al. 2015; Sugiarto et al. 2011). (**e**) In the aging brain, the number of NSCs and OPCs stays unchanged (Capilla-Gonzalez et al. 2013, 2014; Daynac et al. 2014). However, aNSCs lengthen their G1 phase and produce less TACs, reducing the number of neuroblasts reaching the olfactory bulbs (Daynac et al. 2016). OPCs also undergo a dramatic age-related cell cycle lengthening (Young et al. 2013), and a slight reduction of OPC ACD was detected with age (Boda et al. 2015). Further studies are needed to elucidate the mechanisms for age-related changes of ACD. *LGE* Lateral ganglionic eminence, *OPC* oligodendrocyte progenitor cell, *IPC* intermediate progenitor cell, *qNSC* quiescent neural stem cell, *aNSC* activated neural stem cell, *TAC* Transit-amplifying cell, *Im. Nbs* immature neuroblasts, *Mig. Nbs* Migrating Neuroblasts, *SVZ* subventricular zone, *CC* corpus callosum, *SGZ* subgranular zone

The generation of new neurons is amplified by a more recently identified type of progenitor, the outer radial glial cells (oRGs) situated in the outer SVZ (oSVZ) of mammals (Fietz et al. 2010; Hansen et al. 2010; Wang et al. 2011a, b, c). In both humans and rodents, oRGs are produced by ACD of radial glia and will continue dividing asymmetrically during the peak of neurogenesis (E13–18) to produce new lineage of progenitor cells and a large number of neurons (Hansen et al. 2010; Wang et al. 2011a, b, c). In summary, neurodevelopment is based on dynamic switches between asymmetric and symmetric cell division mode, the regulation of which remains to be fully unraveled.

17.3 ACD in the Subgranular Zone of the Adult Hippocampus

Whether adult NSCs similar to their embryonic counterparts divide asymmetrically or symmetrically is the topic of ongoing controversial research, which will be discussed here. These investigations are frequently tackled in the context of the questions of whether NSCs descend from embryonic RG divisions and whether adult NSC division patterns change with age. New neurons continue to be produced in the neonatal brain (Luskin 1993) and in the adult rodent brain in two main neurogenic niches: the SVZ along the lateral ventricles (Doetsch et al. 1999; Lois and Alvarez-Buylla 1993) and the subgranular zone (SGZ) of the dentate gyrus of the hippocampus (Gage et al. 1998; Kaplan and Hinds 1977). Several recent studies point to an embryonic origin or adult NSC (Fuentealba et al. 2015; Furutachi et al. 2015; Li et al. 2013). In the developing brain, RGs from the ventral hippocampus migrate from the temporal to septal poles and are retained as NSCs in the adult SGZ (Fig. 17.1a, b). NSCs in the SGZ exhibit a polarized morphology with a long process extending from the SGZ toward the molecular layer (Mignone et al. 2004). Pulse labeling with thymidine analog 5-bromo-2-deoxyuridine, BrdU, in a nestin-cyan fluorescent protein (CFP) reporter transgenic mouse revealed that NSCs proliferate very little and they were therefore called quiescent neural progenitors (QNP). In situ analyses of BrdU-labeled, GFAP-immunostained reporter transgenic mice further revealed that QNPs divide asymmetrically to generate a GFAP-positive NSC and a transient-amplifying cell, also called amplifying neural progenitors (ANPs). The long process and Gfap are inherited by the QNP and not the ANP, which makes it possible to identify the two different daughter cells by morphologic criteria as well as by GFAP immunoreactivity. In subsequent divisions, ANPs generate neuronal progenitor cells or neuroblasts, which will ultimately differentiate into granule neurons (Encinas et al. 2006). A subsequent report by Bonaguidi et al. used a clonal labeling strategy to genetically mark single cells in Nestin-CreER^{T2}; Z/EG mice and to trace the fate of dividing cells (Bonaguidi et al. 2011). In addition, they applied mosaic analysis with double markers (MADM) to label clones with two colors. Cell pairs were identified based on the close proximity of two cells, and cell fate of daughter cells was determined by morphology and costaining for GFAP, neuronal progenitor marker Tbr2, neuronal marker Prox1, and astrocyte marker S100β as well as proliferation marker MCM2. Both approaches gave similar results and mostly corroborated earlier studies, showing that QNPs predominantly divide asymmetrically. Symmetric self-renewing divisions yielding two QNP cells were found at low rates and of the progeny of these divisions, one either differentiated or continued to self-renew or both cells differentiated. The majority of divisions generated QNP/Neuronal lineage cell pairs, whereby the QNP reentered quiescence and the non-QNP cell was either an ANP/intermediated progenitor, which continued to proliferate symmetrically, or an immature neuron. Surprisingly, this study also found QNP/mature astroglia pairs, whereby both the QNP and the astroglia cells cease to proliferate (Bonaguidi et al. 2011). Taken together, several conclusions can be drawn from Bonaguidi et al. and earlier studies, including the conclusion that NSCs in the adult brain are largely quiescent and when they divide, they undergo mostly ACD to generate one proliferative progeny and one NSC that reenters quiescence. Moreover, NSCs exhibit the capacity for multi-lineage differentiation and long-term self-renewal through ACD. These advances in identifying ACD in the dentate gyrus, and correlating it with cell fate, pave the way for further studies of how cell division mode is regulated.

Interestingly, several environmental factors and stimuli have been identified that negatively and positively influence hippocampal neurogenesis. These factors include environmental enrichment in mice (Kempermann et al. 1997), glucocorticoid treatment during Kainate-induced injury in the adult rat brain (Stein-Behrens et al. 1994) and prolonged treatment with the antidepressant drug and serotonin reuptake inhibitor (SSRI) fluoxetine (Malberg et al. 2000). A subsequent study found increased symmetric divisions of ANP in fluoxetine-treated rodents, suggesting that ANPs and not QNPs are the fluoxetine target population in the brain (Encinas et al. 2006). Collectively, these studies underline that adult NSCs in the hippocampus use both asymmetric and symmetric division modes to control their fate and that cell division mode is altered by extrinsic factors, such as medication. A better understanding of how the medication affects ACD and thereby alters neurogenesis is needed.

17.4 Cell Division Mode in the Adult Subventricular Zone

In this paragraph, we discuss division mode of NSC in the adult subventricular zone (SVZ), which encompasses the ventricle with cerebrospinal fluid, ependymal cells, NSC, NSC progeny, and capillaries (Mirzadeh et al. 2008; Shen et al. 2008). In the SVZ, only a fraction of the NSCs (8.6% according to Ponti et al. 2013) are actively proliferating (Furutachi et al. 2013; Ponti et al. 2013).

Adult SVZ NSCs, frequently referred to as B1 cells, divide to generate transit-amplifying progenitors (TACs; Doetsch et al. 1997; Fig. 17.1). TACs in turn produce postmitotic progeny, including neuroblasts, also called type A cells (Doetsch et al.

1999), and, to a lesser extent, oligodendrocytes (Menn et al. 2006). Until recently, it was proposed that RGs in the SVZ retract their processes and differentiate into astrocytes and ependymal cells at embryonic stages (Spassky et al. 2005; Voigt 1989). The origin of adult NSC was elusive until a recent lineage tracing study (Furutachi et al. 2015) and another study using a retroviral library to track the progeny of embryonic NSCs (Fuentealba et al. 2015) identified a slow-dividing population of embryonic NSCs as an origin of the majority of young adult SVZ NSCs. The majority of the future adult NSCs are produced between E13.5 and E15.5 and enter quiescence until their eventual reactivation in the adult brain (Fig. 17.1c, d).

Due to the embryonic origin and some morphologic similarities, such as intrinsic cell polarity (Mirzadeh et al. 2008), it could be assumed that adult NSCs undergo ACD in the adult brain. However, given the complexity of their niche and their high level of quiescence, it has been challenging to assess the cell division mode of adult NSCs in situ. Live imaging of SVZ-derived cells ex vivo has revealed that slow-dividing NSCs from the adult SVZ have an extended cell cycle ($\geq$36 h) and undergo one or two symmetric divisions to give rise to proliferating "activated" NSCs (aNSCs; Fig. 17.1c; Costa et al. 2011). The progeny of this symmetric division went on to generate an asymmetric lineage tree, with one branch retaining astro-glial fate and eventually entering quiescence and the second branch expressing neuronal markers. The slow-dividing founder cells of these asymmetric clones were identified as Gfap+ cells with the help of a transgenic Gfap-RFP line and were considered to be NSCs. The model arising from these data is that NSCs exit quiescence, whereby they upregulate Gfap and divide symmetrically to give rise to two activated, fast-proliferating NSCs (aNSC), which in turn divide asymmetrically into aNSC/TAC pairs. TACs downregulate Gfap expression and divide typically two to three (up to five) times symmetrically to give rise to mostly neurons, whereas aNSCs divide up to two times, before reentering quiescence. This suggests that a reservoir of quiescent NSC is maintained after a bout of divisions. The data also suggest that aNSCs use ACD to undergo unlimited self-renewal and to generate TACs at the same time. However, due to the limitations of the experimental approach, it cannot be ruled out that aNSCs terminally differentiate and become permanently postmitotic in vivo, similar to what has been proposed in QNP or the dentate gyrus (Encinas et al. 2006). Subsequent in situ analyses used serial injections of thymidine analogs into mice, followed by detection of specific cell fate by immunofluorescence in the brain of injected mice. These data for the most part corroborated ex vivo findings of the cell division rate of NSCs and their direct progeny (aNSC: one TAC: three amplifying divisions, Nbs: one or two divisions). These analyses however did not unequivocally distinguish between asymmetric and symmetric cell division mode of activated NSCs (Ponti et al. 2013).

More recently, Calzolari et al. have used clonal analyses with a mouse line targeting NSCs (GLASTCreER^{T2}) crossed with the "Confetti" multicolor reporter to show that a single activated NSC generates multiple subsequent subclones by supposedly undergoing two to three rounds of ACD (Calzolari et al. 2015). These subclone analyses are consistent with the earlier study by Ponti et al. (2013), showing 3–4 rounds of symmetrical divisions of TACs before giving rise to neuroblasts, which symmetrically divide 1–2 times before migrating to the OBs. Long-term clonal analyses carried out over the course of 4–6 months indicated that aNSCs continue to generate new progeny once activated through multiple (2–3) rounds of divisions, and then they terminally differentiate or die, rather than enter quiescence as NSCs (Calzolari et al. 2015). This study together with an earlier report (Balordi and Fishell 2007a) indicates that the self-renewal potential of aNSC is limited. Further studies are needed to unequivocally determine whether the cell division mode of q/aNSC in the adult SVZ is asymmetric or symmetric.

One hallmark of a classic ACD is that cell fate markers and determinants distribute unequally during mitosis. Cell fate determinants that are proposed to segregate asymmetrically during aNSC mitosis in the adult SVZ include epidermal growth factor receptor (EGFR; Sun et al. 2005), dual specificity kinase 1 (Dyrk1; Ferron et al. 2010), and Delta-like ligand1 (Dll1; Kawaguchi et al. 2013). Further proof that cell fate markers and determinants segregate asymmetric may come from time-lapse imaging of mitoses of aNSC, which express fluorescence-tagged cell fate determinants such as Dll1. This approach has so far been unattainable for the adult brain although it is feasible for tissue sections isolated from embryonic neocortex (Kawaguchi et al. 2013; Noctor et al. 2004). It is likely that ACDs of adult NSC are rather rare in the SVZ, and yet it is deemed an important puzzle to solve. One reason for distinguishing between asymmetric and symmetric cell division mode is that they are regulated very differently. A better understanding of the regulatory mechanism behind asymmetric aNSC divisions, should they exist, will pave the way for therapeutic approaches utilizing aNSC for regenerative purposes. On the other hand, aNSCs are putative cells of origin of adult gliomas (Zong et al. 2015). Unraveling the mechanism by which aNSCs are activated and divide might lead to insights into neoplastic transformation and tumorigenesis and is expected to provide novel points of disruption to which anti-glioma therapies can be targeted.

17.5 The Origin of Oligodendrocytes in the Central Nervous System

In the following two paragraphs, we discuss evidence for ACD during gliogenesis, and in particular, during oligodendrogenesis, which starts at late embryonic stages and continues throughout adulthood. Oligodendrocytes are specialized myelin producing cells that form a myelin sheath around axons and thereby enable them to carry action potentials by saltatory conduction. Oligodendrocytes arise from multiple pools of oligodendrocyte precursor cells (OPCs) present in the developing brain and spinal cord. The production of OPC in mice starts in the spinal cord and occurs in three distinct phases. It starts at E12.5 at the ventral neural tube, then at E15.5 at the dorsal neural tube, and after birth at the central canal subependyma (Gallo and Deneen 2014; Lu et al. 2000; Orentas et al. 1999; Rowitch and Kriegstein 2010; Vallstedt et al. 2005). Similarly, genetic fate mapping using three different transgenic mice each expressing Cre recombinase from a different cell type-specific promoter has shown that OPCs arise in three distinct phases in the mouse forebrain, and these are characterized by the selective expression of transcription factors (Kessaris et al. 2006). Upon reaching their final destination in the adult brain at perinatal stages, OPCs will proliferate locally and expand and some will exit the cell cycle and produce myelinating oligodendrocytes, the majority of which are generated within the first 3–4 postnatal weeks (Miller 2002). In addition to their embryonic origin, OPCs arise from NSCs in the dorsal/ventral SVZ in the adult brain (Gonzalez-Perez et al. 2009; Menn et al. 2006; Ortega et al. 2013), whereby around one in ten adult NSCs is capable of giving rise to oligodendrogenic lineage cells (Fig. 17.1c) (Ortega et al. 2013). Single cell tracking in vitro suggested that aNSCs generate distinct neuronal and oligodendroglial clones. Within the clone, TACs undergo two to three rounds of symmetrical expansion. A subpopulation of TACs produces OPCs (Ortega et al. 2013). Whether OPCs can arise from cells other than TACs and whether resident and SVZ-derived OPCs are functionally distinct are open questions, which will be addressed by future research.

17.6 ACD of Oligodendrocyte Progenitor Cells in the Adult Brain

Postnatal and adult OPCs express the chondroitin sulfate proteoglycan NG2 and generate oligodendrocytes. OPCs also proliferate to self-renew in the adult brain and continue to do so with age, albeit at decreased rates (Dimou et al. 2008; Psachoulia et al. 2009; Kang et al. 2010; Zhu et al. 2011). Time-lapse imaging in adult cortex showed that OPCs are lost either during normal turnover or through injury. OPC loss triggered rapid migration and proliferation of adjacent cells to achieve homeostasis (Hughes et al. 2013). Studies of isolated rat optic nerve progenitors, called O-2A, conducted by Wren and Noble first suggested that OPCs might undergo ACD. They discovered that clones derived from single adult O-2A cells and grown under adherent conditions contained both differentiated postmitotic oligodendrocytes and proliferating progenitor cells. These observations led them to speculate that O-2A cells undergo ACD to self-renew and generate mature cells. They also found oligodendrocyte-only and progenitor-only clones, albeit at low frequency, which would indicate that O-2A cells occasionally undergo symmetric, self-renewing as well as symmetric, differentiating divisions (Wren et al. 1992). Almost two decades later, their predictions were experimentally evaluated and validated by subjecting acutely isolated murine adult OPC from the corpus callosum of P60–P90 mice to a detailed analysis of their cell division mode (Sugiarto et al. 2011). In this study, OPCs were subjected to a cell pair assay, which has been initially designed to distinguish between symmetric and asymmetric cell divisions of embryonic neural progenitors (Shen et al. 2002; Sun et al. 2005). Immunocytochemical analyses of cell pairs indicated that the proteoglycan and known OPC marker NG2 itself was inherited unequally between daughter cells (Sugiarto et al. 2011). Collectively, ex vivo and in vivo data resulting from this study indicated that about 50% of mammalian adult OPCs undergo ACD to self-renew and generate mature oligodendrocytes, whereas the remaining 50% undergo symmetric divisions predominantly to self-renew. The rates of ACD versus symmetric self-renewing (SSD) or symmetric differentiating divisions (SDD) may vary, depending on the age of the mice, the technique used to image cell divisions, and the neuro-anatomical location. For example, live imaging of single cells on brain slices from early postnatal stages of NG2CreBAC transgenic mice, followed by immunostaining for differentiation marker CC1, reportedly detected ACD but at a lower frequency than observed by in vivo pair assays (Zhu et al. 2011). Subsequent in vivo cell pair studies, using thymidine analogs in the adult brain, were consistent with the study by Sugiarto et al. (Boda et al. 2015). Since numerous studies have indicated that OPCs are heterogeneous in proliferation and differentiation rate depending on their neuro-anatomical differences (Hill et al. 2014; Kang et al. 2010; Young et al. 2013) and response to growth factor signaling (Hill et al. 2013), it is feasible that cell division mode shows regional variations. Noteworthy, studies which report low rates of ACD (Hughes et al. 2013; Zhu et al. 2011) had some important difference in their technical approach. These studies investigated distribution of a fluorescent reporter protein dsRed, which was expressed under the control of the NG2 promoter, rather than distribution of NG2 protein itself. Detection of OPC markers like NG2 and PDGFRα (Boda et al. 2015; Sugiarto et al. 2011) in conjunction with detection of mitotic markers such as phospho-histone-H3 and visualizing incorporation of EdU and BrdU by immunofluorescence (Sugiarto et al. 2011) are additional suitable methods to detect ACD. Lineage tracing experiments and thymidine analog injections and immunostaining for differentiation markers further allow to conduct fate analyses of cell pairs, and such studies confirmed that single OPCs produce daughter cells of distinct cell fate and that asymmetry in cell fate persists for months. Extrinsic factors such as increased exercise, injury through stab wounds, or lysolecithin injection increased the proportion of symmetric, self-renewing divisions at the expense of symmetric, differentiating divisions, but it is not clear whether they significantly impacted ACD (Boda et al. 2015). Recent studies using cumulative labeling with EdU showed that all OPCs throughout the central nervous system are able to undergo at least one cell division and that they continue to divide throughout adulthood and to generate surviving oligodendrocytes. These newly generated oligodendrocytes contribute to de novo myelination as well as remodeling of existing myelin tracts (Clarke et al. 2012; Young et al. 2013). Important next steps will be to develop advanced reported constructs and transgenic mice to study OPC divisions live in the healthy and diseased rodent brain. With such improved ability to image ACD, it will become feasible to determine the neuroanatomical and temporal aspects and regulations of cell division modes of OPC during homeostasis and in disease.

17.7 Aging and Cell Division Mode

17.7.1 Neural Stem Cells and Aging

In the rodent, the aging brain is subject to a progressive reduction in proliferating progenitor cells from the SGZ and SVZ, leading to a dramatic drop in the number of neurons produced during aging (Blackmore et al. 2009; Bouab et al. 2011; Enwere et al. 2004; Maslov et al. 2004; Tropepe et al. 1997). Accordingly, the number of TACs and Neuroblasts is strongly decreased with aging (Balordi and Fishell 2007b; Blackmore et al. 2009; Daynac et al. 2014; Shook et al. 2012). Unexpectedly, the pools of quiescent and activated NSCs remain stable until middle age in mice (12 months) (Daynac et al. 2014; Piccin et al. 2014), even though activated NSCs lose their proliferative capacities (Ahlenius et al. 2009; Capilla-Gonzalez et al. 2014; Daynac et al. 2014, 2016a) and progressively enter into quiescence (Bouab et al. 2011; Lugert et al. 2010). Clonal analyses with Confetti mice revealed that activated NSCs undergo self-renewing ACDs, but clone size is progressively reduced with age, suggesting a limited self-renewal capacity of activated NSCs (Calzolari et al. 2015). Further recent evidence suggests that activated NSC cell cycle is altered as early as 6 months of age in mice, due to a G1 phase lengthening in response to an age-related vascular overproduction of TGFβ1 (Fig. 17.1e) (Daynac et al. 2014, 2016a; Pineda et al. 2013). The activation process of quiescent NSCs is controlled by Shh signaling (Daynac et al. 2016b). Indeed, long-term activation of Shh signaling through Ptc inactivation in NSCs provokes quiescent NSCs to switch their division mode from ACD to SCD and accumulate at a quiescent, inactive state (Ferent et al. 2014). Further studies are required to address whether these long-lived quiescent NSCs can be activated by external stimuli and reproduce a functional niche, as observed with anti-TGFβ1 treatment in old and irradiated mice (Pineda et al. 2013). Such studies will have important implications for the human brain, where both the SGZ of the hippocampus (Knoth et al. 2010) and to a much lesser extent the SVZ of the lateral ventricle (Sanai et al. 2011) contain proliferating neuronal precursors. A few mitotically active Dcx+ cells were found in the SVZ which frequently appeared as doublets, which suggests that neural progenitors divide symmetrically to generate neuroblasts in the adult brain (Wang et al. 2011a, b, c). Adult neurogenesis appears to be more relevant in the hippocampus, where breakthrough studies using ^{14}C level analyses in genomic DNA have proved that nearly all dentate granule neurons turn over in the adult human hippocampus, with ≈700 new neurons per day in each hippocampus (Spalding et al. 2013). Adult neurogenesis is important for learning, emotions, and memory, and abnormal adult neurogenesis has been linked to brain diseases, including Alzheimer's disease, demyelinating disease, depression, and brain tumorigenesis (Liu and Song 2016). A better understanding of cell division mode of adult NSC will therefore provide important insights into the normal and abnormal functions of the brain.

17.7.2 OPC and Aging

The aged brain reportedly has decreased remyelination which slows recovery from demyelinating diseases such as multiple sclerosis (Sim et al. 2002). The underlying causes for this impairment are partially attributable to age-related changes in the oligodendrocyte lineage, and whether they are caused by loss or impaired differentiation of OPCs is currently being investigated. A switch between cell division modes could potentially alter OPC numbers and/or alter the rates of differentiation, since asymmetrical distribution of NG2, EGFR, GPR17, and PDGFRα is required for the long-term generation of distinct cell fate (Boda et al. 2015; Sugiarto et al. 2011). OPC numbers were found to be stable during aging in mouse, which suggested that there is no exhaustion of the progenitor pool (Capilla-Gonzalez et al. 2013; Rivers et al. 2008; van Wijngaarden and Franklin 2013). Importantly, while OPC numbers are stable, their capacity to differentiate into myelinating oligodendrocytes significantly decreases with age (from 66% in P62 to only 39% in P120 mice) (Zhu et al. 2011). This age-related decrease in differentiation capacity is potentially the underlying cause for impaired remyelination and repair in old rat brains (Sim et al. 2002). To address whether the ability to generate daughter cells through ACD changes with age, Boda et al. quantified the asymmetric and symmetric distribution of PDGFRα in OPC cell pairs in mice at different ages and for up to 1 year. While the rate of ACD of OPC increases significantly when mice transition from early postnatal to adult stages (between P20 and P60), it subsequently decreases again in old mice (12 months of age). This age-related decrease is accompanied by an increase in symmetric, self-renewing divisions and a significant decline in newly generated OPC pairs (Boda et al. 2015). A separate study addressed whether OPCs change cell cycle dynamics with age and found a dramatic increase in cell cycle lengths of NG2+ OPC in postnatal versus middle-aged mice (Fig. 17.1e) by around 2/3 of a day, every day after birth (cell cycle lenghts <2 days at P6; $= 36$ days at P60 and >70 days in P240 mice) (Psachoulia et al. 2009; Simon et al. 2011; Young et al. 2013). Collectively, these data suggest that OPCs overall lower their proliferation rate by extending their cell cycle lengths. With age, OPCs also differentiate less due to increases in symmetric, self-renewing divisions at the expense of symmetric, differentiating and asymmetric divisions.

Hypothetically, these changes in cell division mode are the underlying cause for age-related decrease in differentiation and myelin repair, but more research will be needed to test this notion and to fully elucidate the dynamics of cell division of OPCs and its functional implications during aging.

17.8 Regulation of ACD by Cell Polarity Proteins

In the final paragraphs, we will discuss how cell polarity is regulated in the model system *Drosophila melanogaster* embryonic neuroblasts, how it links to ACD and cell fate control, and how it compares to regulation of polarity and ACD in mammalian neural stem cells. Prokaryotes as well as unicellular and multicellular eukaryotes utilize an evolutionary conserved mechanism to undergo ACD and thereby generate cell fate diversity within daughter cells. We glean many hypotheses on ACD in mammalian stem and progenitor cells from the studies in *Drosophila. Drosophila melanogaster* embryonic neuroblasts (NBs) are stem-like cells, which undergo several rounds of ACD. During ACD, determinants of differentiating fate concentrate at the basal cell cortex before mitosis and segregate unequally during cytokinesis, to generate each time another NB and a more restricted progenitor called ganglion mother cell (GMC). NBs break symmetry in a stepwise process, which exemplifies the classic, cell-intrinsic type of ACD. Prior to mitosis, NBs establish cell polarity and as a critical next step align the mitotic spindle along the axis of polarity and simultaneously organize cell fate determinants in subcellular compartments near the mitotic spindle poles, such that those cell fate determinants with equal functions segregate together and only into one of the future daughter cells (Gomez-Lopez et al. 2013). Here, we will focus on the mechanisms by which cell polarity is established and how it links to the subsequent steps of ACD.

17.8.1 How Is Cell Polarity Established?

First-generation *Drosophila* NBs delaminate from the polarized neuro-ectoderm and from it inherit the partition-defective-3 (Par3) homolog Bazooka (Baz) as an apical polarity cue. Baz/Par3 assembles a polarity complex in NBs during late interphase/early prophase (Schaefer et al. 2000). Baz/Par3 binds to and activates Cdc42, a GTPase of the Rho family (Atwood et al. 2007), which in turn recruits atypical protein kinase C (aPKC) (Wodarz et al. 2000) and the aPKC inhibitory subunit Par6 (Atwood et al. 2007; Peterson et al. 2004). The apical complex binds the adaptor protein Inscuteable (Insc) (Wodarz et al. 1999) and thereby initiates the assembly of partner of Insc (Pins) and the heterotrimeric G protein-coupled subunits $G\alpha i$ and $G\beta\gamma$. The WD40 protein lethal giant larvae (Lgl) binds to aPKC/Par6, which leads to a mutual antagonism of Lgl and aPKC, as demonstrated by loss-of-function experiments. When bound to the polarity complex, Lgl restricts aPKC localization to the apical cortex and thereby helps to maintain cell polarity. Genetically enforced reduction of aPKC levels suppresses the NB overproliferation phenotype of an Lgl hypomorphic mutation. Moreover, aPKC loss-of-function mutant larval NBs lose the polarized localization of Lgl and cell fate determinant Miranda, indicating that aPKC—despite its antagonism of Lgl function—is required for apical binding of Lgl and subsequent asymmetric distribution of cell fate determinants (Rolls et al. 2003). After first-generation NBs have undergone cytokinesis, the NB daughter first inherits and subsequently downregulates cell apical polarity components in interphase. Several studies indicated that to the contrary the centrosome remains stable and provides spatial memory for proper spindle axis formation in subsequent NB divisions (Januschke et al. 2011; Rebollo et al. 2007; Siller and Doe 2009). The centrosome also shows asymmetry, when it duplicates into a newer “daughter” and an older “mother” centriole and distributes between daughter cells in a nonrandom fashion, with the daughter centriole consistently remaining in the NB (Januschke et al. 2011; for a review, see Conduit 2013). Centrosome asymmetry is regulated by Polo kinase phosphorylation of human CENTROBIN homolog CnB which promotes the formation of aster microtubules and anchors the centrosome and provides directionality for the mitotic spindle (Singh et al. 2014; Januschke et al. 2013). More recent studies have shown that NBs mutant for centrosome asymmetry undergo seemingly normal ACD (Singh et al. 2014), suggesting that centrosome asymmetry is not the dominant mechanism to establish proper spindle alignment. Noteworthy, mammalian radial glial cells non-randomly segregate their centrosomes during ACD and centrosome components control cell fate (Wang et al. 2009), which suggest that centrosome asymmetry is a conserved feature of asymmetrically dividing cells.

17.8.2 Cortical Cell Polarity as Dominant Mechanism to Control Mitotic Spindle Alignment

A dynamic cross talk between the apical complexes (Par/aPKC and Pins/G protein) with centrosomes and astral microtubules is crucial for proper spindle orientation in *Drosophila* NBs. In mitosis, the centrioles duplicate and the daughter centrosome generates large aster microtubules, which bind and link the Drosophila NuMa homolog and coiled-coiled domain protein mushroom body-defective (Mud) to Pins/Gαi to form a large apical complex which tethers the mitotic spindle to the cortex and aligns it along the apicobasal axis (Izumi et al. 2006; Siller et al. 2006; Bowman et al. 2006). Cortical–microtubule interactions such as binding of the MAGUK Disc large (Dlg) to kinesin Khc-73 at the microtubule plus ends (Siegrist and Doe 2005, 2007) further feed back to the cortex and maintain polarity. Spindle positioning is linked to global control of mitosis by Polo and AurA kinase activities (Lee et al. 2006a; Wang et al. 2006, 2007). In mammalian neurogenesis, RG divisions are biased toward aligning the mitotic spindles within 30° of the ventricular surface (Konno et al. 2008). LGN, an ortholog of Pins, is highly expressed in RG and localizes to the lateral membrane of dividing RG (Du et al. 2001). Genetic ablation (Konno et al. 2008) or siRNA-mediated knockdown (Shitamukai et al. 2011) of *LGN* alters the position of the mitotic spindle, which leads to asymmetric inheritance of the Par polarity complex and impaired differentiation. The function of mammalian Inscuteable is less clear and appears to be opposite from LGN, which is different from *Drosophila* and does not match a conserved function of Inscuteable as a cellular linker of the Par complex and LGN/Gαι (Wang et al. 2011a; Postiglione et al. 2011). The phosphoprotein Treacle at the centrosome is critical not only for mitosis but also for correct spindle positioning and mitotic progression potentially through its interaction with Polo-kinase 1 (Plk1), a mammalian Polo homolog (Sakai et al. 2012). A Treacle/Plk1-controlled checkpoint that links cell polarity with mitotic progression has been proposed for RG. A potentially similar polarity checkpoint is relevant in cancer stem cells where it has been shown the Plk1 activity is critical for ACD and vice versa cell polarity is required for progression through mitosis (Lerner et al. 2015).

17.8.3 How Does Cell Polarity Link to Cell Fate Control?

Cell polarity is tightly linked with cell fate control. In metaphase, the mitotic kinase Aurora A (AurA) phosphorylates Par6 and thereby activates aPKC (Ogawa et al. 2009; Wirtz-Peitz et al. 2008). Active aPKC phosphorylates Lgl (Betschinger et al. 2003) and Numb (Smith et al. 2007), thereby releasing those two proteins from the membrane. Phosphorylated Numb localizes to the basal side of the cell (Smith et al. 2007) by interacting with Polo kinase-activated Partner of Numb (PON) (Lu et al. 1998; Wang et al. 2007) and once inherited by the GMC prevents self-renewal mainly through antagonizing Notch. A mouse Numb homolog (m-Numb) is asymmetrically localized during cortical neurogenesis and physically interacts with mouse Notch-1 (Zhong et al. 1996). Genetically enforced deletion of m-Numb leads to severe defects in the neural tube and brain, resulting in embryonic lethality (Zhong et al. 2000). Asymmetric distribution of m-Numb is critical for ACD of mouse cerebral cortical stem cells (Shen et al. 2002). Deletion of the two mammalian homologs of *Drosophila* Numb, m-Numb and Numblike (Numbl), during development depletes neural stem cells, suggesting that both promote self-renewal (Petersen et al. 2002). Deletion of m-Numb and Numbl in the postnatal SVZ resulted in defects in lateral wall integrity and impaired neurogenesis. The results of the mouse double knockout study indicate a conserved role for Numb in promoting neuronal differentiation, rather than self-renewal (Kuo et al. 2006).

These data suggest that developmental stage and the microenvironment potentially affect Numb function. Interestingly, in vitro studies have shown that in the murine brain a fraction of adult SVZ-derived NSCs asymmetrically distribute activated Notch and EGFR between daughter cells. Activated Notch is unequally distributed during NSC division, and it has been shown to directly regulate *Egfr* transcription (Andreu-Agulló et al. 2009). Moreover, Dyrk1a kinase is asymmetrically segregated to the $EGFR^{high}$ daughter, where it prevents EGFR degradation (Ferron et al. 2010). Notch ligand delta-like 1 (Dll1) was found to be expressed in aNSC and TACs, and Dll1 deletion promotes the activation of qNSC in the adult SVZ. Dll1 protein segregates asymmetrically in embryonic and adult NSC in vitro and associates with neuronal progenitor fate. Given that Notch is expressed in qNSC, the data suggest that Dll1 expression in an NSC and TACs provides a signal to maintain qNSC by activating Notch receptor (Kawaguchi et al. 2013).

Drosphila NBs have a second adaptor-cell fate determinant system consisting of Miranda (Mira) and its binding partners which localize asymmetrically and segregate to the basal GMC (Shen et al. 1998). APKC also phosphorylates Mira

(Atwood and Prehoda 2009), but the functional significance of aPKC phosphorylation for Mira asymmetry is not clear (Lee et al. 2006a; Wang et al. 2006), and studies imply a range of alternative mechanisms of Mira localization acting upstream or in parallel to aPKC. These include proteasome-independent mechanism of ubiquitylation and the anaphase-promoting complex (Slack et al. 2007), MyoVI-directed vesicle transport (Petritsch et al. 2003), and passive diffusion (Erben et al. 2008). Loss of Miranda function leads to aberrant cell fate specification due to mislocalization of cell fate determinants, such as the transcriptional regulator Prospero (Pros) (Ikeshima-Kataoka et al. 1997; Shen et al. 1997), the double-stranded (ds) RNA-binding protein Staufen (Stau) (Matsuzaki et al. 1998; Shen et al. 1998), and the NHL-domain protein Brain tumor (Brat) (Lee et al. 2006b; Betschinger et al. 2006).

Staufen2 (Stau2), the mouse homolog of Staufen, is apically localized in RGs and preferentially segregates to the progenitor daughter (Kusek et al. 2012; Vessey et al. 2012). The Stau2 loss-of-function phenotypes cannot entirely be reconciled with each other in the two studies, making it difficult to pinpoint the exact role of Stau in the mammalian brain. Importantly, similar to *Drosophila* Stau, Stau2 is part of a ribonucleoprotein complex that cargos mRNAs for mammalian homologs of *prospero* and *brat*, *Prospero-related homeobox 1* (*Prox1*) (Vessey et al. 2012) and *Tripartite-motif containing 32* (*Trim32*) (Kusek et al. 2012), respectively. Prox1 is known to mediate cell cycle exit and neurogenesis in the neural retina (Dyer et al. 2003) and promotes oligodendrocyte identity of adult NSC in the SVZ (Bunk et al. 2016) and inhibits proliferation and promotes differentiation of NG2+ OPC (Kato et al. 2015). Trim32 activates miRNA and prevents self-renewal in mouse neural progenitors (Schwamborn et al. 2009). These data collectively suggest that, similar to its *Drosophila* homolog, Stau2 functions by binding and localizing determinants of differentiation away from the self-renewing daughter. It will be exciting to see how future studies elucidate the exact function of mRNA localization in ACD and cell fate specification of mammalian NSC and OPC.

Genome-wide RNAi screens for regulators of NB self-renewal identified subunits of the chromatin-remodeling SIW/SNF complex (Neumüller et al. 2011), and depletion of these subunits in larval brains leads to hyperplasia upon transplantation. Further functional studies reveal that SWI/SNF components do not regulate cell polarity but rather prevent dedifferentiation of intermediate neural progenitors, a transit-amplifying and asymmetrically dividing cell type in the type II lineage of larval NB. They prevent hyperproliferation in part through transcriptional upregulation of target gene Hamlet (Eroglu et al. 2014). The SWI/SNF complex has important functions during mammalian neurogenesis and SWI/SNF components are mutated in cancer. These data indicate that chromatin remodeling regulates ACD-associated cell fate decisions.

In summary, *Drosophila* NB cell polarity is established through two major mechanisms: the first is the dynamic physical association of scaffold proteins that activate GTPase and kinase activities. Moreover, the centrosome provides positional memory during cytokinesis that determines the orientation of subsequent cell divisions. The integrity of the polarity complex is essential for activating G protein signaling in a transmembrane receptor-independent manner (Willard et al. 2004) and in the absence of nucleotide exchange (Yu et al. 2005). As a result, it restricts proliferation and proper cell fate determination. Centrosome positioning is a backup mechanism for cortical polarity to establish proper spindle alignment, which is the second major mechanism that controls ACD. Noteworthy, the activation of G protein signaling through Pins occurs cell intrinsically and not only stabilizes apical polarity but also positions the nascent mitotic spindle along the apicobasal axis and determines its size asymmetry. Because of cell polarization, the mitotic spindle aligns such that cell fate determinants with opposing functions are localized to opposite poles during mitotis and segregate to one daughter cell only. The unequal inheritance of cell fate determinants probably induces chromatin remodeling controlled by SWI/SNF in one daughter cell, thereby promoting it to differentiate rather than proliferate. The specification of fates is irreversible, and no cases of spontaneous dedifferentiation have been observed in the NB lineage. Many polarity proteins originally identified in *Drosophila* are conserved in the mammalian genome, which raises the possibility that they may have similar functions in mammalian cells.

The comparison of cell fate determinants during ACD in drosophila Nbs and adult mouse NSC is schematized in Fig. 17.2.

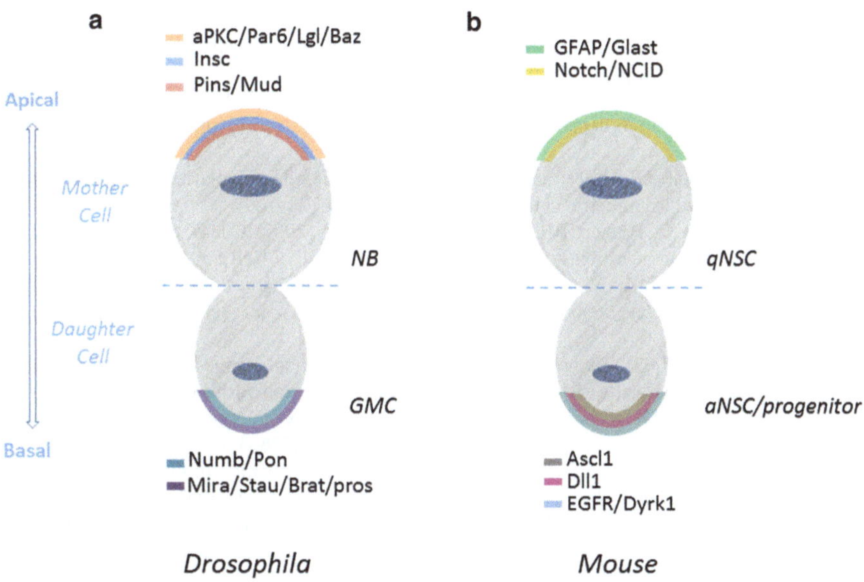

Fig. 17.2 Asymmetric distribution of cell fate determinants in *Drosophila* neuroblasts and *Mouse* adult neural stem cells. (**a**) During metaphase and telophase in drosophila neuroblasts, the mitotic spindle is anchored along to the apicobasal polarity, and cell fate determinants are asymmetrically distributed in the self-renewing neuroblast (NB) and the differentiating ganglion mother cells (GMC). (**b**) The asymmetric cell division of adult mouse NSCs is not well characterized; still there is evidence of asymmetrically segregated cell fate determinants between quiescent NSCs and their proliferating progeny

17.9 Concluding Remarks

Founding studies in Drosophila have paved the way to a mechanistic understanding of neural progenitor cell division mode in higher organisms. The elucidation of neural and glial progenitor cell division mode in mammals is challenging to a large part due to functional redundancy, neuro-anatomical heterogeneity within stem and progenitor cell populations (Marques et al. 2016), and dramatic changes during developmental and with aging. We expect that future studies will further investigate the existence and function of ACD in adult neural stem cells and will determine whether deregulation of ACD is causal to diseases, such as brain tumorigenesis. Improving the understanding of the regulation of ACD such as polarity, and how it links to cell fate determination, holds the promise to develop new therapeutics targeting tumors and other degenerative diseases of the brain.

References

Ahlenius H, Visan V, Kokaia M, Lindvall O, Kokaia Z (2009) Neural stem and progenitor cells retain their potential for proliferation and differentiation into functional neurons despite lower number in aged brain. J Neurosci 29:4408–4419

Alvarez-Buylla A, Garcia-Verdugo JM, Tramontin AD (2001) A unified hypothesis on the lineage of neural stem cells. Nat Rev Neurosci 2:287–293

Andreu-Agulló C, Morante-Redolat JM, Delgado AC, Fariñas I (2009) Vascular niche factor PEDF modulates Notch-dependent stemness in the adult subependymal zone. Nat Neurosci 12:1514–1523

Atwood SX, Prehoda KE (2009) aPKC phosphorylates Miranda to polarize fate determinants during neuroblast asymmetric cell division. Curr Biol 19:723–729

Atwood SX, Chabu C, Penkert RR, Doe CQ, Prehoda KE (2007) Cdc42 acts downstream of Bazooka to regulate neuroblast polarity through Par-6 aPKC. J Cell Sci 120:3200–3206

Balordi F, Fishell G (2007a) Mosaic removal of hedgehog signaling in the adult SVZ reveals that the residual wild-type stem cells have a limited capacity for self-renewal. J Neurosci 27:14248–14259

Balordi F, Fishell G (2007b) Hedgehog signaling in the subventricular zone is required for both the maintenance of stem cells and the migration of newborn neurons. J Neurosci 27:5936–5947

Betschinger J, Mechtler K, Knoblich JJA (2003) The Par complex directs asymmetric cell division by phosphorylating the cytoskeletal protein Lgl. Nature 422:326–330

Betschinger J, Mechtler K, Knoblich JA (2006) Asymmetric segregation of the tumor suppressor brat regulates self-renewal in Drosophila neural stem cells. Cell 124:1241–1253

Blackmore DG, Golmohammadi MG, Large B, Waters MJ, Rietze RL (2009) Exercise increases neural stem cell number in a growth hormone-dependent manner, augmenting the regenerative response in aged mice. Stem Cells 27:2044–2052

Boda E, Di Maria S, Rosa P, Taylor V, Abbracchio M, Buffo A (2015) Early phenotypic asymmetry of sister oligodendrocyte progenitor cells after mitosis and its modulation by aging and extrinsic factors. Glia 63:271–286

Bonaguidi MA, Wheeler MA, Shapiro JS, Stadel RP, Sun GJ, Ming GL, Song H (2011) In vivo clonal analysis reveals self-renewing and multipotent adult neural stem cell characteristics. Cell 145:1142–1155

Bouab M, Paliouras GN, Aumont A, Forest-Berard K, Fernandes KJL (2011) Aging of the subventricular zone neural stem cell niche: evidence for quiescence-associated changes between early and mid-adulthood. Neuroscience 173:135–149
Bowman SK, Neumüller RA, Novatchkova M, Du Q, Knoblich JA (2006) The Drosophila NuMA Homolog Mud regulates spindle orientation in asymmetric cell division. Dev Cell 10:731–742
Bunk EC, Ertaylan G, Ortega F, Pavlou MA, Gonzalez Cano L, Stergiopoulos A, Safaiyan S, Völs S, van Cann M, Politis PK, Simons M, Berninger B, Del Sol A, Schwamborn JC (2016) Prox1 is required for oligodendrocyte cell identity in adult neural stem cells of the subventricular zone. Stem Cells 34(8):2115–2129
Calzolari F, Michel J, Baumgart EV, Theis F, Götz M, Ninkovic J, Gotz M, Ninkovic J (2015) Fast clonal expansion and limited neural stem cell self-renewal in the adult subependymal zone. Nat Neurosci 18:490–492
Capilla-Gonzalez V, Cebrian-Silla A, Guerrero-Cazares H, Garcia-Verdugo JM, Quinones-Hinojosa A (2013) The generation of oligodendroglial cells is preserved in the rostral migratory stream during aging. Front Cell Neurosci 7:147
Capilla-Gonzalez V, Cebrian-Silla A, Guerrero-Cazares H, Garcia-Verdugo JM, Quinones-Hinojosa A (2014) Age-related changes in astrocytic and ependymal cells of the subventricular zone. Glia 62:790–803
Cicalese A, Bonizzi G, Pasi CE, Faretta M, Ronzoni S, Giulini B, Brisken C, Minucci S, Di Fiore PP, Pelicci PG (2009) The tumor suppressor p53 regulates polarity of self-renewing divisions in mammary stem cells. Cell 138:1083–1095
Clarke LE, Young KM, Hamilton NB, Li H, Richardson WD, Attwell D (2012) Properties and fate of oligodendrocyte progenitor cells in the corpus callosum, motor cortex, and piriform cortex of the mouse. J Neurosci 32:8173–8185
Conduit PT (2013) The dominant force of Centrobin in centrosome asymmetry. Nat Cell Biol 15:235–237
Costa MR, Ortega F, Brill MS, Beckervordersandforth R, Petrone C, Schroeder T, Gotz M, Berninger B (2011) Continuous live imaging of adult neural stem cell division and lineage progression in vitro. Development 138:1057–1068
Daynac M, Pineda JR, Chicheportiche A, Gauthier LR, Morizur L, Boussin FD, Mouthon MA (2014) TGFbeta lengthens the G1 phase of stem cells in aged mouse brain. Stem Cells 32:3257–3265
Daynac M, Morizur L, Chicheportiche A, Mouthon MA, Boussin FD (2016a) Age-related neurogenesis decline in the subventricular zone is associated with specific cell cycle regulation changes in activated neural stem cells. Sci Rep 6:21505
Daynac M, Tirou L, Faure H, Mouthon MA, Gauthier LR, Hahn H, Boussin FD, Ruat M (2016b) Hedgehog controls quiescence and activation of neural stem cells in the adult ventricular-subventricular zone. Stem Cell Reports 7:735–748
Dimou L, Simon C, Kirchhoff F, Takebayashi H, Götz M, Gotz M (2008) Progeny of Olig2-expressing progenitors in the gray and white matter of the adult mouse cerebral cortex. J Neurosci 28:10434–10442
Doetsch F, Garcia-Verdugo JM, Alvarez-Buylla A (1997) Cellular composition and three-dimensional organization of the subventricular germinal zone in the adult mammalian brain. J Neurosci 17:5046–5061
Doetsch F, Caillé I, Lim DA, García-Verdugo JM, Alvarez-Buylla A, Caille I, Lim DA, Garcia-Verdugo JM, Alvarez-Buylla A (1999) Subventricular zone astrocytes are neural stem cells in the adult mammalian brain. Cell 97:703–716
Du Q, Stukenberg PT, Macara IG (2001) A mammalian partner of inscuteable binds NuMA and regulates mitotic spindle organization. Nat Cell Biol 3:1069–1075
Dyer MA, Livesey FJ, Cepko CL, Oliver G (2003) Prox1 function controls progenitor cell proliferation and horizontal cell genesis in the mammalian retina. Nat Genet 34:53–58
Encinas JM, Vaahtokari A, Enikolopov G (2006) Fluoxetine targets early progenitor cells in the adult brain. Proc Natl Acad Sci USA 103:8233–8238

Enwere E, Shingo T, Gregg C, Fujikawa H, Ohta S, Weiss S (2004) Aging results in reduced epidermal growth factor receptor signaling, diminished olfactory neurogenesis, and deficits in fine olfactory discrimination. J Neurosci 24:8354–8365

Erben V, Waldhuber M, Langer D, Fetka I, Jansen RP, Petritsch C (2008) Asymmetric localization of the adaptor protein Miranda in neuroblasts is achieved by diffusion and sequential interaction of Myosin II and VI. J Cell Sci 121:1403–1414

Eroglu E, Burkard TR, Jiang Y, Saini N, Homem CCF, Reichert H, Knoblich JA (2014) SWI/SNF complex prevents lineage reversion and induces temporal patterning in neural stem cells. Cell 156:1259–1273

Ferent J, Cochard L, Faure H, Taddei M, Hahn H, Ruat M, Traiffort E (2014) Genetic activation of Hedgehog signaling unbalances the rate of neural stem cell renewal by increasing symmetric divisions. Stem Cell Rep 3:312–323

Ferron SRSSR, Pozo N, Laguna A, Aranda S, Porlan E, Moreno M, Fillat C, de la Luna S, Sánchez P, Arbonés ML, Fariñas I, Sanchez P, Arbones ML, Farinas I (2010) Regulated segregation of kinase Dyrk1A during asymmetric neural stem cell division is critical for EGFR-mediated biased signaling. Cell Stem Cell 7:367–379

Fietz SA, Kelava I, Vogt J, Wilsch-Brauninger M, Stenzel D, Fish JL, Corbeil D, Riehn A, Distler W, Nitsch R, Huttner WB (2010) OSVZ progenitors of human and ferret neocortex are epithelial-like and expand by integrin signaling. Nat Neurosci 13:690–699

Florio M, Huttner WB (2014) Neural progenitors, neurogenesis and the evolution of the neocortex. Development 141:2182–2194

Fuentealba LC, Rompani SB, Parraguez JI, Obernier K, Romero R, Cepko CL, Alvarez-Buylla A (2015) Embryonic origin of postnatal neural stem cells. Cell 161:1644–1655

Furutachi S, Matsumoto A, Nakayama KI, Gotoh Y (2013) p57 controls adult neural stem cell quiescence and modulates the pace of lifelong neurogenesis. EMBO J 32:970–981

Furutachi S, Miya H, Watanabe T, Kawai H, Yamasaki N, Harada Y, Imayoshi I, Nelson M, Nakayama KI, Hirabayashi Y, Gotoh Y (2015) Slowly dividing neural progenitors are an embryonic origin of adult neural stem cells. Nat Neurosci 18:657–665

Gage FH, Kempermann G, Palmer TD, Peterson DA, Ray J (1998) Multipotent progenitor cells in the adult dentate gyrus. J Neurobiol 36:249–266

Gallo V, Deneen B (2014) Glial development: the crossroads of regeneration and repair in the CNS. Neuron 83(2):283–308

Gao P, Postiglione MP, Krieger TG, Hernandez L, Wang C, Han Z, Streicher C, Papusheva E, Insolera R, Chugh K, Kodish O, Huang K, Simons BD, Luo L, Hippenmeyer S, Shi SH (2014) Deterministic progenitor behavior and unitary production of neurons in the neocortex. Cell 159:775–788

Gomez-Lopez S, Lerner RG, Petritsch C, Gómez-López S, Lerner RG, Petritsch C (2013) Asymmetric cell division of stem and progenitor cells during homeostasis and cancer. CMLS 71:575–597

Gonzalez-Perez O, Romero-Rodriguez R, Soriano-Navarro M, Garcia-Verdugo JM, Alvarez-Buylla A (2009) Epidermal growth factor induces the progeny of subventricular zone type B cells to migrate and differentiate into oligodendrocytes. Stem Cells 27:2032–2043

Gotz M, Huttner WB (2005) The cell biology of neurogenesis. Nat Rev Mol Cell Biol 6:777–788

Greig LC, Woodworth MB, Galazo MJ, Padmanabhan H, Macklis JD (2013) Molecular logic of neocortical projection neuron specification, development and diversity. Nat Rev Neurosci 14:755–769

Hansen DV, Lui JH, Parker PRL, Kriegstein AR (2010) Neurogenic radial glia in the outer subventricular zone of human neocortex. Nature 464:554–561

Hill R, Patel K, Medved J, Reiss A, Nishiyama A (2013) NG2 cells in white matter but not gray matter proliferate in response to PDGF. J Neurosci 33:14558–14566

Hill R, Patel K, Goncalves C, Grutzendler J, Nishiyama A (2014) Modulation of oligodendrocyte generation during a critical temporal window after NG2 cell division. Nat Neurosci 17:1518–1527

Hughes EG, Kang SH, Fukaya M, Bergles DE (2013) Oligodendrocyte progenitors balance growth with self-repulsion to achieve homeostasis in the adult brain. Nat Neurosci 16:668–676

Ikeshima-Kataoka H, Skeath JB, Nabeshima Y, Doe CQ, Matsuzaki F (1997) Miranda directs Prospero to a daughter cell during Drosophila asymmetric divisions. Nature 390:625–629

Izumi Y, Ohta N, Hisata K, Raabe T, Matsuzaki F (2006) Drosophila Pins-binding protein Mud regulates spindle-polarity coupling and centrosome organization. Nat Cell Biol 8:586–593

Januschke J, Llamazares S, Reina J, Gonzalez C (2011) Drosophila neuroblasts retain the daughter centrosome. Nat Commun 2:243

Januschke J, Reina J, Llamazares S, Bertran T, Rossi F, Roig J, Gonzalez C (2013) Centrobin controls mother-daughter centriole asymmetry in Drosophila neuroblasts. Nat Cell Biol 15:241–248

Kang S, Fukaya M, Yang J, Rothstein J, Bergles D (2010) NG2+ CNS glial progenitors remain committed to the oligodendrocyte lineage in postnatal life and following neurodegeneration. Neuron 68:668–681

Kaplan MS, Hinds JW (1977) Neurogenesis in the adult rat: electron microscopic analysis of light radioautographs. Science 197(4308):1092–1094

Kato K, Konno D, Berry M, Matsuzaki F, Logan A, Hidalgo A (2015) Prox1 inhibits proliferation and is required for differentiation of the oligodendrocyte cell lineage in the mouse. PLoS One 10:1–19

Kawaguchi D, Furutachi S, Kawai H, Hozumi K, Gotoh Y (2013) Dll1 maintains quiescence of adult neural stem cells and segregates asymmetrically during mitosis. Nat Commun 4:1880

Kempermann G, Kuhn HG, Gage FH (1997) More hippocampal neurons in adult mice living in an enriched environment. Nature 386:493–495

Kessaris N, Fogarty M, Iannarelli P, Grist M, Wegner M, Richardson WD (2006) Competing waves of oligodendrocytes in the forebrain and postnatal elimination of an embryonic lineage. Nat Neurosci 9:173–179

Knoth R, Singec I, Ditter M, Pantazis G, Capetian P, Meyer RP, Horvat V, Volk B, Kempermann G (2010) Murine features of neurogenesis in the human hippocampus across the lifespan from 0 to 100 years. PLoS One 5

Konno D, Shioi G, Shitamukai A, Mori A, Kiyonari H, Miyata T, Matsuzaki F (2008) Neuroepithelial progenitors undergo LGN-dependent planar divisions to maintain self-renewability during mammalian neurogenesis. Nat Cell Biol 10:93–101

Kuo CT, Mirzadeh Z, Soriano-Navarro M, Rasin M, Wang D, Shen J, Sestan N, Garcia-Verdugo J, Alvarez-Buylla A, Jan LY, Jan Y-NN (2006) Postnatal deletion of Numb/Numblike reveals repair and remodeling capacity in the subventricular neurogenic niche. Cell 127:1253–1264

Kusek G, Campbell M, Doyle F, Tenenbaum SA, Kiebler M, Temple S (2012) Asymmetric segregation of the double-stranded RNA binding protein Staufen2 during mammalian neural stem cell divisions promotes lineage progression. Cell Stem Cell 11:505–516

Lee C-YY, Andersen RO, Cabernard C, Manning L, Tran KD, Lanskey MJ, Bashirullah A, Doe CQ (2006a) Drosophila Aurora-A kinase inhibits neuroblast self-renewal by regulating aPKC/Numb cortical polarity and spindle orientation. Genes Dev 20:3464–3474

Lee C-YY, Wilkinson BD, Siegrist SE, Wharton RP, Doe CQ (2006b) Brat is a Miranda cargo protein that promotes neuronal differentiation and inhibits neuroblast self-renewal. Dev Cell 10:441–449

Lerner RG, Grossauer S, Kadkhodaei B, Meyers I, Sidorov M, Koeck K, Hashizume R, Ozawa T, Phillips JJ, Berger MS, Nicolaides T, James CD, Petritsch CK (2015) Targeting a Plk1-controlled polarity checkpoint in therapy-resistant glioblastoma-propagating cells. Cancer Res 75:5355–5366

Li G, Fang L, Fernandez G, Pleasure S (2013) The ventral hippocampus is the embryonic origin for adult neural stem cells in the dentate gyrus. Neuron 78:658–672

Liu H, Song N (2016) Molecular mechanism of adult neurogenesis and its association with human brain diseases. J Cent Nerv Syst Dis 8:5–11

Lois C, Alvarez-Buylla A (1993) Proliferating subventricular zone cells in the adult mammalian forebrain can differentiate into neurons and glia. Proc Natl Acad Sci USA 90:2074–2077

Lu B, Rothenberg M, Jan LY, Jan YN (1998) Partner of Numb colocalizes with Numb during mitosis and directs Numb asymmetric localization in Drosophila neural and muscle progenitors. Cell 95:225–235

Lu QR, Yuk D, Alberta JA, Zhu Z, Pawlitzky I, Chan J, McMahon AP, Stiles CD, Rowitch DH (2000) Sonic hedgehog–regulated oligodendrocyte lineage genes encoding bHLH proteins in the mammalian central nervous system. Neuron 25:317–329

Lugert S, Basak O, Knuckles P, Haussler U, Fabel K, Gotz M, Haas CA, Kempermann G, Taylor V, Giachino C, Götz M, Haas CA, Kempermann G, Taylor V, Giachino C, Gotz M, Haas CA, Kempermann G, Taylor V, Giachino C (2010) Quiescent and active hippocampal neural stem cells with distinct morphologies respond selectively to physiological and pathological stimuli and aging. Cell Stem Cell 6:445–456

Luskin MB (1993) Restricted proliferation and migration of postnatally generated neurons derived from the forebrain subventricular zone. Neuron 11:173–189

Malberg JE, Eisch AJ, Nestler EJ, Duman RS (2000) Chronic antidepressant treatment increases neurogenesis in adult rat hippocampus. J Neurosci 20:9104–9110

Marques S, Zeisel A, Codeluppi S, van Bruggen D, Mendanha Falcão A, Xiao L, Li H, Häring M, Hochgerner H, Romanov RA, Gyllborg D, Muñoz-Manchado AB, La Manno G, Lönnerberg P, Floriddia EM, Rezayee F, Ernfors P, Arenas E, Hjerling-Leffler J, Harkany T, Richardson WD, Linnarsson S, Castelo-Branco G (2016) Oligodendrocyte heterogeneity in the mouse juvenile and adult central nervous system. Science 352:1326–1329

Maslov AY, Barone TA, Plunkett RJ, Pruitt SC (2004) Neural stem cell detection, characterization, and age-related changes in the subventricular zone of mice. J Neurosci 24:1726–1733

Matsuzaki F, Ohshiro T, Ikeshima-Kataoka H, Izumi H (1998) miranda localizes staufen and prospero asymmetrically in mitotic neuroblasts and epithelial cells in early Drosophila embryogenesis. Development 125:4089–4098

Menn B, Garcia-Verdugo JM, Yaschine C, Gonzalez-Perez O, Rowitch D, Alvarez-Buylla A (2006) Origin of oligodendrocytes in the subventricular zone of the adult brain. J Neurosci 26:7907–7918

Mignone JL, Kukekov V, Chiang A-S, Steindler D, Enikolopov G (2004) Neural stem and progenitor cells in nestin-GFP transgenic mice. J Comp Neurol 469:311–324

Miller RH (2002) Regulation of oligodendrocyte development in the vertebrate CNS. Prog Neurobiol 67:451–467

Mirzadeh Z, Merkle FT, Soriano-Navarro M, Garcia-Verdugo JM, Alvarez-Buylla A (2008) Neural stem cells confer unique pinwheel architecture to the ventricular surface in neurogenic regions of the adult brain. Cell Stem Cell 3:265–278

Neumüller RA, Richter C, Fischer A, Novatchkova M, Neumüller KG, Knoblich JA (2011) Genome-wide analysis of self-renewal in Drosophila neural stem cells by transgenic RNAi. Cell Stem Cell 8:580–593

Noctor SC, Martinez-Cerdeno V, Ivic L, Kriegstein AR, Martínez-Cerdeño V, Ivic L, Kriegstein AR, Martinez-Cerdeno V, Ivic L, Kriegstein AR (2004) Cortical neurons arise in symmetric and asymmetric division zones and migrate through specific phases. Nat Neurosci 7:136–144

Noctor SC, Martinez-Cerdeno V, Kriegstein AR, Martínez-Cerdeño V, Kriegstein AR, Martinez-Cerdeno V, Kriegstein AR (2008) Distinct behaviors of neural stem and progenitor cells underlie cortical neurogenesis. J Comp Neurol 508:28–44

Ogawa H, Ohta N, Moon W, Matsuzaki F (2009) Protein phosphatase 2A negatively regulates aPKC signaling by modulating phosphorylation of Par-6 in Drosophila neuroblast asymmetric divisions. J Cell Sci 122:3242–3249

Orentas DM, Hayes JE, Dyer KL, Miller RH (1999) Sonic hedgehog signaling is required during the appearance of spinal cord oligodendrocyte precursors. Development 126:2419–2429

Ortega F, Gascon S, Masserdotti G, Deshpande A, Simon C, Fischer J, Dimou L, Chichung Lie D, Schroeder T, Berninger B, Gascón S, Masserdotti G, Deshpande A, Simon C, Fischer J, Dimou L, Chichung Lie D, Schroeder T, Berninger B, Gascon S, Masserdotti G, Deshpande A, Simon C, Fischer J, Dimou L, Chichung Lie D, Schroeder T, Berninger B (2013) Oligodendrogliogenic and neurogenic adult subependymal zone neural stem cells constitute distinct lineages and exhibit differential responsiveness to Wnt signalling. Nat Cell Biol 15:602–613

Petersen PH, Zou K, Hwang JK, Jan YN, Zhong W (2002) Progenitor cell maintenance requires numb and numblike during mouse neurogenesis. Nature 419:929–934

Peterson FC, Penkert RR, Volkman BF, Prehoda KE (2004) Cdc42 regulates the Par-6 PDZ domain through an allosteric CRIB-PDZ transition. Mol Cell 13:665–676

Petritsch C, Tavosanis G, Turck CW, Jan LY, Jan YN (2003) The Drosophila myosin VI Jaguar is required for basal protein targeting and correct spindle orientation in mitotic neuroblasts. Dev Cell 4:273–281

Piccin D, Tufford A, Morshead CM (2014) Neural stem and progenitor cells in the aged subependyma are activated by the young niche. Neurobiol Aging 35:1669–1679

Pineda JR, Daynac M, Chicheportiche A, Cebrian-Silla A, Sii Felice K, Garcia-Verdugo JM, Boussin FD, Mouthon MA (2013) Vascular-derived TGF-beta increases in the stem cell niche and perturbs neurogenesis during aging and following irradiation in the adult mouse brain. EMBO Mol Med 5:548–562

Ponti G, Obernier K, Guinto C, Jose L, Bonfanti L, Alvarez-Buylla A (2013) Cell cycle and lineage progression of neural progenitors in the ventricular-subventricular zones of adult mice. Proc Natl Acad Sci USA 110:E1045–E1054

Postiglione MP, Jüschke C, Xie Y, Haas GA, Charalambous C, Knoblich JA (2011) Mouse inscuteable induces apical-basal spindle orientation to facilitate intermediate progenitor generation in the developing neocortex. Neuron 72:269–284

Psachoulia K, Jamen F, Young KM, Richardson WD (2009) Cell cycle dynamics of NG2 cells in the postnatal and ageing brain. Neuron Glia Biol 5:57–67

Rakic P (1995) A small step for the cell, a giant leap for mankind: a hypothesis of neocortical expansion during evolution. Trends Neurosci 18:383–388

Rebollo E, Sampaio P, Januscke J, Llamazares S, Varmerk H, Gonzalez C, Januschke J, Llamazares S, Varmark H, González C (2007) Functionally unequal centrosomes drive spindle orientation in asymmetrically dividing Drosophila neural stem cells. Dev Cell 12:467–474

Rivers LE, Young KM, Rizzi M, Jamen F, Psachoulia K, Wade A, Kessaris N, Richardson WD (2008) PDGFRA/NG2 glia generate myelinating oligodendrocytes and piriform projection neurons in adult mice. Nat Neurosci 11:1392–1401

Rolls MM, Albertson R, Shih H-PP, Lee C-YY, Doe CQ (2003) Drosophila aPKC regulates cell polarity and cell proliferation in neuroblasts and epithelia. J Cell Biol 163:1089–1098

Rowitch DH, Kriegstein AR (2010) Developmental genetics of vertebrate glial-cell specification. Nature 468:214–222

Sakai D, Dixon J, Dixon MJ, Trainor PA (2012) Mammalian neurogenesis requires Treacle-Plk1 for precise control of spindle orientation, mitotic progression, and maintenance of neural progenitor cells. PLoS Genet 8:e1002566

Sanai N, Nguyen T, Ihrie RA, Mirzadeh Z, Tsai H-H, Wong M, Gupta N, Berger MS, Huang E, Garcia-Verdugo J-MM, Rowitch DH, Alvarez-Buylla A (2011) Corridors of migrating neurons in the human brain and their decline during infancy. Nature 478:382–386

Schaefer M, Shevchenko A, Knoblich JA (2000) A protein complex containing Inscuteable and the Galpha-binding protein Pins orients asymmetric cell divisions in Drosophila. Curr Biol 10:353–362

Schwamborn JC, Berezikov E, Knoblich JA (2009) The TRIM-NHL protein TRIM32 activates microRNAs and prevents self-renewal in mouse neural progenitors. Cell 136:913–925

Shen CP, Jan LY, Jan YN (1997) Miranda is required for the asymmetric localization of Prospero during mitosis in Drosophila. Cell 90:449–458

Shen CP, Knoblich JA, Chan YM, Jiang MM, Jan LY, Jan YN (1998) Miranda as a multidomain adapter linking apically localized Inscuteable and basally localized Staufen and Prospero during asymmetric cell division in Drosophila. Genes Dev 12:1837–1846

Shen Q, Zhong W, Jan YN, Temple S (2002) Asymmetric Numb distribution is critical for asymmetric cell division of mouse cerebral cortical stem cells and neuroblasts. Development 129:4843–4853

Shen Q, Wang Y, Kokovay E, Lin G, Chuang S-MM, Goderie SK, Roysam B, Temple S (2008) Adult SVZ stem cells lie in a vascular niche: a quantitative analysis of niche cell-cell interactions. Cell Stem Cell 3:289–300

Shitamukai A, Konno D, Matsuzaki F (2011) Oblique radial glial divisions in the developing mouse neocortex induce self-renewing progenitors outside the germinal zone that resemble primate outer subventricular zone progenitors. J Neurosci 31:3683–3695

Shook BA, Manz DH, Peters JJ, Kang S, Conover JC (2012) Spatiotemporal changes to the subventricular zone stem cell pool through aging. J Neurosci 32:6947–6956

Siegrist SE, Doe CQ (2005) Microtubule-induced Pins/Galphai cortical polarity in Drosophila neuroblasts. Cell 123:1323–1335

Siegrist SE, Doe CQ (2007) Microtubule-induced cortical cell polarity. Genes Dev 21:483–496

Siller KH, Doe CQ (2009) Spindle orientation during asymmetric cell division. Nat Cell Biol 11:365–374

Siller KH, Cabernard C, Doe CQ (2006) The NuMA-related Mud protein binds Pins and regulates spindle orientation in Drosophila neuroblasts. Nat Biotechnol 8:594–600

Sim FJ, Zhao C, Penderis J, Franklin RJM (2002) The age-related decrease in CNS remyelination efficiency is attributable to an impairment of both oligodendrocyte progenitor recruitment and differentiation. J Neurosci 22:2451–2459

Simon C, Goetz M, Dimou L, Gotz M, Dimou L (2011) Progenitors in the adult cerebral cortex: cell cycle properties and regulation by physiological stimuli and injury. Glia 59:869–881

Singh P, Ramdas Nair A, Cabernard C (2014) The centriolar protein Bld10/Cep135 is required to establish centrosome asymmetry in Drosophila neuroblasts. Curr Biol 24:1548–1555

Slack C, Overton PM, Tuxworth RI, Chia W (2007) Asymmetric localisation of Miranda and its cargo proteins during neuroblast division requires the anaphase-promoting complex/cyclosome. Development 134:3781–3787

Smith CA, Lau KM, Rahmani Z, Dho SE, Brothers G, She YM, Berry DM, Bonneil E, Thibault P, Schweisguth F, Le Borgne R, McGlade CJ (2007) aPKC-mediated phosphorylation regulates asymmetric membrane localization of the cell fate determinant Numb. EMBO J 26:468–480

Spalding KL, Bergmann O, Alkass K, Bernard S, Salehpour M, Huttner HB, Boström E, Westerlund I, Vial C, Buchholz BA, Possnert G, Mash DC, Druid H, Frisén J (2013) Dynamics of hippocampal neurogenesis in adult humans. Cell 153:1219–1227

Spassky N, Merkle FT, Flames N, Tramontin AD, Garcia-Verdugo JM, Alvarez-Buylla A, García-Verdugo JM, Alvarez-Buylla A, Garcia-Verdugo JM, Alvarez-Buylla A (2005) Adult ependymal cells are postmitotic and are derived from radial glial cells during embryogenesis. J Neurosci 25:10–18

Stein-Behrens B, Mattson MP, Chang I, Yeh M, Sapolsky R (1994) Stress exacerbates neuron loss and cytoskeletal pathology in the hippocampus. J Neurosci 14:5373–5380

Sugiarto S, Persson A, Munoz E, Waldhuber M, Lamagna C, Andor N, Hanecker P, Ayers-Ringler J, Phillips J, Siu J, Lim D, Vandenberg S, Stallcup W, Berger M, Bergers G, Weiss W, Petritsch C (2011) Asymmetry-defective oligodendrocyte progenitors are glioma precursors. Cancer Cell 20:328–340

Sun Y, Goderie S, Temple S (2005) Asymmetric distribution of EGFR receptor during mitosis generates diverse CNS progenitor cells. Neuron 45:873–886. doi:10.1016/j.neuron.2005.01.045

Tropepe V, Craig CG, Morshead CM, VanderKooy D (1997) Transforming growth factor-alpha null and senescent mice show decreased neural progenitor cell proliferation in the forebrain subependyma. J Neurosci 17:7850–7859

Vallstedt A, Klos JM, Ericson J (2005) Multiple dorsoventral origins of oligodendrocyte generation in the spinal cord and hindbrain. Neuron 45:55–67

van Wijngaarden P, Franklin RJ (2013) Ageing stem and progenitor cells: implications for rejuvenation of the central nervous system. Development 140:2562–2575

Vessey JP, Amadei G, Burns SE, Kiebler MA, Kaplan DR, Miller FD (2012) An asymmetrically localized staufen2-dependent RNA complex regulates maintenance of Mammalian neural stem cells. Cell Stem Cell 11:517–528

Voigt T (1989) Development of glial cells in the cerebral wall of ferrets: direct tracing of their transformation from radial glia into astrocytes. J Comp Neurol 289:74–88

Wang H, Somers GW, Bashirullah A, Heberlein U, Yu F, Chia W (2006) Aurora-A acts as a tumor suppressor and regulates self-renewal of Drosophila neuroblasts. Genes Dev 20:3453–3463

Wang H, Ouyang Y, Somers WG, Chia W, Lu B (2007) Polo inhibits progenitor self-renewal and regulates Numb asymmetry by phosphorylating Pon. Nature 449:96–100

Wang X, Tsai J-W, Imai JH, Lian W-N, Vallee RB, Shi S-H (2009) Asymmetric centrosome inheritance maintains neural progenitors in the neocortex. Nature 461:947–955

Wang C, Liu F, Liu Y-Y, Zhao C-H, You Y, Wang L, Zhang J, Wei B, Ma T, Zhang Q, Zhang Y, Chen R, Song H, Yang Z (2011a) Identification and characterization of neuroblasts in the subventricular zone and rostral migratory stream of the adult human brain. Cell Res 21:1534–1550

Wang X, Lui J, Kriegstein A (2011a) Orienting fate: spatial regulation of neurogenic division. Neuron:191–193

Wang X, Tsai J, LaMonica B, Kriegstein A (2011b) A new subtype of progenitor cell in the mouse embryonic neocortex. Nat Neurosci 14:555–561

Willard FS, Kimple RJ, Siderovski DP (2004) Return of the GDI: the GoLoco motif in cell Division. Annu Rev Biochem 73:925–951

Wirtz-Peitz F, Nishimura T, Knoblich JA (2008) Linking cell cycle to asymmetric division: Aurora-A phosphorylates the Par complex to regulate Numb localization. Cell 135:161–173

Wodarz A, Ramrath A, Kuchinke U, Knust E (1999) Bazooka provides an apical cue for Inscuteable localization in Drosophila neuroblasts. Nature 402:544–547

Wodarz A, Ramrath A, Grimm A, Knust E (2000) Drosophila atypical protein kinase C associates with Bazooka and controls polarity of epithelia and neuroblasts. J Cell Biol 150:1361–1374

Wren D, Wolswijk G, Noble M (1992) In vitro analysis of the origin and maintenance of O-2Aadult progenitor cells. J Cell Biol 116:167–176

Wu M, Kwon HY, Rattis F, Blum J, Zhao C, Ashkenazi R, Jackson TL, Gaiano N, Oliver T, Reya T (2007) Imaging hematopoietic precursor division in real time. Cell Stem Cell 1:541–554

Young KM, Psachoulia K, Tripathi RB, Dunn SJ, Cossell L, Attwell D, Tohyama K, Richardson WD (2013) Oligodendrocyte dynamics in the healthy adult CNS: evidence for myelin remodeling. Neuron 77:873–885

Yu F, Wang H, Qian H, Kaushik R, Bownes M, Yang X, Chia W (2005) Locomotion defects, together with Pins, regulates heterotrimeric G-protein signaling during Drosophila neuroblast asymmetric divisions. Genes Dev 19:1341–1353. doi:10.1101/gad.1295505.eration

Zhong W, Feder JN, Jiang MM, Jan LY, Jan YN (1996) Asymmetric localization of a mammalian numb homolog during mouse cortical neurogenesis. Neuron 17:43–53

Zhong W, Jiang MM, Schonemann MD, Meneses JJ, Pedersen RA, Jan LY, Jan YN (2000) Mouse numb is an essential gene involved in cortical neurogenesis. Proc Natl Acad Sci USA 97:6844–6849

Zhu X, Hill RA, Dietrich D, Komitova M, Suzuki R, Nishiyama A (2011) Age-dependent fate and lineage restriction of single NG2 cells. Development 138:745–753

Zong H, Parada LF, Baker SJ (2015) Cell of origin for malignant gliomas and its implication in therapeutic development. Cold Spring Harb Perspect Biol 7

Chapter 18
Molecular Programs Underlying Asymmetric Stem Cell Division and Their Disruption in Malignancy

Subhas Mukherjee and Daniel J. Brat

Abstract Asymmetric division of stem cells is a highly conserved and tightly regulated process by which a single stem cell produces two unequal daughter cells. One retains its stem cell identity while the other becomes specialized through a differentiation program and loses stem cell properties. Coordinating these events requires control over numerous intra- and extracellular biological processes and signaling networks. In the initial stages, critical events include the compartmentalization of fate determining proteins within the mother cell and their subsequent passage to the appropriate daughter cell in order to direct their destiny. Disturbance of these events results in an altered dynamic of self-renewing and differentiation within the cell population, which is highly relevant to the growth and progression of cancer. Other critical events include proper asymmetric spindle assembly, extrinsic regulation through micro-environmental cues, and non-canonical signaling networks that impact cell division and fate determination. In this review, we discuss mechanisms that maintain the delicate balance of asymmetric cell division in normal tissues and describe the current understanding how some of these mechanisms are deregulated in cancer.

18.1 Introduction

Stem cells are a small, specialized cell population that supports tissue and organ development, growth, homeostasis, and regeneration through their ability to self-renew and eventually differentiate along multiple paths into highly specialized cells

S. Mukherjee
Departments of Pathology and Laboratory Medicine, Emory University, Atlanta, GA, USA

D.J. Brat, M.D., Ph.D. (✉)
Departments of Pathology and Laboratory Medicine, Emory University, Atlanta, GA, USA
Winship Cancer Institute, Emory University, 1701 Uppergate Drive, Building C, Rm#C5038, Atlanta, GA, USA
e-mail: dbrat@emory.edu

J.-P. Tassan, J.Z. Kubiak (eds.), *Asymmetric Cell Division in Development, Differentiation and Cancer*, Results and Problems in Cell Differentiation 61,
DOI 10.1007/978-3-319-53150-2_18

that perform end organ-specific functions. Stem cells are strategically located within specialized biological tissue niches of developing and mature organisms in order to carry out this critical role. One fundamental capability is critical to the identity and proper function of stem cells: the ability to undergo a mitotic division in which one resultant daughter cell maintains stemness and the capacity to self-renew, while the other acquires the capability to differentiate. This complex process is known as asymmetric cell division (Betschinger et al. 2006), a highly specialized stem cell property that, for good reasons, is tightly regulated by intricate networks of molecular events. Disruption of these regulatory cascades leads to the generation of two daughter cells that retain their stemness and cannot fully differentiate (Knoblich 2010; Yong and Yan 2011), with the ensuing disequilibrium of growth dynamics causing abnormal tissue growth properties with relevance to cancer that is quickly being recognized (Mukherjee et al. 2015).

Earliest evidence of asymmetric mitotic division came in 1905 by Edwin Conklin (Conklin 1905) when he observed that certain cells of the invertebrate Ascidian egg produced daughter cells that were phenotypically distinct. Since then, asymmetric cell division has been recognized in all living organisms, including invertebrates and vertebrates. More recently, the understanding of molecular events guiding asymmetric division has improved dramatically, mostly due to studies on specific developing cell populations in *Drosophila*, including neuroblasts, intestinal stem cells, and germline stem cells (Knoblich 2010). Advances have also come from mammalian systems, through the study of asymmetric cell division in mouse radial glial progenitors during neurogenesis (Chenn and McConnell 1995; Miyata et al. 2004; Noctor et al. 2004, 2008; Bultje et al. 2009) and muscle satellite cells (Troy et al. 2012). The tumor-like growths that arise when asymmetric cell division is disrupted in these models naturally sparked curiosity regarding the role of asymmetric cell division in cancer. At the same time, the stem cell paradigm derived from normal development and regeneration has been increasingly applied to the study of neoplasms, leading to a burgeoning field of cancer stem cell biology. Unlike normal cells, cancerous populations have oncogenic alterations, disturbed stem–nonstem equilibria, and differentiation of cells with aggressive biologic properties. Here, we summarize the current understanding of the molecular cascade that normally directs asymmetric cell division during development and physiologic homeostasis and also describe how its pathological disruption may be related to the development and progression of cancer.

18.2 Mechanisms of Asymmetric Cell Division

Asymmetric cell division and its regulatory mechanisms have been documented in numerous model systems ranging from *C. elegans* to mammals, but investigations of *Drosophila* have advanced the furthest (Knoblich 2010). Four critical and interdependent events need to occur properly for this intricate process to succeed: (1) correct localization and function of protein complexes at apical and basal sides of the division plane to ensure appropriate activation of fate determinants, (2) proper asymmetric spindle orientation and function both to regulate division and to place daughter cells in the correct orientation to the niche, (3) extrinsic regulation through the stem cell niche, and (4) influences from noncanonical signaling pathways (Chia et al. 2008). This process begins at interphase and ends with cytokinesis.

18.2.1 Regulation of Asymmetry Through Localization of Fate Determinants

Our current understanding of cell-intrinsic regulators of asymmetric cell division is derived mostly from studies of developing neuroblasts of the *Drosophila* brain. Before cell division starts in this cell population, a specific set of proteins form complexes that localize to either the apical or basal side of the cell division plane. On the apical side, these proteins include atypical protein kinase C (aPKC), Partition defective 6 (Par-6), and Lethal Giant Larvae [L(2)GL]. On the basal side, a second protein complex forms from Miranda, Brat, and Prospero. It is still unclear how these protein complexes are initiated or guided to the apical or basal side; the ultimate result is that their positioning directs the identity of the daughter cell containing them.

18.2.1.1 Apical Protein Complex and Localization of Apical Determinants

Normal Conditions

Once aPKC, Par-6, and L(2)GL have formed a complex on the apical side (Fig. 18.1), Aurora A, a serine-threonine protein kinase, then initiates signaling, first by phosphorylating Par-6 and then aPKC. Co-immunoprecipitation and [^{32}P]ATP labeling experiments first indicated that Par-6 is phosphorylated at Ser34 (Wirtz-Peitz et al. 2008), while in vitro kinase assays confirmed the specific phosphorylation and activation of aPKC (Wirtz-Peitz et al. 2008) by Aurora A. aPKC can also phosphorylate itself in the presence of Par-6, L(2)GL, and Aurora A (Wirtz-Peitz et al. 2008). Activated aPKC then proceeds to phosphorylate L(2)GL, thereby reducing L(2)GL affinity to the aPKC–Par-6–L(2)GL complex. Reduced phosphorylation of L(2)GL by aPKC is responsible for delayed cortical release of L(2)GL, whereas constitutively active aPKC$^{\Delta N}$ redistributes L(2)GL into the cytoplasm during cell cycle (Wirtz-Peitz et al. 2008), ensuring asymmetric fate of neuroblasts. Under normal conditions, when L(2)GL dissociates, Par-3 is able to join the complex (Kalmes et al. 1996; Betschinger et al. 2003; Yamanaka et al. 2003), leading to the critical event of phosphorylation of the fate determinant Numb. Once phosphorylated, Numb becomes inactive and dissociates from the plasma membrane (Fig. 18.1). Since Numb acts as a major suppressor of Notch signaling, its inactivation has the effect of activating Notch signaling, providing

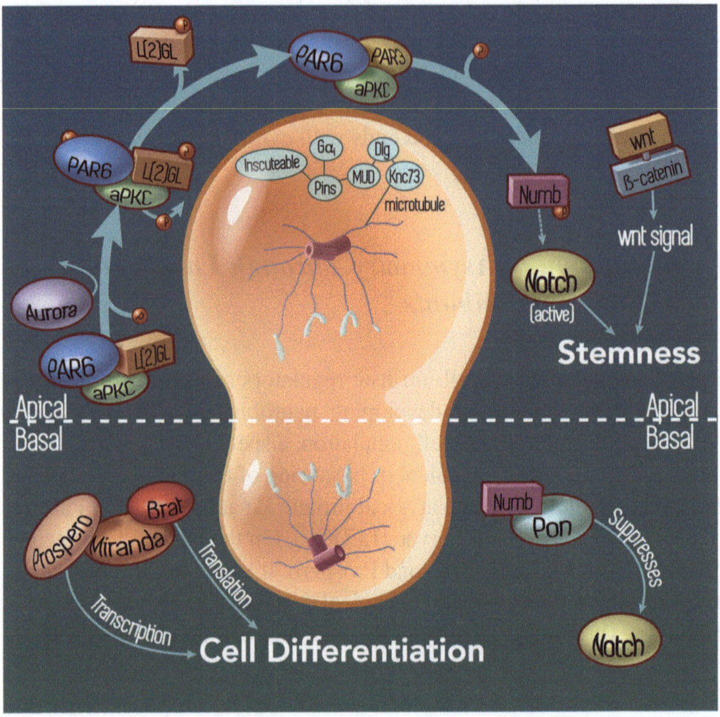

Fig. 18.1 Distinct apical and basal molecular programs regulating asymmetric cell division. *Apical pole*: On the apical side, aPKC/PAR6/PAR3 complex formation initiates during interphase, giving apical pole the identity of self-renewal. Aurora A protein kinase phosphorylates aPKC leading to its activation. Active aPKC phosphorylates L(2)GL which releases L(2)GL from the complex and PAR3 enters. This complex phosphorylates Numb releasing it from the apical membrane and increasing Numb concentration at the basal side. This keeps the Notch signal active on the apical side and promotes stemness. Wnt signal also takes part in the self-renewal, though the detailed mechanism is not known. In metaphase and telophase, apical microtubule arrangement is maintained by Inscuteable/Pins/Gα_i complex. Cytoskeletal adapter protein Inscuteable binds with Pins and Gα_i, whereas Mud forms complex with Dlg and Knc73. These two complexes come together that arranges the microtubules attached to Knc73. *Basal pole*: Adapter protein Miranda on the basal pole binds with Prospero and Brat. Degradation of Miranda releases transcription factor Prospero that turns on the genes driving differentiation. Brat, on the other hand, acts as a translational repressor, possibly suppressing protein production needed for proliferation. Accumulation of Numb on the basal membrane leads to the degradation of Notch. This suppresses Notch signal and its proliferative effect on the basal side

self-renewal properties to the daughter cell emerging from the apical side of division (Wirtz-Peitz et al. 2008).

Mechanisms concordant with those in *Drosophila* neuroblasts have also been established in mammalian systems. During mammalian neurogenesis, neural progenitor cells undergoing asymmetric cell division also show localization of the aPKC/PAR6/PAR3 complex to the apical side (Manabe et al. 2002), where it is concentrated at the adherens junctions, with the help of a small GTP binding protein, CDC42 (Joberty et al. 2000). This complex formation guarantees apical adherens junction integrity and establishes apicobasal polarity (Gomez-Lopez et al. 2013). Protein complexes on the apical side also ensure NUMB inactivation and NOTCH activation. One function of NOTCH in the maintenance of the stem cell fate is to bind with active β-catenin and prevent it from entering the nucleus to initiate its transcriptional program. In this manner, NOTCH is antagonistic to β-catenin pathway and NOTCH knockdown in embryonic stem cells (ESCs) leads to transcriptional activation by β-catenin (Kwon et al. 2011); others have clearly defined the regulation of asymmetric cell division by β-catenin (Habib et al. 2013).

In Neoplasia

In *Drosophila* tumor models, disruption of the tightly regulated mechanisms that control asymmetric division usually generates a highly proliferative population of stem-like cells with limited capacity to differentiate. One of the earliest findings was that loss-of-function mutants of Numb, which normally antagonizes Notch in neuroblasts, result in a neoplastic proliferation of stem-like cells in the developing brain (Lee et al. 2006a; Wang et al. 2006). Similarly, a delocalized and active form of *Drosophila* aPKC protein (Lee et al. 2006b), which activates Notch signaling throughout the neuroblast by suppressing active Numb, also promotes tumor formation (Knoblich 2010; Liu et al. 2010). Suppression of aPKC, on the other hand, reduces the number of neuroblasts. Aurora A is a mitotic kinase that promotes centrosome localization and its maturation. Besides its phosphorylation of aPKC and Par-6, Aurora A is also required for Numb's asymmetric localization on the basal side. In Aurora A mutants, inactive Numb is released from basal cortex and drives overproliferation to generate tumors through the final common pathway of activated Notch signaling (Chia et al. 2008). Mutants of L(2)GL cause the formation of nonfunctional aPKC/Par-6/Par-3 complex that promotes symmetric division. In general, those alterations that lead to inactivation of Numb or promote Notch signaling will cause tumors.

Similar to *Drosophila*, disruption of apical protein complex aPKCi/PAR6/LGL1 in mammalian systems also promotes a neoplastic state. aPKC iota (aPKCi), the human ortholog of *Drosophila* aPKC of the apical complex, is both amplified and overactive in ovarian cancer cells, promoting self-renewal, and is associated with aggressive growth (Eder et al. 2005). In mice, mutated PAR6 attenuates TGF-**β** signaling and suppresses lung metastasis of mammary tumors (Viloria-Petit et al. 2009). In humans, PAR6 is known to promote epithelial–mesenchymal transition (EMT) in breast cancer through TGF-**β** signaling. Loss of LGL1, a mammalian ortholog of fly L(2)GL, promotes overproliferation of neural progenitor cells in the ventricular region of the mouse cerebellum at the expense of their differentiation (Hou et al. 2014). Overexpression of PAR3, another partner of apical protein complex, keeps NUMB inactive, thereby enhancing Notch signaling (Gaiano et al. 2000) in radial glial cell population. This results in symmetric division in these cell populations, resulting in retained stem-like properties of both daughter cells (Bultje et al. 2009). Along these lines, Musashi-2 is an inhibitor of NUMB that promotes Notch signaling, and activation of Musashi-2 in hematopoietic stem cells leads to leukemia, demonstrating neoplastic growth in myeloid cells following activation of this pathway (Ito et al. 2010).

There are numerous examples of human cancers that harbor mutations or copy number alterations (CNA) of genes related to asymmetric cell division. For example, human epithelial neoplasms show frequent deletions in *PARD6* (PAR6), *PARD3* (PAR3), and *DLG2*. Lung and adrenal cancer shows *PARD6* deletion, which disrupts cell polarity and causes neoplastic growths. *PARD3* deletion leads to the loss of cell polarity, symmetric cell division, and enhanced proliferation in glioblastomas and squamous cell carcinomas. *DLG2* is deleted most frequently in cervical and lung cancer and phenocopies *PARD3* deletion (Kunnev et al. 2009; Rothenberg et al. 2010). Although in *Drosophila*, loss-of-function mutations of Aurora kinase drive overproliferation of neuroblasts, many human neoplastic diseases such as pancreatic cancer and prostate cancer show overexpression of its human orthologs AURKA and AURKB (Li et al. 2003; Chieffi et al. 2006). On the other hand, TCGA provisional data analysis shows that more than 75% head and neck squamous cell carcinomas carry high frequency deletion of *AURKB*(Mukherjee et al. 2015), indicating a unique role of Aurora kinase in neoplastic growth that demands further studies. Nevertheless, all these data strongly support a link between neoplastic disease and the disruption of asymmetric cell division.

18.2.1.2 Basal Protein Complex and Localization of Basal Determinants

Normal Conditions

The combination of Numb phosphorylation and inactivation on the apical side, together with Aurora A, mediated localization of unphosphorylated active Numb on the basal side, leading to inhibition of Notch signaling and directing the emerging basal daughter toward differentiation. At the same time, during interphase, a highly coil-coiled adapter protein Miranda also appears specifically on basal side. Immunocytochemical studies have demonstrated that Miranda binds with two critical fate determining proteins that guide the transcriptional and translational fate of the basal daughter: (a) Prospero, a transcription factor, and (b) Brain Tumor (Brat), a translational repressor that may have other functions as well (Lee et al. 2006c). Once proper localization of Prospero and Brat is accomplished, Miranda is proteolytically degraded, releasing Prospero to initiate a transcriptional program that stimulates differentiation (Slack et al. 2007). Brat, on the other hand, likely works through multiple mechanisms, including activity as a translational repressor and by its ability to suppress self-renewing wnt signaling through β-catenin/Armadillo inhibition (Komori et al. 2014).

The incomplete understanding of Brat mechanisms on the basal side (Fig. 18.1) during asymmetric cell division is due in part to its complex structure (Arama et al. 2000). One of its first described functions was as a translational repressor of Hunchback protein during embryonic development (Sonoda and Wharton 2001). Brat belongs to the Tripartite Motif Containing Protein family, which includes almost 70 members of the mammalian TRIM protein family. Based on sequence homology and similar protein domains, the human ortholog TRIM3 has the closest resemblance (48% identity) to *Drosophila* Brat. TRIM3 has already been established as an E3 ubiquitin ligase and regulates asymmetric cell division in human neural progenitor cells (Chen et al. 2014). TRIM32, another ubiquitin ligase closely related to TRIM3, has been reported to drive neurogenesis in a mouse model (Schwamborn et al. 2009). TRIM32 also has several functions including activation of microRNAs, such as *Let-7*, and binding of c-MYC to promote its degradation (Schwamborn et al. 2009).

In parallel to findings in *Drosophila*, the mammalian ortholog of Prospero, PROX1, acts as a fate determinant in asymmetric cell division. During neurogenesis in mouse models, PROX1 (Dyer et al. 2003; Kaltezioti et al. 2010; Vessey et al. 2012) transcriptionally activates differentiation factors that lead to the generation of the differentiated cell from the basal side after division.

In Neoplasia

Loss of proper asymmetric cell division in *Drosophila* favors an increase in cell division that is more symmetric, enhancing the stem cell pool and causing neoplastic growth. Activated Numb serves as a lynchpin on the basal side in its role of suppressing Notch signaling (Fig. 18.1) and promoting differentiation during asymmetric division. This critical event also depends on the inactivation of Numb on the apical side through the actions of aPKC and L(2)GL. When aPKC is inhibited, phosphorylation of L(2)GL does not occur (Betschinger et al. 2003, 2005) on the apical side, thereby reducing Numb phosphorylation on that side and leading to the failure of the unphosphorylated Numb localization to the basal membrane (Wirtz-Peitz et al. 2008). Loss of functional Numb in this setting as well as the Numb null mutant allele causes inappropriate Notch signaling after division (Bowman et al. 2008), leading to disruption in asymmetric cell division.

Mutations in *miranda, prospero*, or *brat* lead to a brain tumor phenotype during *Drosophila* development in which there is a massive brain enlargement consisting of neuroblasts that fail to differentiate. The remarkable similarity of tumor growth following the mutation of any of these three genes points to a common regulatory mechanism. In addition to local growth properties, mutant neuroblasts also show invasive and metastatic property when transplanted into the abdomen of wild-type *Drosophila*, indicating they are malignant, rather than benign growths (Caussinus and Gonzalez 2005; Beaucher et al. 2007; Chia et al. 2008; Janic et al. 2010). Recently, Mukherjee and Brat used *brat-RNAi* in *Drosophila* neuroblasts to cause Brat protein downregulation and generated a brain tumor phenotype in adult flies, which showed a shortened life expectancy. Brat suppression in these cells led to increased Musashi protein, a translational inhibitor of Numb, which in turn generated active Notch. Brat may regulate the nuclear transport of activated Notch through Ketel (mammalian ortholog, Importin β), suggesting a novel mechanism by which Notch nuclear transport is regulated in neoplastic diseases (Mukherjee et al. 2016).

Mammalian PROX1, the ortholog of Prospero, is mutated in a large number of human cancers, including pancreatic, esophageal, and colon cancer, and its functional consequences are thought to cause progression. When PROX1 protein was ectopically expressed in mice xenograft experiments with pancreatic cancer cells that harbored PROX1 mutations, a reduction in self-renewal and proliferation was observed (Takahashi et al. 2006).

In hematologic malignancies, DNA methylation of *PROX1* gene is frequent (56.3%), suggesting another mechanism for *PROX1* transcript suppression (Nagai et al. 2003). In breast cancer, epigenetic silencing has also been noted, and in hepatocellular malignancies, *PROX1* transcription is altered in a manner that promotes disease progression (Versmold et al. 2007; Dudas et al. 2008). Mutation and deletion of *NUMB* gene are hallmarks of many neoplastic diseases, including endometrial carcinoma and adenoid cystic carcinoma. NUMB protein is also reduced in some subsets of breast cancer (Colaluca et al. 2008; Ito et al. 2010; Gomez-Lopez et al. 2013).

Human orthologs of Brat are also gaining attention in their relation to human pathological diseases, such as cancer, and especially in brain tumors. Human TRIM3, the closest ortholog of *Drosophila* Brat, is exclusively expressed in the brain. *TRIM3* is deleted in about 25% of human glioblastomas with similar deletion frequency in lower grade gliomas, signifying this mutation as an early event in gliomagenesis (Boulay et al. 2009). TRIM3 has E3 ubiquitin ligase activity and functions as a tumor suppressor at least partially through ubiquitination and degradation of p21. However, in oligodendrogliomas, TRIM3 overexpression suppresses proliferation in both presence and absence of p21, suggesting contribution of different signaling pathways (Liu et al. 2014). Although 25% of GBMs have at least one *TRIM3* allele deleted, recent work has suggested that non-deleted gliomas have reduced expression of TRIM3 and have suggested that other mechanisms, such as *TRIM3* promoter methylation, could account for its lower RNA expression (Chen et al. 2014). Low TRIM3 protein expression was linked to upregulated Notch signaling in GBM neurospheres, possibly through reduced expression of the negative regulator NUMB. Reconstitution of TRIM3 in *TRIM3*-null neurospheres led to Musashi1, the mammalian ortholog of *Drosophila* Musashi, which also attenuates NUMB translation. Thus, it has been suggested that activation of NUMB is the ultimate mechanism responsible for suppressing Notch signaling. A recent study demonstrated a novel role of TRIM3, in which the protein binds with Importin β and regulates nuclear transportation of proteins including NOTCH1 (Mukherjee et al. 2016). TRIM3 reduction was also tightly coupled with elevated expression and activity of c-MYC in both glioma cell lines and neurospheres. Functional assays measuring asymmetric cell division have also showed that TRIM3 expression can rescue asymmetric cell division phenotype in GBM neurospheres. Altogether, it is clear that the Brat mammalian ortholog TRIM3 regulates self-renewal in human gliomas (Chen et al. 2014). Other TRIM family members, including the Brat ortholog TRIM32, have also been shown to attenuate proliferation and self-renewal during asymmetric cell division in mice through ubiquitination of MYC protein (Betschinger et al. 2006; Schwamborn et al. 2009).

18.2.2 Asymmetric Spindle Assembly

18.2.2.1 Apical Spindle Orientation

Normal Conditions

After localization of fate determinants during interphase, mitotic spindle formation in metaphase and telophase takes center stage to support asymmetric division that leads to unequal daughters and also results in the correct localization of daughter cells within the niche following division. Inscuteable, a cytoskeletal adapter protein (Wodarz et al. 1999), first localizes apically and binds with Partner of Inscuteable (Pins) and the G protein receptor substrate $G\alpha_i$, in a receptor independent manner (Siegrist and Doe 2005). Another protein, microtubule-bound kinesin Khc73, binds with Disc Large (Dlg) and forms a complex with $G\alpha_i$, Pins, and Mud (Siller et al. 2006). This large protein complex together binds with the apical cytoskeleton through Inscuteable, establishing the correct spindle alignment and the orthogonal cleavage plane along the apical/basal axis (Fig. 18.1).

The mammalian ortholog of *Drosophila* Pins is LGN. Studies suggested that LGN binds with a microtubule-associated protein NuMA as well as $G\alpha_i$ to establish asymmetric microtubule orientation (Du and Macara 2004). When LGN is absent, PAR3 exclusively localizes to one cell, which ultimately causes excessive production of outer radial glia (Konno et al. 2008).

In Neoplasia

Drosophila mutants of Disc large (*Dlg*) have a disrupted microfilament and microtubule network that results in loss of localization of basal determinants. Disruption of the apicobasal polarity in imaginal discs during development drives cellular enlargement, making Dlg a well-known tumor suppressor (Woods et al. 1996; Knoblich 2010). The mitotic spindle anchoring protein Inscuteable helps in the localization of basal determinants, such as Miranda and Prospero. Loss of *inscuteable* causes mislocalization of Miranda and Prospero and disrupts their ability to direct asymmetric cell division (Chia et al. 2008). The Partner of Inscuteable (Pins) and Mud act together to establish proper spindle orientation. Absence of either (or both) results in a lack of spindle alignment with the asymmetric fate determinants, causing their random localization and driving the formation of proliferative neuroblast-like cells (Wodarz and Nathke 2007).

Mouse LGN is an ortholog of *Drosophila* Pins that rescues Pins mutants to maintain asymmetric division. siRNA experiments as well as genetic mutations have shown that functionally impaired LGN disrupts asymmetry in radial glial cell division and drives predominant oblique division as opposed to vertical division (Shitamukai et al. 2011). NuMA, a mammalian ortholog of Mud, can cause aneuploidy in mouse myeloid cell (Ota et al. 2003) when overexpressed.

LGN protein is expressed in a large number of human breast cancer cell lines (Fukukawa et al. 2010), signifying a potential role in tumor development and progression. Studies have shown that PBK/TOPK phosphorylates and activates LGN, ultimately favoring symmetric spindle orientation, leading to proliferation and tumor formation. Adenomatous polyposis coli (APC) is another tumor suppressor that binds to human DLG1 protein and negatively regulates cell cycle progression (Matsumine et al. 1996). Mutations in this gene are well described in both familial and sporadic forms of colorectal cancer. APC also initiates downstream β-catenin ubiquitination and degradation by binding with β-catenin through GSK-3β and axin complex. This whole process has the end result of inhibiting the WNT signal and suppressing cell proliferation (Li et al. 2012).

18.2.2.2 Basal Spindle Orientation

In contrast to apical spindle organization, full descriptions of proper basal spindle arrangement are not yet available. Studies indicate that Dlg and Pins may promote basal spindle orientation since basal determinants do not localize in *dlg* and *pins* mutants; however, during telophase, asymmetric cell division can still occur in these mutants (Ohshiro et al. 2000; Peng et al. 2000).

18.3 Extrinsic Regulators of Asymmetric Cell Division

Recent findings indicate that the stem cell niche, the microenvironment surrounding stem cell, also significantly participates in the regulation of asymmetric cell division and the determination of cellular fate. *Drosophila* germline stem cells have been a prime model for exploring niche effects, since the surrounding epithelial cells appear to regulate apical–basal spindle orientation and asymmetric division (Siegrist and Doe 2005; Knoblich 2008). Stem cells in *Drosophila* ovary and testis have their own niche known as cap and hub, respectively (Roeder and Lorenz 2006). Within this niche, stem cells maintain their self-renewal properties as long as they are physically attached; detachment from the niche initiates differentiation. Signals from Notch ligand Delta and Jagged and β-catenin ligand Wnt from adjacent cells present at the niche are delivered to the apical side of the germline stem cells in order to support stemness to that side; the basal side that does not contact the niche generates a daughter cell that will detach and eventually differentiate.

Drosophila germline stem cells are the best studied model for establishing signaling pathways from the niche to the cells undergoing division. In the ovary, there are two distinct sets of cells, namely, Germline stem cells (GSCs) and Escort stem cells (ESCs), that are regulated by the stem cell niche. In each germarium, 2–3 GSCs are surrounded by the stem cell niche, which consists of an equal number of cap cells. Cellular adhesion between GSCs and cap cells is the major regulator of asymmetric division in this niche. GSCs attached to cap cells retain their stemness, whereas the other daughter cell moves away from the niche and differentiates. Both β-catenin and E-cadherin play vital roles in maintaining cellular adhesion between GSCs and cap cells and their absence leads to stem cell loss (Song et al. 2002). Cap cells drive asymmetric division to GSCs by releasing short burst of BMP ligands Decapentaplegic (Dpp) and Glass bottom boat (Gbb) (Fuller and Spradling 2007), which together activate BMP receptors on stem cells and repress differentiation by inducing Bag of marbles (Bam). After division, one daughter cell, called the cytoblast, moves away from the niche and activates Bam transcription, resulting in differentiation. Bam mutant cytoblasts fail to differentiate, giving rise to an ovarian tumor.

Escort stem cells (ESCs) surround germarium and are in contact with cap cells and GSCs. When the cystoblast moves away from the niche after differentiation, Escort cells (ECs) wrap them with their cytoplasmic projections. Eventually, ECs undergo apoptosis. ESCs undergo asymmetric division by the influence of Unpaired, a ligand of the JAK–STAT pathway. Studies have shown that a lack of STAT depletes the ESC pool, whereas overexpression of Unpaired occasionally causes germline stem cell tumor (Decotto and Spradling 2005). In Drosophila male germline stem cells (GSCs), the stem cell niche also regulates the mother cell and daughter cell centrosomal orientation during asymmetric division. In this process, the male germline stem cell (GSC) niche called "hub" anchors the astral microtubules present in the mother GSC. These microtubules are already anchored to the mother centrosomes by a centrosomal integral protein called Centrosomin (Cnn). This hub-microtubule-centrosome anchoring ensures mother centrosomes to stay close to the hub than the daughter centrosomes. The homozygous mutant of centrosomin (cnn) causes random positioning of both mother and daughter centrosomes. Anchoring of the mother centrosome with the hub helps it stay close to the niche, whereas it allows the movement of the daughter cell away from the hub (Yamashita et al. 2007).

Interestingly, in the germarium of aging flies, stem cell loss is a common finding, but the role of the niche and its signaling in this aging process has not been fully explored. It has been determined that overexpression of superoxide dismutase (SOD) can prevent stem cell loss in these aging flies, suggesting that levels of oxidative stress likely play a major role in maintaining the germline stem cell pool (Pan et al. 2007).

While Wnt is a well-established determinant of cell polarity in *C. elegans* and *Drosophila* (Hilliard and Bargmann 2006; Korkut et al. 2009), more recent studies have also suggested that it is a regulator of asymmetric cell division. In the case of embryonic stem (ES) cells, when WNT protein is brought into close proximity of the external surface of ES cells undergoing division, the daughter cell closest to WNT will continue to display pluripotency, whereas the distant daughter will eventually differentiate (Habib et al. 2013). Dysregulation of Wnt signaling has been well documented in human cancers, and *wnt* is an established proto-oncogene frequently linked to breast cancer, specific forms of medulloblastoma, and other forms of neoplasia. Upregulated Wnt drives nuclear localization of β-catenin, initiating target gene expression that favors tumor progression and is associated with poor prognosis in some cancers. Wnt signaling (Fig. 18.1) is also a key regulator of epithelial–mesenchymal transition (EMT) (Lamouille et al. 2014), which could at least partially be an outcome of deregulated asymmetric cell division.

Neural and epidermal stem cells have been ideal models to study comparable niche phenomena in in vivo mammalian systems. In both models, cells receive intrinsic and extrinsic (niche driven) cues that leads to asymmetric cell division (Knoblich 2008). Neural stem cells are attached to ependymal cells that line the ventricular surface within the subventricular zone of lateral ventricle (Doetsch 2003), serving as the niche to the neural stem cells. The daughter cell that loses contact with the ependymal niche after division proceeds to differentiate. Nearby blood vessels and cerebrospinal fluid provide addition input to the niche by supplying growth factors that promote neural stem cell proliferation (Shen et al. 2004; Andreu-Agullo et al. 2009; Lehtinen et al. 2011). Similar phenomena can be observed in embryonic epidermal cells, where daughter cells detach themselves from the basement membrane following division and become differentiated suprabasal cells (Miyata et al. 2004; Noctor et al. 2004). Both EGFR and Notch signaling play roles in directing self-renewal in these settings (Andreu-Agullo et al. 2009). Parallels with cancer biology are evident, since both activating EGFR alterations and dysregulated Notch signaling are common in cancer (Espinoza et al. 2013).

18.4 Noncanonical Signaling Pathway Influence on Asymmetric Cell Division

Apical–basal localization of protein complexes and fate determinants along with the unequal spindle arrangement on two sides of the division plane are central cell-intrinsic drivers of asymmetric stem cell division. While mechanisms have been partially established, signals that initiate and regulate the cell-intrinsic asymmetric division are only partially understood. Noncanonical signaling pathways are such signaling cues that are not a part of conventional apical and basal complexes and fate determinants during asymmetric division but play a large role in the initiation or modification protein localization and fate determination, either directly or indirectly. Here, we will discuss Myc signaling, self-renewal maintaining transcription factors, and mi-RNA signaling.

18.4.1 Myc Regulation

MYC signaling pathways are key regulators of cell growth, proliferation, and development and have a strong impact on stem cell maintenance. Constitutively active c-MYC has been linked with many neoplastic diseases (Zheng et al. 2008).

dMyc (*Drosophila* Myc) levels are high in stem cells and drop significantly in differentiated daughter cells that arise following asymmetric cell division. Additionally, dMyc overexpression prevents differentiation of daughter cells in the ovary (Rhiner et al. 2009). Together these data indicate that dMyc could be a regulator of asymmetric cell division in *Drosophila* neuroblasts and germline stem cells (Quinn et al. 2013), as dMyc is linked to stem cell maintenance and is not expressed in the differentiated daughter. During asymmetric cell division, *Drosophila* Brat suppresses dMyc expression posttranscriptionally in neuroblasts (Betschinger et al. 2006) and in germline stem cells (Harris et al. 2011). This direct control of dMyc by fate determinant Brat is critical for maintaining a homeostatic balance during asymmetric division (Betschinger et al. 2006).

In mice models, mammalian ortholog of *Drosophila* Brat, TRIM32, ubiquitinates c-MYC for its degradation in neural progenitors in neocortex (Fig. 18.2a) (Schwamborn

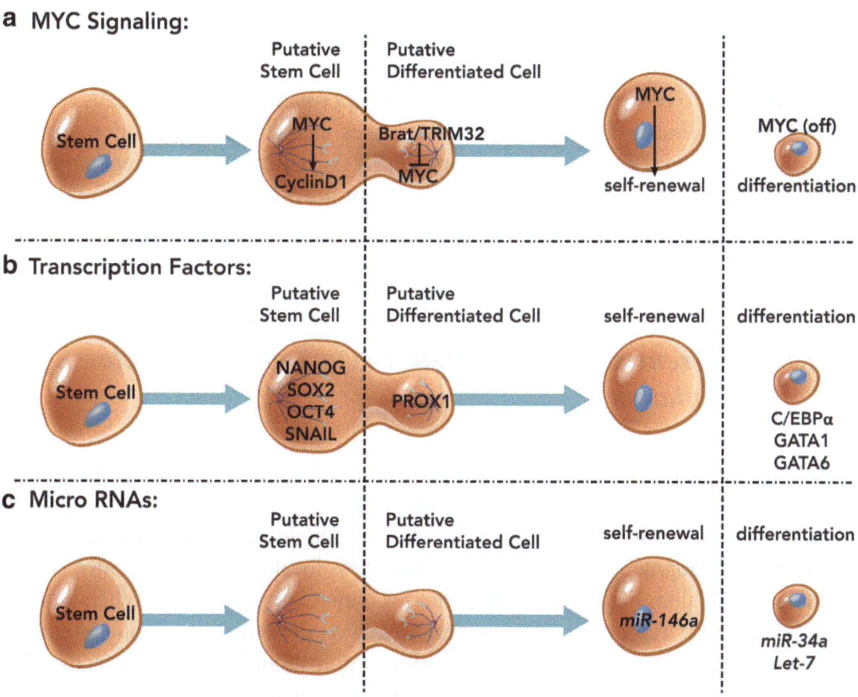

Fig. 18.2 Noncanonical signaling pathways regulate asymmetric cell division. (**a**) Various growth signaling pathways affect the asymmetric cell division balance. Pro-growth regulators such as MYC shift the asymmetric cell division balance toward self-renewal, through activating CYCLIN D1. On the basal side, *Drosophila* Brat or mammalian TRIM32 attenuates MYC function to start differentiation signal. (**b**) Dysregulation of pluripotency and self-renewal transcription factors such as NANOG, SOX2, OCT4, and SNAIL also initiates symmetric stem cell proliferation. Transcription factors such as PROX1, C/EBP alpha, GATA1, and GATA6 express in the differentiated daughter cell. (**c**) MicroRNAs also regulate asymmetric cell division by affecting stem cell renewal and differentiation. *miR-146a* drives stem cell renewal, whereas *miR-34a* and *Let-7* drive differentiation

et al. 2009) initiating differentiation. c-MYC regulates an array of cell growth pathways including cell cycle genes such as *Cyclin D1* (Fig. 18.2a) (Mateyak et al. 1999; Kapinas et al. 2013), chromatin modification, and energy metabolism that promote cell division and stemness. Both *Drosophila* Myc and mammalian MYC bind to the promoter region of TORC1 target genes (Quinn et al. 2013) and exert control over Insulin/TOR metabolic axis and ribosome synthesis driving self-renewal. Another potential mechanism is that MYC represses a master regulator of endoderm differentiation, GATA6 (Quinn et al. 2013) and drives proliferation.

Upon APC (antigen-presenting cell)-mediated activation of T-lymphocytes, these cells undergo asymmetric division where c-Myc asymmetrically segregates to the proximal daughter cell and is maintained by mTORC1 (Verbist et al. 2016). In neuroblastoma cells, MYCN is abundant, especially in neuroblastoma stem cells that lack asymmetric division. Suppressing MYCN in those MYCN expressing cells induces asymmetric division, proving that MYC can regulate asymmetric cell division (Izumi and Kaneko 2012). The excessive activity of proto-oncogene c-MYC in many cancers supports a strong relationship between self-renewal program and carcinogenesis (Nesbit et al. 1999). Evidently, normal cellular safeguards that save cells from aberrant cMYC activation seem to be broken in cancer. In non-neoplastic follicular basal cells of mice for example, aberrant *c-MYC* activation diminishes the stem cell population and initiates differentiation (Perez-Losada and Balmain 2003), indicating an inherent protective mechanism against uncontrolled proliferation in normal cells (Perez-Losada and Balmain 2003).

18.4.2 Self-renewal Transcription Factors and Cell Cycle Regulation

Transcription factors involved in retaining pluripotency of a stem cell include NANOG, SOX2, OCT4 (Fig. 18.2b) (Kapinas et al. 2013). These transcription factors maintain activity in the mother stem cell and in daughter stem cell but not in the differentiated daughter. It's unclear whether they localize asymmetrically during asymmetric division, but they are important for maintaining stem fate. When induced by WNT3a, these transcription factors accumulate on the self-renewal side of embryonic stem cells during asymmetric cell division (Habib et al. 2013). These factors also regulate cell cycle and microRNA genes to maintain pluripotency. Many human neoplastic growths harbor copy number alteration of these pluripotent transcription factors, suggesting a contribution of these factors in the disease.

Cell cycle regulators are the final arbiters in a cell's commitment to divide and create progeny, and their defects are known to drive specific diseases, including cancer. Cyclins and CDKs, the major regulators of cell cycle, determine cell cycle entry and exit. Active CDC42 localizes on the apical side during asymmetric cell division, keeping cyclin D1 active (Besson et al. 2004) by the aPKC/PAR6/PAR3 complex. Cyclin D1 interacts with CDK4 and regulates G1–S cell cycle transition through active E2F in stem cells (Kondo et al. 2006). High levels of Cyclin-CDK guarantee short cell cycle length and self-renewal.

18.4.3 microRNA Regulation

In recent years, the role of microRNAs (mi-RNA) in self-renewal (Fig. 18.2c), pluripotency, asymmetric cell division, and cancer has been a topic of intense investigation. These short, noncoding RNAs target mRNAs in a sequence-specific manner and attenuate their expression (Kapinas and Delany 2011). Self-renewal transcriptional factors, such as NANOG, SOX2 and OCT4 are known regulators of mi-RNAs (Kapinas et al. 2013). In the context of germline asymmetric cell division in *Drosophila*, protein Mei-P26 inhibits mi-RNA pathways and mutant *mei-P26* flies show germline overproliferation (Page et al. 2000). A novel microRNA *miR-34a* suppresses Notch signaling in differentiated daughter cells by localizing asymmetrically (Lerner and Petritsch 2014). Another microRNA *miR-146a* has also been described as an initiator of self-renewal signal in colorectal cancer stem cells (Lerner and Petritsch 2014) and asymmetrically distributed in only 10–20% of these cancer stem cells. In those same types of cells, *miR-146a* activation phenocopies SNAIL-dependent colorectal cancers, possibly because SNAIL regulates *miR-146a* through the Wnt signaling pathway. *Let-7*, another class of microRNA, regulates Ras oncogenic activity. In lung cancer, downregulation of *Let-7* is evident, thus driving the proliferation (Esquela-Kerscher and Slack 2006).

18.5 Summary

Molecular programs of asymmetric cell division are complex, and much of our current understanding stems from *Drosophila* (Knoblich 2008).

- Localization of apical and basal protein complexes along with fate determinants, such as Numb, Brat, and Prospero, during early stages of cell cycle directs asymmetric division, and disruption of these complexes disturbs the bipolar fate, generating stem cells with potential to form neoplastic masses.
- Asymmetric arrangement of spindles and microtubules during mitosis also plays a critical role in maintaining apical–basal polarity and ensuring appropriate asymmetric cell division.
- Both *Drosophila* and mammalian systems have cell-extrinsic (microenvironmental) cues that impact asymmetric cell division.
- Noncanonical pathways that directly and indirectly influence asymmetric cell division include transcription factors, cell cycle regulators, signaling pathways, and microRNAs.

Many cancers display a reduced frequency of asymmetric cell division (Lathia et al. 2011; Gomez-Lopez et al. 2013), which may be explained by disruption in the mechanisms above. Further understanding of stem cell division in cancer will provide opportunities for the development of novel diagnostics and therapies.

Author Contributions Subhas Mukherjee: Concept and design, chapter writing. Daniel J. Brat: Chapter writing, financial support.

Grant Acknowledgement This work was supported by US Public Health Service National Institutes of Health (NIH) grant R01CA149107 (DJB).

Conflict of Interest Statement The authors have no conflicts of interest to disclose.

References

Andreu-Agullo C, Morante-Redolat JM, Delgado AC, Farinas I (2009) Vascular niche factor PEDF modulates Notch-dependent stemness in the adult subependymal zone. Nat Neurosci 12 (12):1514–1523

Arama E, Dickman D, Kimchie Z, Shearn A, Lev Z (2000) Mutations in the beta-propeller domain of the Drosophila brain tumor (brat) protein induce neoplasm in the larval brain. Oncogene 19 (33):3706–3716

Beaucher M, Hersperger E, Page-McCaw A, Shearn A (2007) Metastatic ability of Drosophila tumors depends on MMP activity. Dev Biol 303(2):625–634

Besson A, Assoian RK, Roberts JM (2004) Regulation of the cytoskeleton: an oncogenic function for CDK inhibitors? Nat Rev Cancer 4(12):948–955

Betschinger J, Mechtler K, Knoblich JA (2003) The Par complex directs asymmetric cell division by phosphorylating the cytoskeletal protein Lgl. Nature 422(6929):326–330

Betschinger J, Eisenhaber F, Knoblich JA (2005) Phosphorylation-induced autoinhibition regulates the cytoskeletal protein Lethal (2) giant larvae. Curr Biol 15(3):276–282

Betschinger J, Mechtler K, Knoblich JA (2006) Asymmetric segregation of the tumor suppressor brat regulates self-renewal in Drosophila neural stem cells. Cell 124(6):1241–1253

Boulay JL, Stiefel U, Taylor E, Dolder B, Merlo A, Hirth F (2009) Loss of heterozygosity of TRIM3 in malignant gliomas. BMC Cancer 9:71

Bowman SK, Rolland V, Betschinger J, Kinsey KA, Emery G, Knoblich JA (2008) The tumor suppressors Brat and Numb regulate transit-amplifying neuroblast lineages in Drosophila. Dev Cell 14(4):535–546

Bultje RS, Castaneda-Castellanos DR, Jan LY, Jan YN, Kriegstein AR, Shi SH (2009) Mammalian Par3 regulates progenitor cell asymmetric division via notch signaling in the developing neocortex. Neuron 63(2):189–202

Caussinus E, Gonzalez C (2005) Induction of tumor growth by altered stem-cell asymmetric division in *Drosophila melanogaster*. Nat Genet 37(10):1125–1129

Chen G, Kong J, Tucker-Burden C, Anand M, Rong Y, Rahman F, Moreno CS, Van Meir EG, Hadjipanayis CG, Brat DJ (2014) Human Brat ortholog TRIM3 is a tumor suppressor that regulates asymmetric cell division in glioblastoma. Cancer Res 74(16):4536–4548

Chenn A, McConnell SK (1995) Cleavage orientation and the asymmetric inheritance of Notch1 immunoreactivity in mammalian neurogenesis. Cell 82(4):631–641

Chia W, Somers WG, Wang H (2008) Drosophila neuroblast asymmetric divisions: cell cycle regulators, asymmetric protein localization, and tumorigenesis. J Cell Biol 180(2):267–272

Chieffi P, Cozzolino L, Kisslinger A, Libertini S, Staibano S, Mansueto G, De Rosa G, Villacci A, Vitale M, Linardopoulos S, Portella G, Tramontano D (2006) Aurora B expression directly

correlates with prostate cancer malignancy and influence prostate cell proliferation. Prostate 66 (3):326–333
Colaluca IN, Tosoni D, Nuciforo P, Senic-Matuglia F, Galimberti V, Viale G, Pece S, Di Fiore PP (2008) NUMB controls p53 tumour suppressor activity. Nature 451(7174):76–80
Conklin EG (1905) The mutation theory from the standpoint of cytology. Science 21(536):525–529
Decotto E, Spradling AC (2005) The Drosophila ovarian and testis stem cell niches: similar somatic stem cells and signals. Dev Cell 9(4):501–510
Doetsch F (2003) A niche for adult neural stem cells. Curr Opin Genet Dev 13(5):543–550
Du Q, Macara IG (2004) Mammalian Pins is a conformational switch that links NuMA to heterotrimeric G proteins. Cell 119(4):503–516
Dudas J, Mansuroglu T, Moriconi F, Haller F, Wilting J, Lorf T, Fuzesi L, Ramadori G (2008) Altered regulation of Prox1-gene-expression in liver tumors. BMC Cancer 8:92
Dyer MA, Livesey FJ, Cepko CL, Oliver G (2003) Prox1 function controls progenitor cell proliferation and horizontal cell genesis in the mammalian retina. Nat Genet 34(1):53–58
Eder AM, Sui X, Rosen DG, Nolden LK, Cheng KW, Lahad JP, Kango-Singh M, Lu KH, Warneke CL, Atkinson EN, Bedrosian I, Keyomarsi K, Kuo WL, Gray JW, Yin JC, Liu J, Halder G, Mills GB (2005) Atypical PKCiota contributes to poor prognosis through loss of apical-basal polarity and cyclin E overexpression in ovarian cancer. Proc Natl Acad Sci USA 102(35):12519–12524
Espinoza I, Pochampally R, Xing F, Watabe K, Miele L (2013) Notch signaling: targeting cancer stem cells and epithelial-to-mesenchymal transition. Onco Targets Ther 6:1249–1259
Esquela-Kerscher A, Slack FJ (2006) Oncomirs—microRNAs with a role in cancer. Nat Rev Cancer 6(4):259–269
Fukukawa C, Ueda K, Nishidate T, Katagiri T, Nakamura Y (2010) Critical roles of LGN/GPSM2 phosphorylation by PBK/TOPK in cell division of breast cancer cells. Genes Chromosomes Cancer 49(10):861–872
Fuller MT, Spradling AC (2007) Male and female Drosophila germline stem cells: two versions of immortality. Science 316(5823):402–404
Gaiano N, Nye JS, Fishell G (2000) Radial glial identity is promoted by Notch1 signaling in the murine forebrain. Neuron 26(2):395–404
Gomez-Lopez S, Lerner RG, Petritsch C (2013) Asymmetric cell division of stem and progenitor cells during homeostasis and cancer. Cell Mol Life Sci 71(4):575–597
Habib SJ, Chen BC, Tsai FC, Anastassiadis K, Meyer T, Betzig E, Nusse R (2013) A localized Wnt signal orients asymmetric stem cell division in vitro. Science 339(6126):1445–1448
Harris RE, Pargett M, Sutcliffe C, Umulis D, Ashe HL (2011) Brat promotes stem cell differentiation via control of a bistable switch that restricts BMP signaling. Dev Cell 20(1):72–83
Hilliard MA, Bargmann CI (2006) Wnt signals and frizzled activity orient anterior-posterior axon outgrowth in *C. elegans*. Dev Cell 10(3):379–390
Hou C, Ding L, Zhang J, Jin Y, Sun C, Li Z, Sun X, Zhang T, Zhang A, Li H, Gao J (2014) Abnormal cerebellar development and Purkinje cell defects in Lgl1-Pax2 conditional knockout mice. Dev Biol 395(1):167–181
Ito T, Kwon HY, Zimdahl B, Congdon KL, Blum J, Lento WE, Zhao C, Lagoo A, Gerrard G, Foroni L, Goldman J, Goh H, Kim SH, Kim DW, Chuah C, Oehler VG, Radich JP, Jordan CT, Reya T (2010) Regulation of myeloid leukaemia by the cell-fate determinant Musashi. Nature 466(7307):765–768
Izumi H, Kaneko Y (2012) Evidence of asymmetric cell division and centrosome inheritance in human neuroblastoma cells. Proc Natl Acad Sci USA 109(44):18048–18053
Janic A, Mendizabal L, Llamazares S, Rossell D, Gonzalez C (2010) Ectopic expression of germline genes drives malignant brain tumor growth in Drosophila. Science 330(6012):1824–1827
Joberty G, Petersen C, Gao L, Macara IG (2000) The cell-polarity protein Par6 links Par3 and atypical protein kinase C to Cdc42. Nat Cell Biol 2(8):531–539
Kalmes A, Merdes G, Neumann B, Strand D, Mechler BM (1996) A serine-kinase associated with the p127-l(2)gl tumour suppressor of Drosophila may regulate the binding of p127 to nonmuscle

myosin II heavy chain and the attachment of p127 to the plasma membrane. J Cell Sci 109 (Pt 6):1359–1368

Kaltezioti V, Kouroupi G, Oikonomaki M, Mantouvalou E, Stergiopoulos A, Charonis A, Rohrer H, Matsas R, Politis PK (2010) Prox1 regulates the notch1-mediated inhibition of neurogenesis. PLoS Biol 8(12):e1000565

Kapinas K, Delany AM (2011) MicroRNA biogenesis and regulation of bone remodeling. Arthritis Res Ther 13(3):220

Kapinas K, Grandy R, Ghule P, Medina R, Becker K, Pardee A, Zaidi SK, Lian J, Stein J, van Wijnen A, Stein G (2013) The abbreviated pluripotent cell cycle. J Cell Physiol 228(1):9–20

Knoblich JA (2008) Mechanisms of asymmetric stem cell division. Cell 132(4):583–597

Knoblich JA (2010) Asymmetric cell division: recent developments and their implications for tumour biology. Nat Rev Mol Cell Biol 11(12):849–860

Komori H, Xiao Q, McCartney BM, Lee CY (2014) Brain tumor specifies intermediate progenitor cell identity by attenuating beta-catenin/Armadillo activity. Development 141(1):51–62

Kondo T, Ezzat S, Asa SL (2006) Pathogenetic mechanisms in thyroid follicular-cell neoplasia. Nat Rev Cancer 6(4):292–306

Konno D, Shioi G, Shitamukai A, Mori A, Kiyonari H, Miyata T, Matsuzaki F (2008) Neuroepithelial progenitors undergo LGN-dependent planar divisions to maintain self-renewability during mammalian neurogenesis. Nat Cell Biol 10(1):93–101

Korkut C, Ataman B, Ramachandran P, Ashley J, Barria R, Gherbesi N, Budnik V (2009) Trans-synaptic transmission of vesicular Wnt signals through Evi/Wntless. Cell 139(2):393–404

Kunnev D, Ivanov I, Ionov Y (2009) Par-3 partitioning defective 3 homolog (*C. elegans*) and androgen-induced prostate proliferative shutoff associated protein genes are mutationally inactivated in prostate cancer cells. BMC Cancer 9:318

Kwon C, Cheng P, King IN, Andersen P, Shenje L, Nigam V, Srivastava D (2011) Notch post-translationally regulates beta-catenin protein in stem and progenitor cells. Nat Cell Biol 13 (10):1244–1251

Lamouille S, Xu J, Derynck R (2014) Molecular mechanisms of epithelial-mesenchymal transition. Nat Rev Mol Cell Biol 15(3):178–196

Lathia JD, Hitomi M, Gallagher J, Gadani SP, Adkins J, Vasanji A, Liu L, Eyler CE, Heddleston JM, Wu Q, Minhas S, Soeda A, Hoeppner DJ, Ravin R, McKay RD, McLendon RE, Corbeil D, Chenn A, Hjelmeland AB, Park DM, Rich JN (2011) Distribution of CD133 reveals glioma stem cells self-renew through symmetric and asymmetric cell divisions. Cell Death Dis 2:e200

Lee CY, Andersen RO, Cabernard C, Manning L, Tran KD, Lanskey MJ, Bashirullah A, Doe CQ (2006a) Drosophila Aurora-A kinase inhibits neuroblast self-renewal by regulating aPKC/Numb cortical polarity and spindle orientation. Genes Dev 20(24):3464–3474

Lee CY, Robinson KJ, Doe CQ (2006b) Lgl, Pins and aPKC regulate neuroblast self-renewal versus differentiation. Nature 439(7076):594–598

Lee CY, Wilkinson BD, Siegrist SE, Wharton RP, Doe CQ (2006c) Brat is a Miranda cargo protein that promotes neuronal differentiation and inhibits neuroblast self-renewal. Dev Cell 10 (4):441–449

Lehtinen MK, Zappaterra MW, Chen X, Yang YJ, Hill AD, Lun M, Maynard T, Gonzalez D, Kim S, Ye P, D'Ercole AJ, Wong ET, LaMantia AS, Walsh CA (2011) The cerebrospinal fluid provides a proliferative niche for neural progenitor cells. Neuron 69(5):893–905

Lerner RG, Petritsch C (2014) A microRNA-operated switch of asymmetric-to-symmetric cancer stem cell divisions. Nat Cell Biol 16(3):212–214

Li D, Zhu J, Firozi PF, Abbruzzese JL, Evans DB, Cleary K, Friess H, Sen S (2003) Overexpression of oncogenic STK15/BTAK/Aurora A kinase in human pancreatic cancer. Clin Cancer Res 9 (3):991–997

Li VS, Ng SS, Boersema PJ, Low TY, Karthaus WR, Gerlach JP, Mohammed S, Heck AJ, Maurice MM, Mahmoudi T, Clevers H (2012) Wnt signaling through inhibition of beta-catenin degradation in an intact Axin1 complex. Cell 149(6):1245–1256

Liu J, Sato C, Cerletti M, Wagers A (2010) Notch signaling in the regulation of stem cell self-renewal and differentiation. Curr Top Dev Biol 92:367–409

Liu Y, Raheja R, Yeh N, Ciznadija D, Pedraza AM, Ozawa T, Hukkelhoven E, Erdjument-Bromage H, Tempst P, Gauthier NP, Brennan C, Holland EC, Koff A (2014) TRIM3, a tumor suppressor linked to regulation of p21(Waf1/Cip1.). Oncogene 33(3):308–315

Manabe N, Hirai S, Imai F, Nakanishi H, Takai Y, Ohno S (2002) Association of ASIP/mPAR-3 with adherens junctions of mouse neuroepithelial cells. Dev Dyn 225(1):61–69

Mateyak MK, Obaya AJ, Sedivy JM (1999) c-Myc regulates cyclin D-Cdk4 and -Cdk6 activity but affects cell cycle progression at multiple independent points. Mol Cell Biol 19(7):4672–4683

Matsumine A, Ogai A, Senda T, Okumura N, Satoh K, Baeg GH, Kawahara T, Kobayashi S, Okada M, Toyoshima K, Akiyama T (1996) Binding of APC to the human homolog of the Drosophila discs large tumor suppressor protein. Science 272(5264):1020–1023

Miyata T, Kawaguchi A, Saito K, Kawano M, Muto T, Ogawa M (2004) Asymmetric production of surface-dividing and non-surface-dividing cortical progenitor cells. Development 131 (13):3133–3145

Mukherjee S, Kong J, Brat DJ (2015) Cancer stem cell division: when the rules of asymmetry are broken. Stem Cells Dev 24(4):405–416

Mukherjee S, Tucker-Burden C, Zhang C, Moberg K, Read R, Hadjipanayis C, Brat DJ (2016) Drosophila Brat and human ortholog TRIM3 maintain stem cell equilibrium and suppress brain tumorigenesis by attenuating Notch nuclear transport. Cancer Res 76(8):2443–2452

Nagai H, Li Y, Hatano S, Toshihito O, Yuge M, Ito E, Utsumi M, Saito H, Kinoshita T (2003) Mutations and aberrant DNA methylation of the PROX1 gene in hematologic malignancies. Genes Chromosomes Cancer 38(1):13–21

Nesbit CE, Tersak JM, Prochownik EV (1999) MYC oncogenes and human neoplastic disease. Oncogene 18(19):3004–3016

Noctor SC, Martinez-Cerdeno V, Ivic L, Kriegstein AR (2004) Cortical neurons arise in symmetric and asymmetric division zones and migrate through specific phases. Nat Neurosci 7(2):136–144

Noctor SC, Martinez-Cerdeno V, Kriegstein AR (2008) Distinct behaviors of neural stem and progenitor cells underlie cortical neurogenesis. J Comp Neurol 508(1):28–44

Ohshiro T, Yagami T, Zhang C, Matsuzaki F (2000) Role of cortical tumour-suppressor proteins in asymmetric division of Drosophila neuroblast. Nature 408(6812):593–596

Ota J, Yamashita Y, Okawa K, Kisanuki H, Fujiwara S, Ishikawa M, Lim Choi Y, Ueno S, Ohki R, Koinuma K, Wada T, Compton D, Kadoya T, Mano H (2003) Proteomic analysis of hematopoietic stem cell-like fractions in leukemic disorders. Oncogene 22(36):5720–5728

Page SL, McKim KS, Deneen B, Van Hook TL, Hawley RS (2000) Genetic studies of mei-P26 reveal a link between the processes that control germ cell proliferation in both sexes and those that control meiotic exchange in Drosophila. Genetics 155(4):1757–1772

Pan L, Chen S, Weng C, Call G, Zhu D, Tang H, Zhang N, Xie T (2007) Stem cell aging is controlled both intrinsically and extrinsically in the Drosophila ovary. Cell Stem Cell 1(4):458–469

Peng CY, Manning L, Albertson R, Doe CQ (2000) The tumour-suppressor genes lgl and dlg regulate basal protein targeting in Drosophila neuroblasts. Nature 408(6812):596–600

Perez-Losada J, Balmain A (2003) Stem-cell hierarchy in skin cancer. Nat Rev Cancer 3(6):434–443

Quinn LM, Secombe J, Hime GR (2013) Myc in stem cell behaviour: insights from Drosophila. Adv Exp Med Biol 786:269–285

Rhiner C, Diaz B, Portela M, Poyatos JF, Fernandez-Ruiz I, Lopez-Gay JM, Gerlitz O, Moreno E (2009) Persistent competition among stem cells and their daughters in the Drosophila ovary germline niche. Development 136(6):995–1006

Roeder I, Lorenz R (2006) Asymmetry of stem cell fate and the potential impact of the niche: observations, simulations, and interpretations. Stem Cell Rev 2(3):171–180

Rothenberg SM, Mohapatra G, Rivera MN, Winokur D, Greninger P, Nitta M, Sadow PM, Sooriyakumar G, Brannigan BW, Ulman MJ, Perera RM, Wang R, Tam A, Ma XJ, Erlander M, Sgroi DC, Rocco JW, Lingen MW, Cohen EE, Louis DN, Settleman J, Haber DA (2010) A genome-wide screen for microdeletions reveals disruption of polarity complex genes in diverse human cancers. Cancer Res 70(6):2158–2164

Schwamborn JC, Berezikov E, Knoblich JA (2009) The TRIM-NHL protein TRIM32 activates microRNAs and prevents self-renewal in mouse neural progenitors. Cell 136(5):913–925

Shen Q, Goderie SK, Jin L, Karanth N, Sun Y, Abramova N, Vincent P, Pumiglia K, Temple S (2004) Endothelial cells stimulate self-renewal and expand neurogenesis of neural stem cells. Science 304(5675):1338–1340

Shitamukai A, Konno D, Matsuzaki F (2011) Oblique radial glial divisions in the developing mouse neocortex induce self-renewing progenitors outside the germinal zone that resemble primate outer subventricular zone progenitors. J Neurosci 31(10):3683–3695

Siegrist SE, Doe CQ (2005) Microtubule-induced Pins/Galphai cortical polarity in Drosophila neuroblasts. Cell 123(7):1323–1335

Siller KH, Cabernard C, Doe CQ (2006) The NuMA-related Mud protein binds Pins and regulates spindle orientation in Drosophila neuroblasts. Nat Cell Biol 8(6):594–600

Slack C, Overton PM, Tuxworth RI, Chia W (2007) Asymmetric localisation of Miranda and its cargo proteins during neuroblast division requires the anaphase-promoting complex/cyclosome. Development 134(21):3781–3787

Song X, Zhu CH, Doan C, Xie T (2002) Germline stem cells anchored by adherens junctions in the Drosophila ovary niches. Science 296(5574):1855–1857

Sonoda J, Wharton RP (2001) Drosophila Brain Tumor is a translational repressor. Genes Dev 15 (6):762–773

Takahashi M, Yoshimoto T, Shimoda M, Kono T, Koizumi M, Yazumi S, Shimada Y, Doi R, Chiba T, Kubo H (2006) Loss of function of the candidate tumor suppressor prox1 by RNA mutation in human cancer cells. Neoplasia 8(12):1003–1010

Troy A, Cadwallader AB, Fedorov Y, Tyner K, Tanaka KK, Olwin BB (2012) Coordination of satellite cell activation and self-renewal by Par-complex-dependent asymmetric activation of p38alpha/beta MAPK. Cell Stem Cell 11(4):541–553

Verbist KC, Guy CS, Milasta S, Liedmann S, Kaminski MM, Wang R, Green DR (2016) Metabolic maintenance of cell asymmetry following division in activated T lymphocytes. Nature 532 (7599):389–393

Versmold B, Felsberg J, Mikeska T, Ehrentraut D, Kohler J, Hampl JA, Rohn G, Niederacher D, Betz B, Hellmich M, Pietsch T, Schmutzler RK, Waha A (2007) Epigenetic silencing of the candidate tumor suppressor gene PROX1 in sporadic breast cancer. Int J Cancer 121(3):547–554

Vessey JP, Amadei G, Burns SE, Kiebler MA, Kaplan DR, Miller FD (2012) An asymmetrically localized Staufen2-dependent RNA complex regulates maintenance of mammalian neural stem cells. Cell Stem Cell 11(4):517–528

Viloria-Petit AM, David L, Jia JY, Erdemir T, Bane AL, Pinnaduwage D, Roncari L, Narimatsu M, Bose R, Moffat J, Wong JW, Kerbel RS, O'Malley FP, Andrulis IL, Wrana JL (2009) A role for the TGFbeta-Par6 polarity pathway in breast cancer progression. Proc Natl Acad Sci USA 106 (33):14028–14033

Wang H, Somers GW, Bashirullah A, Heberlein U, Yu F, Chia W (2006) Aurora-A acts as a tumor suppressor and regulates self-renewal of Drosophila neuroblasts. Genes Dev 20(24):3453–3463

Wirtz-Peitz F, Nishimura T, Knoblich JA (2008) Linking cell cycle to asymmetric division: Aurora-A phosphorylates the Par complex to regulate Numb localization. Cell 135(1):161–173

Wodarz A, Nathke I (2007) Cell polarity in development and cancer. Nat Cell Biol 9(9):1016–1024

Wodarz A, Ramrath A, Kuchinke U, Knust E (1999) Bazooka provides an apical cue for Inscuteable localization in Drosophila neuroblasts. Nature 402(6761):544–547

Woods DF, Hough C, Peel D, Callaini G, Bryant PJ (1996) Dlg protein is required for junction structure, cell polarity, and proliferation control in Drosophila epithelia. J Cell Biol 134 (6):1469–1482

Yamanaka T, Horikoshi Y, Sugiyama Y, Ishiyama C, Suzuki A, Hirose T, Iwamatsu A, Shinohara A, Ohno S (2003) Mammalian Lgl forms a protein complex with PAR-6 and aPKC independently of PAR-3 to regulate epithelial cell polarity. Curr Biol 13(9):734–743

Yamashita YM, Mahowald AP, Perlin JR, Fuller MT (2007) Asymmetric inheritance of mother versus daughter centrosome in stem cell division. Science 315(5811):518–521

Yong KJ, Yan B (2011) The relevance of symmetric and asymmetric cell divisions to human central nervous system diseases. J Clin Neurosci 18(4):458–463

Zheng H, Ying H, Yan H, Kimmelman AC, Hiller DJ, Chen AJ, Perry SR, Tonon G, Chu GC, Ding Z, Stommel JM, Dunn KL, Wiedemeyer R, You MJ, Brennan C, Wang YA, Ligon KL, Wong WH, Chin L, dePinho RA (2008) Pten and p53 converge on c-Myc to control differentiation, self-renewal, and transformation of normal and neoplastic stem cells in glioblastoma. Cold Spring Harb Symp Quant Biol 73:427–437

MIX
Papier aus verantwortungsvollen Quellen
Paper from responsible sources
FSC® C105338

If you have any concerns about our products,
you can contact us on
ProductSafety@springernature.com

In case Publisher is established outside the EU,
the EU authorized representative is:
Springer Nature Customer Service Center GmbH
Europaplatz 3, 69115 Heidelberg, Germany

Printed by Libri Plureos GmbH
in Hamburg, Germany